Stauber / Vollrath
Plastics in Automotive Engineering

Rudolf Stauber
Ludwig Vollrath (Eds.)

Plastics in Automotive Engineering

Exterior Applications

Hanser Publishers, Munich • Hanser Gardner Publications, Cincinnati

HANSER

The Editors:
Prof. Dr. Rudolf Stauber, BMW Group, EG-5, Knorrstr. 147, 80788 München, Germany
Dr. Ing. Ludwig Vollrath, VDI Gesellschaft Kunststofftechnik, Graf-Recke-Str. 84, 40239 Düsseldorf, Germany

Distributed in the USA and in Canada by
Hanser Gardner Publications, Inc.
6915 Valley Avenue, Cincinnati, Ohio 45244-3029, USA
Fax: (513) 527-8801
Phone: (513) 527-8977 or 1-800-950-8977
www.hansergardner.com

Distributed in all other countries by
Carl Hanser Verlag
Postfach 86 04 20, 81631 München, Germany
Fax: +49 (89) 98 48 09
www.hanser.de

The use of general descriptive names, trademarks, etc., in this publication, even if the former are not especially identified, is not to be taken as a sign that such names, as understood by the Trade Marks and Merchandise Marks Act, may accordingly be used freely by anyone.
While the advice and information in this book are believed to be true and accurate at the date of going to press, neither the authors nor the editors nor the publisher can accept any legal responsibility for any errors or omissions that may be made. The publisher makes no warranty, express or implied, with respect to the material contained herein.

Library of Congress Cataloging-in-Publication Data
 Plastics in automotive engineering : exterior applications / R. Stauber, L. Vollrath (Eds.). – 1st ed.
    p. cm.
  ISBN-13: 978-1-56990-406-0 (hardcover)
 1. Plastics in automobiles. I. Stauber, R. (Rudolf) II. Vollrath, L.
  TL154.P5435 2007
  629.2'32–dc22
                    2007001235

Bibliografische Information Der Deutschen Bibliothek
Die Deutsche Bibliothek verzeichnet diese Publikation in der Deutschen Nationalbibliografie; detaillierte biblio-grafische Daten sind im Internet über <http://dnb.d-nb.de> abrufbar.

ISBN: 978-3-446-41120-3

All rights reserved. No part of this book may be reproduced or transmitted in any form or by any means, electronic or mechanical, including photocopying or by any information storage and retrieval system, without permission in writing from the publisher.

© Carl Hanser Verlag, Munich 2007
Production Management: Oswald Immel
Typeset by Manuela Treindl, Laaber, Germany
Coverconcept: Marc Müller-Bremer, Rebranding, München, Germany
Coverdesign: MCP • Susanne Kraus GbR, Holzkirchen, Germany
Printed and bound by Firmengruppe APPL, aprinta druck, Wemding, Germany

Preface

Today, polymer materials play leading roles in advanced automotive engineering.

Lightweight design, active and passive safety features, appealing industrial designers' concepts, optimized properties for noise abatement or aerodynamics, as well as an impressive overall image of high quality and advanced automotive engineering have long profited from the use of polymer materials and their associated composites.

Today's automotive industry is challenged by ever more stringent demands to reduce fuel consumption and exhaust emissions and to design products that can comply with environmental legislation today and in the future.

Engineers and manufacturers who develop and produce polymer-based components for automotives need to focus on reducing development times and optimizing production processes for quality and economic viability. Computer-aided selection of polymer materials, mathematical simulation of both the production and mechanical properties of plastic components are tools that can help the industry arrive at innovative and economical solutions when designing polymer applications for cars.

Every member in the polymer processing chain is working hard on new concepts to meet these challenges as best as they can.

For more than thirty years, the VDI Gesellschaft Kunststofftechnik has supported German and European automobile and commercial vehicle manufacturers by organizing international meetings featuring experts and presentations for participants from the feed stock, polymer processing and automotive industries as well as from renowned scientific institutions for polymer engineering. The aim is to provide a joint platform for the worlds of polymer and automobile technologies to report on recent polymer developments and applications for passenger cars and commercial vehicles at regular intervals and to discuss technological issues of polymer engineering with fellow members of these communities.

This volume entitled "Plastics in Automotive Engineering" – Exterior Applications" gives an overview of novel polymer applications for automotive engineering. Case studies illustrate current polymer applications in the exterior of passenger cars and commercial vehicles "made in Germany and Europe". They describe component-specific and vehicle-specific solutions, providing a sweeping insight into current developments in the polymer producing industry, novel production methods and the property profiles of advanced polymers.

This volume will be followed by another two volumes currently in preparation which will address polymer applications in the car interior and in the engine compartment and assemblies.

The editors wish to express their personal gratitude to the authors of the technical papers and to the editorial staff of Hanser Publishers.

Rudolf Stauber   *Ludwig Vollrath*

Contents

Preface V

List of Contributors XIII

1 Introduction to the Application of Plastics in Vehicle Design 1

1.0 Introduction 1

1.1 Plastics in Lightweight Vehicle Engineering 1

1.2 Plastics Applications for the Exterior 3
1.2.1 Injection Molding with Polymers 4
1.2.2 Textile-reinforced Plastics 5

1.3 Outlook 6

2 Automotive Concepts and Lightweight Design 7

2.1 Increased Use of Plastics in Body Applications: Opportunities and Risks 7
2.1.1 Current Situation of Lightweight Materials 7
2.1.2 Future Regulatory and Market Requirements 9
2.1.3 Lightweight Construction for Rigid Components 12
2.1.4 Use of Plastic as a Structural Material 13
2.1.5 Competition Between Plastic and Stainless Steel 15
2.1.6 Acoustics 16
2.1.7 Summary 16

2.2 Use of Plastics in Body Construction in Future Automobile Design and Production ... 19
2.2.1 Introduction 19
2.2.2 Structural Plastics 21
2.2.3 The Body Tub Assembly 21
2.2.4 SMC at Ford Motor Company 24
2.2.5 Implications of the End of Life Vehicle Directive 26
2.2.6 Summary 29

2.3 Lightweight Design with Plastics in the New BMW 6 Series 30
2.3.1 Material Mixture Saving Weight and Increasing Collision Safety 30
2.3.2 Requirements for Thermoplastic Side Panels 31
2.3.3 Criteria for the Use of a Thermoplastic Side Panel 31
2.3.4 Material Requirements 33
2.3.5 Product Requirements for Component Design 34
2.3.6 Manufacturing and Assembly Process 36
2.3.7 Outlook and Future Requirements 38

3	**Material Concepts and Process Technologies**	41
3.1	Plastics Engineering in Automotive Exteriors	41
3.1.1	Introduction	41
3.1.2	History	41
3.1.3	Processing Methods and Applications for Thermoplastics	43
3.1.4	Processing Methods and Applications for Polyurethane	46
3.1.5	Computer-Aided Engineering (CAE)	49
3.1.6	Outlook	50
3.2	Plastics for the Car of Tomorrow – Invisible Contribution: Visible Success	53
3.2.1	Introduction	53
3.2.2	Innovation Processes	53
3.2.3	Making the "Invisible" Visible: Applying Nanotechnology	54
3.2.4	Making the Virtual Real: Applied Simulation Technology	56
3.3	CFRP – from Motor Sports to Automotive Series – Challenges and Opportunities Facing Technology Transfer	59
3.3.1	Introduction	59
3.3.2	CFRP at the Transition Between Motor Sports and Road Cars	59
3.3.3	Current Applications of CFRP	59
3.3.4	The Trends: a Closer Look	61
3.3.5	Challenges	61
3.3.6	Opportunities	64
3.3.7	Outlook	65
3.4	Further Developments in SMC Technology	68
3.4.1	Reasons for Using SMC	68
3.4.2	The Road to Premium SMC	68
3.4.3	Lightweight Potential of SMC Technology	70
3.4.4	Outlook for SMC	74
3.4.5	Acknowledgements	74
3.5	New TPO Grades with Reduced Linear Thermal Expansion for Exterior Parts Painted Offline	75
3.5.1	Introduction	75
3.5.2	Requirements and Basic Conditions	75
3.5.3	Analysis and Simulation Techniques	77
3.5.4	New TPO Grades and their Properties Compared with Other Technical Thermoplastics	78
3.5.5	Component Test Results	80
3.6	Lightweight Bodywork – Use of Structural Foams in Hollow Sections	82
3.6.1	Introduction	82
3.6.2	Requirements Applicable to Manufacture	83
3.6.3	Selection of the Material System	84
3.6.4	Process Conditions and Implementation	87

3.6.5	Outlook	88
3.7	**Body-in-White turns Black**	**89**
3.7.1	Introduction	89
3.7.2	FRP Manufacturing Technologies in the Mercedes-Benz SLR Mclaren	90
3.7.3	Prepreg Technologies (Classical and RFI)	90
3.7.4	Preform Technologies	91
3.7.5	Summary	94
3.7.6	Acknowledgements	94
3.8	**Innovative Manufacturing Processes Adapted for the Production of Structural Modules**	**95**
3.8.1	Introduction	95
3.8.2	The Evolution of Production Technologies for Structural Parts	95
3.8.3	Developments of New Applications	97
3.8.4	Future Potential of a New Process and Machine Technology	98
3.9	**Innovations in Injection Molding for the Automobile of the Future**	**102**
3.9.1	Introduction	102
3.9.2	Innovative Areas for New Process Technologies	103
3.9.3	Current Process Innovations in Injection Molding	112
4	**Modelling and Rapid Prototyping**	**123**
4.1	**Innovative Solutions for Prototype-Making in Automobile Manufacturing – Prototype Component Requirements and Practical Examples**	**123**
4.1.1	Passenger Car Development	123
4.1.2	Prototype Requirements	125
4.1.3	Selection of Suppliers	129
4.1.4	Examples	129
4.1.5	Summary	133
4.2	**Rapid Prototyping of Plastic Components for Efficient Vehicle Development**	**135**
4.2.1	Introduction	135
4.2.2	The Importance of Rapid Prototyping	135
4.2.3	Rapid Prototyping – a Definition	136
4.2.4	Rapid Prototyping (RP), Rapid Tooling (RT), Rapid Manufacturing (RM): Some Differences	138
4.2.5	The Prototype Assembly Department as Internal RP Service Provider	138
4.2.6	Systems and Applications in the Prototype Assembly Department	140
4.2.7	Outlook	142
4.3	**Calculating the Fatigue Life of Short-Glass-Fiber-Reinforced Thermoplastics**	**143**
4.3.1	Introduction	143
4.3.2	Material Behavior under Cyclic Loading	143
4.3.3	Formulation for Calculated Fatigue Life Estimation	145
4.3.4	Examples of Calculation	147
4.3.5	Summary and Outlook	151

4.4	Calculated Estimate of Service Life for Components Made of Short-glass-fiber-reinforced Plastics	153
4.4.1	Introduction	153
4.4.2	Initial Situation and Objectives	153
4.4.3	Theoretical Considerations	154
4.4.4	Procedure for Estimating Service Life	158
4.4.5	Experimental Investigations	160
4.4.6	Components and Component Tests	163
4.4.7	Summary and Further Procedure	164
4.4.8	Acknowledgments	165
4.5	Material Data for Imaging Thermoplastic Plastics in Crash Simulation	167
4.5.1	Introduction	167
4.5.2	Input Data for Crash Calculation	167
4.5.3	Representation of Non-Linear Behavior in the Simulation	169
4.5.4	Fiber-Reinforced Thermoplastics	171
4.5.5	Validation of Material Data and Material Models	172
4.5.6	Failure Analysis	172
4.5.7	Outlook	172
5	**Joining**	173
5.1	Bonded Joints in and on the Vehicle Structure	173
5.1.1	Introduction	173
5.1.2	Assembly Bonding	173
5.1.3	Bonded Joints in and on the Body	175
5.1.4	Testing, Development, Dimensioning	178
5.1.5	Summary and Outlook	180
5.2	Joining Techniques Used for Integrating Fiber Composite Plastic Components in Metallic Body Structures	181
5.2.1	Introduction	181
5.2.2	Lightweight Design Using Body Structures	181
5.2.3	Joining Techniques for Hybrid Aluminum-FRP Designs	182
5.2.4	Study: Audi A2 with FRP Body Components	189
5.2.5	Summary and Outlook	190
5.3	Bonding Automotive Bodies	192
5.3.1	Bonding – Often the Only Feasible Joining Technology	192
5.3.2	New Adhesives for SMC Components Meet Maximum Requirements	195
5.3.3	Chemical Thixotropic Adhesives	197
5.3.4	Summary	197
5.4	Warm Reactive Bonding	198
5.4.1	Headlight Bonding	198
5.4.2	Current Status of the Technology	198

5.4.3		Market Requirements and Wishes	199
5.4.4		Just-in-Time Production Feasible	199
5.4.5		Influence on Handling Strength	201
5.4.6		Pump and Application System for Production	203
5.4.7		Utilization in Series Production	203
6		**Case Studies – Design, Production, Performance**	**205**
6.1		Structures and Body Panels	205
6.1.1	Case Study 1:	Film Technology in Automobile Manufacturing – a Comparison	205
6.1.2	Case Study 2:	Hybrid Tailgate: Getting the Best From Thermosetting and Thermoplastic Materials	218
6.1.3	Case Study 3:	Experience with On-Line Paintable Plastics on Vehicle Exteriors	227
6.1.4	Case Study 4:	Application of Plastic Bodywork Components in a Modern Bus	234
6.1.5	Case Study 5:	Cab Body Panels and Parts: from Thermosets To Thermoplastics	243
6.2		Front Modules, Crash Elements, Safety Concepts	253
6.2.1	Case Study 6:	Development of a Thermoplastic Lower Bumper Stiffener for Pedestrian Protection	253
6.2.2	Case Study 7:	Crash Simulation with Plastic Components	264
6.2.3	Case Study 8:	SLR Crash Element: from Concept to Volume Production	271
6.2.4	Case Study 9:	Thermoplastic Crashboxes	280
6.3		Roof Modules, Hardtops	289
6.3.1	Case Study 10:	Roof Module: As Exemplified by the New Opel Zafira	289
6.3.2	Case Study 11:	The Z4 Hardtop Made of SMC – a Self-Supporting Automotive Part Made of Thermosetting Material in the Context of CARB Legislation	299
6.3.3	Case Study 12:	Innovative Noise-Optimized Folding Top for the New BMW 6 Series Convertible	306
6.3.4	Case Study 13:	Roof Module for Commercial Vehicles Made with SMI Technology	312
6.4		Automotive Glazing	327
6.4.1	Case Study 14:	Organic Glazing in the Automobile	327
6.4.2	Case Study 15:	Polycarbonate Automobile Glazing: Automotive Industry Requirements and Solutions	335
6.4.3	Case Study 16:	Plastic Automobile Windows	346
6.5		Acoustics and Aerodynamics	354
6.5.1	Case Study 17:	Specific Requirements on Aeroacoustic Development for Convertibles	354
6.5.2	Case Study 18:	Automotive – Compression Molding – LWRT Technique for Car Underbody Covers	365
6.5.3	Case Study 19:	Noise-Reducing Coatings in Buses: the Mercedes-Benz Citaro	369
Appendix			**383**
Subject Index			**393**

List of Contributors

Editors

Stauber, Rudolf, Prof. Dr., BMW Group, Munich, Germany

Vollrath, Ludwig, Dr.-Ing., VDI-Gesellschaft Kunststofftechnik, Düsseldorf, Germany

Authors

Adam, Frank, Dr.-Ing., Institut für Leichtbau und Kunststofftechnik, TU Dresden, Dresden, Germany, Chapter 1

Ader, Stephanie, Plastic Omnium, Sainte-Julie, France, Chapter 6.1.2

Aengenheyster, Gerald, Dr.-Ing., Freeglass GmbH & Co. KG, Schwaikheim, Germany, Chapter 6.4.2

Albers, Hartmut, Dipl.-Ing., DaimlerChrysler AG, Sindelfingen, Germany, Chapter 4.1

Arntz, Hans-Detlef, Dr., Bayer Material Science AG, Leverkusen, Germany, Chapter 3.1

Balika, Werner, Polymer Competence Center, Leoben, Austria, Chapter 4.4

Bangel, Martin, Dr.-Ing., AUDI AG, Neckarsulm, Germany, Chapter 5.2

Bechtold, Michael, Dipl.-Ing., Dipl. Wirt.-Ing., formerly DaimlerChrysler AG, Sindelfingen, Germany, Chapter 3.7, 6.2.3

Borne, Peter, Wilhelm Karmann GmbH, Osnabrück, Germany, Chapter 6.5.1

Brambrink, Roland, Bayer Material Science AG, Leverkusen, Germany, Chapter 3.1

Brune, Martin, Dr., BMW Group, Munich, Germany, Chapter 4.3, 4.4

Buchholz, Udo, Dr., Huntsman Advanced Materials GmbH, Basel, Switzerland, Chapter 5.3

Bürkle, Erwin, Dr.-Ing., Krauss Maffei Kunststofftechnik GmbH, Munich, Germany, Chapter 3.8

Büthe, Ingolf, Dr., BASF Aktiengesellschaft, Ludwigshafen, Germany, Chapter 3.2

Buron, Marie-Pierre, Dr., FAURECIA DF-Audincourt Cedex, France, Chapter 3.8

Corsi, Jacopo, IVECO SpA., Torino TO, Italy, Chapter 6.1.5

Davis, Roger, Ford of Europe, Laindon, UK, Chapter 2.2

Deinzer, Günter H., Dipl.-Ing., AUDI AG, Ingolstadt, Germany, Chapter 5.2

Derks, Martin, Dipl.-Ing. (FH), BMW Group, Munich, Germany, Chapter 3.6

Dumazet, Philippe, Dr., Hagenbach Faurecia, Hagenbach, Germany, Chapter 3.8

Elseberg, Dirk, Wilhelm Karmann GmbH, Osnabrück, Germany, Chapter 6.5.1

Erzgräber, Matthias, Dipl.-Ing., Adam Opel GmbH, Rüsselsheim, Germany, Chapter 6.2.1

Fleischer, Harald, BMW Group, Munich, Germany, Chapter 4.4

Forster, Jan, Dipl.-Ing., formerly Institut für Kunststoffverarbeitung der RWTH Aachen, Aachen, Germany, Chapter 6.4.3

Frank, Uwe, Dr.-Ing., BMW Group, Munich, Germany, Chapter 6.3.2

Frik, Steffen, Dr.-Ing., Adam Opel GmbH, Rüsselsheim, Germany, Chapter 6.2.1,

Funkhauser, Steffen, Dipl.-Ing., BASF Aktiengesellschaft, Ludwigshafen, Germany, Chapter 3.2

Geisler, Wolfgang, Dr., Volkswagen AG, Wolfsburg, Germany, Chapter 4.2

Ginsberg, Lutz, Dipl.-Ing. (FH), Neoplan Bus GmbH, Stuttgart, Germany, Chapter 6.1.4

Glaser, Stefan, Dr., BASF Aktiengesellschaft, Ludwigshafen, Germany, Chapter 6.2.1

Göbel, Sebastian, Dipl.-Ing., Braun GmbH, Kronberg, Germany, Chapter 6.4.3

Guster, Christoph, Montanuniversität Leoben, Austria, Chapter 4.4

Haldenwanger, Hans-Günther, Prof. Dr.-Ing., formerly AUDI AG, Ingolstadt, Germany, Chapter 5.1

Harrison, Allan, Ford of Europe, Laindon, UK, Chapter 2.2

Hopmann, Christian, Dr.-Ing., RKW AG, Petersaurach Germany, Chapter 6.4.3

Horstmann, Jürgen, formerly Hella GK Hueck & Co., Lippstadt, Germany, Chapter 5.4

Hufenbach, Werner, Prof. Dr.-Ing. habil., Institut für Leichtbau und Kunststofftechnik, TU Dresden, Dresden, Germany, Chapter 1

Kalinke, Peter, Dr., formerly Wilhelm Karmann GmbH, Osnabrück, Germany, Chapter 6.5.1

Kampke, Manfred, Dr.-Ing., Roechling Automotive AG & Co. KG, Ingolstadt, Germany, Chapter 6.5.2

Kempf, Jürgen, Dipl.-Ing., BMW Group, Munich, Germany, Chapter 3.6

Kim, Patrick, Dr., formerly DaimlerChrysler AG, Sindelfingen, Germany, Chapter 3.3

Kölle, Olaf-Björn, Volkswagen AG, Wolfsburg, Germany, Chapter 4.2

Kühfusz, Rudi, Menzolit-Fibron GmbH, Kraichtal-Gochsheim, Germany, Chapter 6.3.2

Lehner, Erich, Dr.-mont, DaimlerChrysler AG, Sindelfingen, Germany, Chapter 6.4.2

Lepper, Martin, Dr.-Ing., Institut für Leichtbau und Kunststofftechnik, TU Dresden, Dresden, Germany, Chapter 1

Leroy, Alain, Dipl.-Ing., formerly Webasto AG, Stockdorf, Germany, Chapter 6.3.1

Liebertz, Helge, Dr., Volkswagen AG Wolfsburg, Germany, Chapter 4.5

Liebold, Rolf, Bretten, Germany, Chapter 6.3.2

Ludwig, Hans-Joachim, Dr.-Ing., Decoma Europe, Altbach, Germany, Chapter 6.1.1

MacKenzie, Paul, McLaren Composite, Portsmouth, UK, Chapter 3.7

Maier, Anja, Dipl.-Ing., BMW Group, Munich, Germany, Chapter 2.3

McDonagh, Michael, Ford of Europe, Laindon, UK, Chapter 2.2

Mederle, Günther, Dipl.-Ing. (FH), MAN Nutzfahrzeuge AG, Munich, Germany, Chapter 6.3.4

Merz, Peter W., Dr., Sika Technology AG, Zürich, Switzerland, Chapter 5.4

Meyr, Wolfgang, Dipl.-Ing., BMW Group, Munich, Germany, Chapter 2.3

Michalowski, Simon, Wilhelm Karmann GmbH, Osnabrück, Germany, Chapter 6.5.1

Möltgen, Bruno, Dipl.-Ing., DaimlerChrysler AG, Sindelfingen, Germany, Chapter 6.2.3

Moos, Egon, Dr.-Ing., Roechling Automotive AG & Co. KG, Worms, Germany, Chapter 6.5.2

Moulin, Jean-Paul, Plastic Omníum, Sainte-Julie, France, Chapter 6.1.2

Müller, Reinhard, Adam Opel GmbH, Rüsselsheim, Germany, Chapter 6.2.2

Paul, Reiner, Dr., Bayer Material Science AG, Leverkusen, Germany, Chapter 3.1

Protte, Rainer, Bayer Material Science AG, Leverkusen, Germany, Chapter 3.1

Radunz, Herbert, Bayer Material Science AG, Leverkusen, Germany, Chapter 3.1

Rau, Walter, Dipl.-Ing., BASF Aktiengesellschaft, Ludwigshafen, Germany, Chapter 3.2

Riepenhausen, Holm, Dipl.-Ing., REHAU AG & Co., Rehau, Germany, Chapter 6.2.4

Roth, Günther, BMW Group, Munich, Germany, Chapter 6.3.3

Sauer, Jochem, Dr.-Ing., Huntsman Advanced Materials (Switzerland) GmbH, Basel, Switzerland, Chapter 5.3

Schelisch, Lutz, formerly Adam Opel GmbH, Rüsselsheim, Germany, Chapter 6.3.1

Schmid, Georg, Dipl.-Ing. (FH), AUDI AG, Ingolstadt, Germany, Chapter 5.2

Schmidt, Frank, Wilhelm Karmann GmbH, Osnabrück, Germany, Chapter 6.5.1

Schuh, Thomas, Dr., DaimlerChrysler AG, Sindelfingen, Germany, Chapter 3.3, 3.4. 6.1.3

Seufert, Martin, Dipl.-Ing. (FH), AUDI AG, Ingolstadt, Germany, Chapter 2.1

Siener, Edmund, chem. Dipl.-Ing. (FH), EvoBus GmbH, Mannheim, Germany, Chapter 6.5.3

Starke, Joachim, Dr.-Ing., formerly Roechling Automotive AG & Co. KG, Worms, Germany, Chapter 6.5.2

Steuer, Ulrich, Dipl.-Ing., AUDI AG, Ingolstadt, Germany, Chapter 2.1

Steinbichler, Georg, Dipl.-Ing., Engel Austria Gesellschaft mbH, Schwertberg, Austria, Chapter 3.9

Sullivan, John, Ford of Europe, Laindon, UK, Chapter 2.2

Ullmann, Falk, Dr.-Ing., DaimlerChrysler AG, Sindelfingen, Germany, Chapter 6.4.1

Vogel, Julius, Institut für Kunststofftechnologie, Universität Paderborn, Germany, Appendix B

Walther, Ulrich, Dr.-Ing., formerly AUDI AG, Ingolstadt, Germany, Chapter 5.1

Wazula, Fritz, DaimlerChrysler AG, Sindelfingen, Germany, Chapter 6.1.3

Wegmann, Ulf, Wilhelm Karmann GmbH, Osnabrück, Germany, Chapter 6.5.1

Wehner, Joachim, Dipl.-Ing. (FH), BMW Group, Munich, Germany, Chapter 2.3

Wentzien, Heino, Dipl.-Ing. (FH), formerly Adam Opel GmbH, Rüsselsheim, Germany, Chapter 3.5

Wetter, Dr.-Ing., AUDI AG, Ingolstadt, Germany, Chapter 5.2

Windpassinger, Martin, Dipl.-Ing. (FH), Parat Automotive, Neureichenau, Germany, Chapter 6.3.3

Witek, Wolfgang, Dr.-Ing., BMW Group, Munich, Germany, Chapter 6.3.2

Wittig, Wolfgang, Dr., formerly DaimlerChrysler AG, Stuttgart, Germany, Chapter 3.3

Wüst, Andreas, Dipl.-Ing., BASF Aktiengesellschaft, Ludwigshafen, Germany, Chapter 6.2.1

Zängerl, Franz, Dipl.-Ing., Borealis GmbH, Linz, Austria, Chapter 3.5

Zago, Alessandro, Dr., formerly BMW Group, Munich, Germany, Chapter 4.3

Zöllner, Olaf, Bayer MaterialScience, Leverkusen, Germany, Chapter 3.1

1 Introduction to the Application of Plastics in Vehicle Design

WERNER HUFENBACH

1.0 Introduction

Tomorrow's cars should be lighter, safer, more economical, more comfortable and, at the same time, ecologically compatible – all in the face of global competition and with high net product. This guarantees that there will be not only conflicting goals, but also a continuing trend to weight-saving multi-material design and its inherent advantages and drawbacks. Above and beyond these, multifarious emotional und socio-dynamic factors have to be taken into consideration.

State-of-the-art automobile designs stand out for the way their structure is beautifully adapted to their use requirements. In addition to technical and technological restrictions, design concepts are influenced especially by economic-ecological guidelines and international specifics. All previous efforts in light-weight design have only been partially successful at compensating weight increases necessitated by legislative demands for greater safety in vehicles, as well as customer wishes for greater comfort.

Since the technical issues in vehicle engineering are so complex, the competition among numerous materials and technologies is loud and obvious. It is the driving force behind innovative solutions in materials and processes tailored to specific applications. Any integrated concept has to involve the entire chain of value creation, as well as the entire service life of the product right down to its recycling.

In materials development, the goal is materials compositions suitable for loading and recycling. Plastics and composites, as well as sandwich and polymer-metal hybrid composites, are being included ever more often in the developmental search for innovative component solutions.

1.1 Plastics in Lightweight Vehicle Engineering

To start with, it should be made clear that there are no patent recipes for developing optimized lightweight products. Instead, the choice of a design strategy is largely dictated by the case of application, each time presenting a new and considerable engineering challenge. This does not make it easy to conceive modern vehicle structures, since there are hardly any pertinent guidelines and catalogues to fall back upon. Development is always a novel task, requiring a high level of creativity and an interdisciplinary team of colleagues covering the entire engineering chain, starting with materials development and processing, through virtual product development and the manufacturing process, all the way to quality assurance and recycling in one holistic lightweight concept.

For vehicle construction to be cost-efficient, it requires a detailed analysis of weight ratios based on the selected chassis and drive concepts. For instance, it should be obvious that, given front-wheel drive, especially the entire vehicle front, not just the front and rear crash zones, has to be weight-optimized due to the weight concentration there (power pack, drive train). This achieves a balanced weight distribution on front and rear axles, with a positive effect on handling. Lightweight construction at the rear of the vehicle is more effective on a rear-drive vehicle due to the greater weight there (differential etc.). In the safety cell, lightweight measures in the roof area aim at lowering vehicle center of gravity. In the floor area, it makes sense to use lightweight materials only when body shell and power train can prevent the vehicle center of gravity from shifting upward.

When vehicle components are being developed, great importance is attached to the determination of loading state, as well as to implementing a corresponding design concept with suitable materials and structures – taking all production restrictions into consideration. Already in the incipient stage of development, appropriate decisions have to taken affecting technical-economic, repair and recycling aspects of the specific application, as to which basic type of construction will provide optimum results in the case at hand:

- differential
- integral or
- integrating hybrid.

Every new or further development should adhere to the guiding principle that choices of materials always have to be made directly in connection with the choice of construction and the choice of production method!

Integral components have a high level of function integration achievable in production by modern injection molding and compression technologies. By contrast, individual components in differential construction are often complex structures bonded together from different materials by suitable jointing techniques.

Here, gluing is a cold-jointing technique ($T \leq 200\ °C$) that is especially suited for laminar bonding any number of materials combinations that either cannot be joined by other methods, or only at considerable expense. As gluing techniques continue to be refined and perfected, we will see an increasing number of structurally glued supporting connections. A further trend can be seen to combining gluing with other jointing techniques in order to compensate the specific disadvantages of each technique. Combination jointing will help achieve notably higher dynamic bonding strength, since the result is relatively uniform force transmission over a wide area, and since gluing inhibits crack initiation.

Well-adapted jointing concepts are required in order to exploit the potential for lightweight supporting structures from plastics and/or polymer matrix composite materials. For example, in the 2005 VW Passat, Remform screws are used to fasten add-ons directly into the mounting panel (KMU) made from PP-GF30, while lines and cables are held in place by integrated clip functions [1].

On the other hand, metal-plastic hybrid technology has been increasingly applied in many supporting body components. This hybrid technology is based on the combination of two proven production methods – metal deep-drawing and polymer injection molding – and produces ready-to-mount hybrid components with high function integration. Considerable remaining lightweight reserves can be mobilized by these methods thanks to the adaptability of their stiffness and strength to loading conditions.

Load transmission areas and cutouts are especially fracture-prone zones in the thin-walled constructions desired, since thin-walled profiles and panels are relatively sensitive to buckling and denting. Such types of instability failure over a large loading area can be largely eliminated by using calculated plastics rib structures. For instance, steel can be loaded up to its yield point in GFC-steel hybrid components configured to application if the thin-walled sheets can be prevented from collapsing prematurely by including a ribbed structure made from glass-fiber reinforced polyamid (PA6-GF).

Moreover, a number of functions, in addition to the structural stiffening achieved by ribbing, can be integrated into the component in one operation by defined coatings on sheet metal parts. For instance, necessary functional elements (e.g., inserts) are bonded by one-shot injection molding into integral front-end beams on the Ford Focus, Audi A6, to name a few, while simultaneously improving the functional properties of the component. Doing so reduces both the number of individual parts and processing steps, as well as the expense of assembling the component.

As this is written, 20 million vehicles containing such hybrid components have been produced worldwide [2]. Current research is concentrating especially on the production of door modules using this innovative technology (see Figure 1-1).

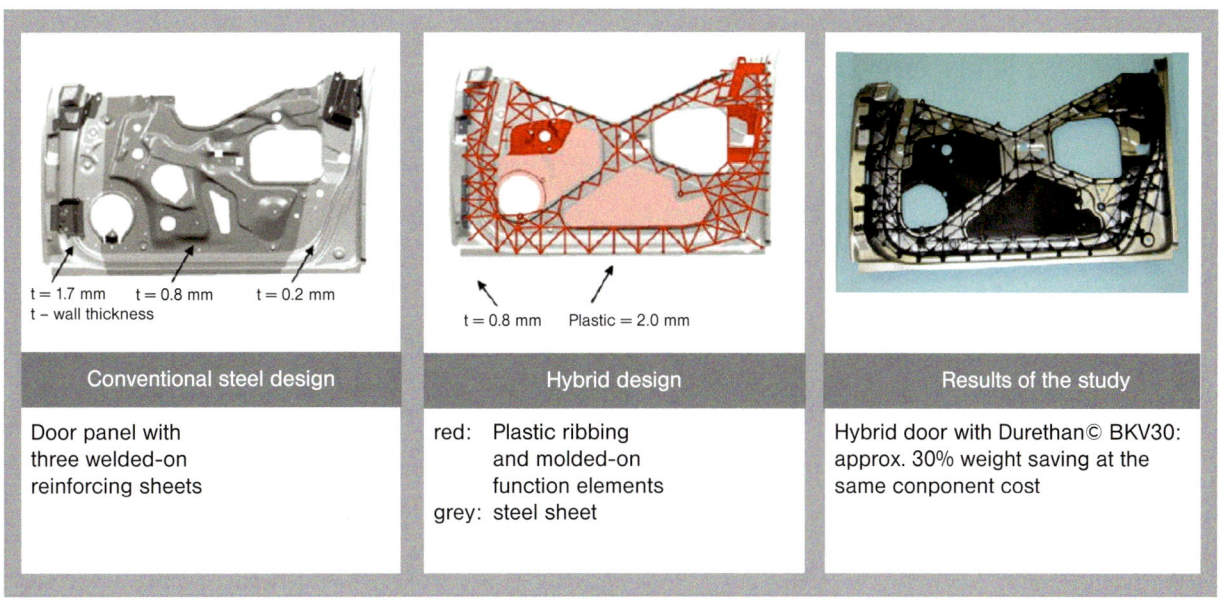

Figure 1-1 Door module study using metal-plastics hybrid technology [2]

Thanks to their low density, plastics are especially suited for lightweight applications in vehicle structures. Out of the wide variety of available plastics, almost every technical application can find a polymer that satisfies its particular geometrical, thermo-mechanical, medial, physiological and esthetic requirements. Combining the polymer with additives and reinforcers enables far-going modification of technical properties, such as strength, elasticity, damping, electrical, thermal and acoustic insulation/damping, including friction and wear behavior.

The multi-faceted properties spectrum of polymers has secured their position in vehicle engineering. Applications range from interior to engine room, from structural components to exterior body parts [3]. Whereas plastics applications in interiors and small parts have hardly any room for growth, since the possibilities have been largely exhausted (e.g., mirror frames, hubcaps), the increasing trend to specialty vehicles indicates that there is considerable development potential for the use of shape-stable high-performance plastics under the hood, as well as for hybrid technology (metal-plastics composites) especially for large body panels, as well as structural components.

1.2 Plastics Applications for the Exterior

The development of plastics for car bodies is targeted at their utilization on the basis of particular, complex requirement profiles as to strength, stiffness, class A surface and color matching. In addition to high stiffness and shape stability, low water-absorption in the enameling line is a must for success in mass production. Modern thermoplastics, such as mineral-reinforced polyamid-ABS blends, are an obvious choice for such applications. Thanks to their low thermal linear expansion coefficient, they keep the gaps narrow between neighboring assemblies [4].

The technological properties of plastics make them economical and highly productive to process, providing extensive freedom for shaping components creatively, while ensuring dimensional and shape stability, as well as high function integration. The highly automated processing methods widely available are characterized by ideal material utilization with relatively low energy requirements. Short cycle times, especially when thermoplastics are processed, are another advantage.

1.2.1 Injection Molding with Polymers

Processing methods are subject to constant further development. The current trend is tending toward combined technologies for producing function-integrating components that are close to final contour. Besides essential shaping, final surface quality (in-mold decorating in injection, compression or blow molding) or functional elements are formed from a second polymer (Figure 1-2).

Polymers can be reinforced with fibers to increase their strength and stiffness properties. Their thermomechanical properties can be adapted to loading via fiber orientation. Fiber-polymer composites (FPC) can exhibit load-adapted properties by optimum use of injection molding technology, thus offering remarkable lightweight potential.

Some of the decisive advantages of, for example, a glass-fiber reinforced thermoplastic component (e.g. trunk lid) produced by low-pressure injection-compression molding and equipped with a coextruded multi-layer foil are:

- high component stiffness
- good crash behavior even at low temperatures
- low thermal elongation
- low structural weight
- class A surface enamel quality
- high creative freedom with functional integration
- good materials recycling
- production cost savings by direct compounding on an injection molding compounder.

The elimination of subsequent enameling by using coextruded foils in conjunction with lower materials costs achieves cost advantages vis-à-vis classic production methods.

Whereas processing short-fiber polymers with 0.2 to 0.5 mm fiber lengths has become standard technology, current further research involves the development of injection molding for long-fiber thermoplastics (LFT). Short fibers alone are incapable of exhausting the potential that fiber reinforcing holds for increasing strength and stiffness properties of unreinforced thermoplastics. Whereas fiber-glass mat reinforced thermoplastics (GMT) have been compression molded

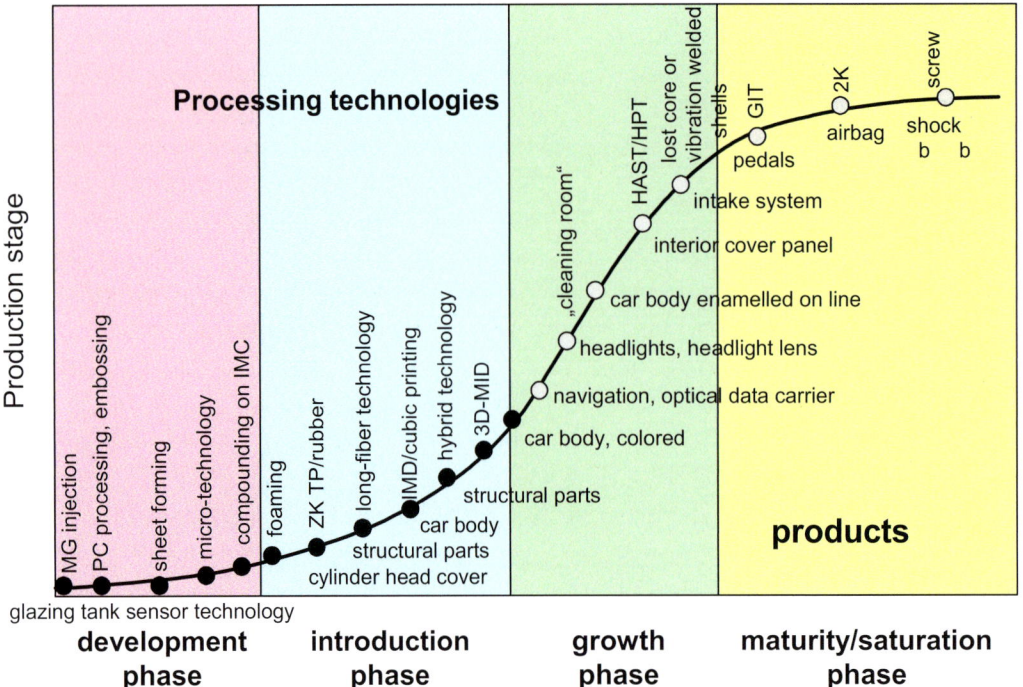

Figure 1-2 Special injection molding methods for plastics parts in vehicle engineering [5]

for some time, recent processing developments involve compounding thermoplastics directly with additives and glass fiber roving, chopped strands (LFT-D) or long-fiber pellets (LFT-G). Depending on the raw materials supplier, fiber lengths range from 10 to 25 mm.

For the matrix material, economical polypropylene is used, in addition to polyamid, für engine-room structures. Some current applications of PP-LFT involve injection molded, highly adapative interior and exterior components, such at the instrument panel of the Škoda Fabia or the front assembly beam on the VW Touran und VW Tuareg. Underlaid car body components of the Audi A2 are a further example for the utilization of PP-GF. Here the reinforced thermoplastic is elastomer-modified by addition of EPDM to increase its impact strength.

Recently, SMC (sheet molding compound) and RTM (resin transfer molding) technologies have been increasingly applied in the production of lightweight components from fiber-reinforced thermosets. Especially the reinforcing in RTM consists of carbon fibers (CF) whose remarkably high strength and stiffness potential suits them for utilization in external structural parts (car body segments) that must sustain extreme loads. Structural and/or class A components on the basis of glass-fiber reinforced SMC are exemplified by the trunk lid of the BMW 600 series convertible and coupe or the Mercedes-Benz models CL and CLK. The goal was to exploit the high level of functional integration in addition to the variety of shapes feasible in production. Thanks to the favorable electromagnetic properties of SMC materials, various antenna systems (radio, GPS, etc.) could be integrated directly into the suspension system. Design-technological measures achieved, for example, approx. 17% weight-savings with an SMC trunk lid for the BMW 600 series coupe [6].

The RTM process finds application especially in the production of large-area CFK structures, e.g., producing the roof of the BMW M6 coupe. By reducing roof weight with CFK, the vehicle center of gravity is lowered and driving dynamics improved. Moreover, the visible CF weave serves as a styling feature with a high-tech look. Further large-area high-tech applications for CFK in vehicle engineering can be found in the Mercedes-Benz SLR coupe or Porsche Carrera GT, whose entire monocoque is manufactured from this composite material.

For the near future, polyurethane will be applied more and more for exterior car body components. The application potential has not yet been exhausted for the basic polyole and isocyanate materials. Their wide properties spectrum is adjustable from highly elastic to hard-brittle behavior. The component-oriented combination of PUR methods, such as long-fiber injection (LFI) with in-mold coating, enables economical production of complex structures with high surface quality.

1.2.2 Textile-reinforced Plastics

Compared to conventional fiber-composites, these materials possess the greatest flexibility for adapting the structure of the material to loading. That predestines them for optimizing the materials mix in the hybrid technology required by the complex requirements of lightweight vehicle engineering. The structural properties of components from textile composites have variable, directionally dependent structural properties formed in a simultaneous process involving materials design and component configuration. This requires all process stages to be especially tightly meshed, something unknown with conventional materials. Therefore, all R&D work for vehicle applications has to focus on a thoroughgoing investigation of the entire process for developing hybrid lightweight structures with thermoplastic textile composites. This means working out essential information for all links in the value-creation chain, starting with the filament, through hybrid yarn, semi-finished material and textile preform all the way to consolidated, function-integrating hybrid components, including reproducible quality and short cycle times.

For structural reasons, composites reinforced with conventional weave exhibit low stiffness under flexural and tensile loading. Therefore, multi-layer hybrid yarn weaves (MLW) were developed. They combine the advantages of weave, such as excellent drapability and good impact behavior, with those of biaxial reinforced weaves and can be realized in a product. This structure can be combined in well-known engineering techniques for the flexible production of flat or spatial weave-preforms. The resulting advantages of MLW can find application in three-dimensional reinforcement of polymers, for example, in safety-relevant thermoplastic components for vehicle engineering.

The application of this still nascent materials group seems to lie in the development of innovative smart materials. Textile preforms seem ideally suited for integrating actor and sensoric systems directly into the material. Textile reinforced plastics thus offer enormous future potential for tomorrow's vehicles.

1.3 Outlook

The trend to ever more sharply differentiated customer needs has to be met by innovative derivative strategies in vehicle engineering. Multi-material design will thereby gain considerable significance. Modern plastics and composites have extraordinary market potential due to their wide properties spectrum and great creative freedom.

This requires meshing materials development, component development and process development ever more tightly and efficiently into intelligent system development. The value creation chain has to be thoroughly mapped over the entire product growth process and product service life including recycling. In the long term, this is the only way to secure and expand competitive advantages in the global market.

References

[1] Hillmann, J., u. a.: Karosserie – Perfektion in Anmutung und lightweight. Der neue Passat, ATZ-Sonderausgabe April 2005
[2] Koemm, U.: LANXESS Medientag "Innovation – LANXESS liefert Antworten" LANXESS AG, 14. September 2006
[3] Stauber, R.; Cecco, Ch.: Moderne Werkstoffe im Automobilbau. Werkstoffe im Automobilbau, ATZ-Sonderausgabe, November 2005
[4] Erbstößer, U.: Innovative Lösungen für den Kunststoffsektor. Presseinformation: LANXESS auf der Plastics Design & Moulding, 2006
[5] Bürkle, E.: Kunststoffe 90 (2000) 1
[6] Grün, R., u. a.: Die Kunst des Karosseriebaus. Der neue BMW 6er, ATZ-Sonderausgabe Mai 2004

2 Automotive Concepts and Lightweight Design

2.1 Increased Use of Plastics in Body Applications: Opportunities and Risks

Ulrich Steuer

2.1.1 Current Situation of Lightweight Materials

The plastics industry forecasts that the amount of plastics used in each automobile will increase to 18% by 2008 [1], while the aluminum industry predicts an increase in the amount of aluminum used in the body from 16 to 26 kg and in the entire vehicle from 101 to 134 kg by 2005 [2]. Audi has already reached this scenario. The proportion of polymers used in current models ranges from 17.4 to 22.5%, while the amount of light alloy varies between 12.3 and 34.3% or 181 and 602 kg, respectively [3].

Neither one of the materials evaluated had an advantage because of market-related conditions. A comparison of aluminum and polypropylene prices shows that price fluctuations very much run parallel. This means that there are no special economic advantages or disadvantages for lightweight materials [4]. The comparative prices for thin sheet metal are subject to considerably smaller changes [5] and characterize the current competitive pressure on steel.

The increase in the amount of plastic used in automotives can be attributed to innovative developments but is also related to the increase in vehicle weight and production quantities. The percentage of steel has decreased as far as the vehicle itself is concerned but at the same time, total quantities have increased. This is related to stricter requirements on safety, driving comfort, compliance with relevant legislation, vehicle equipment and driving performance.

Optimizing running resistances and efficiencies in the drive system has – particularly with the introduction of direct-injection diesel engines, 6-gear transmissions, aluminum bodies and Cw values approaching 0.25 – limited the once dominating influence of weight-reduction measures in the vehicle on fuel consumption.

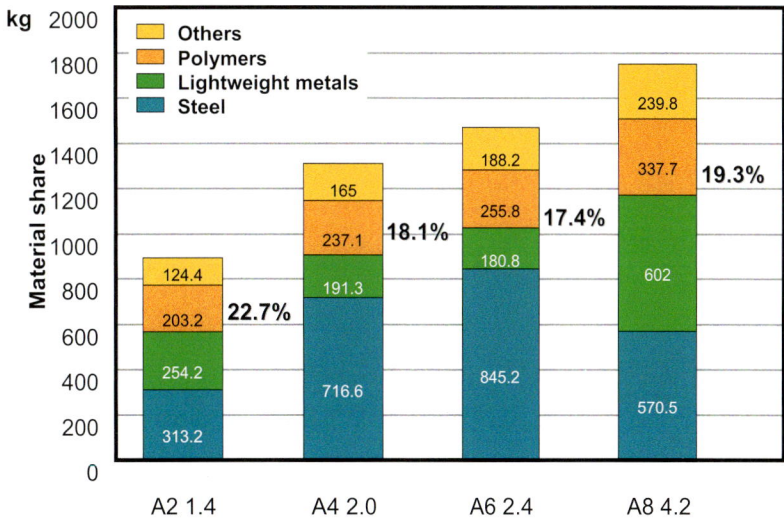

Figure 2-1 Proportions of materials in Audi models

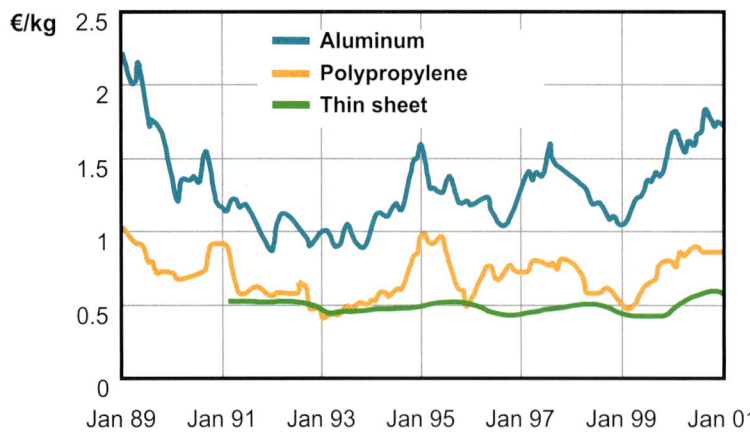

Figure 2-2 Course of raw materials prices on the world market

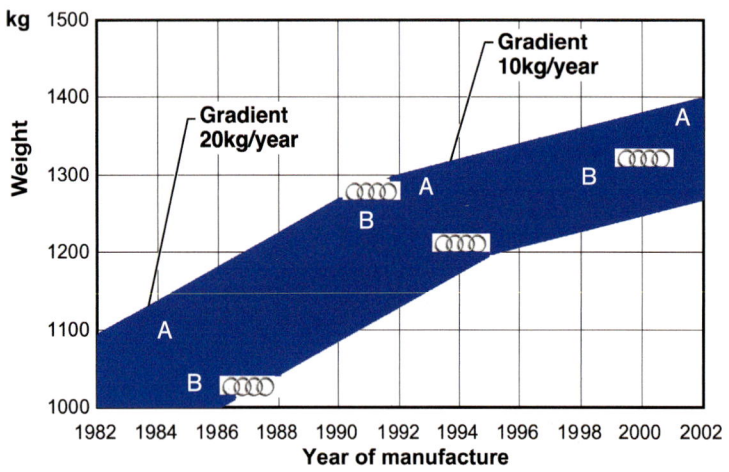

Figure 2-3 Trends in weight in the B segment

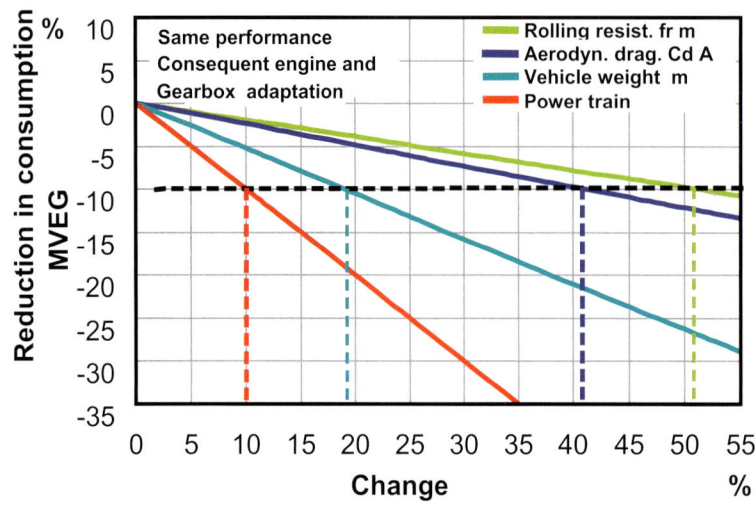

Figure 2-4 Performance and fuel consumption

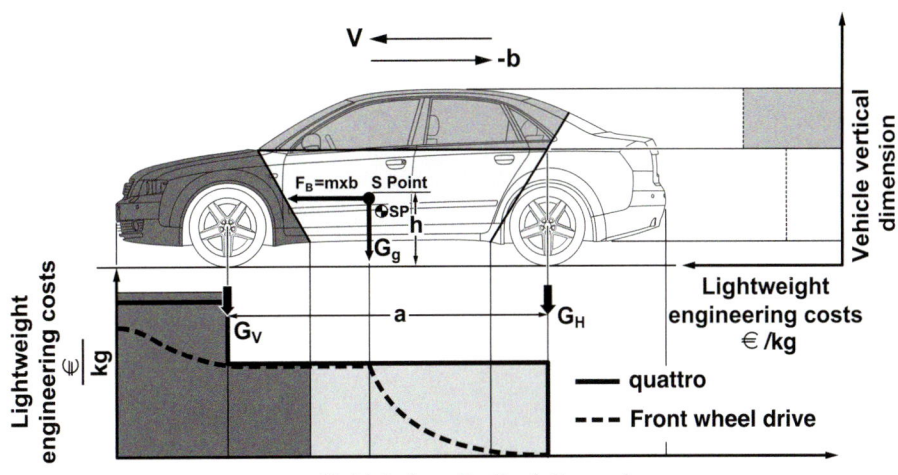

Figure 2-5 Areas of lightweight construction in the vehicle

An additional 10% reduction in weight in a state-of-the-art vehicle such as the Audi A2 Öko only causes a decrease in fuel consumption of 0.15 litres per 100 km [6, 7]. In the case of mass-produced vehicles it therefore makes more sense to invest in the development of direct-injection gasoline engines, such as Audi's FSI engine with a reduction of up to 15% in fuel consumption, than in efforts to reduce weight.

Nonetheless, weight reduction still remains very important but is being given a different priority. In the future, the focus will be on achieving weight reduction and a better distribution of weight – in other words, the load on the front axle *and* the height of the centre of gravity will need to be reduced if further improvements in vehicle handling and performance are to be achieved.

2.1.2 Future Regulatory and Market Requirements

2.1.2.1 HC Emissions

Since the 2004 model year new HC limits for evaporative and refuelling emissions apply on the American market to 40% of the vehicles. As of the 2007 model year, the limits must be met by all vehicles.

The permitted daily HC evaporative emission loss during a 3-day parking period at ambient temperatures from 18 to 40 °C (SHED test) is being reduced from 2.0 to 0.5 grams per test. Since the total amount of organic compound emissions is being measured, emissions from all components have to be taken into consideration. The permissible quantity of emissions

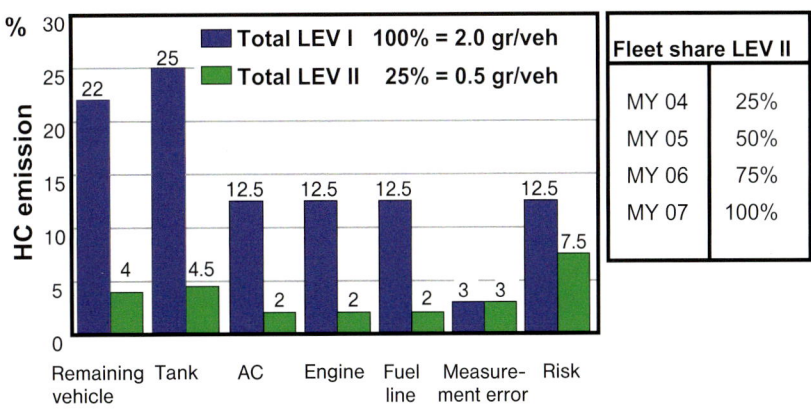

Figure 2-6 HC limits for evaporative and refuelling emissions in the SHED test

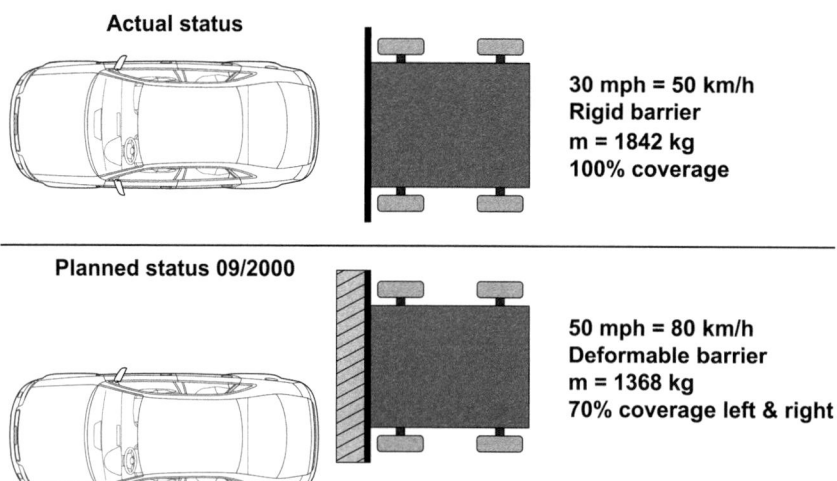

Figure 2-7 New US rear impact legislation (FMVSS 301)

is independent of the size of the vehicle. Finding a solution for this design task is therefore much more difficult for a high-end car than for a compact car. A calculation of the distribution indicates that HC emissions will have to be reduced by a factor of 4 to 5 for all components. This will involve all fuels, paints, adhesives, elastomers and thermoplastics. Developmental work is currently under way with these goals in mind. The greatest source of emissions, of course, is the fuel tank itself. A reduction of the diffusion losses to zero is only feasible with a hermetically sealed fuel tank made of high-quality shapable stainless steel.

2.1.2.2 Rear Impact at 80 km/h

Since September 2004 new requirements apply for rear and side impacts according to FMVSS 301. This tightens the requirement that the fuel system be left intact. For the rear impact test, vehicle overlap, barrier design, barrier weight and also the impact velocity are all changed.

For the side impact text, the overlap remains unchanged. However, the deformable barrier is used with an increased impact velocity: up from 32 to 54 kph. The requirement for an intact fuel system still applies. This affects the current state of development for plastic spare-wheel wells and the plastic fuel tank.

2.1.2.3 Pedestrian Protection

New requirements for the front module structure will arise resulting from regulations for the protection of pedestrians struck by vehicles. Impact protection for pedestrians will be improved in two stages. Stage A is a registration requirement for new vehicles since July 2005, while stage B will apply as of 2010. After 2015, all new vehicles must provide pedestrians with protection according to the regulations in stage B. In addition to the test criteria shown in the diagram, the compliance criteria, such as reduction of the permitted HIC values, will also be tightened.

As long as there are no intelligent, anticipatory systems for accident prevention available, active (airbags) or passive (deformation zones) elements will have to be integrated into the vehicle front end structure. These changes could affect not only the vehicle design, the engine compartment package (in the event that the engine bears on the front axle) but also the selection of materials for components and assemblies.

2.1.2.4 Recycling of Old Vehicles Directive

In future, environmental compatibility, safety and functionality aspects will demand equal priority in the development of automobiles. A holistic approach

2.1 Increased Use of Plastics in Body Applications: Opportunities and Risks

Figure 2-8 Pedestrian protection commitment ACEA/EC

Leg impactor				
Part	m [kg]	v [km/h]	Knee bend angle	Tibia accel.
A	13.4	40	21°	< 200g
B	13.4	40	15°	< 150g

Thigh impactor				
Part	m [kg]	v [km/h]	Total force	Bending torque
A	----	----	----	----
B	12-20	20-40	5 kN	300 Nm

Child head impactor			
Part	m [kg]	v [km/h]	HIC
A	3.5	35	<1000 (2/3 of area) <2000 (1/3 of area)
B	2.5	40	<1000

Adult head impactor			
Part	m [kg]	v [km/h]	HIC
A	4.8	35	----
B	4.8	40	<1000

Windshield: test not evaluated

to these issues will not only minimize the generation or release of harmful substances during vehicle manufacture and during vehicle service but also the consumption of resources such as fuels and raw materials [8]. The avoidance of waste is also gaining increasing importance.

While today there is no mandatory quota for material or thermal recycling, a material recycling quota of > 80 wt.% and an energy recycling quota of > 5 wt.% will be mandatory as of 2006 according to the EC directive. As of 2015, the amount for material recycling will be raised to 85 wt.% and that for energy recycling to 10 wt.%. Confirmation that these recycling quotas can be reached were required by 2005. This means that end-of-life vehicles will be recycled up to 95 wt.%. The remaining amount for landfill can be further reduced with other types of recycling. In general, the principle of 'material recycling rather than thermal recycling rather than landfill' applies.

These requirements will influence not only material selection but also the actual design of components and assemblies. Vehicles of the future must be not only easy to assemble but also easy to disassemble, thus minimizing recycling costs. Depending on their separability, materials within the same module or assembly should be mutually compatible. A useful approach to reach this goal is to contract entire assemblies out to development suppliers who will be able to develop financially more attractive recycling concepts for these assemblies rather than for individual parts.

In addition to the stricter recycling requirements, the future will also see the continued replacement of those materials with questionable health or safety (biocompatibility) records. Accordingly, the use of lead, mercury, cadmium and chrome VI in all vehicles was banned by Appendix II of EC Directive 2000/53/EC in July 2003.

Figure 2-9 Present and future recycling quotas

2.1.3 Lightweight Construction for Rigid Components

Exterior body parts such as exterior door panels, lids and fenders are dimensioned for flexural strength and buckling resistance. Therefore, the modulus of elasticity of the selected materials plays a dominant role. To obtain the same rigidity as with sheet steel with a thickness of 0.75 mm, the use of other materials leads to wall thicknesses up to 4.5 mm as shown in Figure 2-10.

The consideration of weights per unit area provides a drastically different picture, see Figure 2-11. Using plastic materials such as PP-TV, PPO/PA blends and RRIM, approximately the same weight per unit area as for wrought aluminum alloys can be reached. However, the most favorable material regarding weight is magnesium. Although SMC has the same specific weight, its modulus of elasticity is still lower by a factor of 2 to 3 than that of magnesium and the very low wall thicknesses are technically difficult to achieve.

The current price for wrought magnesium alloys is still above € 15 per kg. Magnesium has an elongation at break of approx. 22% [9] and is only suitable for technically less demanding parts unless they are thermoformed at 220 °C [10]. In addition, an economic solution for corrosion protection needs to be found. Until these issues are resolved, magnesium will be more likely to find applications in vehicle interior. However, considering the volume of production for wrought aluminum alloys with prices of € 5 per magnesium will be competitive with aluminum and plastics, once this price is reached, see Figure 2-12.

Series production applications already exist for magnesium die castings with a material price of € 2.80 per kg. Examples of the use of magnesium include an interior door component of the Mercedes SLC, the roof frame of a Porsche convertible, the top cover of a BMW 3 Series convertible and the interior frame of the tailgate of the 3-liter Lupo.

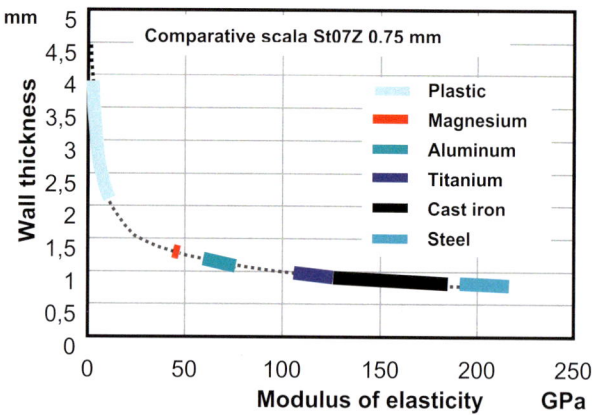

Figure 2-10 Wall thicknesses at the same rigidity

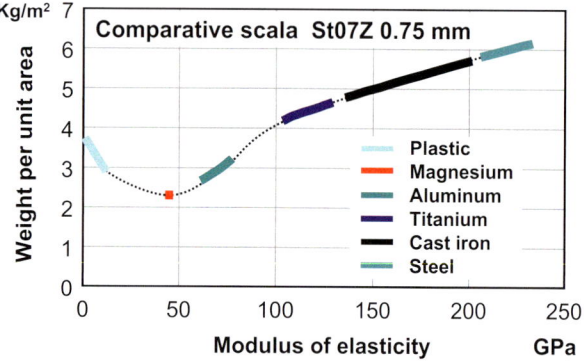

Figure 2-11 Weight per unit area at the same rigidity

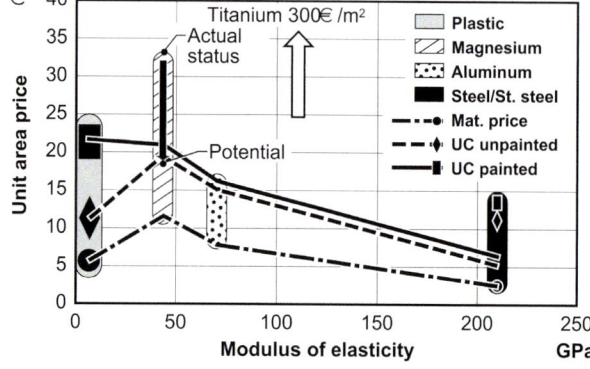

Figure 2-12 Price per unit area at the same rigidity

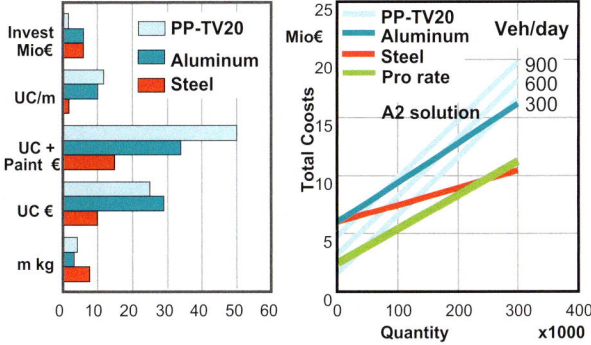

Figure 2-13 Cost comparison for fenders

In mass-produced vehicles, flat plastic components on the outer skin cannot compete with aluminum on a pure material substitution basis.

Plastic fenders will only be compatible – even in midrange production quantities – once integration possibilities result in savings in both individual costs and investment. An example is the integrated fender extension of the Audi A2 made of PP-TV. If the required fender shape is too difficult to draw from sheet metal, plastic may be considered an alternative. However, in general plastic fenders will not become established as an accepted solution.

Presently, the cost structure for painted exterior body parts is not satisfactory. Although injection molded parts made of polypropylene materials are relatively inexpensive to manufacture, the proportion of the costs attributable to painting, transportation and packaging makes up more than half of the total cost. The relationship between costs and weight is even more unfavorable in the case of the more expensive materials such as RRIM. Although on-line painting is aimed at for extremely expensive high-temperature-resistant materials such as mineral-filled PPO/PA, most are made by costly in-line painting, requiring prior priming by the supplier.

Newly developed materials will therefore only be competitive if they can also withstand the even higher future drying temperatures used in electrophoretic dip coating. This measure is being introduced at Audi to achieve cost reduction, since it eliminates the step of separate hardening of the aluminum.

The same problems can be found with the vehicle bottom line; for example, off-line painted sill panels made of PP-TV or RRIM. Here, a solution focusing on wear resistance and higher-strength materials such as PA6-GF would be appropriate. Delivering a conductive component (that has not already passed through the supplier's paint plant) will permit inexpensive on-line painting and thus a cost reduction potential of more than 50%. This solution is not yet available on the market but does make sense from both the technical and economic points of view. The limited surface quality of materials with a high degree of glass-fibre reinforcement is tolerable in the sill area.

2.1.4 Use of Plastic as a Structural Material

Before plastics can be used in the bodyshell structure, they must be on-line paintable. Trials with the hybrid front-end of the Audi A6 and with plastic spare-wheel wells have demonstrated that PA6-GF30 meets these requirements. A fender brace made of PA6.6-GF30 is in series production for the Audi A4 and reliably passes through the entire painting process (see Figure 2-14). By integrating clips and retainers a weight reduction of 40% and a cost reduction of 30% were achieved for this application.

To a limited extent glass fiber reinforced plastics are suitable for structural components. Care must be taken to meet requirements for fatigue strength. Taking PA6-GF35 as an example, it is seen that temperature, moisture and fiber orientation have an extraordinarily great influence on fatigue life [11] (see Figure 2-15).

In addition, the damping characteristics of thermoplastics result in a marked dependency of fatigue life on frequency, see Figure 2-16. A classic fatigue failure is to be expected with low exciting frequencies. As frequency increases, the test piece heats up markedly and in extreme cases failure due to thermal damage will occur [12, 13].

These material properties prevent a service life calculation using the well-known damage accumulation

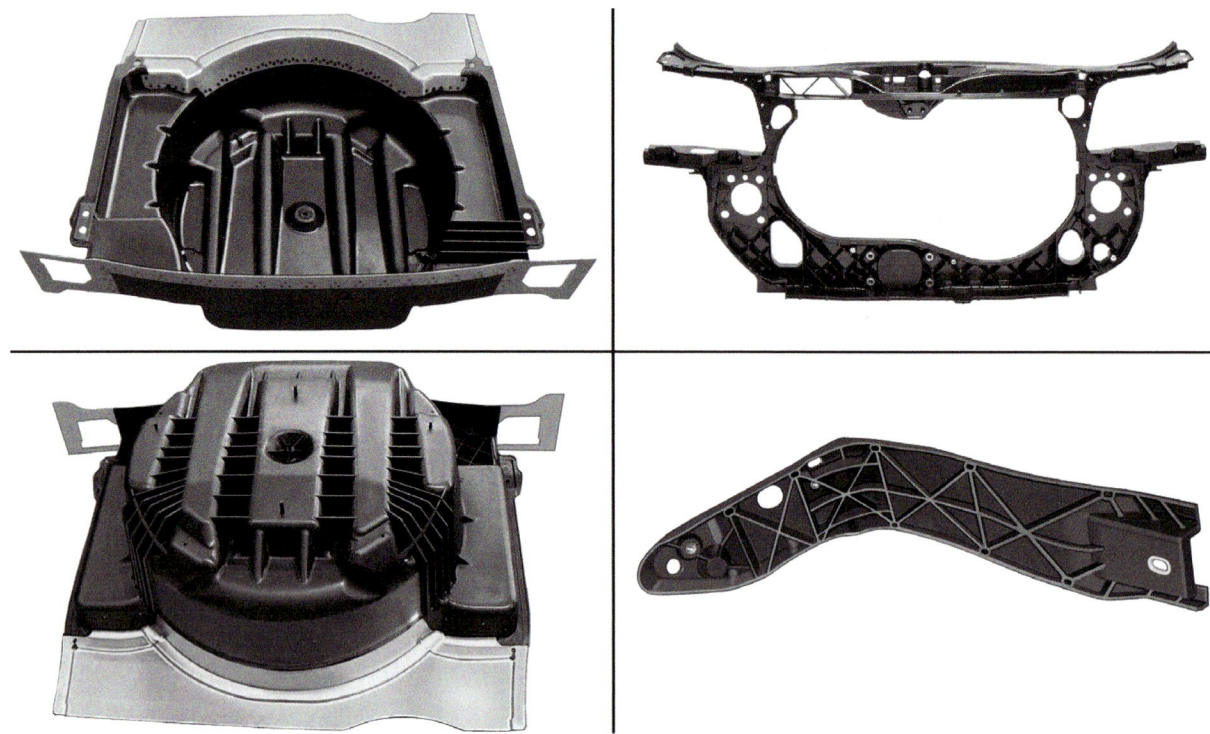

Figure 2-14 Parts made of PA6-GF30 and PA6.6-GF30 with cathodic dip painting capability

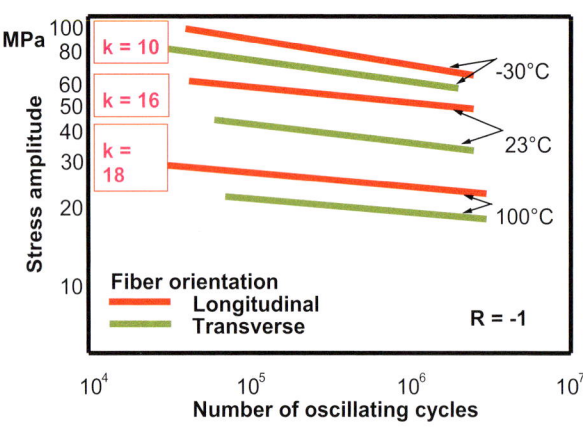

Figure 2-15 Wöhler diagram for PA-GF35

Figure 2-16 The influence of exciting frequency on the fatigue life of PA6-GF30 with sinusoidal excitation [12]

hypotheses. Instead, service life can only be calculated by relative comparison and a weak-point analysis as part of FE calculations, while fatigue life is determined experimentally under extreme temperature conditions [14].

However, economic considerations show that the range of applications is limited. Due to their lower stress tolerances compared to metals, large wall thicknesses are required. This results in uneconomically long cycle times in the injection molding process. Economically it is more viable to determine those areas of the bodyshell that experience low stress levels and to equip these areas with hybrid components. This approach would offer good opportunities for weight reduction at acceptable cost [15].

2.1.5 Competition Between Plastic and Stainless Steel

The last two decades of development in wrought steel alloys resulted in an increase in strength while ensuring the lowest possible reduction in forming properties. Since 1975 an uninterrupted succession of further improved materials for the bodyshell has reached the market. The current state of the art, as far as cold-formed impact-relevant parts are concerned, is a complex-phase steel with a tensile strength of 800 MPa in conjunction with 10 to 20% ultimate elongation. In addition, hot-formed components with tensile strengths ranging up to 1300 MPa with ultimate elongations of 6 to 9% are used in applications where side impact is a consideration, see Figure 2-17.

Although there are no significant increases in material prices, the outlay for investment and in particular for production costs is rising. Whereas around 40% of production costs is taken up by low-alloy deep-drawing steels, an increase to around 75% of product costs is noted when hot-formed parts are used. The demand for low values for elongation at break rule out the production of complicated parts and result in increased expenses for secondary joining operations. These problems can be avoided by using a material new to bodyshell construction, stainless steel. The technical potential of this approach is shown in Figure 2-18.

The elongation at break for stainless steel of more than 50% is better than the one for special extra-deep-drawing steels. Due to outstanding strain-hardening properties, a tensile strength of 1000 MPa can be achieved with a degree of deformation of 30%. The residual elongation of 20% is unrivaled among other metals.

The material price is expected to establish at approx. € 1.50 per kg [16] for mass production requirements. This means that the future holds no opportunities in larger-scale production for carbon fiber reinforced unsaturated polyester resins at or above € 10 per kg, not to mention epoxy resins. Stainless steel offers weight reduction at additional costs of € 3–4 per kg for the bodywork structure, see Figure 2-19.

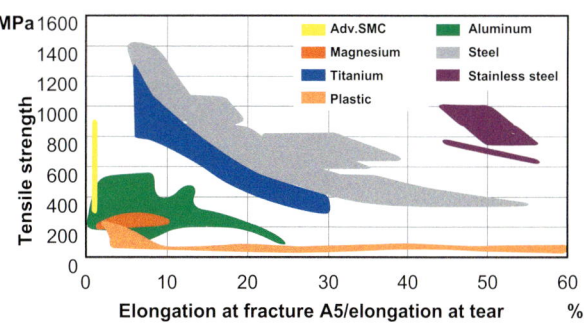

Figure 2-17 Comparison of strength and ductility for different materials

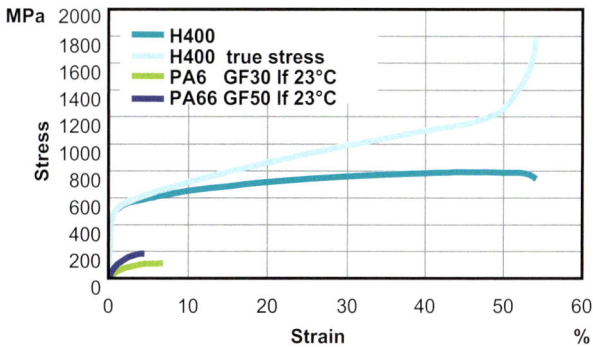

Figure 2-18 Comparison of strength and ductility for stainless steel and plastic

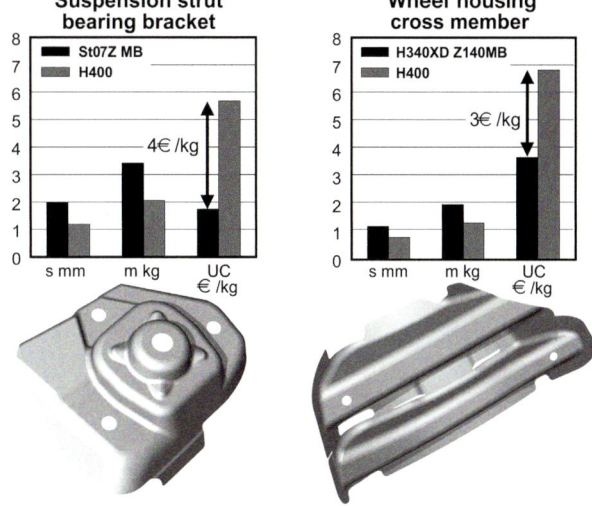

Figure 2-19 Comparison of costs between steel and stainless steel

2.1.6 Acoustics

Passenger compartment acoustics are of overwhelming importance for high-end vehicles. Until recently organic materials had an advantage, because they have a higher dissipation factor than metals and are thus better suitable to dampen structure-borne sound. To respond to the superior properties of these organic materials, the steel industry introduced steel-plastic composites that feature a thin layer of plastic embedded between outer steel layers, see Figure 2-20. With these composite materials, dissipation factors as high as 0.6 can be achieved, depending on temperature and frequency. Components can be shaped and joined using the methods usually employed in the automotive industry. Since bitumen damping sheets are no longer required, weight reductions of 10% without cost increases can be achieved.

To achieve an improvement in airborne sound absorption it is necessary to fit open-celled foam systems to the noise-source side of the surface of the surrounding shell. Melt-bonded sound-deadening materials should only be used in locations where they are unavoidable because of space considerations. In the future, self-supporting, thermally stable, acoustically highly-effective materials will be required that also have to be recyclable.

In the next few years we expect the legislature to demand a reduction in exterior noise, as determined by the drive-by test, from 74 to 71 dBA. Further increases in the thermal efficiency of engines will result in louder combustion noise. This in turn will mean increased demand for sound-damping, -insulating and -absorbing materials. One such example is the textile rear wheel arch panel of the Audi [17]. In the future, all new Audis will be equipped with textile wheel arch panels. In addition, the front wheel arch panels of the A4 and A6 were also changed over to textile panels in 2002. An assessment of acoustically effective components is shown in Figure 2-21.

2.1.7 Summary

Over the last twenty years there has been a continuous increase in the proportion of plastics used in automotives as a whole. Any further increase beyond the current state of the art in series production is problematic because of the limited ductility of high-strength plastics, high material prices and high painting costs. The corresponding opportunities and risks are summarized in Figure 2-22.

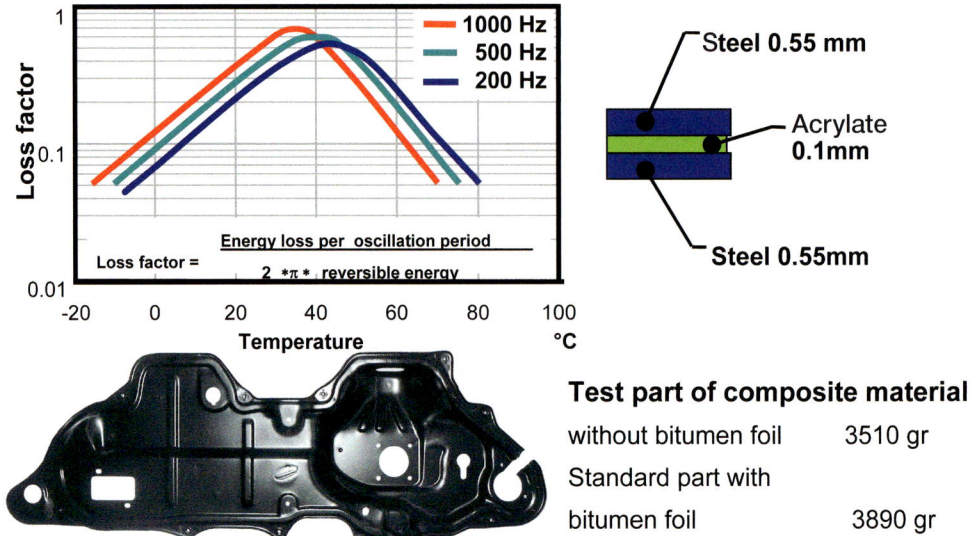

Figure 2-20
Mounting plate for Audi A3 bulkhead

Test part of composite material
without bitumen foil   3510 gr
Standard part with
bitumen foil   3890 gr

Component	Expectation
Textile wheel housing liners	Sharp growth in volume Cost reduction from scale effects
Insulation tray absorber	Volume optimization of foam and cover non-woven
Open-cell (sponge) foams	Elastic PUR and melamine foams Elimination of cover non-wovens
Alu insulating foils	Use only for temperature reasons
Sprayable insul. compounds	Costs not yet competitive
Steel-plastic sandwich	First use planned, if further development of forming properties successful, sharp growth in volume
Bulkhead parts in cavities	Growth due to red. in airborne sound
Thermal insulation (endless glass fibre material)	Competition for sponge foams with alu insulating foil or sheets
Noise insulation measures in force inflow zones	Avoidance of structure-borne sound

Figure 2-21 Assessment of acoustically effective components

Component	Risk for plastic	Opportunity for plastic
Bumper	Very slight	Market saturation reached
Front end	Magnesium strategy	Growth in volume with hybrid technology
Hood Fender	Costs, weight, Al/Mg strategy	Pedestrian protection Parts integration
Outer side part	Costs, strength, ductility	None initially
Door outer part	Costs, surface quality	Only in event of strat. decision against aluminum
Roof	Costs, quality	Modular design
Trunk lid	Costs, creep properties of SMC	Integration of antennas
Wheel housing liners	With PP inj. moldings	If textile materials
Insulation tray	Costs for PP-GM	Wall thickness red. with LFT
Aero floor	Costs for PP-GM	Growth in volume
Inner sheets	Al, St-plastic sandwich	Very slight
Structural parts	Al, St, stainless steel	Only if hybrid technology
Fuel tank	High, HC emission	Market saturation reached
Acoustic material	Component temperatures	Legisl., market demands

Figure 2-22 Summary

References

[1] Klein, G.: Nachhaltige Innovationen, Kunststoffe 91 (2001) 8 S. 72

[2] Information from Alusuisse (www.alusuisse-automotive.com)

[3] Stich, A.: Aluminium und Magnesiumwerkstoffe und Formgebungsverfahren [Aluminum and magnesium materials and production processes], Audi AG internal report

[4] London Metal Exchange (www.ime.co.uk) and www.kunststoffweb.de

[5] German Federal Bureau of Statistics, Technical Series 17, Series 2, 'Preise und Preisindices für gewerblich Produkte' [Prices and price indices for commercial products]

[6] Rummel, J.: Leck vor Bruch [Leak before failure], Automobilentwicklung, Jan. 2001

[7] Hollerweger, H., Kaufmann, D., Schade, M.: Der Einfluss von Gesamtfahrzeugparametern auf den Kraftstoffverbrauch von PKW [The influence of total vehicle parameters on car fuel consumption], HDT Nov. 2000

[8] Vehicles environmental standard, VW 91100

[9] Juchmann, P., et al.: Neue Magnesium-Blechprodukte für den Automobilbau [New magnesium sheet products for automobile manufacturing], Automobiltechnische Zeitschrift 103 (2001) 2

[10] Friedrich, H., Schuhmann, S.: Forschungsstrategien für eine zweites 'Magnesium-Zeitalter' im Fahrzeugbau [Research strategies for a second 'age of magnesium' in automobile manufacture], Volkswagen Distributionsservice Vermold

[11] Ermittlung vom Schwingfestigkeits-Kennwerten an glasfaserverstärkten Polyamiden [Determination of fatigue strength parameters in glass fiber reinforced polyamides], Audi AG internal report 1990

[12] Oberbach, K., Hesse G.: Der Einfluss von Beanspruchungsfrequenz und -ablauf auf das Schwingverhalten von Kunststoffen [The influence of stress frequency and development on the fatigue characteristics of plastics], Materialprüfung 14 (1972) 6

[13] Oberbach, K.: Schwingfestigkeit von Thermoplasten – ein Bemessungskennwert? [Fatigue strength of thermoplastics: a design parameter?] Kunststoffe 77 (1987) 4

[14] Steuer, U., Sponheim, K.: Entwicklung eines Frontendträgerteils aus einem Metall-Kunststoffverbund [Development of a front-end support made of a metal-plastic composite], DVM Report 128, Oct. 2001

[15] Steuer, U., Seufert, M., Mederle, K.: Das Hybridfrontend des Audi A6 [The hybrid front end of the Audi A6], paper at VDIK conference, Mannheim 1998

[16] Krautschik, J.: Information courtesy of Thyssen Krupp Edelstahl

[17] Steuer, U., Puschmann, O.: Gewichtsreduzierung an akustisch wirksamen Bauteilen am Beispiel der hinteren Radlaufschale des Audi A6 [Weight reduction in acoustically effective components as exemplified by the rear wheel arch panel of the Audi A6], paper at the VDIK conference, Mannheim 1999

2.2 Use of Plastics in Body Construction in Future Automobile Design and Production

John Sullivan

2.2.1 Introduction

The pressures on the automotive industry in terms of cost and weight have been, and remain extreme, with continually intensifying technological deployment in response to market needs.

The progressive implementation of European End of the Life Vehicles Directive (ELVD), places the responsibility for end-of-life vehicle disposal with the Original Equipment Manufacturers (OEM's). Vehicle design of the future will need to increasingly focus not just on assembly, but also on final disassembly and consequently, require the number of differing materials in any given assembly to be minimized.

The total life cycle environmental impact of vehicles must also be managed. Environmental legislation often exerts conflicting pressures on vehicle manufacturers. The requirement of the European ELVD for a maximum of 5% disposable waste by 2015 is one example. Such restrictions raise many interesting questions; for example, is it better to employ a steel leaf spring which can be recycled at ELV, or employ a composite leaf spring which cannot be recycled, but delivers significantly improved fuel economy through lower weight and hence reduced CO_2 and No_x over the whole life of the vehicle? Which then is the better equation for the environment, the total lifecycle of a vehicle or the end of life disposability?

Functional performance and variable cost continue to be a challenge. Customers expect our contribution to environmental improvements without additional cost. As we migrate our vehicles to recyclable materials, such as polypropylene or TPO, we need to absorb the incremental weight and cost of these materials, or find viable offsets within the total vehicle, and still maintain profitable margins.

As fuel prices rise, the necessity to improve fuel economy in order to maintain current levels of customer operating costs is clear. However, there is also an expectation to improve fuel economy with noticeable benefits from prior products. New technologies including fuel cell vehicles, such as Focus FCV, offer good long-term solutions to reducing the future cost of fuel. Many advances are also being made in mass production of internal combustion technology, which will result in significant short-term improvements for diesels and medium term benefits for gasoline engines. Considering that, the major gains in powertrain technologies will be realized in the medium-to-long term, weight reduction actions through new and advanced material applications will be the key element in the delivery of improved fuel economy.

Figure 2-23 Ford Focus hybrid FCV

Aluminum was once reserved for lower volume niche production. However, recent trends have seen the increased use of aluminum as an alternative to steel with significant benefits. The new Jaguar XJ vehicle has realized roughly a 200 kg weight improvement over a comparable steel-bodied car –, this is the equivalent to the weight of a baby elephant. However, with the price variance of steel vs. aluminum, this trend will remain in the foreseeable future on premium cars only.

Significant advances have been made in low production performance/premium products, such as the Aston Martin Vanquish. Weight reduction has been achieved through the use of advanced composite materials giving low weight but great structural rigidity. However, these technologies remain low volume and the challenge for the future is to identify the elements, which can be modified to mass production but still support an appropriate recycling and ELV strategy.

Design methodology and strategies have to be modified to integrate new material usage, such as plastics and composites, where previously steel was used. The operational characteristics of the new materials components will have a significant impact on the level of usage, particularly in terms of their physical relationship with 'traditional' materials. For example, a plastic fender can expand up to 17 mm in X-axis when going through the paint process and will not shrink back to the nominal position post process. These parameters need to be considered in the part design and manufacturing process, accommodating increased variability, without restricting the design or affecting quality for the customer.

Plastic utilization has increased significantly for the interior of products. Lower entry products tend to display more visible plastics with foil and additional finishes, such as paint being added, progressing up the vehicle segments. Customers tend to perceive the use of plastic negatively. However, this perception is often driven by component styling, the grain selection, gloss level and applied finish.

We need, through component design, to make the use of plastic more acceptable to our customers. In 1999, Ford launched the Ford Focus. This car has a painted polypropylene instrument panel. However, through the part design, application of soft touch paint and selection of a suitable grain, the panel does not give the appearance of an abundance of plastic. Similarly, when designing the Ka, the use of material was required to meet the business case, but also underlined what the car was trying to convey. The result is a car with exposed interior sheet metal and polypropylene door trims and instrument panel, which through their curvaceous design deliver the character of the car.

We rely very heavily on our supply base and the key to successful plastic application begins with the highest standards of raw material and raw material matching.

Where perceived quality requires that plastics have to be less obvious, the increasing use of foils and finishes can be employed, emphasizing decorative areas of the product, and in most cases, offering opportunities for series differentiation. However, this will drive careful selection of base materials and needs to be integrated into the design process.

Foils, films and finishes will deliver other major benefits with advances in nano technology; we are close to mass production capability of replacing selected glazing with polycarbonate, realizing significant weight savings. Long-term development of water resistant films could eventually allow the deletion or reduced use of windshield wipers. Further development of foils will see increased exterior usage, replacing traditionally painted components, such as door/liftgate handles.

Figure 2-24 Ford Focus instrument panel

This introduction has only touched on some of the major issues facing the automotive industry in respect to plastic utilization. In the following, will discuss in more detail some of these issues and in particular the requirements of our industry to meet ELV legislation.

2.2.2 Structural Plastics

The use of fiber-reinforced plastics to carry significant structural loads has been feasible for many years. Lotus pioneered this technology and the manufacturing processes to make the technology viable. With the exception of Formula I vehicles and very low volume vehicles, such as the McLaren F1, most vehicles simply could not make the technology viable in the long term.

However, several vehicles have used aspects of the technology to their advantage. Lotus, TVR, Matra have manufactured body assemblies or closures where the economics made sense, or where low investment opportunity or design freedom was required. One of the most surprising advantages for structural plastics is the potential for energy absorption, graphically illustrated by the high survival rates now seen in F1 vehicles (even during very high energy accidents).

One of the most innovative vehicles to use structural plastics is the Aston Martin Vanquish. The body/chassis construction is a hybrid of aluminum and composites.

2.2.3 The Body Tub Assembly

The body tub comprises an assembly of aluminum extrusions and sheet sections that are bonded together with rigid epoxy adhesives to form a very stiff compartment. The vehicle has a carbon/glass composite tub tunnel, which is bonded into the hybrid tub. Most of the body/chassis is bonded together with structural adhesives. Benefits are reduced weight, high stiffness and zero corrosion.

All the external body panels are formed from aluminum sheet (except the cant rail). Most of the body panels are manufactured by the super forming process where certain grades of aluminum sheet can be slowly hot formed over single sided tooling with some air pressure. The panels are then hand finished and adhesively bonded to the car. Benefits are reduced weight, low cost tooling, flexible assembly, easy design changes and ultra low corrosion

2.2.3.1 Energy Absorbing Composite Front Ends

The vehicle front and rear ends are assemblies of composite moldings that are designed to absorb energy in a unique way. Metals are tough and generally absorb energy by yield and deformation but brittle fibers and epoxy resins generally do not. However, brittle fibers and resins in composite materials, in certain combinations and angles, can be persuaded to absorb large amounts of energy in a stable manner (larger than metals) by microcracking, friction and total destruction to dust.

Figure 2-25 Aston Martin Vanquish

Figure 2-26 Aston Martin Vanquish – cutaway

2.2.3.2 Composite Body Sides

These composite components are produced with glass fiber and polyester resin using the Resin Transfer Molding (RTM) process. The fiber preforms are made using the robotic spraying down of fibers to a fiber length and distribution that is fully programmable. This process enables a waste-free process to a wide range of body shapes. These preforms are then compacted and molded using RTM.

2.2.3.3 Adhesive Usage

The Vanquish uses a high level of structural adhesives over the tub, body panels and chassis components even in relatively hot areas in the engine compartment. A surprising feature is the use of low stiffness (rubbery) adhesives with large bonding areas to bond aluminum to composites. The reductions in stress concentrations in load transfer areas from the use of rubbery adhesives results in high toughness and excellent durability of the body assembly. Testing was conducted from −40 °C to +70 °C and up to 90% relative humidity.

2.2.3.4 Summary

The Aston Martin Vanquish showcases a combination of materials to produce a unique design with beauty and high performance. Some aspects of these structural composites and lessons learned will be used even in high volume vehicles in time.

2.2.3.5 Mixed Materials

Existing vehicle applications for mixed material structures at Ford have been integrated into the building process for a steel-structured vehicle. This strategy will continue with selection of the optimum mix of materials, which can be integrated, together providing the most cost-effective overall solution.

However, there are many conflicting challenges when using plastic materials in any phase of the build. Key issues with manufacturing process compatibility are summarized in Table 2-1.

The selection and design of joining methods for mixed materials has to be integrated into the selection process to optimize material systems, especially as welding mixed materials is not feasible. Therefore, processes, such as adhesive bonding or mechanical fastening, have to be used which always require additional material/parts, resulting in increased complexity and costs. If mixed material parts can be mechanically interlocked, then this additional complexity can be avoided, or joints can be eliminated altogether by integrating components into castings or moldings. Other key issues with joining methods of mixed materials are listed in Table 2-2.

Further differences in joints between mixed materials are the joining methods feasible for joining a module to a structure compared to joints feasible for fixing sub-components to the module itself and how this is affected by the properties of different materials. For example, plastic parts may have sufficient global stiffness to meet service load requirements, but then need additional local reinforcement to withstand stress concentration from mechanical fasteners.

Major challenges for mixed materials face our industry, one being the analytical method used for CAE simulation. This is increasingly critical because of the need both to compress timing for vehicle development programs, and to increase robustness of new designs. This drives the need for more rapid and more accurate CAE simulation. For plastic materials in non-structural applications, where analysis requirements are mostly concerned with stiffness, simulation can be carried out with existing technology. However, for critical structural applications, suitable material models for plastics are essential. This will cover material behavior to yield point, through the hardening phase to crack initiation and fracture, including, of course, all temperature, strain rate and environmental degradation effects. Such models are starting to become available for local analysis of some materials, but to include this detail in a complete vehicle model with steel for durability analysis is not yet possible.

A successful example of a hybrid part, which utilizes the properties of mixed materials, is the front end

Table 2-1 Summary of key process issues for mixed materials during manufacture of a steel intensive structure

Unpainted Body (BIW)	Paintshop	Assembly
• Handling methods for non-magnetic materials • Fixturing systems • Contamination • Odor • Dimensional tolerances sufficient to allow thermal expansion • Repair/rework of mixed materials	• Degradation at maximum electrocoat paint oven temperatures (208 °C) • Electrostatic paint spray deposition on non-conductive materials • Colored primer required • Part degrease/treated • Differential thermal expansion • Plant emissions	• Color match mixed materials • Limited joining methods • Body tolerances

Table 2-2 Summary of key issues with joining of mixed materials during manufacture of a steel intensive structure

Unpainted Body (BIW)	Paintshop	Assembly
• Adhesives for BIW are too stiff for plastic composites and can cause delamination • Mixed materials joints with steel must allow for electrocoat coverage for corrosion protection or have additional protection before assembly	• Degradation at maximum electrocoat paint oven temperatures (208 °C) • Electrostatic paint spray deposition on non-conductive materials • Drive for automated, lights out paintshop incompatible with assembly processes • Assembly in paintshop requires increased control of contamination	• No pumpable adhesives except for direct glazing because of cleanliness, contamination and health & safety requirements • Color appearance of joints • Limited use of heat activation for joining or solvents for cleaning/pre-treatment because of compatibility with manual assembly processes • Dimensional tolerances and fixturing need to be suitable for adhesive joints • Pressure sensitive adhesives require critical assembly tolerances and part marriage • Robustness of adhesive bonding to surfaces coated in cavity wax corrosion protection

module of the Focus (Figure 2-27). This is an injection molded part with steel reinforcements, which are embedded in the plastic by overmolding. The steel reinforcements also have holes, which are mechanically locked by the plastic filling into them during the molding process. This front end module features a balance between the structural properties of steel (including dimensional and thermal stability) and the component integration of an injection molded part. This module is mechanically fixed to the unpainted body structure.

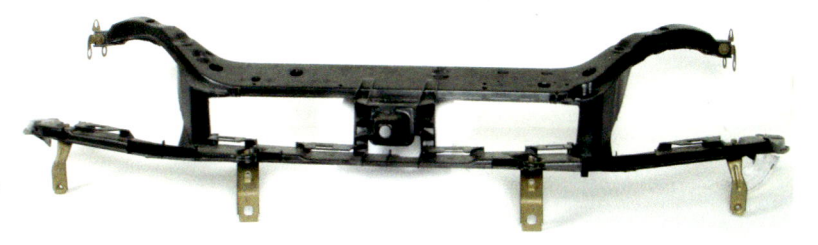

Figure 2-27 Hybrid front end module of Ford Focus

2.2.3.6 Strategic Material Deployment

The choice of interior plastics often comes under scrutiny from customers and the automotive press. Occasionally, articles compare the same raw materials in two different applications with one use being described as 'cheap' and the other 'expensive'.

The perception that a vehicle's interior is 'cheap' or 'plasticy' comes not from the plastic substrate itself but from the surface finishes applied to the substrate.

The question is not necessarily, 'What material do I choose?', as that is often defined by the functional properties required, but rather, 'How do I make this visually attractive and tactile to the customer?'.

There are two key considerations when choosing the surface finish: customer expectation and cost. Rarely are these compatible. Most customers would like matte looking trim; soft to the touch; scratch resistant with the perception of luxury. This is achievable with costly surface treatments. However, on high volume, small vehicles, where variable cost is sensitive, strategic use of surface treatment is vital. In the top-end luxury sector, customer expectations are higher and there is less pressure on variable cost. More expensive finishes can be applied and used in conjunction with more 'traditional' materials such as wood and leather.

Figure 2-28 Jaguar S-Type indicating use of leather, wood and treated plastics on premium vehicles

Given the fact that high volume, small vehicles, such as the Ford Fiesta, are restricted by cost in their choice of surface finishes, it is the design and engineering of the component that is critical to achieve the perception of quality. Careful choice of color, grains and surface style can still make self-color, untreated plastic an attractive material. This approach can go wrong when manufacturers try to simulate a more expensive material such as leather, in a hard plastic. Visually it might be close, but if it doesn't feel like leather, or have a quality sound when tapped, then the customer is not comfortable with it.

Plastics are here for the long term, but the pressures on the automotive industry make it increasingly difficult to justify the business case for some surface treatments on some vehicle lines. Customers demand increased safety, performance and functionality; yet they do not want to pay more. To this end, cleverly designed, self-color grained plastics offer an acceptable strategy for vehicle interiors. However, customers' perceptions cannot be changed easily and manufacturers will continue trying to find a careful balance between cost and expectation.

2.2.4 SMC at Ford Motor Company

SMC has been established and used in the US to a much greater extent than in Europe for body panels. The major issue with SMC has been painting problems caused by porosity. This can now be overcome with the correct use of a vacuum during the mold cycle, which can virtually eliminate the problem. However, very accurate control of the molding operation and the vacuum cycle is required.

The grille opening panel on the Ford Transit van is unique (Figure 2-29). It is a high volume automotive component that has critical requirements both for surface finish and structural demands. This panel spans the vehicle from one fender to the other, supporting body parts and components, including headlamps and fenders, and also meets all the engineering specifications required for such parts (including paint oven temperatures).

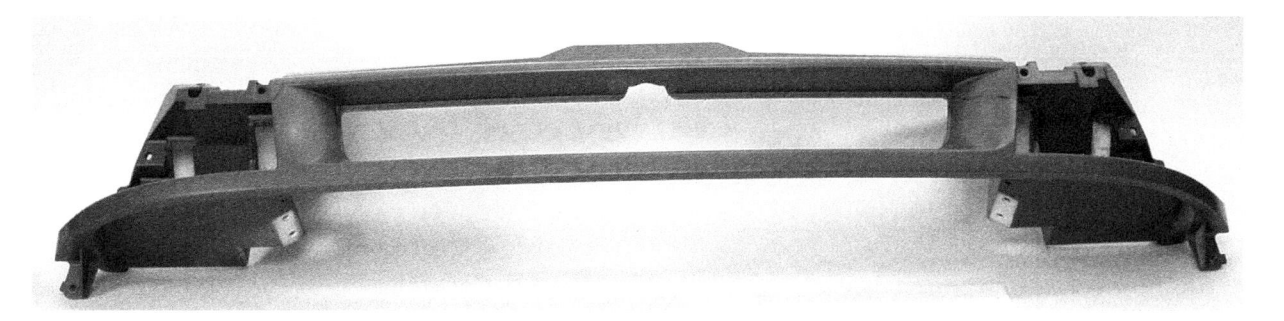

Figure 2-29 Ford Transit grille opening panel

The introduction of a high volume painted SMC part to Ford of Europe has had challenges, but the many advantages have long been recognized, such as low cost tooling. Now that the porosity problem has been solved, the use has been extended to side panels on a new commercial vehicle where the low cost both for tooling and parts were required, along with class 'A' surface finish standards

There are other examples to demonstrate its uses, such as the deck-lid for the Mercedes CL Coupe; other OEM's and are now developing SMC deck lids for premium models.

2.2.4.1 Nano Technology

The development of plasma enhanced vapor deposition of silicate coatings (SiO_2), has had many applicators within the tooling industry (for eliminating friction and wear). Its use as a surface coating for glazing components is predicted to have many applications, for both functional and aesthetic reasons. By using this coating, for example on polycarbonate glazing, it can provide excellent clarity, weight savings and high scratch resistance, way superior to current silicone coatings. Multi-sun roofs have the obvious advantages of weight, once the cost of the process reaches mass production viability (currently three times conventional glazing costs).

2.2.4.2 Foil and Film Technologies

When used in similar applications on interiors, for décor type finishes, problems are rarely encountered with appearance criteria, both color and graphics being feasible (depending on the technology of foil used). The main issues being continuous roll in injection tool, pre-formed sheet foil (with back injection) or printed then formed in the back injection process.

The latest advantageous use is in pre-formed 'printed' applications; where savings can be made by eliminating the number of components yet still achieve style, color and functional specifications. This pre-formed technology allows the replacement of moving switch components with pre-formed surfaces acting on high-tech electronic switchboards behind. Three-dimensional capability can lead to new 'high-tech' looking layouts.

New developments in interlayer resin technology are leading to a potential material for use as a primary bumper or exterior component surface-finish foil. They are formed from four layers within the foil. A carrier layer, coated with colored resin, a layer containing metallic pigments, covered in a high-gloss UV screening clear coat, which when incorporated in the tool and back-injected with the polycarbonate/ABD backing, forms an integral material structure with an extremely high gloss 'A' surface and high durability properties.

Some colors are still in development but show potential for future usage, with savings on painting processes on polymers as used at present. Interior usage combined with polypropylene is also a possibility for facia and others. Painting is probably the highest cost component on a vehicle and occupies up to 50% plant floor space.

2.2.4.3 Raw Materials and Color Matching

To achieve high craftsmanship standards with color matching, an appearance harmony philosophy must be adopted to ensure that raw material selection is correct.

The following criteria for TPO (thermoplastic polyolefins, e.g., PE, PP, EPDM, etc.) use for interior components must be considered:

- More versatility of molding,
- Low gloss finishes equivalent of topcoat finishes,
- Recyclability and
- Cost benefits.

To achieve good matches of interior system components, raw materials must be sourced from suppliers who can guarantee color and gloss consistency. Understanding suppliers' raw materials is essential for meeting agreed Appearance Harmony Targets. This is done by focusing on how different process conditions affect color and gloss, understanding how the pigments in materials behave under different light conditions and understanding the color relationship across different suppliers' materials. Raw material suppliers must be flexible, as pigments often have to be adjusted after processing to ensure that acceptable color match is achieved.

Full service suppliers are advised to use the same raw material supplier within their assemblies to satisfy these criteria. Ford Motor Company's Fusion door trim (Figure 2-30) is a good example for this approach.

This door trim has four components, all made of the same material, all sourced from the same material supplier, thus providing the customer good appearance harmony across this assembly.

Appearance harmony processes must identify color, texture and gloss differences of components, their materials and material suppliers. This process eliminates problems associated with these differences. This must be an ongoing process from concept to production. All systems on the vehicle must be covered to achieve interior, exterior and instrument panel (center stack) harmony.

2.2.5 Implications of the End of Life Vehicle Directive

Over the past few years, legislation has been introduced to further reduce negative impact on the environment. The scope is wide and is growing in many areas, including the automotive sector. The legislation affects the vehicle, design, homologation and marking of materials, and goes as far as to limit the use and landfill of certain materials and the design and operation of recycling infrastructure.

This increase in legislation is not a great surprise since about nine million end life vehicles are removed from service every year and it has been estimated that about two million tons of non-metallic waste is currently landfilled every year. While this can be viewed as pollution, it is also a wasted resource that could be encouraged to be recycled back into the manufacturing process.

The positive news is that about 75% of the average vehicle weight is currently recycled, covering most of the metallic content. Steel, cast iron, copper, aluminum, lead, zinc, etc. are valuable materials and the infrastructure is already in place to return these materials to the market by existing recycling companies.

Figure 2-30 Ford Fusion front door trim

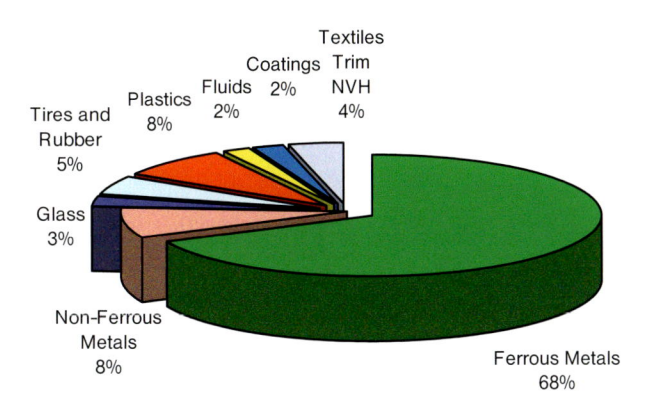

Figure 2-31 Breakdown by material system of the average European vehicle by weight

2.2.5.1 Key Legislation

There are several published directives that have an effect on end of life vehicles and the recycling process. There are eight key steps that need to be considered.

- **Landfill Directive:** Only waste that has been treated may be land filled. It imposes a ban on land filling of whole tires as of 2003 and of shredded tires as of 2006. Directive 1999/31/EC

- **Material Restrictions:** Vehicles' materials and components must not contain lead, mercury, cadmium and hexavalent chromium. Time frames and exceptions are defined. Certain items, e.g., batteries, have to be made identifiable for treatment. Directive 2000/53/EC

- **Pre-treatment:** Defines treatment of specific materials and components including batteries, tires, glazing, catalysts, operating fluids, engine coolants, explosives, lubricants, fuels, mercury and HVAC fluids. As of July 1, 2002 applies for all new vehicles.

- **Cost-Free Take Back:** Requires the acceptance of a vehicle and service parts by an authorized treatment facility at zero cost for the last owner. As of July 1, 2002 for new vehicles and as of January 1, 2007 for all vehicles. Directive 2000/53/EC

- **Parts Marking:** Components must be marked with recognized coding standards to facilitate re-use, recovery and recycling. The Ford objective is to mark all parts over 50 g. Directive 2000/53/EC

- **Recycling Targets:** Member states must ensure that from Jan. 1, 2006 all end of life vehicles, reuse and recycling shall be increased to 80%. Reuse and recovery shall increase to 85% by weight. From Jan 1, 2015 the targets are 85% and 95%, respectively. Directive 2000/53/EC

- **Type Approval:** After 2005, all new vehicle types must be type approved as being designed to be reusable and recyclable to 85% and reusable and recoverable to 95% by weight. Directive 70/156/EEC

- **Published Information:** Producers are to provide dismantling information for each new type of vehicle within 6 months of it being first put on the market. Information is required to help treatment facilities to comply with the ELV Directives. Directive 2000/53/EC

2.2.5.2 Recycling As a New Design Requirement

The implications of the impending ELV legislation is that the current work undertaken to ensure products are recyclable, now becomes a legal requirement, with design criteria including:

- Engineering environmentally friendly aspects and recycling into our products right from the concept phase.

- Reducing complexity by using fewer material types.

- Avoiding mixed materials and marking components with recycling codes where possible.

- Use of materials that can be easily recycled and use of materials containing recyclate where available.

- Design for easy separation.

These requirements inevitably increase cost pressure. The challenge is to find innovative solutions to ensure that systems functionality and quality are achieved without adversely impacting total product cost.

2.2.5.3 Recycling Status

The challenge now is to increase progress with non-metallic materials and in order to do this, the extraction, refinement and transport infrastructure needs to be evolved to build a successful recycling industry for non-metallic materials, comparable to the one existing for metals. The auto industry can facilitate this, by creating large vehicle structures and assemblies that can be quickly and easily disassembled and more economically recycled. As material selections across the industry become more standardized, separation and sorting of materials becomes simpler and preserves the value of the material, thus making recycling more viable and adding residual value to the end of life vehicle.

Ford Motor Company has had considerable success with the use of recycled materials in their products. A pilot disassembly plant at Köln-Niehl, Germany was built near the vehicle plant to benchmark and assess viability of disassembly operations. A Ford European Recycling Action Team (ERAT) has been established for many years that promotes recyclability and encourages the use of recyclates. This team has driven many projects into production and at this time more than 270 polymeric components in Ford vehicles are manufactured from materials that includes recyclate. This progress has resulted in over 24,000 tons of materials being diverted from landfill per year, with this amount increasing every year.

To ensure recyclability, a component has to be recyclable and be easily removable within a specific time period to make the operation viable.

2.2.5.4 Examples of Good Recycling

There are many examples of good recycling in the industry. Not all recycling returns the parts back to the same use. Ford Motor Company uses recycled polypropylene soft drink bottle tops in recycle to make heater units housings. Ford also have used recycled polycarbonate compact disks in recycle to make Ka instrument bezels (Figure 2-32).

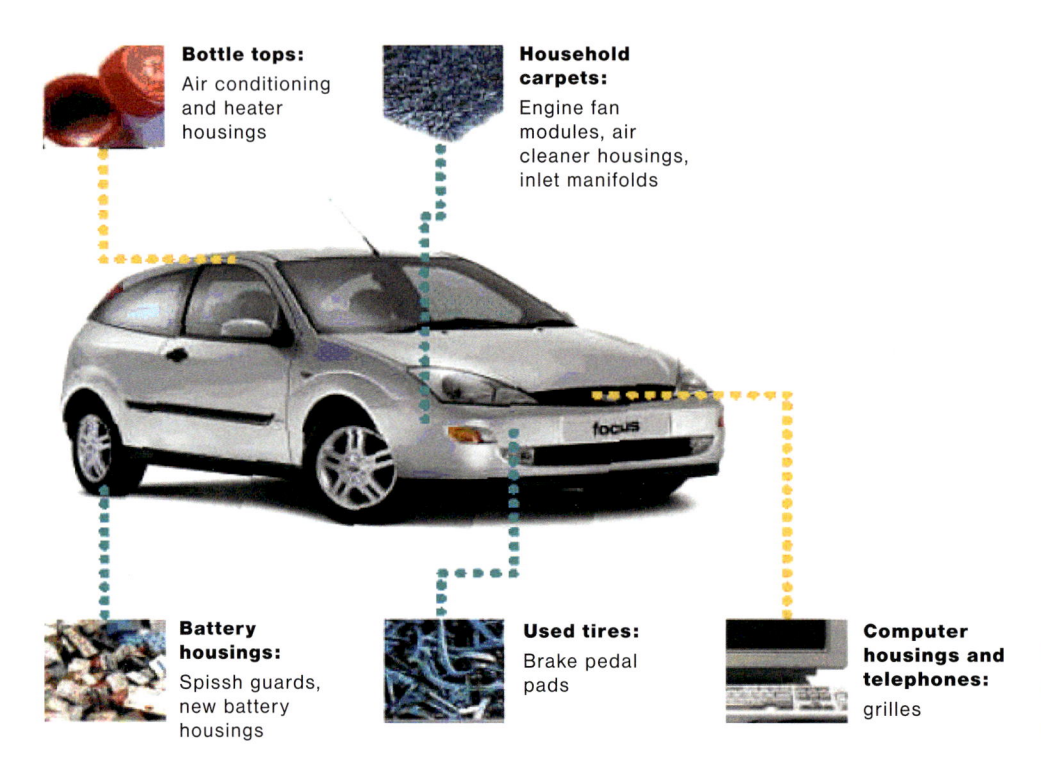

Bottle tops: Air conditioning and heater housings

Household carpets: Engine fan modules, air cleaner housings, inlet manifolds

Battery housings: Spissh guards, new battery housings

Used tires: Brake pedal pads

Computer housings and telephones: grilles

Figure 2-32 Ford's utilization of recyclate material

Automotive window glazing cannot be used to make new automotive window glazing but it can be used to make bottles. Significant amounts of vehicle sound deadener material contain recyclate from the clothing and fabric industry. Further co-operation across industries will see a greater utilization of valuable resources.

2.2.5.5 Roadblocks and Areas for Development

Many materials and components are proving complex and difficult to recycle. These include window glazing, paints and coatings, adhesives and sealers, some foams, thermosetting plastics, elastomers and tires. Mixed and hybrid assemblies also prove difficult to separate.

The ERAT teams established at Ford Motor Company are piloting many studies with partner suppliers to improve recycling performance. One excellent example is the use of Ford transport to return end of life bumpers from our dealers back to a recycling plant which sorts, cleans and regrinds polymers and supplies them to molders to make service bumpers on the Ford Focus. We are learning that the key to success is good organization and logistics.

Automotive industry cooperation with key suppliers will be necessary to solve the issue of recycling, particularly for difficult materials. The use of the International Materials Data System (IMDS) is becoming the accepted standard to record the materials content will help automotive companies control and plan recycling routes. As stated earlier, recycling issues must be resolved without compromise of function, loss of quality, or increase in costs.

2.2.6 Summary

Great progress has been made in engineering vehicles that are safe, efficient, environmentally clean and desirable. We now have to ensure that we meet our recycling obligations and automotive companies can assist this process in helping to create the demand for materials that contain recyclates in automotive products through publication of our successes and the continued design of desirable products.

2.3 Lightweight Design with Plastics in the New BMW 6 Series

ANJA MAIER

Dynamism in the automobile requires not only a powerful engine and excellent running gear but also the lowest vehicle weight possible. "Intelligent lightweight design" is the magic phrase – and this is true of the car body as well. The word "intelligent" means that lightweight components not only reduce the weight of the vehicle – they also add functionality.

2.3.1 Material Mixture Saving Weight and Increasing Collision Safety

When it comes to the vehicle body the name of the game is "materials mixture" (see Figure 2-33): In other words, the bodyshell is a mixed construction of steel, aluminum and plastic components which provides considerable weight advantages. The body structure has nevertheless become more rigid and stronger. This in turn results in improved collision safety and in more precise reactions as far as driving dynamics is concerned [1].

The hood, for example, is made of aluminum but it also benefits from a global innovation which currently is only used by BMW: the inner panel and the outer skin of the hood are bonded not by a one-pack adhesive but by a two-pack adhesive instead. This allows a reduction in the wall thickness of the two aluminum panels, which results in a weight reduction of 9%.

The trunk lid made of SMC is another example: in the beginning, SMC was considered for reasons of styling freedom but then integration became another important issue. Only when using thermoset material made it possible to include the spoiler in the mold and to integrate electronic components, such as the antennae. In all, approx. 10% weight was saved in comparison with a trunk lid made of steel.

Approximately 4 kg of weight was saved by using thermoplastics for the two front side panels, a change that also contributed to the excellent handling characteristics of the new 6 Series. But no less decisive was

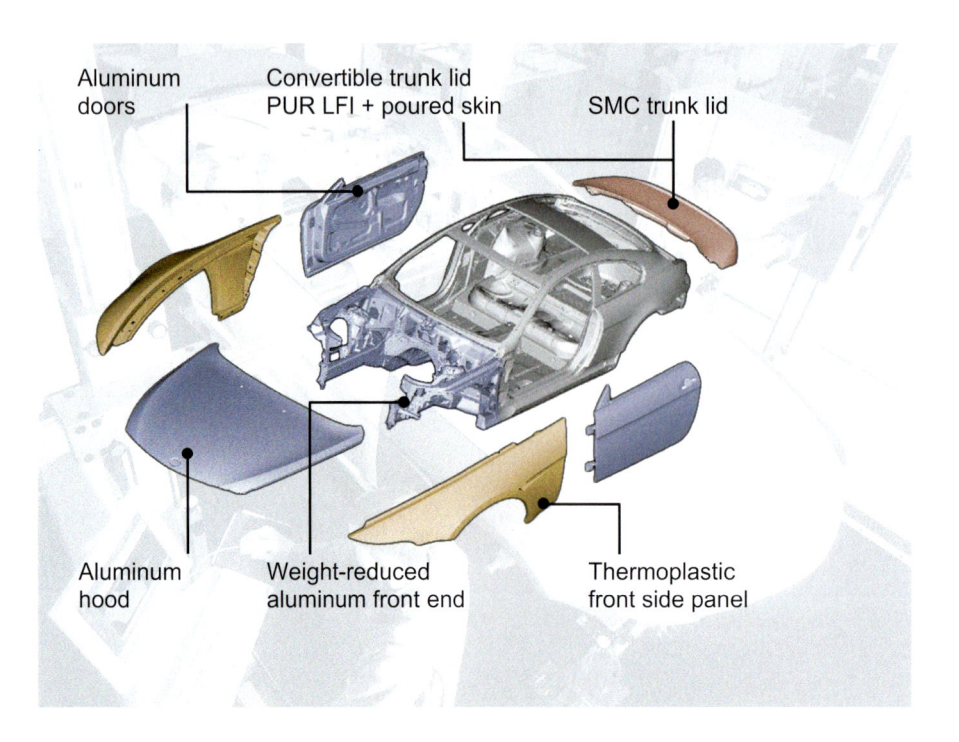

Figure 2-33
Mix of materials in the vehicle body

the fact that the plastic allowed the designers to extend the lines of this large coupé right through into the side panel area. Thus integrating the side trim elements into the side panels would not have been possible using conventional steel design. Other advantages were found in the fact that thermoplastic materials are less sensitive to minor collision damage than steel because of their specific material properties.

2.3.2 Requirements for Thermoplastic Side Panels

In order to capitalize on the advantages of thermoplastic side panels, a suitable combination of the various materials and manufacturing technologies had to be found.

Based on the requirements for side panels (see Figure 2-34), the objective will always be lightweight design. In the light of this, not only plastics but also some metals could be considered for this application, although they would impose some restrictions on styling freedom, integration potential and economic impact.

Another major objective is to create an economically efficient application that will compare favorably to other materials in a break-even analysis. Great emphasis is to be placed on the dynamic fracture behavior in order to meet legislative and product liability requirements. In this respect, light metals should be compared to steel; another major issue to be considered are the changes in properties as a function of temperature, particularly when thermoplastics are used.

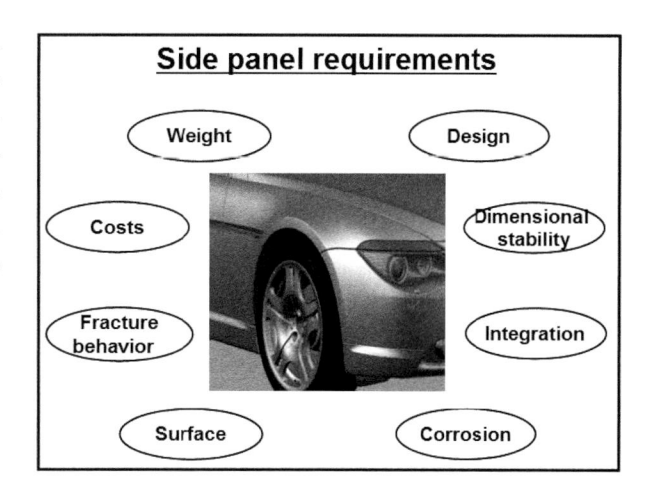

Figure 2-34 Requirements for a side panel

In addition, the look and appearance, especially the design and dimensioning of the joint lines, have considerable influence on the visual impression of a vehicle. Here, a central issue is not only the actual dimensional stability of the material but also its linear expansion behavior.

2.3.3 Criteria for the Use of a Thermoplastic Side Panel

Technical requirements on the one hand and economic considerations on the other required an economical decision from the company.

Evaluation of the economic efficiency, particularly when using thermoplastic materials, is considerably influenced by the problems related to exterior painting (see Figure 2-35). In general, components can be painted off-line (in other words, at the system sup-

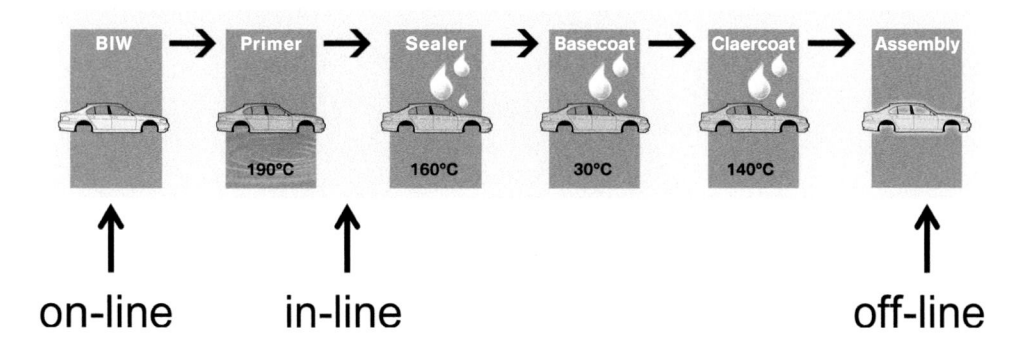

Figure 2-35 Painting scenarios for thermoplastic side panel [2]

plier), in-line or on-line. Components painted in-line and on-line are components painted on the vehicle production line, depending the location where they enter the line.

Side panels that can be painted on-line have already been fitted to the body shell before they pass through the entire vehicle painting process, including electrophoretic dip coating. In this case, the maximum temperature in the dryer of more than 190 °C impose specific requirements on the heat resistance of the material used.

Components that can be painted in-line are not fitted until after they pass the electrophoretic dip coating dryer which reduces the temperature stress on them to about 160 °C.

Alongside the considerable additional costs resulting from off-line painting and the additional plant floor space required for assembly, particular attention must be paid to color-matching.

When BMW had to make a material choice, there were no thermoplastic or polyurethane materials that were suitable for on-line painting. Therefore, the only viable approach from an economic point of view was in-line painting. Off-line painting was also excluded for the same reasons.

A comparison of the various materials (see Table 2-3) showed that the manufacturing costs for a thermoplastic in-line painted side panel differ by only 20% from those of an SMC side panel painted on-line. Therefore, the principle properties of these two materials had to be compared before an informed decision could be made (see Figure 2-36).

Table 2-3 Comparison of materials

Materials	Steel	Aluminum	Thermoplastic	SMC	PRIM
Weight per part [kg]	4.5	2.5	2.5	2.9	3.2
Wall thickness [mm]	0.8	1.2	3	2.1	3.5
Painting scenario	online	online	inline	online	inline
Manufacturing costs [%]	100	170	100	120	150

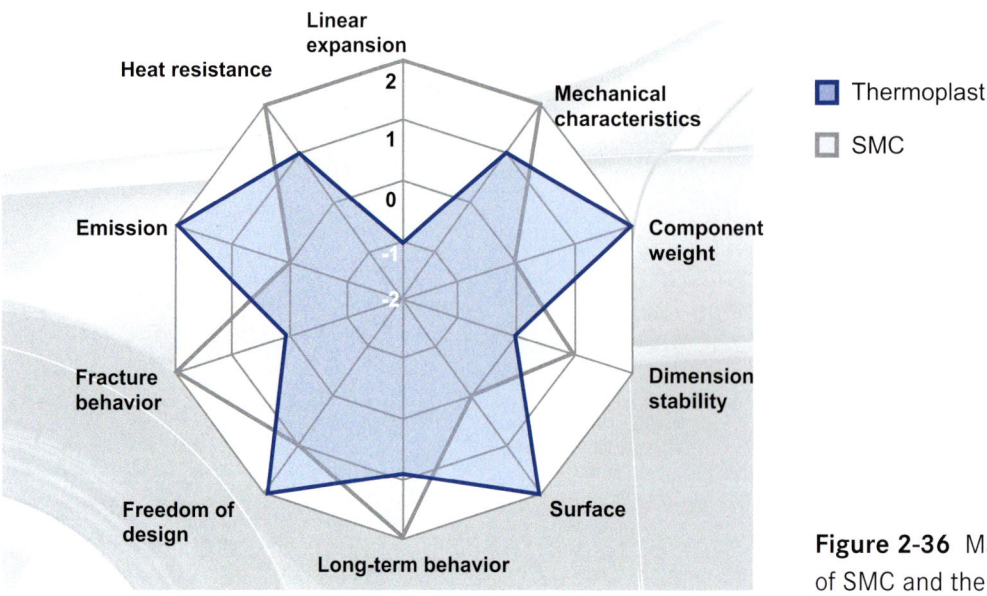

Figure 2-36 Material properties of SMC and thermoplastics

Polyurethane RRIM was not included in further considerations for economic reasons and also due to its low weight-saving potential.

The following advantages and disadvantages are evident:

Advantages of SMC:

- Coefficient of linear expansion
- Rigidity
- Fracture behaviour

Particularly the very low coefficient of linear expansion for the metal materials is of great advantage in component layout and inter-panel gap design.

Advantages of thermoplastics:

- Weight
- Design freedom
- Surface texture

The final decision in favour of in-line painted thermoplastics since this solution best fulfilled the objectives relevant to the vehicle, such as weight reduction, styling implementation, and economic efficiency.

2.3.4 Material Requirements

Of the thermoplastic materials under consideration, an unreinforced PPE+PA blend was finally chosen. However, before the part-specific material properties were examined, the component itself had to pass through the in-line painting process to ensure that it had a certain degree of rigidity at the required paint-drying temperature to emerge free of deformations or other impairments. The evaluation of this so-called heat resistance is carried out via DMA testing (see Figure 2-37).

The component's properties are examined once the vehicle has been produced. Table 2-4 shows an overview of the main mechanical properties of the PPE+PA blends.

Meeting inter-panel gap requirements requires very tight manufacturing tolerances; in addition, only a certain amount of dimensional change is acceptable. Therefore, the linear expansion of the material is of critical importance.

Since the specified material PPE+PA absorbs moisture, which causes changes in its properties and dimensions (see Table 2-4), this behavior must be taken into consideration in component design.

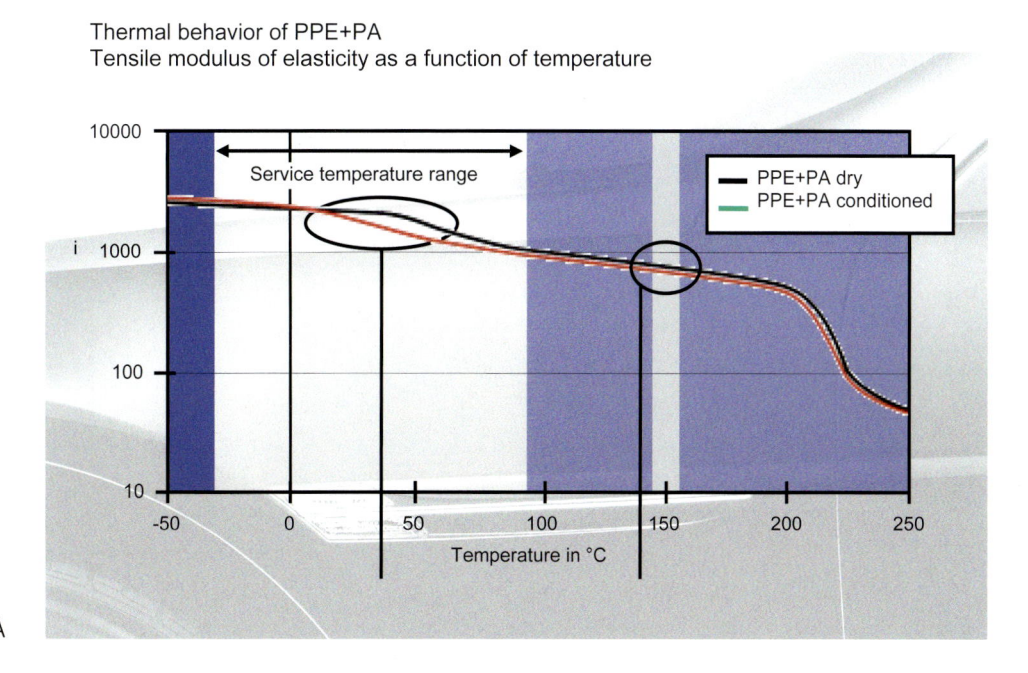

Figure 2-37 Heat resistance of PPE+PA

Table 2-4 Mechanical characteristics of PPE+PA [3]

Property	Dim	Standard	PPE+PA dry	PPE+PA conditioned
Flexural modulus of elasticity	MPa	ISO 178	2000	1600
Tensile modulus of elasticity	MPa	ISO 527-1A 50	2200	1600
Tensile strength	MPa	ISO 527-1A 50	57	49
Elongation at break	%	ISO 527-1A 50	29	142
Impact strength	kJ/m^2	ISO 179/1eU	330 NB	325 NB
Notched impact strength	KJ/m^2	ISO 179/1eA	28 C	27 C
Linear expansion RT $\rightarrow$ +80 °C	$\times 10^{-6}$ 1/K	DIN 53 752	93	102
Linear expansion RT $\rightarrow$ -30 °C	$\times 10^{-6}$ 1/K	DIN 53 752	85	92
Ductile-brittle transition	°C	ISO 6603	> 0	> 0

Another important aspect concerns the safety requirements. All properties relevant to impact/collisions must be delivered over a large temperature range, i.e., the material has to display high degree of impact strength, particularly at temperatures below freezing point. This properties profile cannot be verified until the finished component is mounted, because the geometric design of the component plays a significant role in the determination of its impact strength.

In order to evaluate component behavior under dynamic stress, front impact tests at temperatures down to –20 °C are carried out, during which no fracture should occur. The selected PPE+PA blend met this requirement.

Together with high impact strength, the material must exhibit a certain level of rigidity. This static buckling resistance provides the side panel with a certain inherent strength which is, however, also considerably determined by the component shape. The following describes the series of design and process measures taken to meet these material-specific demands.

2.3.5 Product Requirements for Component Design

In order to produce a high volume of vehicle components made of unreinforced thermoplastic PPE+PA the mandatory quality requirements relating to the total vehicle have to be met.

The side panel as an important component of the body's outer skin must satisfy the following criteria:

- Integration of trim elements and side indicator
- Accommodation of the front indicator
- Uniform joint line between neighbouring components during the complete vehicle lifetime
- Surface quality to the BMW standards
- Simple and process-reliable installation on the bodyshell with adequate adjustment possibilities for the joint and fairing lines according to the interpanel joint plan
- Harmonious overall visual impression of the body.

Possible solutions had to take into account the following important material and process related parameters:

- High coefficient of thermal expansion
- Hygroscopic material behavior
- In-line painting process.

2.3 Lightweight Design with Plastics in the New BMW 6 Series

Due to the considerable size of the side panelling and the high coefficient of expansion, an intelligent fastening concept had to be developed. With an initial longitudinal length of approx. 1400 mm and a maximum use temperature of approx. 85 °C, the component has a measured unconstrained longitudinal range expansion of approx. 6 mm. In order to achieve a harmonious interpanel joint concept, this value is unacceptable.

This problem was solved using a fastening system, which on the one hand only permits a precisely defined thermal expansion of 2.5 mm and on the other hand ensures the necessary assembly tolerance for the threaded connections. Here, the sufficient clearance of the temperature-specific change in length had to be decoupled from the compensation for tolerances.

The contours of the through-holes are integrated into the side wall and appear in the mold as individual inserts. In a preassembly operation, steel sleeves are pressed into the component, with the installation position being uniquely defined by the centering. Any installation tolerances between the body and the side panel are taken up within the sleeves. The component is fastened firmly (force-locked) by means of supports, sleeves and screw heads. The permissible thermal expansion, depending on the size of the through-hole, now can take place by a relative movement of the side panel.

With the help of FEM calculations, the attachment concept was designed in such a way that, following an unconstrained movement of 2.5 mm, the side panel compensates for all remaining expansion by homogeneous "cambering," which does not imply any negative quality impact for the customer.

The indicator is fastened directly to the side panel, in order to prevent any moisture- or temperature-related impact on the joint running along the x-axis between the side panel and the indicator. The design specified an offset between the indicator and the headlight to ensure a uniform appearance of the joint line.

The use of a thermoplastic material allowed the implementation of characteristic design elements. The side

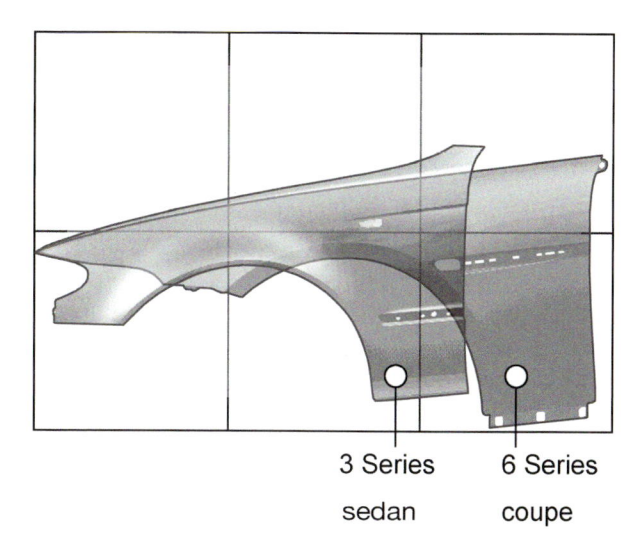

Figure 2-38 Size comparison of the side panel of the 3 Series limousine and the new 6 Series coupé

Figure 2-39 3-D view of an attachment point

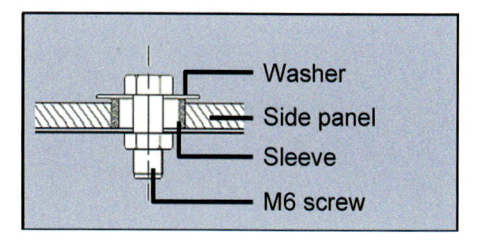

Figure 2-40 Sectional view

indicator with its trim element deserves particular attention: the components are inserted into a recess, whose depth and shape could not be realized as a single part with conventional sheet metal construction. By means of additionally integrated cut-outs, the exterior elements can be fastened directly to the side panel without the need for further fastening elements.

2.3.6 Manufacturing and Assembly Process

The complete production process takes place in the BMW plants at Landshut and Dingolfing in Germany. The individual process steps and their sequence are shown in the following flow diagram. The most important process steps will be explained in more detail.

2.3.6.1 Injection Molding

Due to its size, the component is produced on an injection molding machine with a locking force of 32 000 kN. In order to meet economic requirements, the entire side panel must be molded in a single process step.

A particular challenge with this application is the positioning of the mold parting lines in the nonvisible area. To allow demolding of the undercut areas of the rain channel, the door groove, and the transition to the bumper, a three-plate mold with balanced film gate was used. The adjustment of the different splits – necessary to make sure the surface is of marks – is carried out when the mold is heated. The mold temperature, determined by the flow characteristics of the material, is approx. 120 °C.

Tests showed that the moisture contents of the pellets had an influence on the processing quality. A specific, constant moisture content of the pellets at the time they are fed into the injection molding machine is of decisive importance to ensure a high-quality parts (Figure 2-42).

2.3.6.2 Priming

The next step after injection molding is the application of conductive primer to the components. Due to the hygroscopic properties of the polyamide component, a special two-pack polyurethane primer system was developed. It allows for moisture absorption during conditioning but the release of moisture via the outer surface during the painting process is slowed down.

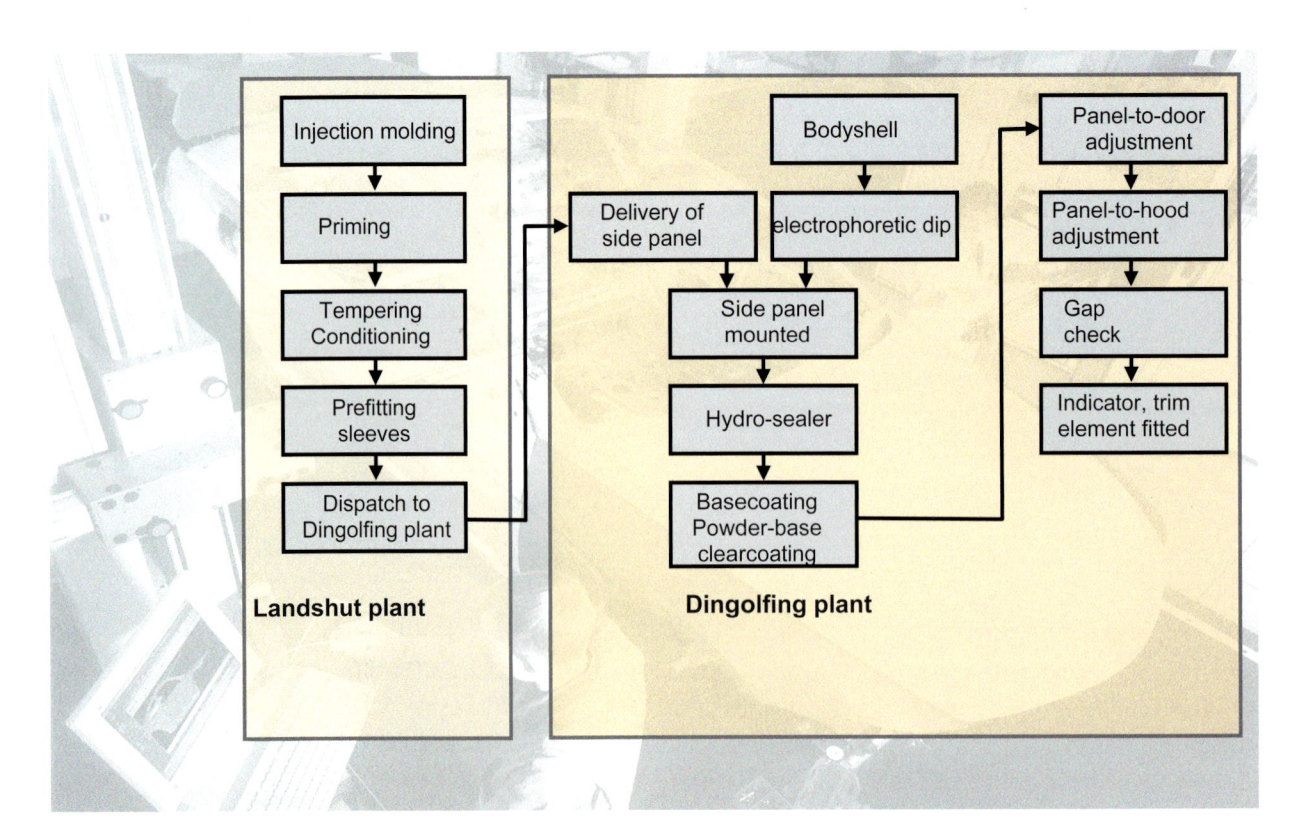

Figure 2-41 Process sequence diagram

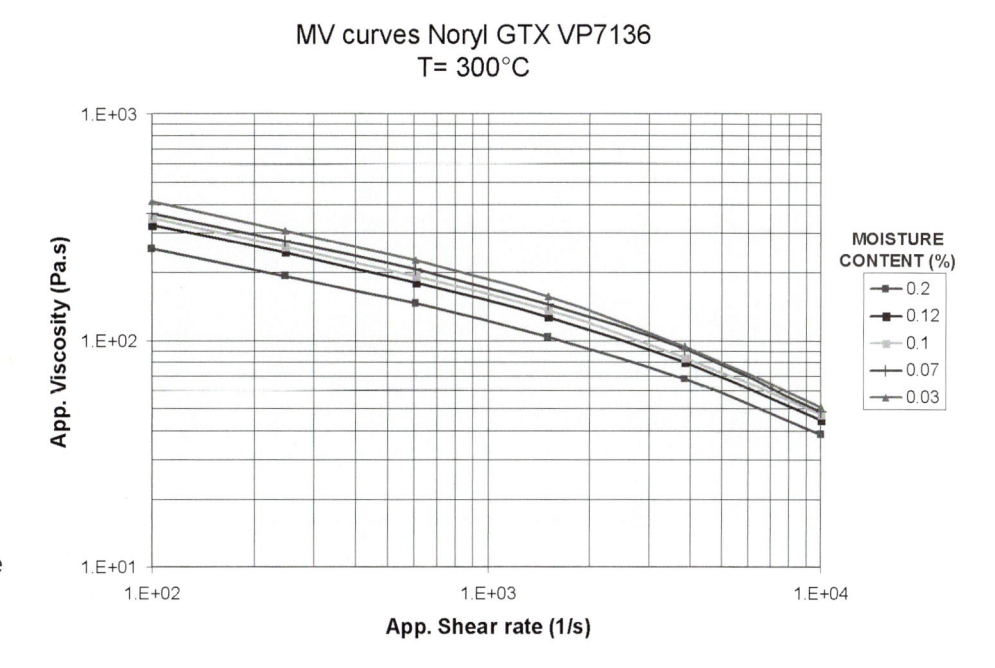

Figure 2-42 Viscosity curve as a function of shear rate and moisture content [4]

2.3.6.3 Tempering and Conditioning

Moisture absorption and release have a direct effect on the dimensions of the component. The equilibrium state in the central European climatic zone is at approx. 1.1 wt.% of the material.

Since a change in moisture content of 1.0% causes a change in length of approx. 0.3%, this aspect must be taken into account. Therefore, before being shipped to Dingolfing, the side panels are conditioned to 2.4% to ensure the required dimensions following the drying processes after painting. To obtain constant moisture content values and to eliminate so-called after-shrinkage, a tempering unit is included upstream.

2.3.6.4 Transportation to the Dingolfing Plant

To make logistics more efficient, special transportation containers are used. The are design so that air in the containers remains more or less sealed off. This means that the conditioned side panels can be stored up to ten days before they are fitted to the vehicle.

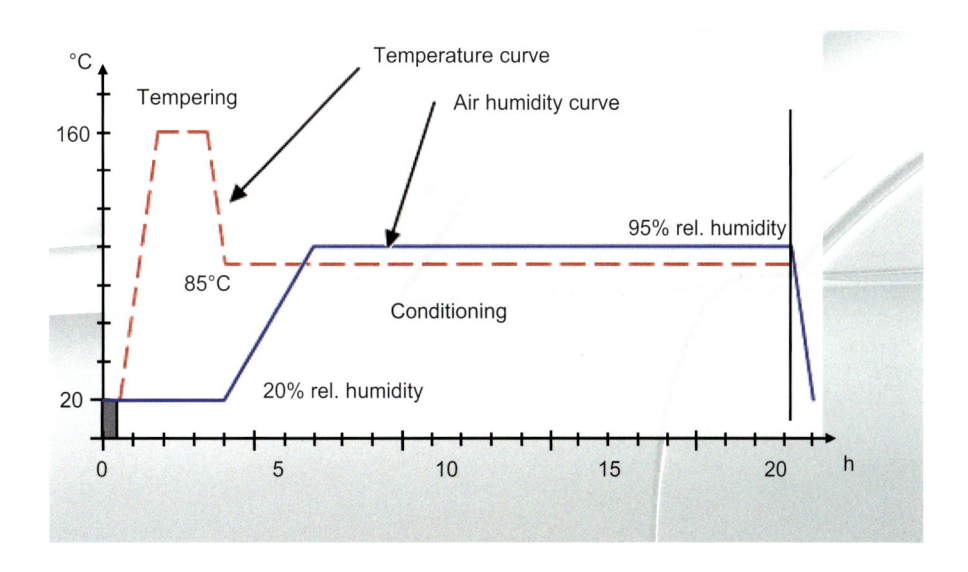

Figure 2-43 Tempering and conditioning curve

2.3.6.5 Attachment of the Side Panel

Compared with the conventional sheet-metal approach, the in-line process calls for a modified manufacturing sequence through the entire process chain. With the vehicle in the body shell state, the hood panel is now mounted on the chassis using corresponding fixtures that act as substitutes for the side panels.

After passing through the electrophoretic dip coating unit, the vehicles are sent into an transfer station. Here, the side wall is fastened to the body shell at the A pillar and its position on the x- and y-axis is set with respect to the door. No further restriction is set on the remaining freedom of movement of the part.

This makes prevents thermal induced stresses and permanent deformations.

On the z-axis, the front of the part rests on a support. After completion of this step in the production process, the vehicle is returned to the line and passes through the remaining painting process used for series production.

2.3.6.6 Assembly

The side panel is the first component to be finally screwed to the vehicle on the assembly line. Here, the joint lines between door and hood are set according to the specified dimensions.

The other add-on body components, such as front indicators, side indicators and trim element are fitted in the subsequent course of the assembly.

2.3.7 Outlook and Future Requirements

Due to the complexity of the manufacturing process, the described concept of a thermoplastic side panel is economically feasible only for smaller production runs, such as for the new 6 Series.

Further material development is needed to reduce the additional costs for design and process in the future.

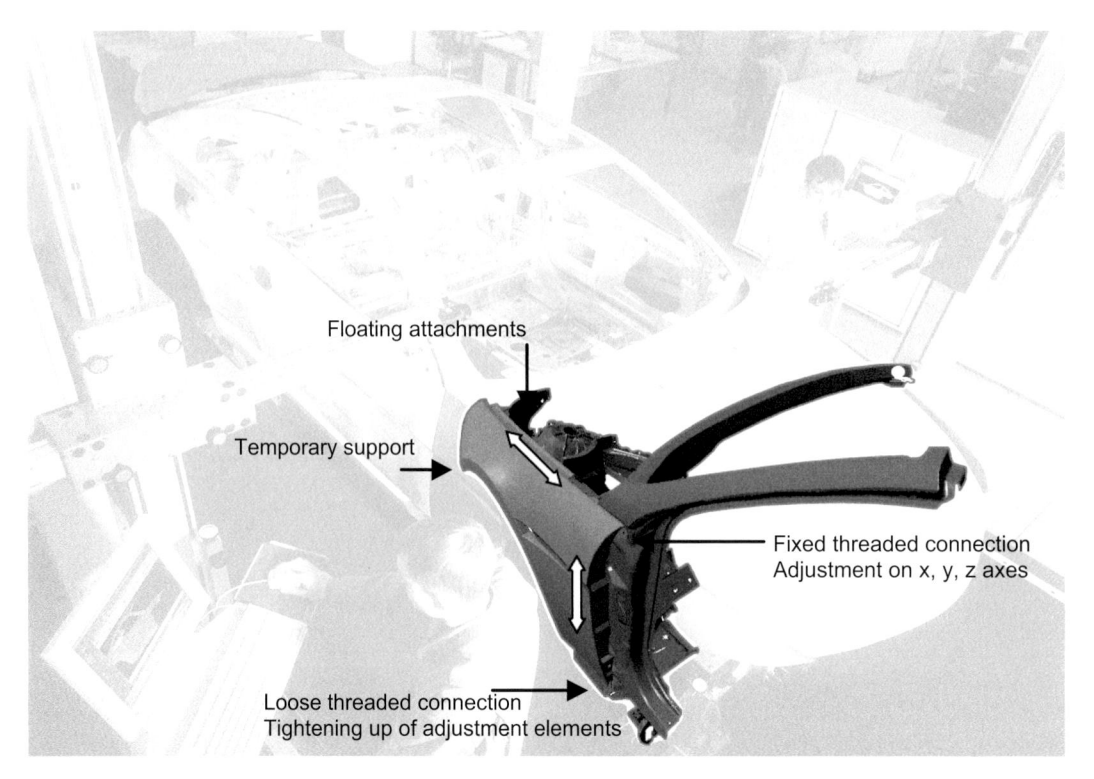

Figure 2-44 Attachment during the painting process

2.3.7.1 On-Line Paintability

As has already mentioned, at the time the decision was made to use a particular material, there was no thermoplastic material on the market that could be painted on-line. Therefore, the first step in material development is to increase heat resistance.

The first possible solutions on the market now are based on PPE+PA blends and PA+ABS blends. Figure 2-45 shows an overview of rigidity as a function of temperature.

However, the on-line paintability of the material is only important at this one point in the vehicle's life and therefore the level of rigidity in the temperature range of the electrophoretic dip coating dryer can be reduced to the necessary minimum. Of equal decisive importance to the deformation behavior of the component is the low-stress, material-friendly injection molding process. This is why the processing behavior of the material plays an important role.

2.3.7.2 Mechanical Properties

Assuming that the material can be painted on-line, material optimization measures are focused on an improvement in properties for the actual service life of the vehicle. Efforts are made to achieve a well-balanced rigidity and toughness profile, while keeping linear expansion to a minimum. The property profile of the PA+ABS blend is shown in Table 2-5.

2.3.7.3 Dimensional Properties and Moisture Absorption

As mentioned earlier, the dimensional changes and a modified property profiles caused by moisture absorption of these materials is a critical aspect, resulting in considerable extra costs in component and vehicle manufacture. Therefore, one of the highest priorities is to develop moisture independent materials. Currently, various possible materials are being developed together with different raw material producers and then tested at BMW.

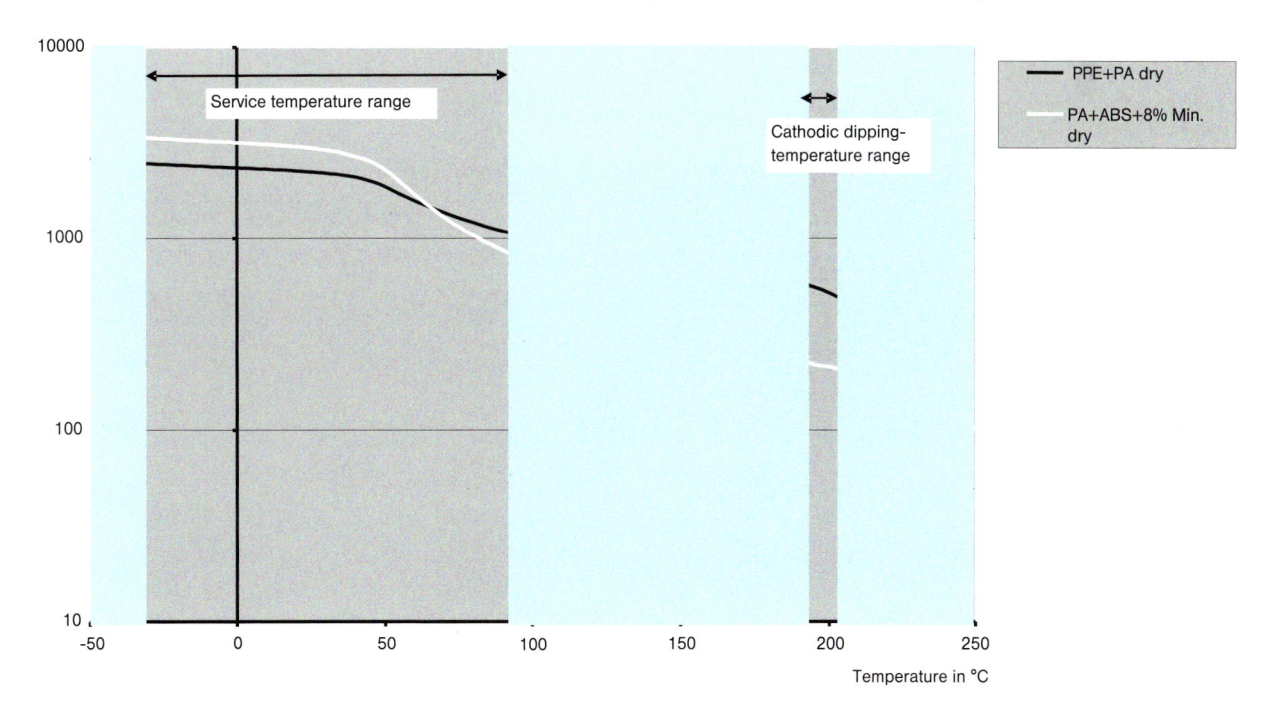

Figure 2-45 Heat resistance of on-line paintable thermoplastics

Table 2-5 Material properties of on-line paintable thermoplastics [3]

Property	Dim	Standard	PPE+PA	PA+ABS + 8% Mineral	Requirements for side panel
Flexural modulus of elasticity	MPa	ISO 178	2000	3000	> 3000
Tensile modulus of elasticity	MPa	ISO 527-1A 50	2200	3150	> 3000
Tensile strength	MPa	ISO 527-1A 50	57	50	> 15
Elongation at break	%	ISO 527-1A 50	29	27	> 20
Impact strength	kJ/m^2	ISO 179/1eU	330 NB	164 C	> 150
Notched impact strength	kJ/m^2	ISO 179/1eA	28 C	11.9 C	> 10 C
Linear expansion RT → +80 °C	×10^{-6} 1/K	DIN 53752	93	68	< 50
Linear expansion RT → -30 °C	×10^{-6} 1/K	DIN 53752	85	62	< 50
Ductile-brittle transition	°C	ISO 6603	> 0	> RT	< RT

2.3.7.4 Conductivity

The next logical step in the direction of greater economic efficiency is to develop conductive materials with the aim of saving on the cost caused by priming. However, such materials can only be used once the interdependency between hygroscopic thermoplastics and the paint system is eliminated or until non-hygroscopic thermoplastics are available.

2.3.7.5 Development of Test Methods

A further major task in the future will be to meet requirements relating to the protection and safety of pedestrians. One focus of development in particular will be finding testing methods which will deliver an early evaluation of dynamic fracture behavior; another is to prepare material cards for computer-aided simulations.

Overall, BMW considers it a very positive step that several raw material producers are now participating intensively in further development of these materials.

References

[1] Innovationen der 6er Baureihe, BMW Group press kit, Munich 2003
[2] Schropp, M.: Paper at Produktionsforum BMW Group, Landshut 2003
[3] BMW AG, Kunststofftechnikum
[4] Viskositätskurve in Abhängigkeit der Schergeschwindigkeit und der Feuchte, product information for Noryl GTX, GE Plastics, 2002

3 Material Concepts and Process Technologies

3.1 Plastics Engineering in Automotive Exteriors

Olaf Zöllner

3.1.1 Introduction

Plastics continue to play a major role in the appearance of automotives. Indeed, the development of automotives as we know them today would have been virtually impossible without plastics. Comfort, safety, weight reduction, corrosion resistance, integration potential and design freedom have all been driven forward by automotive engineers, not least through the use of plastics.

The idea of an all-plastic body was first realized 40 years ago in a number of studies, one of which was the concept car presented at the "K" plastics exhibition of 1967, shown in Figure 3-1. On the right is a concept study from 2005. New materials and processing techniques are responsible for the obvious difference.

What began a few decades ago with a number of concept vehicles has since become an integral part of automotive design. Plastics have gained a permanent place for themselves in vehicle body design and, because the possibilities are still far from exhausted, will continue to play a key and ever-growing role in automotive applications over the coming years. Tailor-made plastics offer much more potential to meet the demands of the automotive industry. It should also be kept in mind that the designing and processing technologies – best described by the word "engineering" – have made continuous advances in recent times. Engineering equipment is now available to realize applications and ideas that would have been inconceivable a few years ago.

3.1.2 History

The proportion of plastics used in automotive engineering has increased several times over in the last 40 years. Figure 3-2 compares the percentages from the 1960s with those of today [1].

One of the main fields of application for plastics is in vehicle interiors, where plastic materials have made a significant contribution to the comfort we know today, while complying with very stringent safety standards. Apart from that, plastics have also become well established in exterior body panel applications.

Figure 3-1 Concepts for an all-plastic body, on the left at K 1967 and on the right from 2005 (from Bayer MaterialScience)

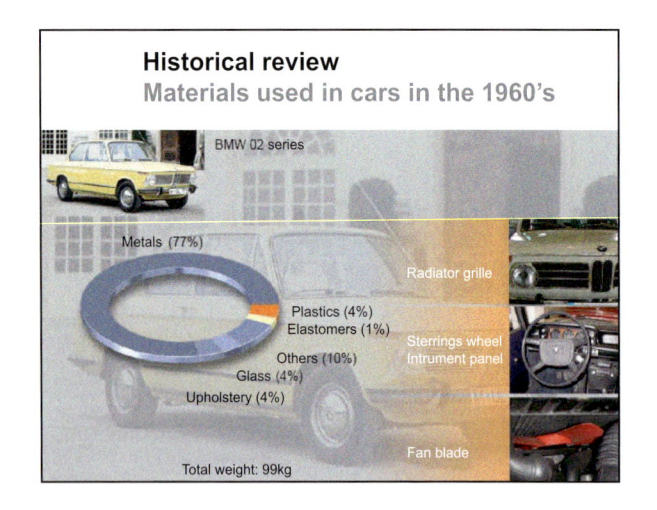

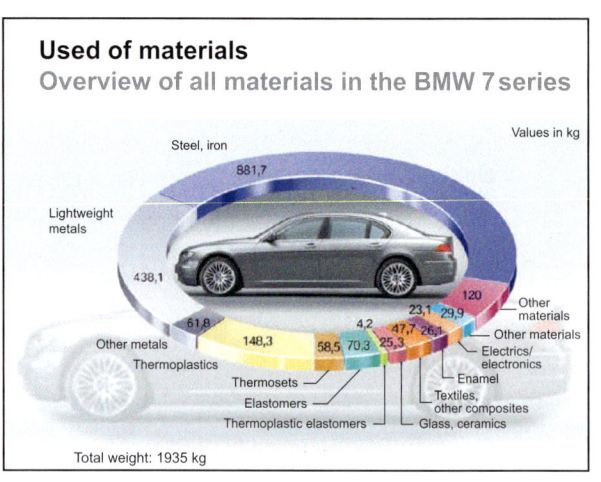

Figure 3-2 Growth in the percentage of plastics in the total weight of a car [1]

In the early days of automotive development, natural rubber was used for manufacturing tires. In the 1960s, the use of plastics for exterior parts began. Today, plastics are found in a wide variety of exterior body applications, yet development in this sector has only really just started and is being driven by the raw material manufacturers to meet specific requirements. Materials used in exterior body applications include:

- PC/ASA   Steel   PC/AES
- PC/PBT   Aluminum   PP
- PC/PET   SMC   PPO/PA
- PU RIM   PU foam   PA/ABS

The heavy steel bumper was one of the first parts to be replaced by a plastic bumper shell, and was accompanied by the introduction of energy-absorbing foam structures. By combining it with an outer plastic skin, development engineers came up with a system that was not only "repair-friendly", since it could elastically absorb small deformations, but was also able to make a significant contribution to "crash management". Plastic radiator grilles soon followed. Today, the bumper and grille are fully integrated into the body and exploit the advantages of plastics for design purposes. Figure 3-3 (left and center) compares an early front-end design with one of today.

In 1993, Hella launched the first polycarbonate headlamp lens onto the market. Today, all new cars are fitted with plastic headlamp lenses and lens modules. Apart from the weight saving compared with glass, the outstanding design freedom provided by plastics has been a key factor in the success of this application. New vehicles demonstrate very clearly how headlights can enhance the aesthetics and the image of a brand, in addition to fulfilling their performance functions (Figure 3-3 center and right).

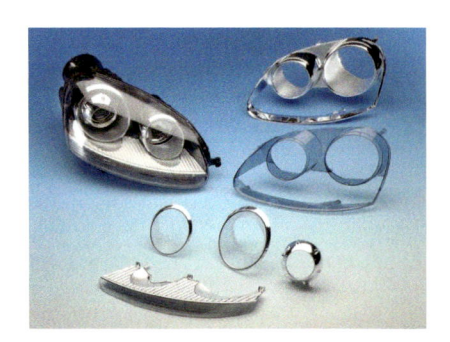

Figure 3-3 Old and new front-end design, and headlight lens made of polycarbonate

3.1 Plastics Engineering in Automotive Exteriors   43

production. Figure 3-4 gives an overview of some of the present-day plastic applications for body panels.

These applications have become possible not only through the development of new and optimized plastics, but also through the enormous innovative momentum in equipment, processing and tool engineering, as well as indispensable CAE technologies.

Figure 3-4 Plastic applications for vehicle body panels

3.1.3 Processing Methods and Applications for Thermoplastics

For all body panel applications, such as side panels, doors, antenna covers and spoilers, a basic requirement is that they have high stiffness, excellent thermal resistance and low thermal expansion. In addition, they must also have adequate toughness at temperatures of −30 to 40 °C. For this purpose, tailor-made blends have been developed such as PC/PBT and PC/PET with special fillers.

Polyurethane materials have a world market share of approx. 7% and can be regarded as a specialty product compared with the commodity plastics. Thanks to their tremendous variability with regard to performance properties, shaping potential and processing, polyurethanes have since found their way into an impressive range of applications.

The industrial application of polyurethane materials began in the 1960s with flexible and rigid foam applications for the furniture, construction and appliance industries. In the automotive sector, a start was made with polyurethane seating foams, which helped to significantly improve comfort and reduce vehicle weight. Today, polyurethane materials have become indispensable for both interior and exterior applications.

Another important specification in the automotive industry is Class A finish. To achieve this, application processes must be used that are precisely geared to the materials and the specifications. Cascade-controlled injection molding (Figure 3-5 left) and sandwich injection molding (Figure 3-5 right) are particularly suitable.

In the last few years, side panels, tailgate liners and the first roof modules made of plastic have all entered mass

In the cascade method, hot runner nozzles are opened and closed in sequence during the filling of the part to avoid weld lines. When optimized, filler separation and

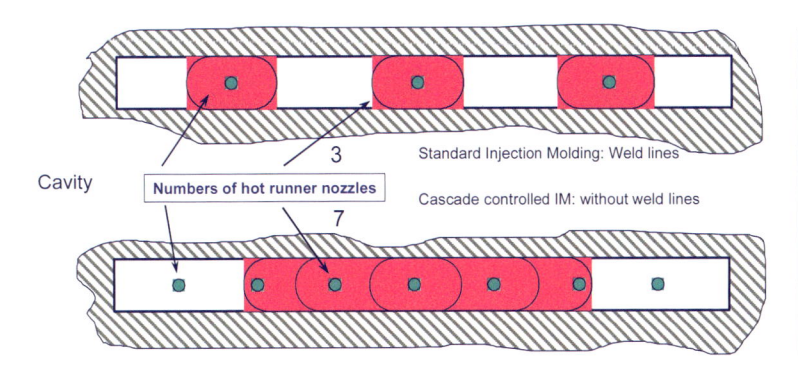

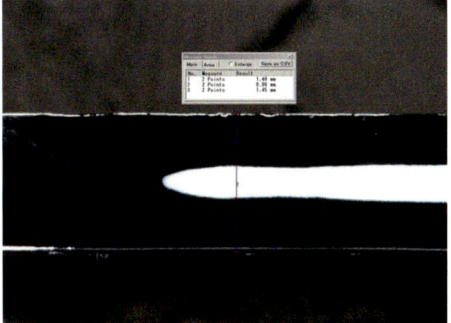

Figure 3-5 Cascade controlled injection molding (left) and sandwich injection molding (right)

the related surface defects can be largely prevented. In this process, each nozzle along the flow path opens as soon as the flow front has passed its aperture. Apart from better surface quality, another major advantage of this process is that injection pressure never acts over the entire mold at once, thus permitting reduced clamp tonnage of the injection molding machine. The injection and compression pressures are reduced via the controlled hot runners [2]. This method has also proved effective for back-molding films and fabrics, since it largely avoids deformation of the substrates along weld lines [3].

In sandwich injection molding, two thermoplastics are injected one after the other into the cavity, forming a core component and a skin component. The core component can consist, for example, of a fiber-reinforced thermoplastic, giving the part high stiffness and low thermal expansion. To ensure a Class A finish, the main component is generally an unfilled thermoplastic. With this process, it is important that the viscosities of the two components are coordinated. The resultant wall thickness and the flow characteristics are strongly influenced by this [4, 5].

Exterior body parts are mostly painted nowadays. Possible exceptions are door sills and bumpers. The coating can either be applied online (after assembly on the body) or offline (coating in a separate process step). In parallel with the coating operation, however, there have been developments in the back-molding of body parts behind film. Thin film composites with tailor-made properties (color, weather protection, etc.) are back-injected or back-foamed using special processing methods. Basically, there are three ways of producing color-matched body parts using film technology [6]:

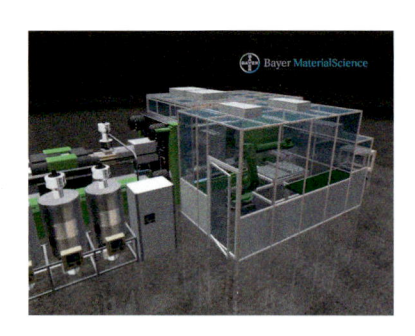

Clean room production cell for back-molding large-area parts behind film

Gripping the film

Brush cleaning the film

Electrostatic charging of the film and its insertion

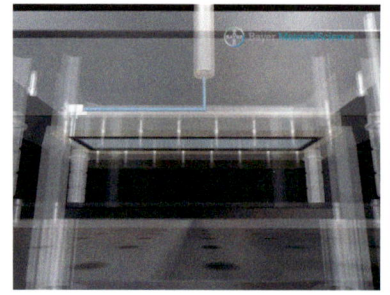

Back-molding the film

Stacking the finished part

Figure 3-6 Injection molding process for back-molding large-area parts with film at Bayer MaterialScience's testing laboratory

3.1 Plastics Engineering in Automotive Exteriors

- Conventional offline coating of the substrate with the film
- Laminated film consisting of a carrier film and paint film (paint film with transparent layer and a colored layer beneath it)
- Coated film with colored basecoat and UV-curing clearcoat. Curing takes place after thermoforming and film insert molding.

In all cases, the films must be inserted into the mold under virtually clean-room conditions. Special robots and handling techniques have been developed to do just this. In Bayer MaterialScience's polycarbonates testing laboratory, the process chain has been developed for back-molding large-area parts with film (Figure 3-6).

Polycarbonate glazing is an impressive example of the technological development in car exteriors. This shows how technological advances have led to new applications and new classes of materials (e.g., polycarbonate replacing glass).

The high quality required for such applications would not have been feasible without the development of a special compression technology in combination with an innovative 2-component injection molding technique using tailor-made swivel-platen injection molding machines [7, 8]. In most 2-component polycarbonate glazing applications, the clear component is injected first (Figure 3-8).

After this mold half opens, the swivel platen rotates the clear component into the "2nd cavity", into which the "black component" is then injected. The actual window cut-out is produced in this way through the injection molded "blackout". Furthermore, a variety of functions can also be integrated into the black component (e.g.,

Figure 3-7 Production-line applications for polycarbonate glazing – left: sun roof and rear fixed side window, right: louver sun roof

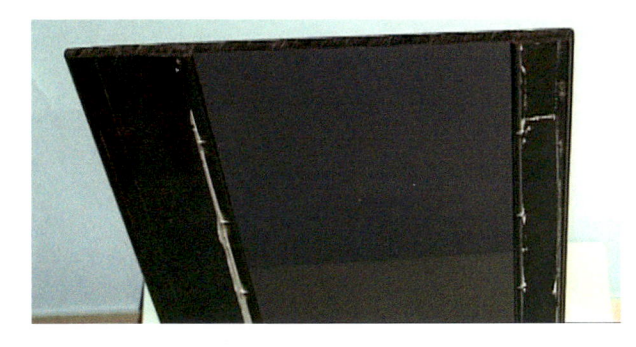

First component

Second Component

Figure 3-8 Left: Cross-section of a 2-component polycarbonate sunroof.
Right: Detailed view: polycarbonate at the top, PC PET blend at the bottom.

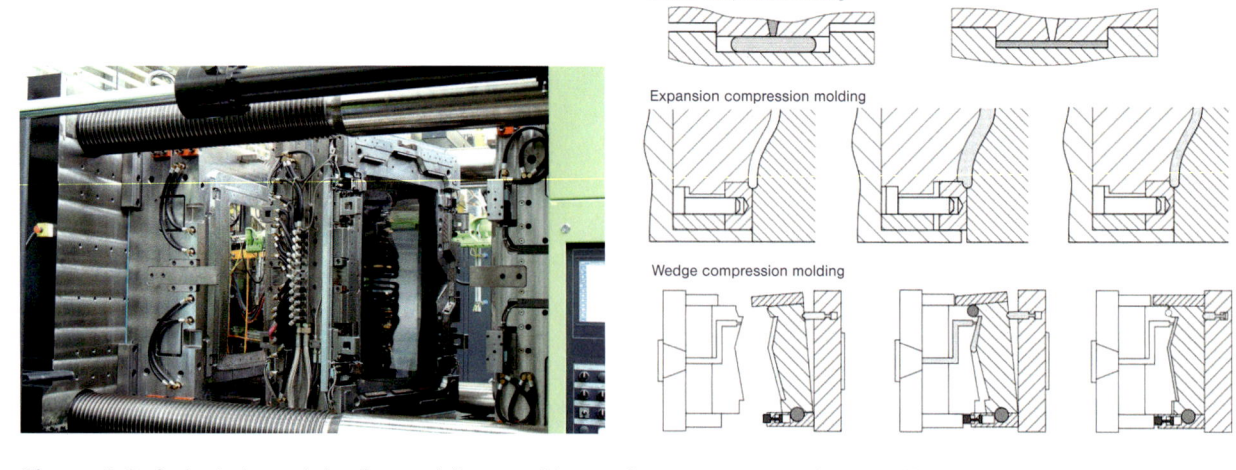

Figure 3-9 Swivel-platen injection molding machine and some compression molding techniques

bosses, fixings, window blind holders, guide rails, etc.). Figure 3-9 (left) shows a swivel-platen injection molding machine for the manufacture of 2-component polycarbonate roofs. The right-hand diagram in Figure 3-9 shows some compression molding processes used to manufacture low-stress PC automotive glazing.

Low-stress, reduced-pressure mold filling can now be used for large roofs measuring up to around 1.5 m^2–1.7 m^2. Optimally coordinated coating systems provide the required scratch protection and the specified UV stability [6, 7, 8, 9].

Although large panoramic roofs have made a successful breakthrough, there is still a great deal of development work to be done. A film technology with optimized film properties, such as IR reflection, blackout printing and other optical effects as well as UV absorption, will further increase the potential of this still young application and will reduce costs. In parallel with recent mold and process developments, work has also been done on the polycarbonate material to tailor it for specific applications. Specially coordinated additive packages and ultra-clean production provide the basic requirements for this application.

In addition to the body parts mentioned so far that are made by thermoplastic injection molding, a large number of other automotive components are made of polyurethane.

3.1.4 Processing Methods and Applications for Polyurethane

Polyurethane parts are produced from liquid reactive components. They can be subdivided into the two main components, A (polyol formulation) and B (polyisocyanate). These two components are conveyed to the mixing head via the metering unit at predefined temperatures and in a fixed stoichiometric ratio. This reaction mix is then fed into the mold, where it reacts to form the finished part.

For the production of polyurethane parts, the RIM process, which stands for **R**eaction **I**njection **M**olding, is generally used. The process involves the rapid metering and mixing of the two liquid polyurethane components by the high-pressure countercurrent injection principle, followed by the injection of the polyurethane reaction mix into either an open or a closed mold, and rapid curing to produce the final polyurethane part. As in the injection molding process for thermoplastic compounds, RIM technology can be used for the economical manufacture of polyurethane parts of virtually any size.

The diagram in Figure 3-10 illustrates the principles of the RIM process. One advantage of polyurethane processing is the comparatively cheap production of the molds, since high-strength steel molds are not generally needed due to the low processing pressures.

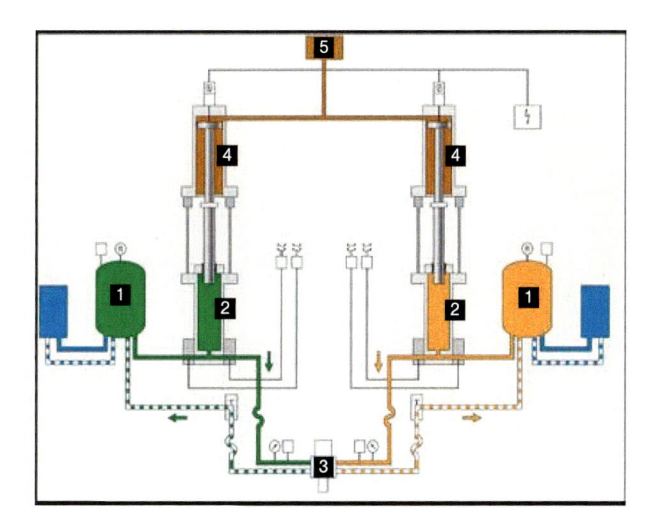

3.1 Plastics Engineering in Automotive Exteriors

The two liquid starting components are conveyed from the two machine tanks (1) through a via the injection mixing head back into the relevant machine tanks. Pressures of 15–30 bar (low-pressure cycle) are generated in the recirculating system. Immediately before triggering the actual injection, the pressure in the system is raised to 150–200 bar (high-pressure cycle) via hydraulically driven piston pumps (4 and 5). During the injection phase, enough energy is therefore available for mixing the two reaction components. To improve processing and enhance surface quality, nitrogen or air can be added to the two starting components via a gas loading unit (6).

Figure 3-10 Diagram of an RIM unit and description of the process [11,12]

For the production of spoilers and mirror frames, use is made of an expanded polyurethane material that is also hard and tough. The non-reinforced polyurethane material is processed in specially designed high-pressure PU units using the RIM process, where the polyurethane reaction mix can be filled via a gating module directly into the closed mold – or poured into the open mold in a set pattern. In combination with a torsion-resistant lightweight core (PU sandwich technology) and an appropriate polyurethane cover, spoilers can be manufactured with exceptionally good properties (Figure 3-11 left). Another polyurethane application is window gasketing (Figure 3-11 right), which makes use of the RIM process and formulations that have been stabilized against UV light. Because of its outstanding flow properties and the low cavity pressures during processing, the material is particularly suitable for large-area convex safety glazing. The liquid reaction mix acts simultaneously as an adhesive and produces a firm bond to glass or polycarbonate windows. Furthermore, inserts can be integrated without problems using RIM technology.

The RRIM process – **R**einforced **R**eaction **I**njection **M**olding – is a special variant of the RIM process. It uses chopped strands or mineral fibers as reinforcing

Figure 3-11 Rear spoilers and rear window gasketing

Figure 3-12 Polyurethane side panel

materials. Depending on the mechanical and physical properties specified for the part, up to 24% by weight of such fibers with a length of 75–200 µm is added to the polyurethane via the polyol component. The wall thicknesses of RRIM parts are generally between 1.6 and 3 mm. Because of the high material feed rate and the comparatively low wall thicknesses with long flow paths of over 1,000 mm, flow pressures of up to 40 bar are built up, which have to be withstood by suitable mold carriers. The normal locking force is between 2.5 and 5 MN.

Fiber-reinforced solid polyurethane materials are used in body add-on parts such as bumper covers, side member panels, side strips, spoilers and wheel arch extensions. In addition to the enormous design freedom with polyurethane systems, which allow long flow paths and complex shapes, parts manufactured from this material also boast outstanding stone chip resistance, good paint adhesion and accurate fitting. Such parts can be coated on- or in-line, either mounted on the body or separate. Two conditions for this are that the design of the parts is suitable for the use of plastics and that the design specifications are coordinated at an early stage.

With the development of high-grade thermoplastic films for exterior car body applications, a new and promising area of application has opened up for polyurethane materials. A fiber-reinforced PU panel is an excellent substrate for thermoplastic film with a high-gloss Class A finish, enabling it to be used as an outer skin for the bodywork. The production of the film/PU modules employs the long-fiber injection process (LFI), in which the polyurethane reaction mix is injected directly into the mold simultaneously with the long-glass fibers (12.5–100 mm). The process uses a specially developed PU mixing head with an integrated glass chopping unit. Different glass fiber contents and glass fiber lengths can be placed specifically at the parts of a component that need to be reinforced in order to attain the specified properties. The stiffness and strength data of the PU substrate materials depend essentially on the glass fiber content, glass fiber length and polyurethane density – see Figure 3-13.

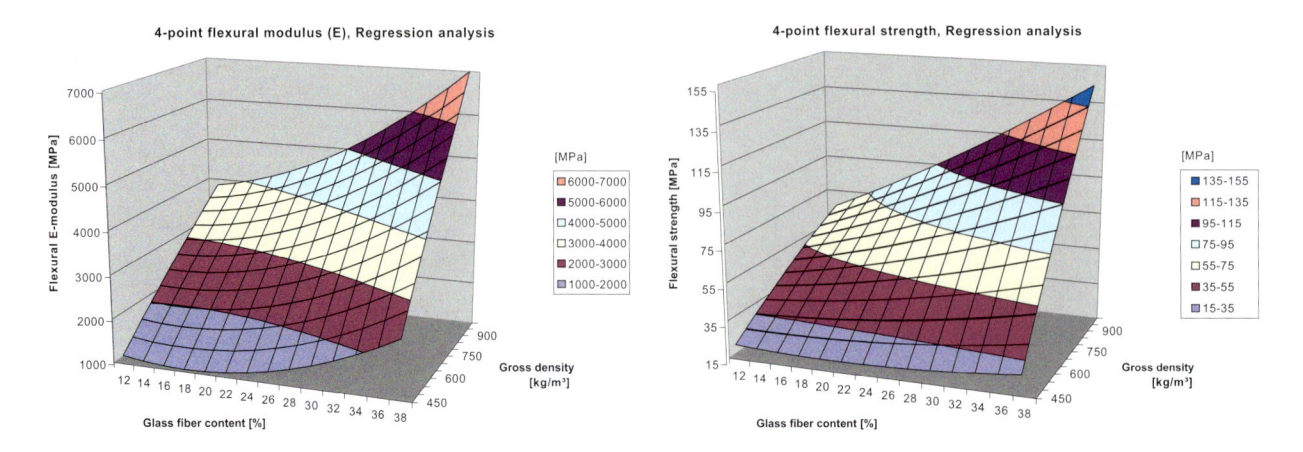

Figure 3-13 Flexural modulus of elasticity and flexural strength of Baydur® STR

In addition, the coefficient of thermal expansion of long-glass fiber-reinforced polyurethane is comparable with that of aluminum, and the supports are characterized by outstanding heat resistance. With this innovative process, it is possible, for example, to produce film composite parts that allow the easy integration of fixing elements, hinges and inserts such as antenna systems. By pouring the fiber-reinforced polyurethane reaction mix into the open mold in a set pattern, large parts can also be economically manufactured.

3.1.5 Computer-Aided Engineering (CAE)

Apart from the rapid ongoing development of the materials, processes, molds and machines, a further influencing factor is the progress being made with CAE tools for attaining the specified properties of the parts. Mold-filling simulations, the calculation of thermal expansion and wind loads, and the simulation of head impact on polycarbonate car glazing all ensure that the optimization process can begin before completion of the first prototypes. Figure 3-14 shows the use of CAE for two applications (left, bumper fascia; right, polycarbonate panoramic roof).

The demands made on the exterior front-end structure, consisting of the bumper panel, radiator grille and headlight lens, have been constantly increasing. Due to regulations regarding pedestrian protection, the entire system has to be designed today in such a way that, in the event of pedestrian impact, certain thresholds relating to acceleration and knee-bending angle are not exceeded [13, 14]. This is achieved through intelligent design and suitable material combinations. Precise harmonization between flexible and rigid areas is particularly important in this respect. Moreover, the front end must contribute significantly to passing the Allianz Crash Test (AZT) or the high-speed crash test. The entire front end must deform appropriately to absorb the impact energy, which calls for ideal interplay between geometry and material properties. Figure 3-15 shows a simulation based on the example of a "lower leg" and the process chain in the CAE calculation, taking the example of pedestrian protection.

With the aid of computer simulations (CAE) and the relevant specific material data, it is possible to tune all components to each other accurately and to optimize their behavior in advance. Through simulation, a variety of different solutions can be examined in terms of their quality and practicability, and the number of practical trials can be reduced to a minimum. As shown in Figure 3-16, deformation behavior and maximum stress can be accurately calculated in advance.

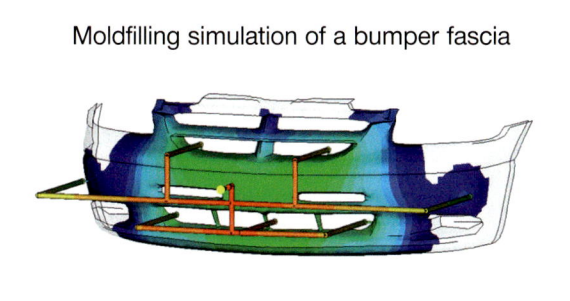

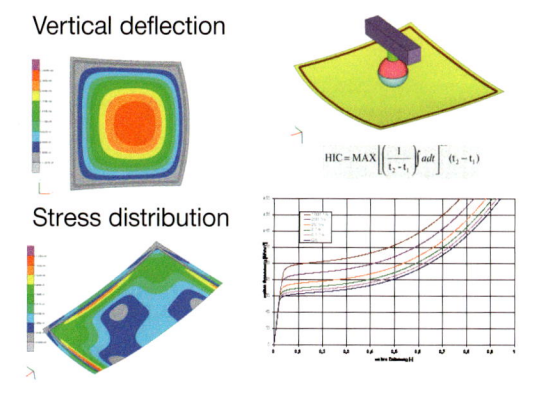

Figure 3-14 Mold-filling simulation for a bumper fascia and the calculation of wind load and head impact for a polycarbonate panoramic roof

Simulation of "lower leg" impact

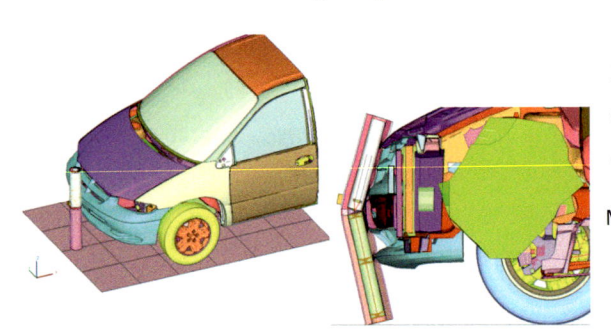

Process chain of pedestrian protection simulation:
1. dynamic material testing
2. material characterization for simulation
3. different impactor models
4. simulation of leg and head impact

Material testing ... characterization ... impactor ... to optimize parts

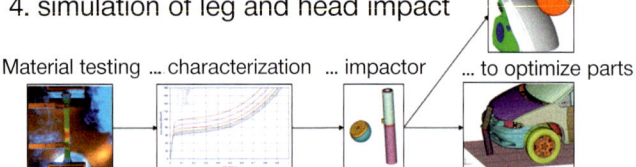

Figure 3-15 Pedestrian protection taking the example of the "lower leg" and the simulation process chain for pedestrian protection testing

Pendulum impact simulation

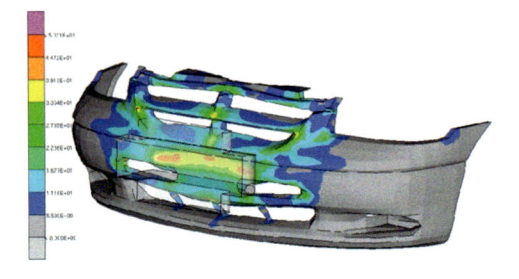

Pendulum side impact on a fender

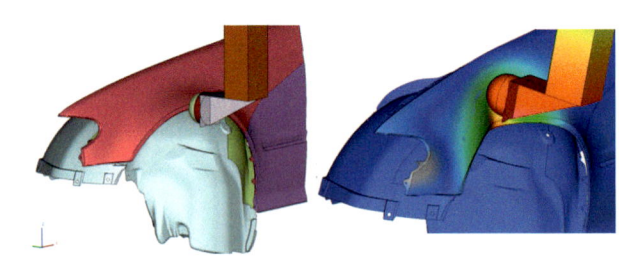

Figure 3-16 Crash calculation for a bumper and a side panel of PC/PBT

3.1.6 Outlook

The ongoing trend towards the development of lightweight exterior car body parts for the automotive industry and their suitability for production-line manufacture ensure that plastics still have outstanding prospects for the future. No one expects completely new classes of plastic based on new raw materials to take the market by storm, because the end-user also wants to see value for his money. This is easier to achieve through the utilization of present materials and economy of scale with higher utilization of plant capacity than through the search for new monomers for new polymers. Future innovations will come about mainly through the intelligent combination of existing materials. Fundamental knowledge of the properties of the individual materials, the processing methods used and the demands made on future components will be of elementary importance for the development and design of new material combinations. Hybrid systems for roof and rear modules, e.g. polycarbonate and PU-RIM, have been thoroughly analyzed in highly advanced project work and design engineering development. There are also highly promising project studies on trunk lids and engine hoods made from a combination of PC and PU. With these parts, the specifications concerning appearance and body mounting are of the utmost importance. Based on its flexibility and design freedom, as well as potential weight savings, the film-insert molding technique will also become increasingly established as an alternative to SMC processing.

Polyurethane is also doing valuable service in the structural sector. By filling the frame with foam, the vehicle body can be significantly reinforced. This enhances both active and passive safety while reducing weight at the same time.

A new technology for the interior is called "direct skinning". Here, for example, a PC/ABS injection molded part is combined with a PU-RIM skin into one unit.

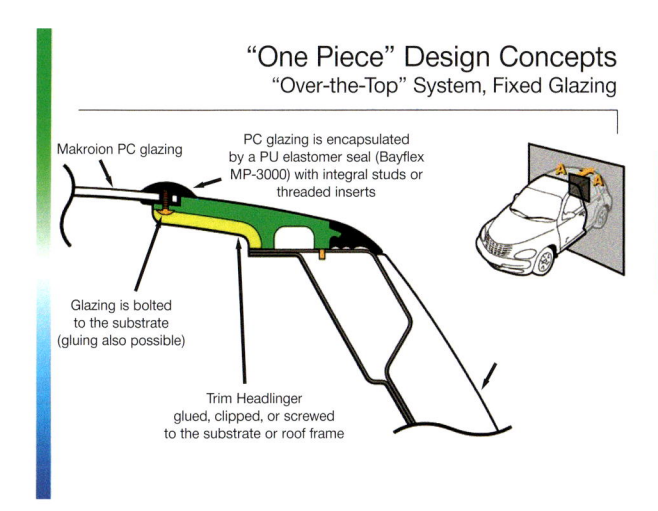

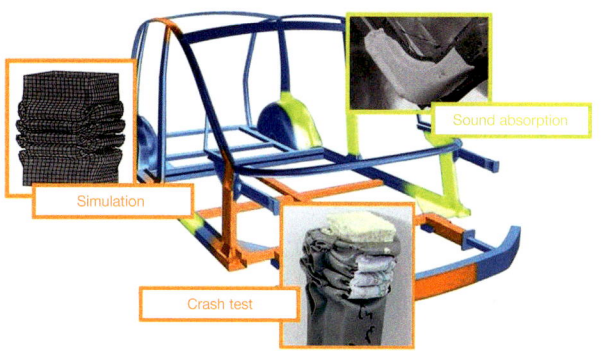

Figure 3-17 Concept for modular structure of a car roof

Figure 3-18 Frame structure filled with foam

The new process, currently in the development stage, is based on the familiar 2-component injection molding process. However, instead of a thermoplastic, a reactive polyurethane system is used as the second component. This involves combining the thermoplastic unit with the PU-RIM unit, from both a mechanical and from a control engineering point of view. The components are produced in two consecutive production steps, switching intelligently from the thermoplastic to the reactive thermoset. The cycle times are very short, making the process particularly economical. The result is a very stable bond that is noted for its high-grade surface and pleasant tactile properties.

The development of technologies, blends and systems for car body applications is still far from complete. New processing methods are being constantly developed and existing methods improved. Thermoplastic and reactive materials are being further optimized and tailored to meet the demands of specific applications. In addition, plastics engineers will constantly advance their knowledge in the handling of new CAE and processing technologies. As a result, car manufacturers will be able to call on advanced plastics that will open the door to many more innovative developments in exterior automotive functions and design in the future.

Figure 3-19 In-mold coating – "direct skinning" technology [15]

References

[1] Heidenreich, C.: SKZ Fachtagung Thermoplaste in Automobilexterieur, 2006

[2] Rothe, J.: Sonderverfahren des Spritzgießens, *Kunststoffe* 11/1997, p. 1564

[3] Schäfer M.: Gleichzeitig, doppelseitig dekorieren, *Plastverarbeiter* 2/1996, p. 62

[4] Johannaber, F., Michaeli, W.: Handbuch Spritzgießen 2002, Carl Hanser Verlag, Munich, p. 427

[5] Rau, S.: Thesis IKV Aachen, 1992

[6] Gestermann, S., Koeppchen, W., Krause, V., Moethraht, M., Popphusen, D., Sandquist, A., Zöllner, O.: Polycarbonates and its Blends for Car Body Parts, *Automobiltechnische Zeitung ATZ*, 11/2005, p. 1010-1016

[7] Zöllner, O., Brambrink, R., Protte, R., Dahmen, H. J., Kohl, W., Krause, V.: KFZ Verscheibung aus Polycarbonate, *Mobiles* 31, 2005-2006, p. 32-34

[8] Giessauf, J., Kralicek, M., Steinbichler, G., Pitscheneder, W.: Große Autoscheiben aus der Spritzgießmaschine, *Kunststoffe* 10/2004, p. 164-170

[9] Aengenheyster, G.: Polycarbonat Fahrzeugverscheibung in der Serienfertigung, *Plastverarbeiter*, 10/2004, p. 146-149

[10] Brockmann, C. M.: Spritzprägen technischer Thermoplaste, IKV- Berichte aus der Kunststoffverarbeitung, Band 84, Verlag Mainz, 1999

[11] Uhlig, K.: Polyurethan Taschenbuch, Carl Hanser Verlag, Munich, 2001

[12] Becker/Braun/Oertel, Kunststoffhandbuch 7, Carl Hanser Verlag Munich, 1993

[13] www.newmaterials.com/Realistic Simulation of Head, Hip and Leg Impact/New Materials International, 15 June 2004, Bayer MaterialScience

[14] www://ec.europa.eu/enterprise/automotive/pagesbackground/pedestrianprotection/
A study on the feasibility of measures relating to the protection of pedestrian and other vulnerable road runners, Enterprise and Industry, 2006

[15] Protte, R., Krull, S., Pohl, T., Fäcke, T.: In Kombination – Spritzgießen und RIM wachsen zusammen, *Plastverarbeiter* 10/2006

3.2 Plastics for the Car of Tomorrow – Invisible Contribution: Visible Success

I. BÜTHE

Deeper understanding of molecular structures, nanotechnology and improved simulation techniques vis-à-vis polymer materials are paving the way to innovative applications in the automotive sector. At the top of the list are improvements in cost-effectiveness, safety and comfort.

3.2.1 Introduction

Anyone wishing to play an active part in shaping the future must look ahead and recognize coming trends and then put them into practice. BASF is working together with its customers to work out innovative and environmentally friendly concepts for automobile manufacturing. An outstanding role is played here by the combination of cost-efficiency and environmental protection, comfort and safety, design and styling. Innovations – springing from a deeper understanding of materials and their properties – are indispensable for the automotive industry and thus are the key to the future success of everyone involved.

BASF's advertising campaign entitled 'Invisible contribution: visible success' is intended to demonstrate how important chemical products are for the automotive industry as well. The actual contributions to success, however, are often made out of sight.

New findings in polymer research, such as in nanotechnology, as well as constant further developments in simulation techniques for component design, are opening up new application possibilities for chemical products and especially for plastics that were inconceivable even a few years ago.

This paper intends to:

- Make the 'invisible' visible
  – Applied nanotechnology
- Make the 'virtual' real
  – Applied simulation technology

3.2.2 Innovation Processes

In the past, innovations were for the most part product-driven. In most cases, first came the development and synthesis of new polymers for which suitable application possibilities then had to be found. In the meantime, most innovations now derive from know-how regarding applications and from an understanding of customer requirements.

New technologies in the manufacture and analysis of properties and in component design make it possible to tailor products to particular applications (Figure 3-20):

Here major roles are played by nanotechnology, molecular modeling and improved simulation techniques. Further developed methods of modeling allow an insight into the properties of polymers and their applications at any order of magnitude. On practically every level, from individual atoms and molecules via polymer chains to calculation of the finished component, we are creating new possibilities for what are considered fully

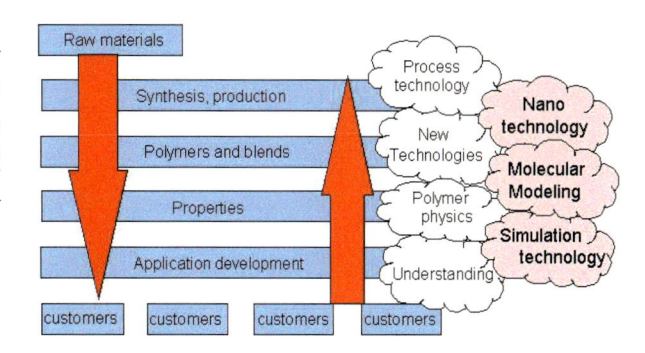

Figure 3-20 Innovation process: product or application driven?

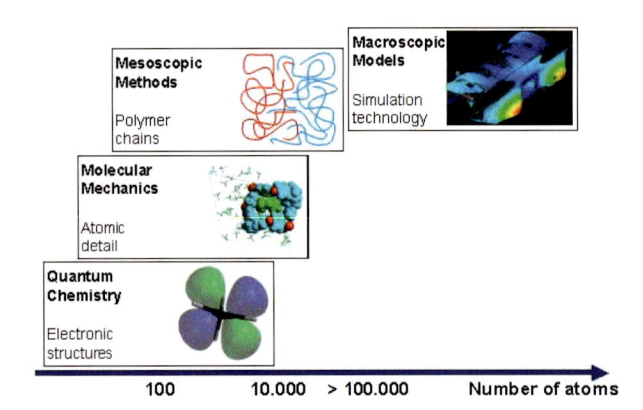

Figure 3-21 Methods in modeling – simulation

matured product classes and doing so with the aid of polymer physics and modern simulation procedures (Figure 3-21).

3.2.3 Making the "Invisible" Visible: Applying Nanotechnology

Nanotechnology is frequently associated with the superhydrophobic properties of exotic plants. This so-called lotus effect (Figure 3-22) can now be produced synthetically with various materials and ultimately results in self-cleaning surfaces.

Nanotechnology is far more than the lotus.

Applied nanotechnology is already found in numerous products and will make further innovations possible in the future.

Figure 3-22 Lotus effect

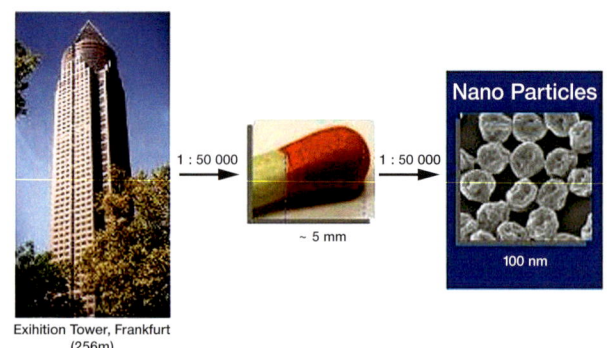

Figure 3-23 Dimensions in the nano world

Figure 3-23 illustrates the dimensions of the nano world: if the Frankfurt Exhibition Tower were shrunk by a factor of 50,000, it would be the same size as the head of a match. Reducing it once more by the same factor would put it in the world of nano particles.

In principle, the nano world is nothing fundamentally new to chemists and physicists. For decades now, polymer dispersions have been examples of applied nanotechnology with some extremely successful applications, such as surface-finishing for paper and also paints and adhesives.

What is special about nano particles is their frequently spectacular properties.

The raisin-like fine structure of the nano particles of dilatant polymer dispersions (Figure 3-24) results in very special flow properties. At low shear rates,

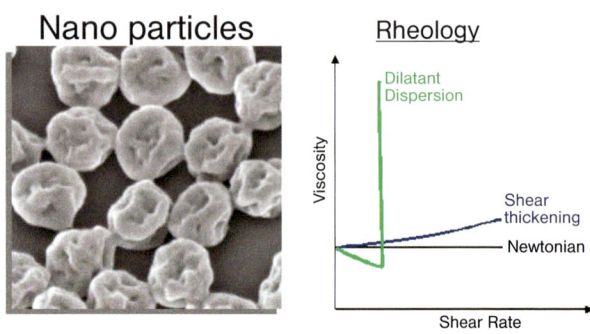

Figure 3-24 Dilatant dispersions

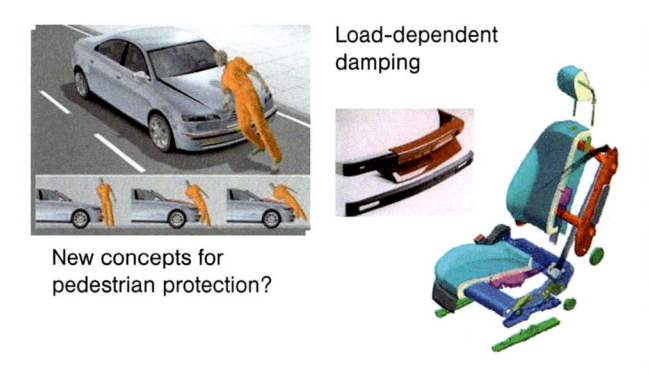

3.2 Plastics for the Car of Tomorrow – Invisible Contribution: Visible Success

Figure 3-25 Possible applications for dilatant dispersions

	PC/ABS	PA/SAN Nano
Heat resistance	++	++
Flowability	+	+
Stiffness	++	++
Toughness	+	++
Chemical resistance	o	++
Weathering resistance	o	++
Scratch resistance	o	+

Figure 3-27 Comparison of properties

the particles slide smoothly past each other. At high shear rates, however, they catch on each other. This leads to a sudden rise in viscosity. Such products are currently being investigated for their potential as an 'intelligent damping device', e.g., for pedestrian protection (Figure 3-25).

Nanotechnology is not simply a laboratory curiosity: it is already applied in practice. With certain plastics, special organic nano particles – even at low concentrations – contribute to considerably improved flow properties (Figure 3-26). With PBT, it was possible to double the flowability while retaining comparable mechanical properties. The processor profits from a reduction in cycle time of up to 30%, due to simpler molds with fewer gates or thinner wall thicknesses.

Even the PA/SAN blend widely used for vehicle interior applications can be further optimized by selective interventions in its microstructure. With this outstanding surface material, the morphology can be influenced by suitable measures to create structures in the nano range. The range of properties is once again considerably improved in comparison with the standard material. The nano-modified material is superior in many respects to the alternatively used ABS/PC blend (Figure 3-27).

In the field of the outer skin of an automobile (Figure 3-28), nanotechnology enables the use of online-paintable plastics with a specially balanced property spectrum of temperature range or dimensional stability.

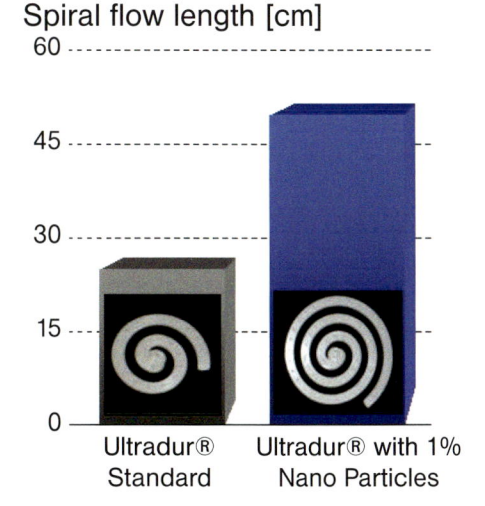

Figure 3-26 Improved flowability of PBT

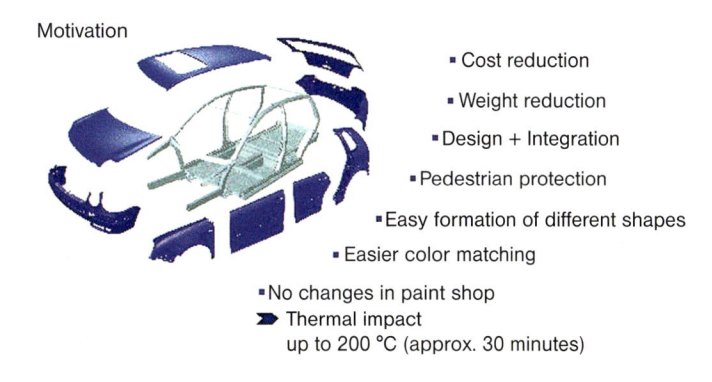

Figure 3-28 Online-paintable plastics for body panels

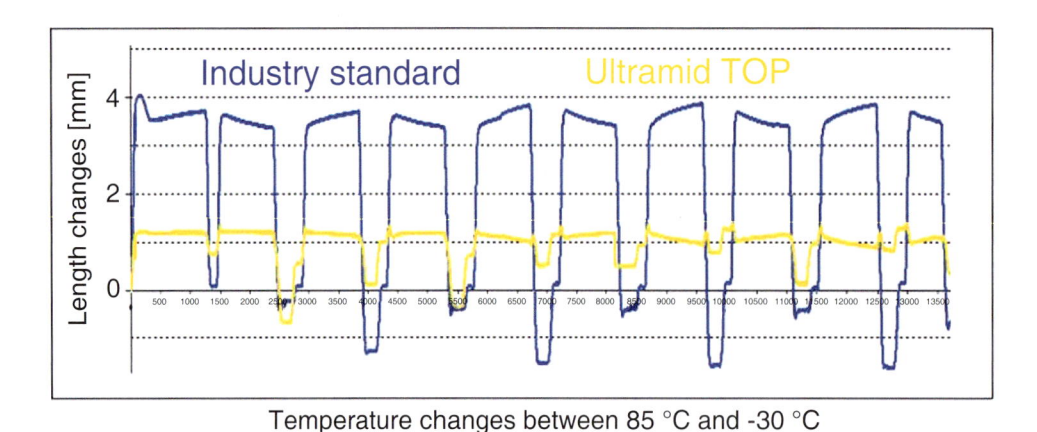

Figure 3-29 This BASF development compared with competition

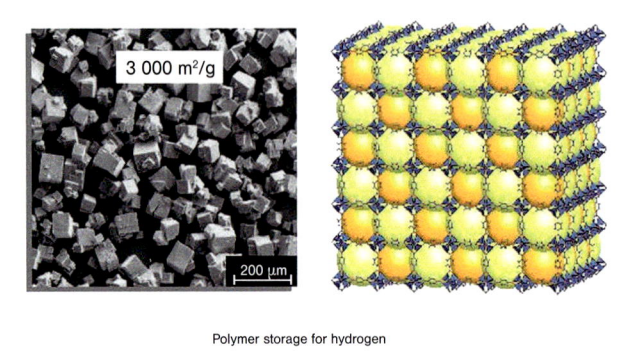

Figure 3-30 Metal-organic framework (MOF)

Figure 3-31 Hydrogen storage

A graph of dimensional stability during temperature changes between 85 °C and −30 °C shows the outstanding properties of this new development (Figure 3-29).

Alternative drive concepts, such as hybrid technology, open up new application possibilities for innovative plastics. Not only fuel cell components, but also intelligent hydrogen storage systems, are ultimately based on nanotechnology. Figure 3-30 shows so-called MOFs (metal-organic frameworks) with cubic lattice structures and extremely large internal surfaces of 3000 m²/g.

Simple raw materials (terephthalic acid and zinc oxide) in a sophisticated combination and molecular arrangement form the basis of this polymer network with a porosity in the nano-range and used for storing hydrogen.

With the aid of these MOFs, about three times the quantity of hydrogen can be stored in a tank at any given pressure.

Nanotechnology is and remains a key technology for the development of highly innovative products.

3.2.4 Making the Virtual Real: Applied Simulation Technology

Development of tomorrow's automobile is increasingly taking place in the virtual world. This saves time-consuming prototype building and cost-intensive testing with real components. Continuous further development of simulation methods is the key to this trend. Integrative simulation now permits more precise predictions of the behavior of components under dynamic loading.

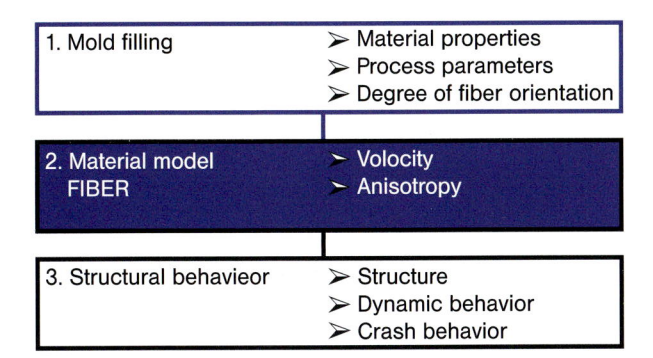

3.2 Plastics for the Car of Tomorrow – Invisible Contribution: Visible Success

1. Mold filling
- Material properties
- Process parameters
- Degree of fiber orientation

2. Material model FIBER
- Volocity
- Anisotropy

3. Structural behavieor
- Structure
- Dynamic behavior
- Crash behavior

Figure 3-32 Integrative simulation

In this way, for example, reliable crash simulations can be carried out for components made of short-glass-fiber-reinforced polymers. The material model "fiber" (Figure 3-32) supplements known tools for simulating mold filling or for calculating component behavior and closes the gap in the simulation of crash behavior.

It has a basic requirement that specific material data be acquired in high-speed measurements and transferred to the material model.

Thanks to these methods, the virtual can be successfully transformed into reality. An example of this is the first thermoplastic engine sump for a truck (Figure 3-33). The following success factors played a considerable role in its development:

- Optimum installation space utilization due to perfect component design
- Optimum damping and toughness
- Weight reduction
- No secondary finishing.

This component marks the starting point for the development of highly integrated oil function modules. In the future, the advantages of plastics can be exploited by integrating filtration or pumping, for example, in the oil sump. The result will be considerable cost and weight savings coupled with improved engine compartment packaging.

Engineering plastics can be used to advantage even in dynamically loaded seat structures thanks to improved simulation techniques. Seat pans made of polyamide have already been put successfully into full-scale production. Figure 3-34 shows an overview of the advantages of a plastic seat pan.

Plastic/metal hybrid solutions for crash-optimized beam structures constitute an enormously wide application sector for modern simulation techniques.

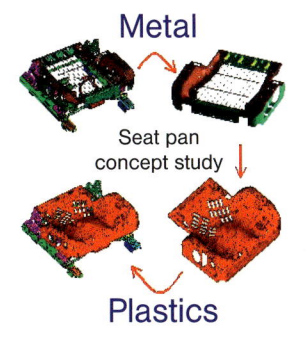

Metal

Seat pan concept study

Plastics

Advantage
- Weight reduction up to 60 %
- Potential for cost reduction
- Reduced tooling costs
  - formation of different shapes
- Common architecture plus vehicle specific options
- Design flexibility - comfort
- Improved damping
- Feature integration

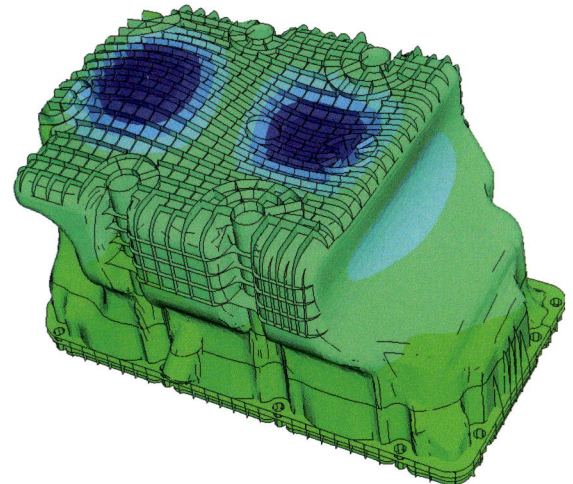

Figure 3-33 Thermoplastic engine oil sump

Figure 3-34 Concept study for a seat pan

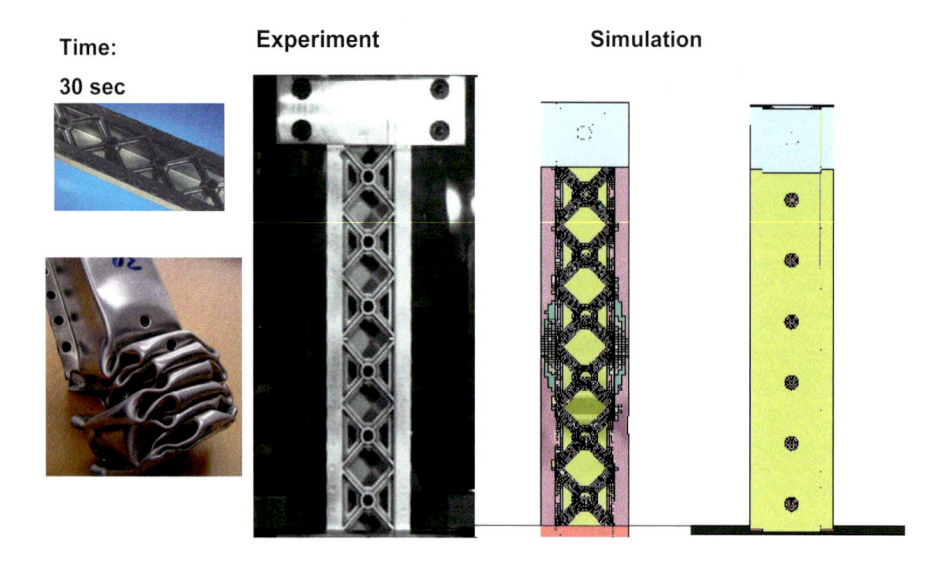

Figure 3-35 Simulation and experiment with a plastic-reinforced steel section

Here plastic inserts can, for example, bring about a significant change in the buckling resistance of metal structures, thus enabling markedly lighter overall component design. Integrative simulation plays a decisive role in this development as well (Figure 3-35).

Optimized products and deeper understanding of their behavior in dynamic loading cases enable many new applications for the intelligent reinforcement of metal structures in vehicle bodies. Innovative hybrid structures that initiate controlled failure upon collision impact will further improve the safety of future vehicles while also contributing to vehicle weight reductions.

The ultimate objective of "better value for the money, lighter and safer" can thus be achieved in many cases.

Increased understanding of the structure and properties of polymer materials in all scales of magnitude will keep the innovation pipeline full for new products and applications in the automobile.

Many of these innovations will, however, remain invisible to the end user. The only thing that is and will remain visible is the shared success of all partners in the innovation process from which a perfect car will ultimately emerge.

3.3 CFRP – from Motor Sports to Automotive Series – Challenges and Opportunities Facing Technology Transfer

Patrick Kim

3.3.1 Introduction

CFRP is the material of choice for demanding motor sports applications. However, to this date, automotive applications in highway vehicles have been predominantly limited to high-end cars. The automotive industry has repeatedly been identified as one of the markets most likely to propel carbon fiber into mass production. This paper discusses the basic issues that will need to be solved in order for CFRP to make the leap to mass-produced primary automobile structures. The principal challenges lie not so much in meeting technical requirements as in reaching volume targets, automotive-level quality, and direct cost competitiveness with lightweight metals such as aluminum and magnesium. A strong vertically integrated supplier chain will be required to achieve this. The goal is to provide materials and process technologies that can ensure required quality and cost competitiveness at volumes in excess of several tens of thousands of cars a year. While the raw materials technologies available can to a great extent be carried over from motor sports, radically new design and manufacturing concepts will be necessary to bring down the cost of CFRP components.

3.3.2 CFRP at the Transition Between Motor Sports and Road Cars

The exploits of modern motor sports owe a great deal to carbon technology. Ever since the first carbon fiber-reinforced plastic monocoque by McLaren in 1981, CFRP has enabled a level of safety and lightweight design in the frame and body that remains unmatched by any other class of materials. The result is that CFRP is now regarded as state-of-the-art, a standard technology throughout motor sports, be it on the track, as with Formula 1 or DTM, or in rally cars.

CFRP has taken some serious steps from the race track to the open road in the past few years. With supercars, such as the Mercedes-Benz SLR McLaren, CFRP has steadily progressed from being used exclusively in professional sports machines to setting performance standards in cars that are also a pleasure to drive in everyday road traffic situations.

Given the proven performance of CFRP, we should still ask ourselves what are the real driving factors behind the use of CFRP in more standard automotive series? CFRP has long been and still remains a status symbol for customers and auto-makers. But as its use progresses into lower-priced vehicles, the question "What are consumers willing to pay?" is becoming central to future developments of the technology for volume models. A different way to express this is: "What are they not willing to give up?" When moving away from the realm of absolute performance machines, customers will make no concessions in space and comfort, reliability, maintenance and insurance costs, and very little in acquisition costs.

CFRP as a material has all the properties needed to make cars with desirable levels of performance and driving pleasure. However, the process technologies which would make this possible are lagging behind. These technologies, along with a more vertically integrated supplier chain, will be a key to bringing CFRP into automotive series. The only technology that can really be "transferred" is the carbon fiber.

3.3.3 Current Applications of CFRP

In highway automobiles, CFRP is currently used primarily in single substitution parts, in upper-end, high-performance GTs and "supersports" cars, and in after-sales market components for the racing enthusiast. Typical substitution components are drive shafts (Mitsubishi Pajero), spoilers (Mitsubishi Lancer Evo8), and hoods (Nissan Skyline GT-R). These are produced in volumes ranging from a couple of thousand to nearly 100,000 cars a year in the case of drive shafts.

In a somewhat higher price range, its use includes body panels (BMW M3 CSL, Honda NSX R-Type). Truly CFRP-specific application in larger structures and monocoques is limited to cars with price tags above €100,000 and volumes below 600 cars a year.

Four cars have recently monopolized the headlines with their strongly publicized use of carbon. A closer look at these applications will give a good idea of trends in the materials and processes and their significance for the automotive industry.

Figure 3-36 Mercedes-Benz SLR McLaren rear scuttle panel compression molded from Advanced SMC. This is a net-shape, no-waste process with a capability of several tens of thousands per year from a single mold.

The hood of the Chevrolet Corvette Commemorative Edition [1] is produced in prepreg/autoclave technology. This has been optimized somewhat to reduce the curing time, but is basically still a process typical of the aerospace sector. The labor-intensive cycle and extensive use of ancillary materials makes this method expensive and precludes its expansion to volumes above a few hundred a year for more complex large parts.

The MG XPower SV body, made of SP Systems' SPRINT CBS [2], breaks with this tradition by working with prepared kits including a tailored syntactic core (to reduce carbon fiber use and thus cost) and surface layer. Most significantly, the material can be processed out of autoclave, although vacuum bagging is still needed for consolidation. SP Systems estimates the economic limit at approx. 2000 cars a year.

Besides its jewel-like autoclave-processed monocoque [3], the Porsche Carrera GT is making some inroads into the use of RTM, even for the fabrication of external parts. Experience with the RTM processing of CFRP in the Opel Speedster shows that there is potential for extending volume toward or even beyond 10,000 cars a year if preforming, handling, and demolding are automated.

Finally, the Mercedes-Benz SLR McLaren within its full CFRP monocoque incorporates a geometrically complex structural part made of carbon fiber-based advanced SMC (Figure 3-36). This process allows for deep-drawn parts reinforced with practically continuous fibers. The production of this component involves a concept for an automated layered preform preparation process with absolutely no waste that enables tailored fiber orientations [4]. The final part is compression molded at cycle times corresponding to standard SMC moldings. Here, for the first time, we have a process technology that, in the short term, is poised to enable the production of structural CFRP parts from a single mold in volumes of several tens of thousands. In combination with other automated processes adapted to the various geometric and functional requirements, as discussed below, it would enable entire body structures to be mass produced.

3.3.4 The Trends: a Closer Look

Several conclusions can be drawn from an analysis of efforts to date on the role of automotive OEMs and suppliers in bringing CFRP to series:

- Developments have focused on optimizing existing processes and are often limited to a single process or an isolated component.

- Development work either has not translated into a concrete series application or is implemented on a markedly reduced scale.

- The projects have often had a pronounced promotional value or were focused on exclusive racing-level products.

Significant recent application profiles:

- Apart from monocoques, CFRP components are designed for a clear, simple interface with the surrounding parts. This simplifies the design of substitution parts, but can result in a sub-optimal layout of the structural system, indicating the need for designs more specifically adapted to CFRP.

- With few exceptions, parts produced in large volumes have a relatively simple geometry, since the processes are not amenable to more complex parts.

- Parts with a complex geometry (e.g. hollow, curved parts with variable cross sections) generally involve joining steps and are produced in very limited volumes and at high costs. The processes used for their production generally have limited potential for expansion to larger series. This means that, in contrast to what recent articles have been propagating, the use of CFRP in supercar monocoques as a rule does NOT "prefigure its application in mass-produced cars" [4].

- CFRP part production relies strongly on experienced, specially qualified labor.

- The suppliers are often small to mid-sized composite molders with limited capabilities in concept and process development, in computational structural analysis, or in experimental validation or testing to automotive standards.

3.3.5 Challenges

Three main areas of CFRP development for series applications present significant challenges: meeting long-term technical requirements, manufacturing capability for an automotive environment and costs.

3.3.5.1 Technical Requirements

There is ample evidence – and little doubt – that the main short-term structural and functional requirements of an automotive body can be met with CFRP, whether alone or combined as a system with other materials. However, many technical questions relative to in-service characteristics remain open. They require concrete, practical and verified answers before the step into large series can be taken.

Beyond the traditional considerations regarding the effective properties of the composites at room temperature, materials suppliers and processors need to address properties in environments typical of vehicle production and operation. Dynamic (fatigue and crash) loads, process and service fluids, high temperatures and humidity cycles are the main loads that need to be taken into consideration. The values for these loads and the requirements derived from them vary according to the type of component and are summarized in Figure 3-37. Material data and road experience documenting the long-term behavior of CFRP car structures under cyclic mechanical and hygrothermal loads (durability, damage tolerance) are needed, but as yet are only available to a limited extent.

Other technical requirements and prerequisites for introducing CFRP into mass production are: systematic, industry-relevant materials and structure qualification certification (those used in aeronautics are too costly and do not address the automotive environment);

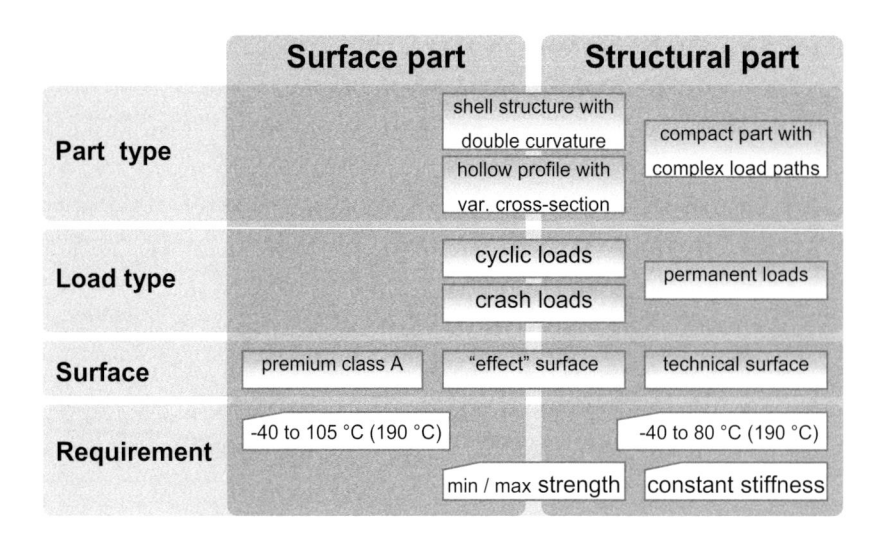

	Surface part		Structural part
Part type		shell structure with double curvature / hollow profile with var. cross-section	compact part with complex load paths
Load type		cyclic loads / crash loads	permanent loads
Surface	premium class A	"effect" surface	technical surface
Requirement	-40 to 105 °C (190 °C)	min / max strength	-40 to 80 °C (190 °C) / constant stiffness

Figure 3-37 Characteristics and loading of FRP structural parts

adapted material properties documentation allowing comparisons between FRP types (databases); and predictive numerical analysis of composite automotive structures under service loads. These are all issues that concern the industry as a whole, and may best be developed in a pre-competitive framework similar to the ACC (Automotive Composites Consortium) in North America.

3.3.5.2 Manufacturing

One conclusion of the Hypercar Revolution project was that, as recently as the late 1990's, "no solution existed that could make high-performance composite structures in the volumes or economies required by the auto industry" [5]. Although CFRP body applications are on the street in the range of up to 2000 cars a year, a real breakthrough will not occur until processes and manufacturing chains are available for annual volumes in the tens of thousands. By comparison, typical volume ranges for Mercedes-Benz are 30,000 a year (SL), 100,000 a year (S Class) and over 250,000 a year (C and E Classes). The process cycle times to achieve the lowest of these volumes lie in the order of 3–5 minutes per tool. Anything longer will decrease the manufacturing cost competitiveness on account of the additional costs associated with multiple tooling and/or equipment.

Beyond the issue of volume, a major challenge will be the integration of CFRP into the predominantly steel-oriented automotive environment, and in particular the introduction of CFRP structural parts into the classic production chain. As a structural material, CFRP is bound to ultimately find its way into primary structures: in other words, into the body-in-white (BIW). An in-line scheme is a likely first step towards this goal: CFRP modules can be integrated into the car after the coating operation. However, this requires significant changes in the BIW concept and design. Faced by the cost of setting up a dedicated plant to accommodate the specifics of composites as well as industry resistance to the modification of existing plants, the alternative would be to provide FRP technologies that allow on-line processing. Since a step directly into a full-CFRP BIW is unlikely, this requires materials and joining technologies adapted to the high temperatures of the paint line – up to 190 °C in the case of the e-coat bake.

Underlying all this is the need for a quality philosophy adapted to automotive series production. The first step involves moving away from the highly specialized manual labor dominating supercar production (e.g. the Porsche Carrera GT body by the hand lay-up/autoclave process), and towards fully automated processes. What is required is not only the capability to meet the production volumes, but also to ensure a quality level and consistency required by an automotive series (for example, precision of the reinforcement structure as well as the magnitude and distribution of production-related defects). Experience with the "Triple C" program supporting the SMC deck lid of the

Mercedes-Benz CL Class [6] has shown that consistent quality is best achieved by an unbroken supply-chain management, starting with raw materials and culminating in the finished component in the car.

3.3.5.3 Cost

Cost is the one factor that will make or break the introduction of CFRP into large-volume automotive series. New lightweight materials must provide solutions that can compete with metals. Once the economical weight-saving potential is taken into account, the main emerging contender is aluminum. A performance/weight/cost comparison with steel or advanced composites, such as autoclaved laminates, is always necessary, often interesting, sometimes impressive, but it does not reflect in full the currently decisive question when selecting a material for a given application: What is the best you can get for your money for a given car? What is good for motor sports is not necessarily good for series automobiles.

A look at the typical cost of CFRP and aluminum BIWs (including parts and assembly) as a function of volume (Figure 3-38) is essential for understanding where costs must – and can – be optimized. At very low volumes, CFRP can be significantly less expensive than aluminum due to the lower investment and tooling costs. The relation changes at large volumes, as the recurring costs – that is, of material (raw and ancillary) and labor – are higher for CFRP. The economical cross-over point depends on the application and generally lies at volumes on the order of 2000 to 5000 units per year. Pushing the cross-over point to the larger volumes of "real" automotive series means reducing these recurring costs, ideally by a factor of 1.5–2.

The price of raw materials, and in particular that of carbon fibers, is usually cited as the main culprit for the high cost of CFRP parts. Although it does make up a significant proportion of the total costs, it is not expected to have the largest potential for cost reduction. Nor is there a magic price at which CFRP technology will suddenly become economically feasible. Much discussion between OEMs and carbon-fiber producers has centered around a chicken-and-egg problem:

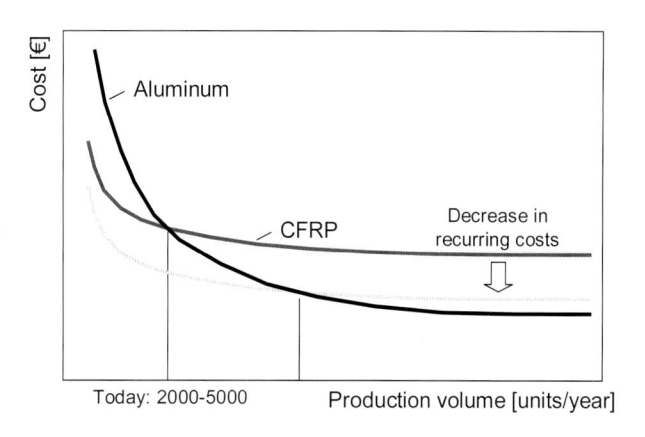

Figure 3-38 Cost comparison between CFRP and aluminum bodies as a function of volume (schematic). The cross-over volume for economical CFRP car body production increases with decreasing recurring costs.

on one hand, a lower fiber price is desired or needed to go into automotive applications, but on the other, the lower price can only come with a commitment to high volumes which OEMs cannot as yet guarantee. The fact is that, even though the price of carbon fibers has decreased from about €16 per kg to less than €11 per kg over the past few years, a correspondingly strong penetration into series application has not occurred.

Considering that a further decrease in the price of carbon fibers to clearly below €10 per kg is unlikely in the short to medium term, even for large volumes, we should turn our attention – and the debate – more strongly towards manufacturing costs as a platform for reaching the necessary affordability of CFRP parts. Here, several goals stand out:

- A high degree of automation at all stages of manufacturing.

- Reducing the number and cost of intermediate products in the process chain.

- Developing processes with strong economies of scale: that is, going beyond scale-up by simple multiplication of existing processes.

- Reducing waste and scrap by means of appropriate net-shape processing routes (it is estimated that a typical textile-based RTM preform generates upwards of 40% waste [7]. The cost of throwing away woven carbon-fiber mat is appalling!)

One argument that is often brought forward when addressing the issue of the cost of composite vs. metal parts is that composites have advantages over the vehicle life-cycle as a result of reduced energy consumption. As has often been demonstrated for aluminum, this type of conclusion is highly dependent on service scenarios; the break-even point often occurs after the first ownership, and is therefore hard to justify to the original owner. Furthermore, this is a flawed argument in practice unless the costs can be properly accounted for and passed on to the affected parties. Until full life-cycle cost analysis becomes an integral part of costing, the initial cost of the component or module will be the measure by which different material solutions are compared [8]. And even then, it should not be forgotten that the competition is not sleeping: being cost-competitive with aluminum must be as true tomorrow as it may be today. This means that there has to be potential for continued cost reductions. The selection of a materials technology for large series is based on the fulfillment of requirements at a competitive cost – leaving life-cycle analysis out of the picture.

Finally, one cost block that cannot be ignored, although it usually is in the current discussion, is maintenance and repair costs. With warrantee costs rising throughout the industry, OEMs place a high priority on reducing the risk of recalls when the service characteristics of a new technology are not fully understood. The availability of standard procedures for damage detection and evaluation as well as economical repair for CFRP will be a key to keeping these costs at an acceptable level.

3.3.6 Opportunities

3.3.6.1 The Stakes

Three 'pillars' have been identified as supporting significant future growth for the carbon-fiber market: wind energy, offshore and automotive. A simple calculation gives a hint at why the carbon-fiber industry has such an avid eye on the automotive sector: Consider, as a volume scenario for automotive series production, a median volume of a million cars a year (this corresponds approximately to the yearly Mercedes-Benz production and is less than 2% of the yearly world production); each kilogram of CFRP (or 0.5 kg of carbon fiber) used per car would represent a carbon-fiber consumption of 500 tons a year (with the simplified assumption of zero waste!). This seems like a drop in the bucket, especially when compared to the glass-fiber market, but it is significant for carbon fiber considering that this represents the annual production of a typical carbon-fiber plant and that current worldwide production capacity is slightly over 30,000 tons per year [9]. Putting this into the perspective of the global automotive industry, carbon-fiber manufacturers will find it challenging to fulfill the demand at even a moderate expansion rate – once CFRP has proven itself. The supply/demand balance and associated potential price (in)stability are to be taken seriously when pondering the introduction of CFRP into automotive products. However, several factors will be in play before this situation obtains.

3.3.6.2 Restricting Factors

The principal factors that tip the balance against CFRP in automotive series are:

- Cost/performance tradeoff: in the current extremely competitive automotive world, there is limited readiness to pay for weight savings and innovation. The typical cost allowance for weight savings lies between zero and €10 per kg of weight reduction, depending on the car model and component type.

The automotive development and production environment: FRPs face an established, mature industrial system dominated by steel. Technology and product development cycles on the order of 5–10 years in total are not conducive to attracting investors to the CFRP cause. This is compounded by the fact that comparatively high up-front investments are needed to establish the technology on the automotive scale, whereas production facilities for steel already exist.

- Risk: the technology jump required with CFRP contrasts with the "safer" incremental improvements achieved with metals – even if new metal applications are a long way from being free of problems. This means that much PR work will be needed industry-wide and within corporations to overcome the attitudes and perceptions of engineers and managers who tend to be especially critical of "failures" of new technologies.

- Supplier structures and capabilities: the FRP industry is fragmented; it consists of many small, highly specialized firms, limited in resources and capabilities. OEMs look for system development partners capable of designing, producing and validating to automotive specifications.

3.3.6.3 Enabling Factors

Offsetting these obstacles are recent developments in composites technology that bring a renewed impetus for the future of automotive CFRP:

- Integrated development tools: linking CAD/CAE and FEA methods for processing and structural analysis allows potentially faster development times. This will be contingent upon the availability of reliable data and thus, initially at least, on the need for costly and time-consuming experimental verification or validation.

- Lowered investments: new rapid tooling and low-pressure processes hold the promise of lower capital investment costs while increasing the speed of development and flexibility in production. This will help compensate the need for new plants to manufacture composites in volume, while established facilities exist for "traditional" metals production.

- Fast processes: compression and flow molding of advanced compounds (Advanced SMC, carbon-fiber SMC: Dodge Viper fender support), fully automated filament winding (Mitsubishi drive shaft), RTM and other processes in development are finally opening the door to volumes ranging up to *several* tens of thousand units per year.

- Vertical process chains: composites competence centers established by certain Tier 1 European suppliers [10] are for the first time bringing together under one roof the full spectrum of composites technologies – from raw materials screening to component validation – needed for a "solutions approach rather than simple material on a roll" [2].

- Finally, recent applications in highway vehicles are providing the basis for gathering important in-service experience. Even though they do not show the way to high volume production, they at least confirm the technical advantages of the CFRP for road-worthy automobiles: improved handling, enhanced crash safety, and a weight-saving potential that goes beyond aluminum.

An initial threshold will have to be overcome in order to leverage these factors, but all in all, conditions and trends have never been more favorable for the introduction of CFRP to automotive series.

3.3.7 Outlook

3.3.7.1 Open Questions

The future of CFRP in automotive series will depend to a large extent on the answers to the many questions besetting it, some of which are:

- What are the "best" steps away from an industry-standard all-metal BIW to body concepts and component designs adapted to CFRP? This question is specific to individual company and product placement strategies.

- How will the core competence of the OEMs (car design, materials and structures specification and qualification, process simulation, predictive crash and fatigue modeling, testing, production systems) evolve in order to gain experience and thus reduce the risk associated with the as yet poorly understood long-term behavior of CFRP under standard operation conditions on the road?

- How will CFRP live up to customer (ab)use without the level of inspection that is standard for motor sports (and commercial) vehicles?

- Considering that it will take time to introduce CFRP and that therefore volume-related carbon-fiber price reductions cannot be expected to make a decisive contribution to decreasing initial costs, how effectively can fast, low-cost processes reduce manufacturing costs?

- Is an industry-wide standardization of CFRP types possible and does it make sense, in order to avoid an uneconomical explosion in the variety of materials?

- How can CFRP material systems and processes for surface parts be made fully reliable and ensure class-A quality without reworking – even after thermal cycling?

- What do suppliers want or need from OEMs? How can they work together better in order to get CFRP into the automobile?

3.3.7.2 The Way There

One key to bringing CFRP into automotive series is to select the "right" development steps from today's BIW towards a full "body-in-black". It should be kept in mind that an all-CFRP solution does not necessarily make sense: today's cars are a perfect example of a function-driven material mix. The most technically and cost-effective solution will most likely be based on hybridization with glass fiber or other materials; even in FRP-specific designs, expensive carbon fibers should be used efficiently, i.e. only where its properties are really needed.

A phased adoption of CFRP technologies will be unavoidable, as the risk of introducing a radically new technology into large series on the scale of a full BIW is considerable. This progressive introduction will not give CFRP optimum conditions to compete directly with established materials. Some "activation costs", in the form of promotional or strategic projects, and an incubation time possibly spanning one product cycle, will thus be needed for taking the initial steps. Bridging this gap can be facilitated by major pre-competitive programs, such as the North American ACC or publicly-funded research initiatives, such as the European CRACTAC and TECABS. As the possibility for setting up the former seems rather illusory in the present European car industry, the latter will remain important platforms for progress in automotive structural composites.

In the meantime, supercars and niche-performance models will provide a proving ground for gaining valuable design and service experience, and for defining best practices for automotive CFRP. The experience will be organizational as well as technical: it will have to lead to a closer coupling between the part and the process developers, and between designers, production engineers, and costing departments. It will have to lead to stronger long-term alliances between OEMs and suppliers, but also among suppliers along the entire value-added chain, in order to provide the capabilities needed for a successful automotive composites project. In the end, leveraging the enabling factors to introduce CFRP into series will take strong, sustained management commitment on the part of the OEMs and suppliers.

3.3.7.3 Conclusions

Just about all major OEMs have carried out CFRP programs with some degree of intensity. Although functional prototypes and ultra-niche series of roofs, floors, frame elements, body panels, and crash elements, to name but a few, exist in large numbers, a significant use of CFRP in series automobiles still lies a good way into the future. It would be foolish to specify a date, although we should indeed fix a time-frame for ourselves. The final step could take the form of a strategic project led by a "heavyweight" decision-maker; it could equally well take place as a more low-key introduction, starting with a limited number of smaller, less spectacular components. A gradual evolution seems much more likely than a revolution-by-program, such as the Volkswagen 1-liter car and the BMW Z22 tried to usher in. This will not be characterized by one single technological trend, but rather by many vectors for the introduction and growth of CFRP in automotive series [11]. In any case, patient, intimate cooperation between OEMs and suppliers will be required in order to ensure the necessary technical and financial means. Everyone must be willing to lean quite far out window on the cost side in order to gain valuable experience on the road.

Coming back to the initial question: apart from the carbon fiber itself, there has been no real technology transfer of CFRP from motor sports (or from aerospace or rail or wind energy, for that matter) into automotive series. That is because the boundary conditions and requirements are all unique to the industry. Looking back some thirty years, it seems little has changed in the outlook for carbon fibers: the material properties, basic fabrication processes, and predicted applications cited today are more or less the same as in 1971 [12]. A closer look shows that much practical progress has really been achieved in the meantime, but the key enabling technologies are only just now starting to come to life, along with a better understanding within the composites industry of what the automotive industry needs. If we are to pave the way for CFRP into 'normal' highway vehicles, it will be on the basis of thoroughly new process technologies, including a high degree of automation, and a newly structured, vertically integrated composites industry.

References/Web sites

[1] Corvette sports a carbon composite bonnet. *Reinforced Plastics* (July/August 2003), p. 4

[2] Summers, J.: Carbon car panels a cost effective reality. *Reinforced Plastics* (Sept. 2003), pp. 40–42

[3] Der GT lehrt den inneren Wert. Automobil-Entwicklung (Sept. 2003), pp. 12–20

[4] Special: Automotive. JEC Composites Vol. 4 (2003), pp. 25–65

[5] Hypercar newsletter. www.hypercar.com (November 2003)

[6] Schuh, T., Wesse, T. Ilzhöfer, K.-H., Hermann, A.: Towards a Premium SMC. www.smc-alliance.com/events/JEC_2002/Schuh_SMC_JEC.pdf (2002)

[7] Friedrich, H., Kopp, J., Stieg, J.: Composites on the way to structural automotive applications. *Materials Science Forum*, Vol. 426–432 (2003), pp. 171–178

[8] Maeder, G., Giocosa, A.: Future applications for plastics in the automotive industries. Kunststoffe im Automobilbau, VDI (2001), pp. 325–334

[9] The global outlook for carbon fiber 2003. Conference proceedings, Intertech (2003)

[10] Plastic Omnium, Sigmatech information. www.plasticomnium.com (2003)

[11] Composites for the Automotive Industry. Conference proceedings, The Engineer (2003)

[12] Jeffries, R.: Prospects for carbon fibres. *Nature*, Vol. 232 (1971), pp. 304–307

3.4 Further Developments in SMC Technology

THOMAS SCHUH

3.4.1 Reasons for Using SMC

SMC (Sheet Molding Compound) is a glass-fiber-reinforced thermoset material that has been used for a long time in automobile manufacturing. Applications in the European and US commercial vehicle sector are predominantly for structural and outer body panels. In the car sector, SMC is mostly used in niche vehicles produced in small or medium quantities. Typical examples include tailgates, fenders, underside panels, and front-end modules. The Renault Espace, where virtually the entire outer skin is made of SMC, is a rather a unique example of SMC applications.

With SMC, even complex component shapes can be manufactured in a single operation in a state close to the final contours. At low production quantities, this yields cost advantages over steel which are frequently the decisive reason for using SMC.

Other typical advantages of SMC technology include:

- Lower component weight compared to steel
- Very high dimensional stability
- A coefficient of thermal expansion similar to steel
- A level of heat resistance that allows on-line painting even with electrophoretic dip coating processes with peak temperatures of 200 °C
- Greater possibilities for design shaping than with aluminum.

3.4.2 The Road to Premium SMC

Commencement in 1999 of large-scale production of the Mercedes S Class coupe opened up a new chapter in the history of SMC technology. The deciding factor for using SMC for the rear lid of this high-end vehicle was, first, the freedom of design afforded by SMC, enabling shapes that cannot be provided by either steel nor, in particular, by aluminum. Second, its permeability for electromagnetic radiation meant that antenna systems could be integrated in the rear lid. The fact that all antennas can be hidden beneath the outer skin offered a further enhancement of design possibilities.

The large horizontal area of the component, the complexity of its overall shape and the requirement that its appearance harmonize with the surrounding steel components placed extremely high demands on the external attractiveness of the component. Under these conditions, the short- and long-wave corrugations typical of SMC show up much more clearly. These requirements on the surface quality of this component far exceeded those on any previous SMC application.

In addition, to exclude color deviation problems right from the start, such as can occur with parts painted off-line, the rear lid was to be painted on-line together with the overall body. Temperature peaks of up to 190 °C in this process constituted an additional challenge, since

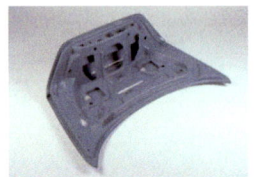

Figure 3-39 Mercedes S class coupe with double-shell SMC rear lid

3.4 Further Developments in SMC Technology

gas releases would need to be prevented. These are much easier to overcome at the lower temperatures components are exposed to in off-line painting.

This new premium standard was finally achieved during the development phase after extensive adjustments had been made to the formulation of the semi-finished SMC. However, once series production started, it became obvious that the SMC process as a whole was also in need of optimization. This optimization process was the focus of a joint project known by the abbreviation "Triple C".

"Triple C" stands for **C**ompound, **C**haracterization and **C**onsistency and aims to ensure consistent quality for components with high surface quality requirements. At the same time, the technological potential of SMC material was to be considered for other vehicles series with greater production volumes.

Partners in this project were all major manufacturers along the SMC process chain: DSM as important suppliers of the raw material, Menzolit-Fibron as developers and manufacturers of the semi-finished product, Peguform as the manufacturer of the component and DaimlerChrysler as the OEM with overall responsibility.

In the first step, a database was set up for process analyses. Along the entire production chain, all data relevant to process and product were collected and recorded for more than 10,000 C215 rear lids, an equivalent to 80 tons of SMC. These data relate to the complete process chain: raw material and machine data, physical index values, logistics and even details as to surface quality. All together, more than 5 million data points were entered into a data bank.

Process analysis consisted of three modules:

- Process monitoring by experts on site
- Calculated process variations as part of design of experiments (DoE)
- A holistic analysis of this complex process chain by data mining.

On-site process monitoring by experts meant that every single process step could be analyzed to identify specific sources of error. Systematic errors were uncovered here, such as incorrectly selected or incorrectly prepared blanks. Since process observation was carried out jointly, everyone involved in the project was more aware of the requirements of the process as a whole and of their individual responsibilities. On the basis of the DoE, important control variables were

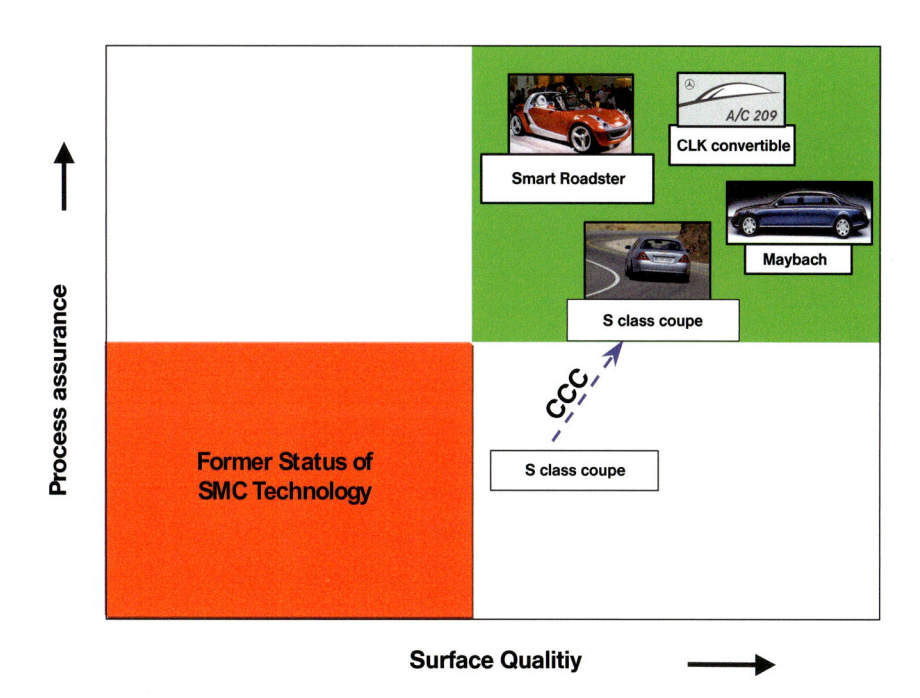

Figure 3-40 "Triple C" as a precondition for applying Premium SMC in the Mercedes Car Group

determined in the manufacture of the semi-finished material and the component itself. Their influence on the quality of the moldings was determined and process parameters optimized.

The large number of interrelating influences on the surface quality of SMC components makes it very difficult to detect less obvious key variables by conventional methods. Therefore, data mining was used in addition – an approach already successfully used for process optimization in other industrial fields, such as chip fabrication. It provides a high degree of transparency into the interrelationships existing in complex data systems. In this way, it was possible, e.g., to detect inadequate temperature levels in the mold which led to an exponential increase in rejects when critical limits were exceeded. These process analyses produced a large number of individual results from which it was possible to derive specific measures whose implementation drastically reduced process fluctuations.

Once implemented, the level of rejects was reduced to less than 2%. Simultaneously, surface quality was again significantly improved. Sustained implementation of the CCC project thus contributed to SMC becoming established on a high quality level and was thereby an important precondition for the high-volume production of new components. Since most of the findings are not specific to a particular component, they can be transferred directly to other projects.

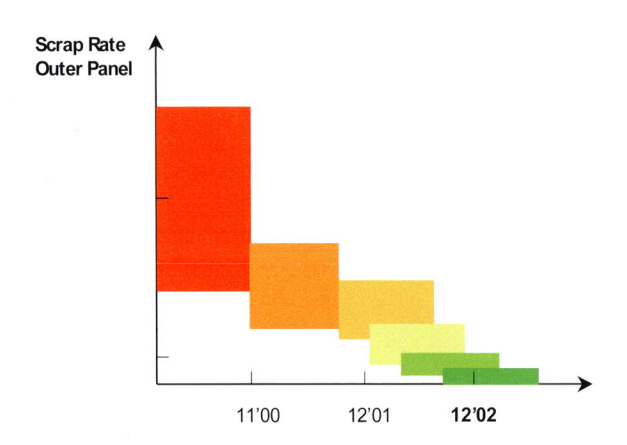

Figure 3-41 SMC technology improvements due to CCC

The decisive factor for the success of this project was a holistic analysis of the process chain as a complex system by means of data mining of a data bank. In addition, this result could not have been achieved without very open collaboration by all parties involved and their declared intent to cooperate to find an optimum solution. Creating this spirit of openness was initially one of the greatest challenges of the project. Today, all of the partners find reassurance in the success achieved. Initial fears of losing out by sharing expertise has turned into its opposite. In addition, the culture of collaboration has undergone a lasting improvement, for which thanks are due to all project partners.

"Triple C" has laid the corner stone for the future manufacture of SMC components at premium quality, both in medium and in large volumes, while maintaining controlled, stable processes.

3.4.3 Lightweight Potential of SMC Technology

Lightweight SMC

Using normal SMC for components, such as a rear lid, results in weight savings of 10% compared to steel. This means that classical SMC is at a disadvantage compared to aluminum which is establishing itself more and more as a body material. New developments referred to as lightweight or low-density (LD) SMC are intended to close this gap, even if the weight-saving potential of aluminum cannot be reached.

The LD SMC material developed by Menzolit-Fibron in collaboration with Daimler-Benz is based on the formulation used in the current series-production material. The substitution of hollow glass microbeads for a portion of the calcium carbonate filler reduces material density by approx. 20% to 1.5 g/cm^3. For the rear lid of the S class coupe this approach provides a reduction in the component weight of approx. 2.0 kg. Although these hollow glass beads are considerably more expensive than calcium carbonate, the additional costs for the component are comparatively moderate, amounting to less than 1 Euro per kilogram of weight saved.

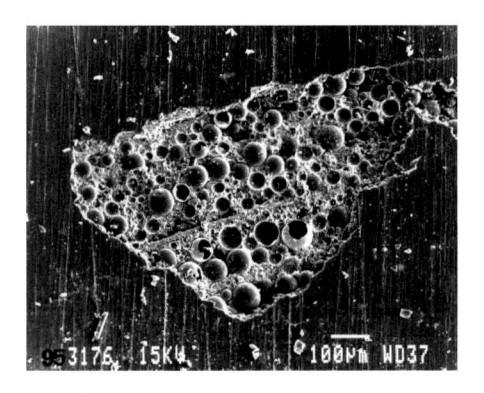

Figure 3-42 LD SMC:
density reduction by using hollow glass beads

As far as plant, tooling and process parameters are concerned, processing LD SMC is no different from standard SMC. Compared with standard SMC, the surface of the LD-SMC parts shows reduced long-wave corrugation, but increased short-wave corrugation. The latter can be improved to a level acceptable to the customer by specific optimization of formulation and application.

Once the level of mechanical properties, bonding capability and collision performance have been secured, the next step is major testing to ensure readiness for series production.

For applications that have no specific demands on surface quality, the proportion of hollow glass beads can be increased further, thus enabling weight reductions of up to 30% compared to standard SMC. Chrysler is already using similar SMC types in structural components in series production of the new Viper.

This potential for lightweight design gives SMC a good chance of establishing itself as a third bodywork material alongside steel and aluminum, particularly if efforts succeed to further improve the economic efficiency of this technology.

Advanced SMC

Heavy-duty fiber composites have demonstrated that they provide the largest potential for lightweight components. They allow for extraordinary weight savings, particularly when load paths can be utilized with the anisotropic properties of these tailor-made materials. Whenever minimal weight has top priority, these materials are used despite their high cost. Not surprisingly, the aerospace sector was the first to use this group of materials. Today, carbon-fiber-reinforced plastics (CRP) have also become the winning material of choice in motor racing – particularly in Formula 1 racing.

Carbon fibers have also gained importance in the sporting goods and leisure industries. The energy industry is on the verge of employing it in the off-shore sector as well as for wind mill rotor blades.

At the present time, two major obstacles prevent CPR from entering into the everyday automobile world. One is the very high price of components made of CPR, the other – and to a certain extent this is directly related to the first obstacle – are the production technologies currently available, because they usually accommodate only small lots.

The availability of relatively inexpensive carbon fibers creates a demand for inexpensive production technologies that also have the potential to accommodate series production. The working principle of SMC technology makes it a good candidate, and for some time research has been conducted to investigate the use of carbon fibers instead of glass fibers in SMC. The first materials of this type are already in series production in semi-structural components. Since the compound contains chopped random fibers, there is a considerable decrease in mechanical properties compared with the composite materials in use in the aerospace industry and in motor racing.

To close this gap, a new technology has been developed as part of a joint project between Menzolit Fibron GmbH, Volkswagen AG and DaimlerChrysler AG. With Advanced SMC, an innovative technology has been created, allowing unidirectional carbon fibers to be processed by a technology similar to SMC.

Figure 3-43 shows a comparison of the lightweight potential of Advanced SMC and other standard bodywork materials.

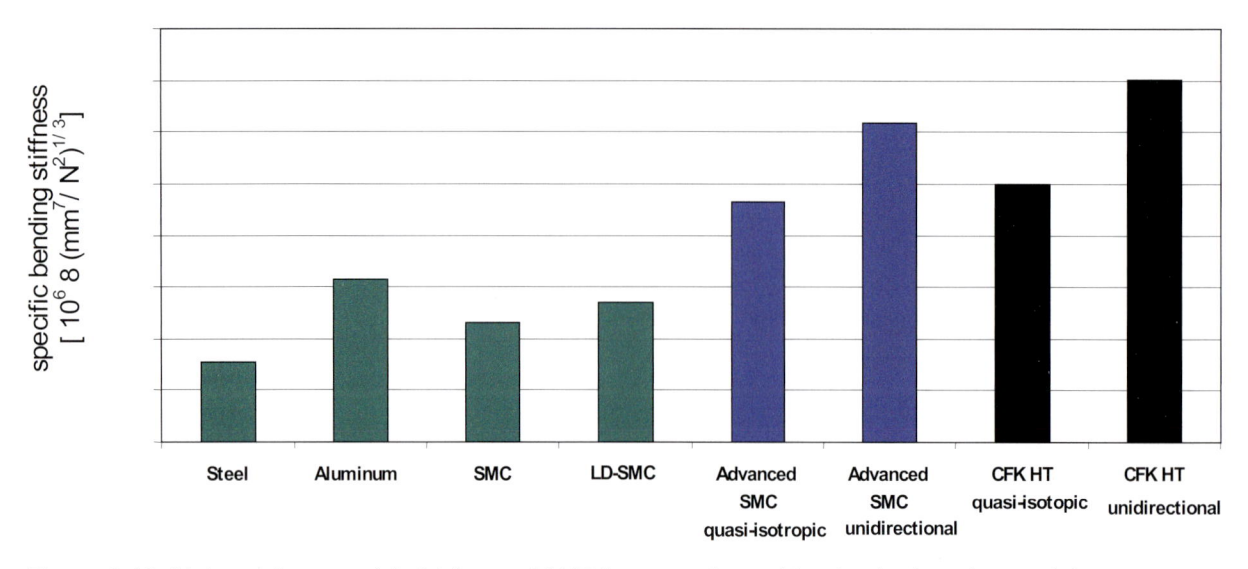

Figure 3-43 Lightweight potential of Advanced SMC in comparison with other bodywork materials

Chart y-axis: specific bending stiffness [$10^6 \cdot 8$ (mm^7/N^2)$^{1/3}$]

Chart categories: Steel, Aluminum, SMC, LD-SMC, Advanced SMC quasi-isotropic, Advanced SMC unidirectional, CFK HT quasi-isotopic, CFK HT unidirectional

Description of the Material

Two main trajectories were pursued in the development of Advanced SMC:

- The development of a material for structural applications with the highest possible fiber content and the highest potential for lightweight design (with approximately the same rigidity level as heavy-duty CPR and slightly lower strength).

- The development of a material with Class-A surface properties, taking slight decreases in fiber content or lightweight potential into account.

Before it was possible to compete with aluminum applications for the vehicle exterior, techniques that allowed the production of very thin wall thicknesses (approx. 1.2 mm) had to be developed. In addition, efforts had to be made to automate major process steps to secure a high level of quality.

Advanced SMC differs from standard SMC in the following important aspects:

- The aim of producing a material with aligned (quasi-)continuous strands requires considerably modified manufacturing technology for the semi-finished material.

- The considerably reduced flowability of this material means that a preform has to be made, representing approx. 85% of the component geometry. Consequently, rectangular blanks can no longer be used for more complex shapes.

- In order to design a component that will hold up to the loads it will encounter, this preform has to be formed from several very thin individual layers.

- Production facilities will have to be adapted for processing conductive carbon fibers.

Production of the Semi-Finished Material

The carbon fibers are supplied as a thin, wide mat between two carrier films coated with resin. It is crucial for the fibers to be evenly arranged and tensed to prevent fiber corrugation and to achieve optimum properties. Once the mat has passed through the impregnation line, the semi-finished material is wound onto reels.

Preform Manufacture

Following a pre-calculated layer design system, multiple layers of semi-finished material with specified differing fiber orientations are assembled into a preform. If the component has a complex shape, a similarly

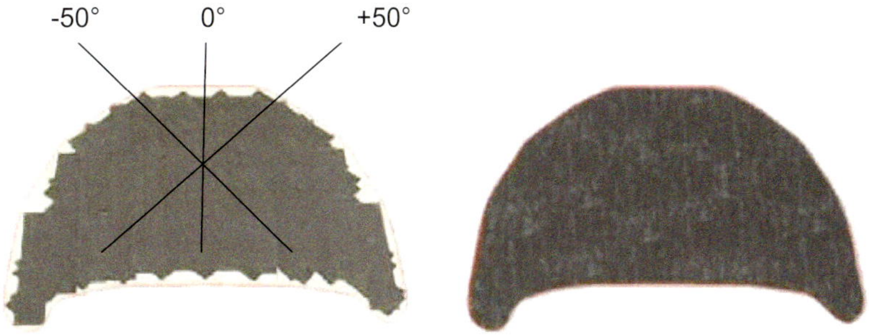

3.4 Further Developments in SMC Technology

Figure 3-44 Waste-free mat cut-out technique

Mat cut close to contour Saturated component

complex preform can be built from a large number of individual strips of different lengths and widths so as to minimize material waste. A planar preform of this kind must cover approx. 85% of the subsequent surface of the component (see Figure 3-44). This important step in preform production should be automated in order to achieve highly uniform quality and also limit costs.

Making the Component

The preforms are cured in a heated compression mold at temperatures and cycle times typical for SMC.

In order to validate the technique as a whole, a rear lid was made at DaimlerChrysler as a reference component. It was used to evaluate production-specific issues. The rear lid consists of an outer shell with a highly contoured inner shell glued to it. Designing this complex shape presented a major challenge.

With the help of these reference parts it was possible to assess the overall level of maturity of the technique as to product concept, component properties and validation of component configuration.

Once the technology was successfully validated by the example of the rear lid, its first series-production application was implemented at DaimlerChrysler.

Even if for economic reasons large-scale production is not yet feasable, Advanced SMC technology has improved the chances for CPR technology to make its entry into small- and medium-lot production in the automotive sector.

Close-to-contour precut blank for inner shell

Complex inner trunk lid shell after compression

Figure 3-45 Making the inner shell of the rear lid

3.4.4 Outlook for SMC

The development of Premium SMC has created the basis for a new range of applications for SMC technology. The excellent suitability of this material for lightweight engineering coupled with its ability to meet the highest requirements regarding surface quality has created an opportunity for a large number of new applications, such as rear lids, hoods, fenders, roofs among others.

The latest developments, which resulted in a marked weight reduction, take the necessity of exploring further areas of lightweight design potential into account. Against this background, light SMC has a good chance of establishing itself as a third bodywork material together with steel and aluminum.

If recognizable potential savings can be realized by further optimizing the overall process, the threshold of economic efficiency can be raised toward higher and higher production volumes, possibly ending SMC technology's niche existence.

This will not, however, be possible unless there is a fundamental and sustained major change in the way the SMC sector defines quality and costs. The cooperation of all partners in this project was a good example for such an effort.

The development of advanced SMC has created a new and innovative technology, which – taking its overall properties into consideration – has the potential to bring about the breakthrough for CPR materials in automotive engineering. It remains to be seen whether this will lead to similar revolutionary changes as those observed with the arrival of CPR technology in the Formula 1 world.

However, there is no doubt that lightweight design will become more and more important in the future. The need for additional components to meet higher demands in vehicle safety, comfort and driving performance initially involve added weight. Future fuel consumption targets will therefore not be attainable without weight reductions in other areas.

SMC technology and its described variations can certainly make a major contribution to this development.

3.4.5 Acknowledgements

Technological development projects seldom conclude with an outlook to as much potential as Advanced SMC. For this reason special thanks are due to the following:

at Menzolit Fibron:
Messrs Ehnert, Schäfer, Stachel and Hörsting;
at Volkswagen AG:
Messrs. Stieg, Becke, Brüdgam and Friedrich;
at DaimlerChrysler:
Messrs. Ilzhoefer, Wittig and Hermann.

3.5 New TPO Grades with Reduced Linear Thermal Expansion for Exterior Parts Painted Offline

Franz Zängerl

3.5.1 Introduction

In the early 1990s, mineral-filled PP/EPDM compounds known as "zero-gap materials" were introduced in bumper panels.

Today, the vast majority of automakers utilize these materials in their vehicles. Most grades only contain about 10% mineral filler and reduce the significant linear thermal expansion coefficient by about 30% compared with unfilled PP/EPDM's. This is illustrated in more detail in Figure 3-46.

Increasing demands on the dimensional stability of components, seamless surfaces at occurring temperatures, production processes as well as mechanical and rheological properties have made it necessary to advance the development of PP further. The results of this intense innovation effort will be presented in this paper by means of materials comparisons, first series production applications and a feasibility study. In addition, technological developments will be presented indicating that this material's potential may soon be realized.

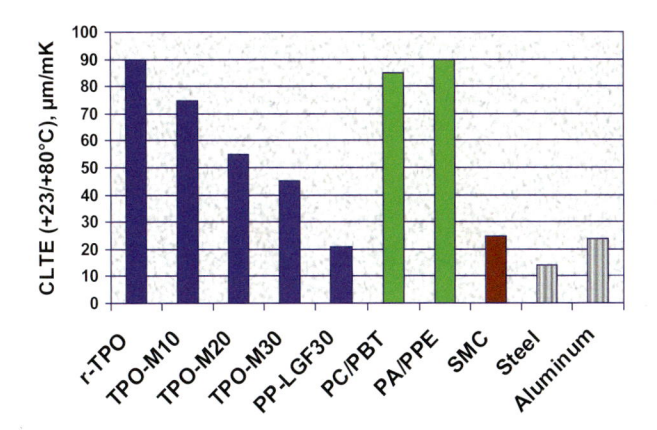

Figure 3-46 Comparison of linear thermal expansion in different materials

3.5.2 Requirements and Basic Conditions

Materials for car bodies must meet many – in part opposing – requirements. The most important are summarized here:

- Mechanical properties (stiffness, toughness, low-temperature viscosity)
- Surface (class-A paintability, scratch-resistance, ultraviolet resistance, homogeneity)
- Dimensional stability (thermal expansion, shrinkage, moisture absorption, warping)
- Process ability (flowability, required pressure, melting stability)
- Thermal behavior (heat resistance, continuous operating temperature)
- Emissions (CARB test).

Today, bumper panels are predominantly made from PP/EPR with up to 10% filler. While the type and quantity of the filler are generally known to play an important role, it is the base polymer that largely determines the properties of the material.

High-impact resistant, heterophase copolymers, defined by a multitude of parameters, are predominantly used.

Only a few of these are listed here:

- Polymerization process
- Molecular weight and molecular weight distribution
- Catalyst
- Additives
- Type, amount and stereochemical distribution of the EPR components.

The last item essentially determines the morphology of the material and has a significant effect on the required properties. The morphology of a heterophase copolymer is shown in Figure 3-47.

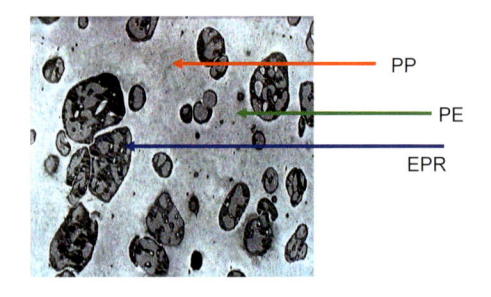

Figure 3-47 Morphology of a heterophase copolymer

The EPR content determines the stiffness/impact strength behavior, as illustrated in Figure 3-48. As the amount is increased, impact strength increases disproportionately from a specific minimum concentration; at the same time, stiffness decreases linearly.

The stereochemical distribution of EPR components in PP matrix is determined, among other things, by their viscosity and composites (C2/C3 ratio). This provides important control variables for product development to achieve an optimum balance of the different properties, up to and including thermal expansion. Figure 3-49 shows the effect of EPR content on linear expansion.

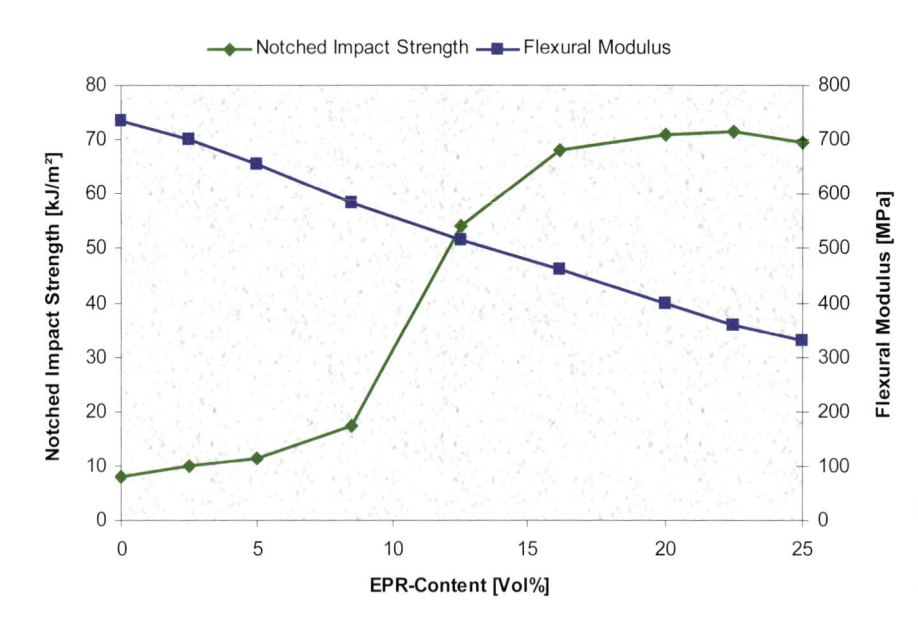

Figure 3-48 Influence of EPR content on impact strength and stiffness

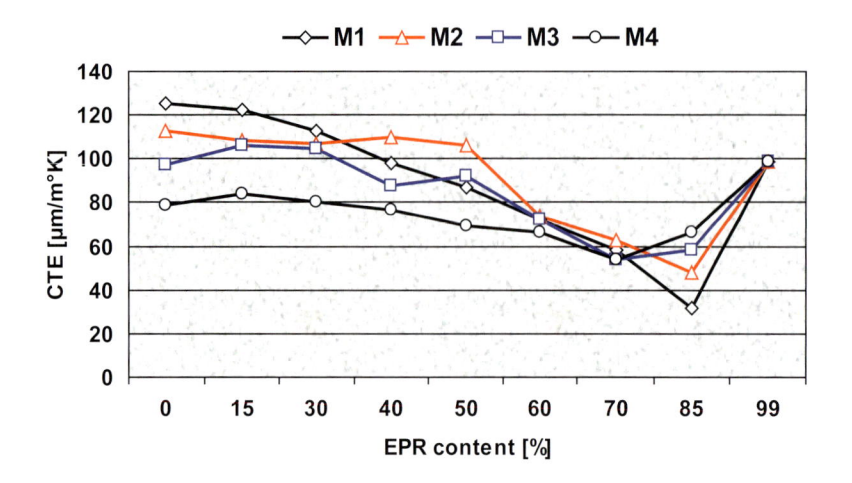

Figure 3-49 Effect of EPR content on the linear thermal expansion of different types of PP

The illustration shows that with the right combination of matrix with the type and concentration of the EPR phase, a relatively low linear thermal expansion coefficient of less than 80 µm/mK can be achieved even in an unfilled state, thereby providing a solid basis for the performance range of the finished compound. The technology for producing the base polymer has to be the technical and economic point of departure. Utilization of various fillers, impact modifiers and other additives in the compound step then paves the way to further optimization.

3.5.3 Analysis and Simulation Techniques

The linear thermal expansion coefficient is defined according to DIN 53 752-A.

The test specimen, measuring only 10 mm, is pretreated in relatively exactly defined process steps before its expansion is measured in a temperature range of 30 to 130 °C in two consecutive cycles. The results are easy to reproduce and enable a meaningful relative comparison of the different materials with respect to their actual linear expansion in the component. Currently, a method with more practical relevance is being developed to further improve this correlation and should become available in the second half of 2006.

The numerical simulation (Abaqus) of the linear thermal expansion of components has already been taken a step further here and deviates by only 5% from the component's physical measurement. Great effort went into identifying suitable material models; in the case of dynamic loading stress, these highly complex models are of paramount importance in order to achieve sound calculations.

The example in Figure 3-50 shows good concurrence almost as far as fracture initiation at approx. 13 mm between the calculated load-displacement diagram and the one obtained during testing in instrumented impact penetration tests.

In the meantime the simulation of visual effects, such as the tiger skin pattern, has also been achieved on components. This will probably lead to a breakthrough in pinpointing the causes of this undesired effect in the entire material/process/design system. In the medium term, calculations will also incorporate the properties of materials based on their formulations; to save time and development costs, a virtual laboratory is under consideration that is similar to the one used by the pharmaceutical industry. It is already possible to calculate simpler material systems with encouraging results.

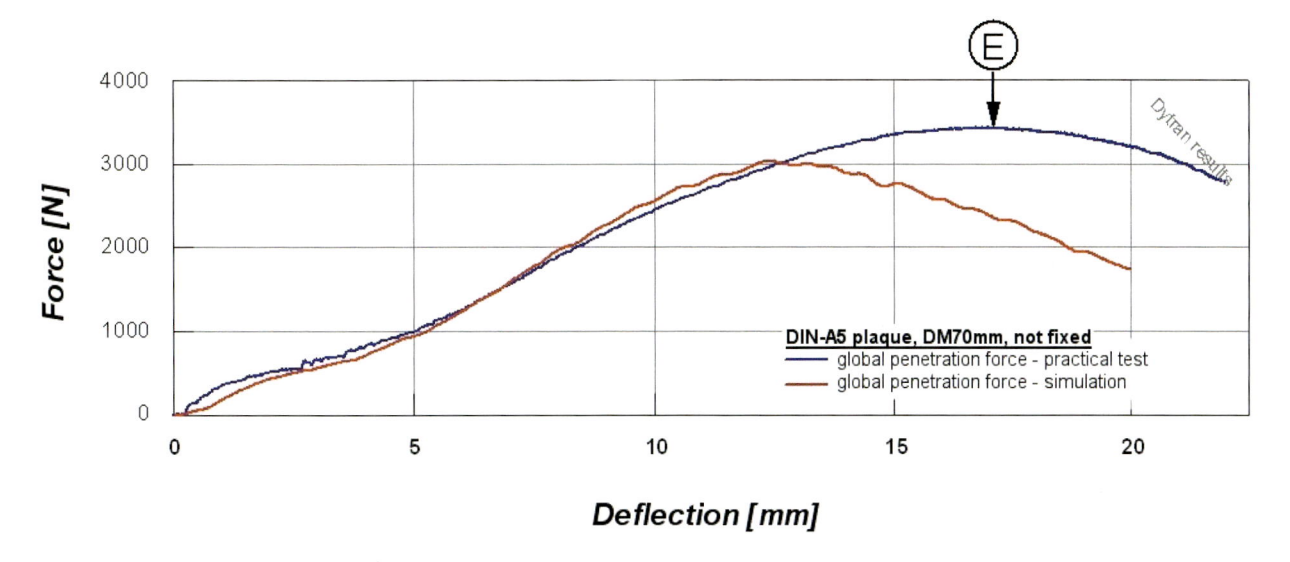

Figure 3-50 Load-displacement diagram from instrumented impact penetration tests on PP-TV30

3.5.4 New TPO Grades and their Properties Compared with Other Technical Thermoplastics

Extensive preparatory work, including the development of new base polymers, has led to interesting innovations, as shown in Figure 3-51.

The specific materials for a component are selected depending on the properties required. While type EE-109AE is recommended for bumper fascias, ED230HP and EF341AE are reserved for offline-painted body panels due to their superior stiffness. Depending on the type, the linear thermal expansion value is between 50 and 60.

Compared with other technical thermoplastics, such as PC/PBT, ED230HP exhibits approx. 1/3 less linear expansion, approx. 14% less density and high flowability with comparable stiffness. Notched impact strength is conspicuously lower. In fact, functional testing (e.g. the pendulum impact test) on components reveals quite the opposite with respect to impact strength. This is probably due to superior flow behavior and the associated lower degradation during processing, and to higher elongation at break.

Properties	Unit	Daplen EE109AE	Daplen ED230HP	Daplen EF341AE
MFI	g/10 min	13	10	17
Filler amount	Wt%	20	20	30
Flexural E – Modulus	MPa	1500	1900	3300
Tensile stress at yield	MPa	16	20	25
Notched impact strength, +23 °C	kJ/m^2	35	29	5
HDT B	°C	93	105	120
LTE (+23 °C/+80 °C)	µm/mK	51	60	52

Figure 3-51 Key properties of new TPO grades

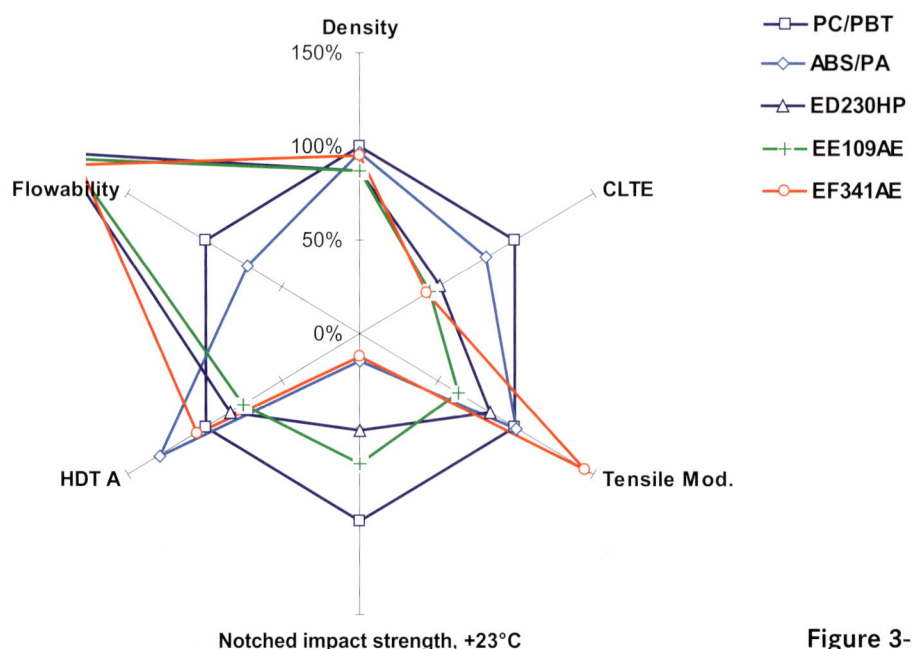

Figure 3-52 Comparison of materials

3.5 New TPO Grades with Reduced Linear Thermal Expansion for Exterior Parts Painted Offline

As with all PP's, hygroscopicity is very low at 0.1% (after 24 h storage) and practically negligible; this, along with low CTE, is an additional factor in the dimensional stability of the components. Chemical resistance to common substances (e.g., petrol, cleaning agents) is very good. Hydrocarbon emissions are tested using the Headspace Method (VW-PV3341) and are conspicuously lower than PA/ABS and comparable with PC/PBT.

This could also become relevant for exterior parts in view of the Californian law stipulating the reduction of non-fuel emissions.

The fall in stiffness at temperatures of between 20 and 80 °C is somewhat faster with the TPO's than with the PC/PBT, but shows the same if not slightly higher stiffness properties at temperatures typical during painting, i.e., approx. 80 °C, depending on the type of TPO.

TPO type ED230HP shows very uniform energy absorption over time/deformation length and no abrupt decrease on reaching elongation at tear.

The ductile fracture behavior so important in practice is already apparent in the fracture pattern of the sample as shown in Figure 3-55.

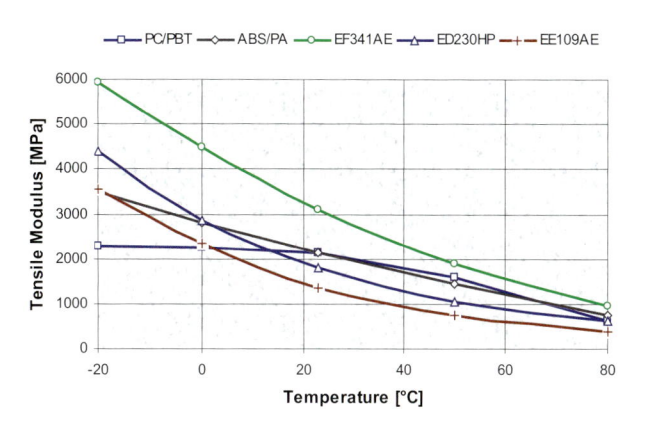

Figure 3-53 Tensile modulus of elasticity plotted against temperature

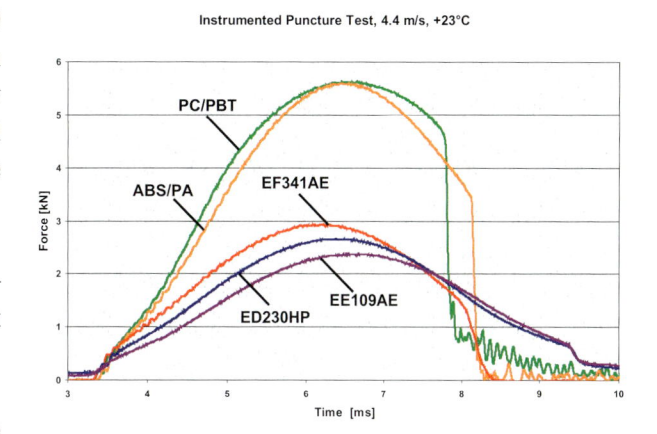

Figure 3-54 Instrumented puncture test, 4.4 m/s, 23 °C

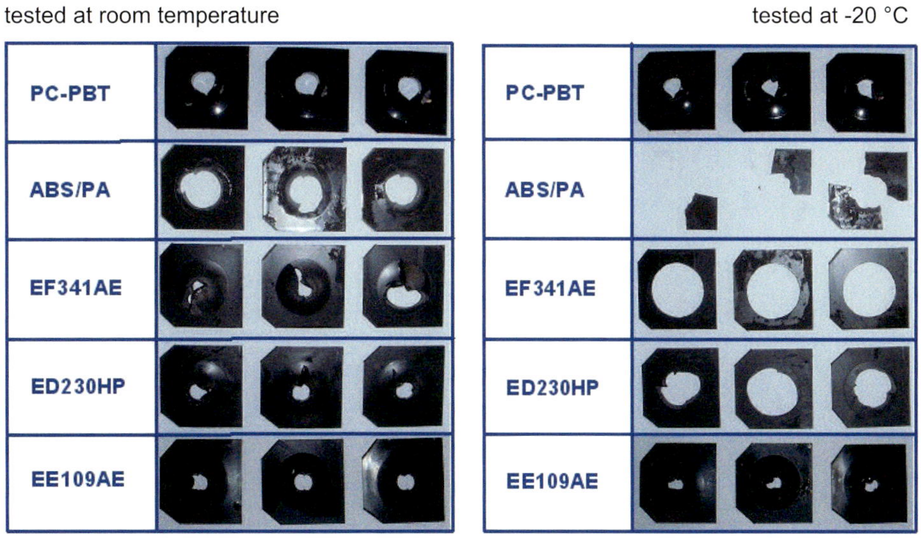

Figure 3-55 Samples after instrumented impact penetration test at RT and –20 °C

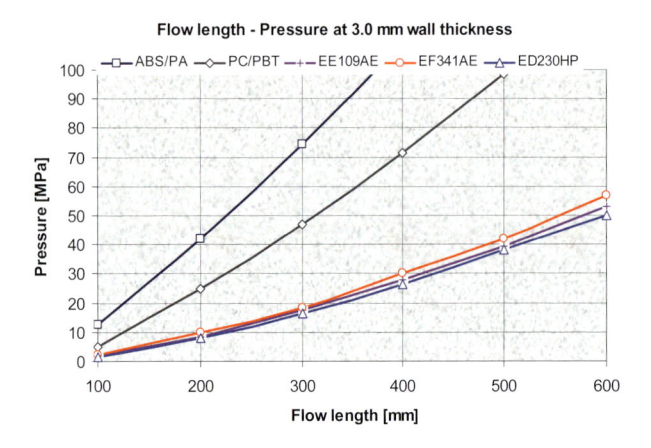

Figure 3-56 Required pressure according to the flow-length of different materials

Figure 3-56 illustrates injection pressure as a function of flow length. The high flowability of TPO materials requires less locking pressure at the same flow length and, due to its potential for use in smaller machines, is also commercially interesting. Considerably longer flow lengths can be realized on a given machine.

3.5.5 Component Test Results

Functional tests carried out on components confirm the laboratory results and will now be briefly described.

Type EE109AE is in series production and is used, among other things, for the bumpers of the new Opel Zafira. These are produced by Dynamit Nobel.

The material's low expansion and high stiffness properties made it possible to eliminate the usual metal backing with the aid of a tilted flange construction.

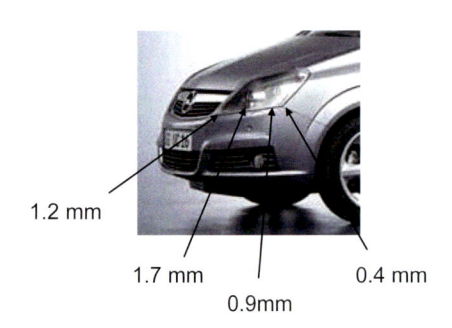

Figure 3-57 Expansion of the bumper panel at 70 °C

At 70 °C the bumper panel expands by only approx. 1.7 mm in the sensitive area around the headlights. Despite the higher density of the bumper material compared with the previous model, it was possible to reduce the weight of the system.

ED230HP, designed for outer skin applications, is being used in the manufacture of the split outer panel in the tailgate of the Renault Modus. The component is painted offline and glued to a bracket made from a LGF-PP. It is manufactured by Plastic Omnium.

This material was additionally tested in the course of an extensive feasibility study using production tooling parts from exterior components of the new Smart ForTwo.

The study's main criteria comprised component functionality, producibility, surface quality, ageing stability, and an impression of quality.

The results were as follows:

- Dimensional stability: OK
- Good processability
- Dimensional accuracy of the components within the tolerance limits
- Reduction of component weight by approx. 14% (compared with PC/PBT)
- Reduction of injection pressure by at least 40% (compared with PC/PBT)

Figure 3-58 Renault Modus tailgate

3.5 New TPO Grades with Reduced Linear Thermal Expansion for Exterior Parts Painted Offline

- ECE R42 pendulum impact test (fender corner impact) at RT: OK
- Fender – cold impact test in compliance with FP regulation MCC at –25 °C: OK
- ECE R42 pendulum impact test (fender corner impact) at –20 °C: qualified OK
- Heat resistance up to 90 °C: OK
- Gap reduction possible due to lower linear thermal expansion.

The performance of the new TPO's can already match – as explained in detail – that of several classical technical thermoplastics. Their potential does not end there, however. The next evolutionary leaps are just around the corner as technology advances (Borstar, reactive compounding), and work on catalysts is also being stepped up. Whereas new catalysts lie the farthest in the future, production technologies can already be expected that will enable new products in the medium term. In reactive compounding, a kind of partial cross-linking of the molecular chains will create more stability at interfaces and layers close to the surface. This means, for example, a tremendous increase in impact strength while maintaining the same stiffness. The rigidity of the substrate in the steam jet test has seen a considerable increase under the most rigorous testing conditions.

New metallocen-based ('single site') PP catalysts open up a whole new range of potential due to improved heat resistance and by incorporating polar comonomers, which puts painting PP without pretreatment in the realm of the feasible.

Borealis is wholly dedicated to these issues in order to create value for its customers through innovation.

Figure 3-59 Fender after corner impact at –25 °C at defined points

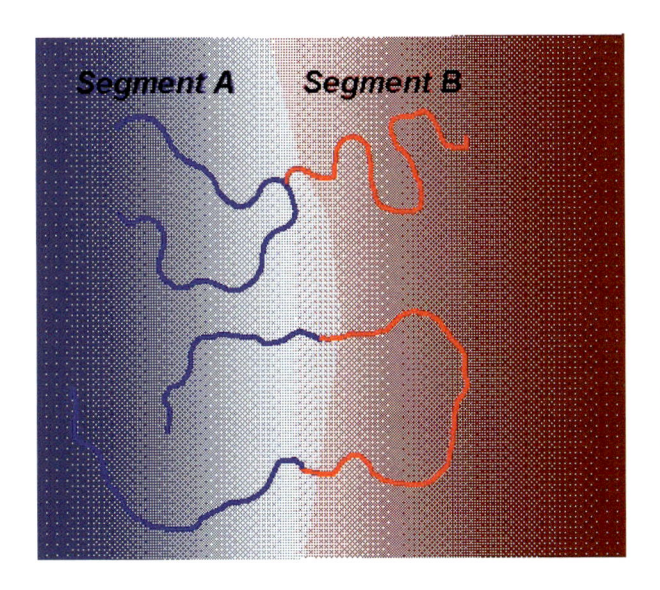

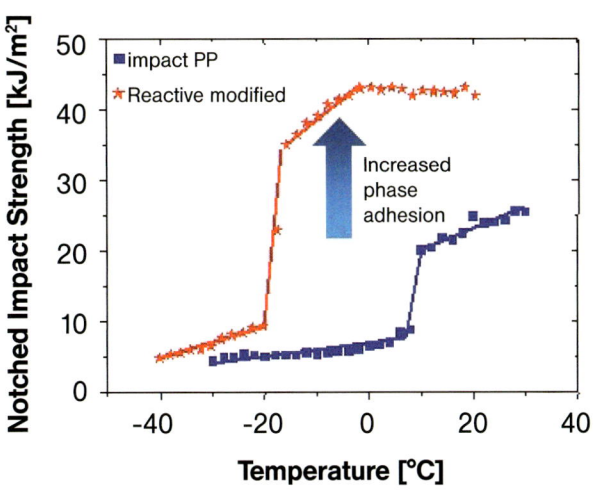

Figure 3-60 and 3-61 Partial crosslinking at the boundaries, shift in temperature/viscosity

3.6 Lightweight Bodywork – Use of Structural Foams in Hollow Sections

MARTIN DERKS

In automobile manufacturing, self-supporting bodies are made from hollow sections in monocoque design. The rigidity of the structure is primarily determined by the size and shape of the cross-sections. Where designs have openings in cross-sectional areas or extended-area geometrical features, they require additional reinforcement. One way of providing this is to use structural foam. This means inserting foamable epoxy resins in the bodywork onto substrate components made of plastic or metal.

We shall describe the requirements applicable to the manufacture of the moldings and the criteria for selecting the material systems. Implementation of the solution takes place under the process conditions obtaining in bodywork production at the BMW Group.

3.6.1 Introduction

In the early days of automobile manufacturing, the body was mounted on the chassis with the chassis taking the entire load when the vehicle was moving. Bodies were originally made of wood and their only job was to protect the occupants and vehicle contents from wind and weather.

During the course of automation and further development of sheet-metalworking and of lightweight design in automobile manufacturing, self-supporting bodies made of metal were developed. The supporting structure of present-day vehicles consists of hollow sections joined together in so-called monocoque design. The hollow sections are connected at joints in the frame. The number and design of these joints is of decisive importance to the rigidity of the body frame. Besides the installation of bulkheads and the use of beads on flat surfaces, the use of structural components thus represents a weight-optimized alternative to local reinforcement of bodywork.

By the term structural foam, we mean compounded (premixed), precured and foamable epoxy resins. These are applied in a two-component injection molding process preferably to polyamide substrates (polyamide 6 with 30% glass-fiber content) or to sheet-metal substrates. The components thus made are then inserted into the hollow section during the bodywork construction process. The epoxy foams expand and cure under the influence of temperature as they pass through the paint shop.

The desired damping and reinforcement effects are achieved by large-area installation of structural foam components inside the hollow section. Local deformation, such as inward or outward bulging, are thus prevented. The weight saving due to structural foam components is assisted by the following conditions:

- Open cross-sectional areas in the car body joints
- Local transfers of force
- Large areas at the joints.

Structural foam   Polyamide support   Expansion

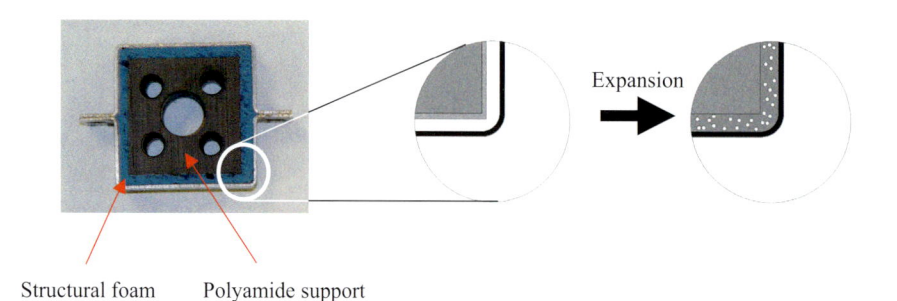

Figure 3-62 Structural foam component inside the sheet-metal top-hat section

3.6 Lightweight Bodywork – Use of Structural Foams in Hollow Sections

Besides the weight savings as compared with the alternatives mentioned, the present solution has the advantage of being implemented faster. This will be described taking the BMW 5 Series Touring as our example.

During the course of model revision, a panorama roof was incorporated into the vehicle. The loss of roof skin and roof bows entailed optimization of the vehicle structure. In the trial phase of the vehicle, test drives also revealed that vibration comfort and rigidity needed adjusting.

For this reason, five structural foam parts were incorporated into each body side frames (Figure 3-63).

Not only was the use of structural foam investigated, but also the incorporation of additional steel reinforcement plates or increasing the wall thickness of the hollow sections near the roof pillar joints.

The solution using structural foam was selected for the following reasons:

- Minimum modification required to the series framework
- Reduction in outlay on tools and equipment
- Weight saving of 10 kg compared with solutions using steel.

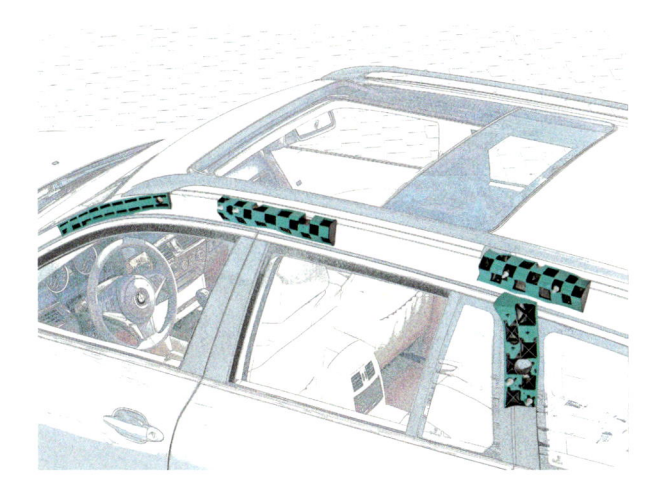

Figure 3-63 Structural foam components in the BMW 5 Touring

3.6.2 Requirements Applicable to Manufacture

Structural foam components are manufactured in a production process with several steps. The necessary raw materials such as resin, filler, catalyst, accelerator and blowing agent are mixed during the compounding process. The compound is then pelletized in the extruder for further processing by injection molding or continuously extruded for direct application to the support components. The polyamide supports are either coated with precured structural foam in the two-component injection molding process or metal components are inserted into a mold and then encapsulated. Due not only to the process conditions in the production line (described in Section 4), but also to their function in the vehicle, the components have to meet strict requirements. It is therefore necessary to optimize manufacture of the moldings with respect to both the processes and their application in the body structure.

A high degree of consistency in the material is required with regard to:

- Adhesion
- Foam structure
- Adjustment of the foaming process with regard to temperature, time and height
- Mechanical aspects and heat resistance.

During fabrication of the structural foam components, specified adhesion requirements must be fulfilled. The foam must not detach from the support while being transported nor when being installed in the hollow sections. In the same way, detachment or slipping during the foaming process cannot be permitted either. Figure 3-64 shows the slippage test with the foam parts installed in a vertical position. The specimens are subjected to the maximum object temperature in the electrophoretic dip coating oven.

Following expansion and curing, there must be an adhesive bond with the hollow section and the support material. Figure 3-49 shows an adhesion test following a road test. Adhesion during the process and in

Faulty   o.k.

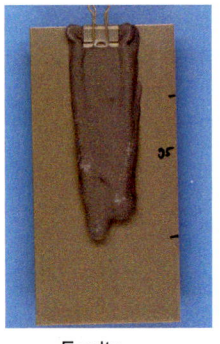

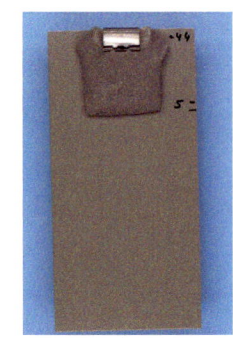

Figure 3-64 Slippage test [4]

Figure 3-65 Adhesion following the road test

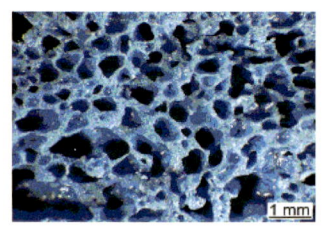

Figure 3-66 Fine and coarse-celled structural foams [1]

the vehicle is determined by the type and quality of resins, catalysts and accelerators selected. Besides batch consistency of the raw materials used, the time elapsing during compounding ensures evenness in the adhesion of the foams. Ageing processes occur during the production of the components. Release agents used during injection molding can impair the adhesion of the foam to the support. Demolding of the support elements should be eased by sufficient tapering so that release agents can be dispensed with. Foam adhesion can be additionally improved by specially texturizing the substrate surface.

Efforts are made to obtain a fine-celled foam structure for optimum mechanical properties in the vehicle. This should be taken into consideration when selecting the blowing agents and during the compounding process. Figure 3-66 shows different cell sizes in epoxy foams:

In order to secure functionality under process conditions on the production line, the foams must have foamed up and reactions have completed at the minimum conditions of 165 °C and 10 minutes. For this reason, production of the foam must be secured within close manufacturing tolerances with regard to the compounding and injection molding processes. The materials must not become damaged beforehand due to shearing processes or curing, or expand excessively.

Measures supporting the process should be taken so as to secure the mechanical properties and temperature stability. In this connection, the glass transition temperature (T_g) should be measured. By the value of the glass transition temperature it is also possible to tell the extent to which the foams have cured. If the structural foams have not cured, this temperature will be approx. 20 °C. Following curing, the value of T_g will rise to approx. 100 °C.

3.6.3 Selection of the Material System

In the epoxy foam group of materials, a distinction is drawn between one- and two-component systems. The present application and the production sequences dealt with in Section 4 require the use of one-component epoxy foams since, due to process-related factors, the curing reaction can only be started by heating. These foams are examined in the plastics development department of the BMW Group, and requirements relating to material properties are defined.

Table 3-1 shows an extract from the BMW standard for structural foams:

3.6 Lightweight Bodywork – Use of Structural Foams in Hollow Sections

Table 3-1 Requirements regarding the mechanical characteristics of structural foams [2, 3]

Property (after curing 10 min at 165 °C)		Requirements on structural foam	Test conditions
Dynamic material behavior			DMA
Modulus of elasticity	at –30 °C	> 800 MPa	
	at RT	> 700 MPa	
	at 80 °C	> 500 MPa	BMW Group standard: GS 97036 Measurement frequency 15 Hz
Damping (tan δ)	at –30 °C	> 0.020	
	at RT	> 0.015	
	at 80 °C	> 0.030	
Glass transition		> 95 °C	
Creep tendency	at RT	0%	Measurement frequency 15 Hz
	at 80 °C	< 25%	Initial load 20 N
	at 95 °C:	< 30%	Load: 2.3 ± 2.0 MPa
			Test duration: 120 s
Quasistatic material behavior			
Compressive strength		> 15 MPa	DIN EN ISO 527-2
Tensile strength		> 5 MPa	DIN EN ISO 527-2
Elongation at break		> 1%	DIN EN ISO 527-2
General parameters			
Density, unfoamed	Raw material after 20 min at 120 °C	< 1.1 g/cm^3	DIN EN ISO 1183-1
Density, foamed	after 10 min at 165 °C	< 0.6 g/cm^3	DIN EN ISO 845
Linear expansion	–30 °C – 80 °C	< 45 · 10^{-6} K^{-1}	DIN 53 752-B

Figure 3-67 shows how the material system is selected and validated. In general, efforts are made to reduce the number of possible materials by means of static and dynamic tests on test specimens before actually carrying out expensive component and vehicle tests. In the first step, the material properties of density, heat resistance and linear expansion are compared. Despite the low density of the foams, coefficients of linear expansion of less than 4×10^{-5} 1/K are reached. Since temperature fluctuations of more than 100 °C occur when the vehicle is operating, it is important to minimize the difference in linear expansion with respect to steel. In the same way, the foams should only shrink to a minimal extent following hardening within the component.

Due to the temperature load in the roof structure, the heat resistance of the materials must be assured at 95 °C. Many structural foams have similar properties in their static material properties and at room temperature. Taking into account the requirements which apply in vehicle operation, a test was for this reason developed in order to determine the creep tendency (retardation) under temperature.

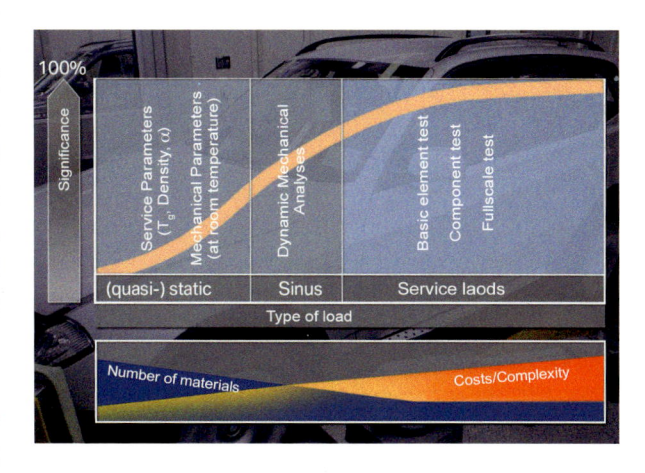

Figure 3-67 Validation of structural foam systems [1]

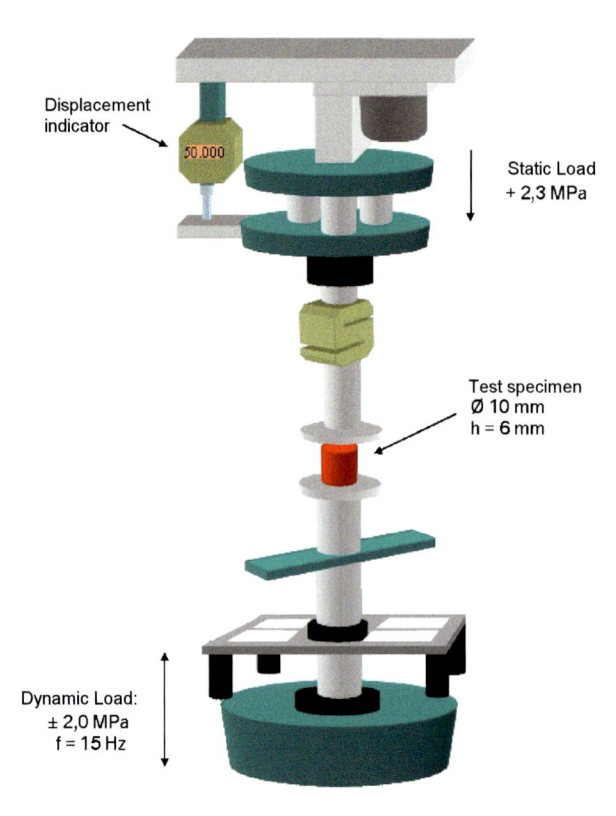

Figure 3-68 Short-term dynamic test [1, 6]

Figure 3-68 shows an example of a test set-up for short-term dynamic testing. The cylindrical specimens are subjected to a dynamic load at a frequency of 15 Hz within the region of repeated compressive stress. Every 30 seconds, the original level is checked again and recorded. Particular attention is paid to examining the influence of the extent of curing, of the fillers and of the foam structure.

Figure 3-69 shows the results of retardation tests on different structural foams at different temperatures.

Since testing uses considerably higher compressive stresses than occur during vehicle operation, it is possible to classify the foams quickly according to creep tendency. Equally important is to consider whether the material properties have been achieved within the process window (production line). Comparison of foaming tests in the production-line sequence and the creep curves measured resulted in the finding that a creep tendency of less than 30% is permissible.

The foam selected is then tested in a basic element (a simple folding, for example) with regard to ageing as well as long-term adhesion and statutory requirements. This means that even emissions standards such as the CARB laws (Californian Air Resources Board) are covered in the requirements profile. In addition, individual special substances are separated out. For example, amine-catalytic structural foams must not have a noticeable odor.

After this come expensive, time-consuming component tests and road tests to confirm the materials selection. In addition to the structural foam components installed, other properties or components of the vehicle are also tested. For this reason it is important to reduce the number of materials to a minimum beforehand.

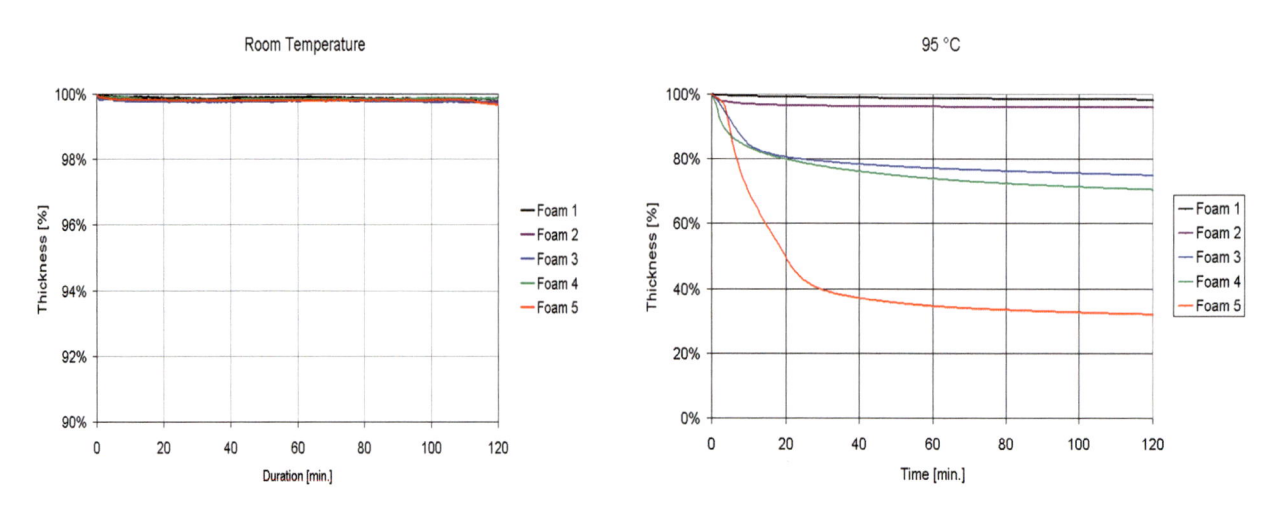

Figure 3-69 Results of short-term dynamic tests at room temperature and at 95 °C [1]

3.6.4 Process Conditions and Implementation

The manufacture of vehicle bodies with structural foam components is an integral part of production-line sequences in the BMW Group. Here these processes secure the corrosion protection, long-term performance and functionality of the structural foam components in the vehicle.

During the course of production, structural foams pass through the following steps:

1. Transportation and storage/logistics chain
2. Installation of the structural foams in the body shop
3. Body shop oven for pre-gelling adhesives
4. Body shop holding area
5. Paint shop pre-treatment/cleaning baths
6. Electrophoretic dip coating
7. Foam curing in the electrophoretic dip coating oven.

The logistics chain for these components covers transportation from the supplier up to arrival at the production network of the BMW Group. Throughout the logistics chain the components are stored and transported at a defined temperature and humidity. Maximum storage time for the components is defined as 6 months below 30 °C. Mechanical loading on the component is prevented by storing in special storage trays.

In the body shop, the structural foam components are inserted or clipped into the hollow sections. Care is taken here that the components are handled in a way appropriate for plastics. Next, the hollow sections are welded together into assemblies. In order to prevent thermal damage to the structural foam component, soldering or inert-gas arc welding work must keep a minimum distance from the foam of 50 mm. In addition to welding and soldering, adhesives are also applied in the body construction shop. Pre-gelling of the applied adhesives then follows – this is to prevent contamination in the paint shop processes downstream. Adhesive pre-gelation takes place in the body shop oven at 120 °C for 12 minutes. As they pass through the oven, the foams must not yet exhibit any reaction or expansion. The vehicle bodies are then held in the body holding facility, after which they are sent to the paint shop. A defined period spent in the holding facility ensures that the structural foam does not absorb too much moisture.

In the paint shop, the car bodies first pass through several pre-treatment baths. These clean the bodies until they are free of metal particles, drawing oils, greases and surface contamination. Here it must be ensured that there is sufficient flushing of the bodywork cavities. This is ensured by a gap of 2 mm between the structural foam component and the hollow section. The design of the components also ensures that no medium is left in the body, as this could subsequently result in bath entrainment and thus to coating faults.

Paint shop pretreatment  | Electrophoretic dip coating

Figure 3-70 Pre-treatment and electrophoretic dip coating in the paint shop [5]

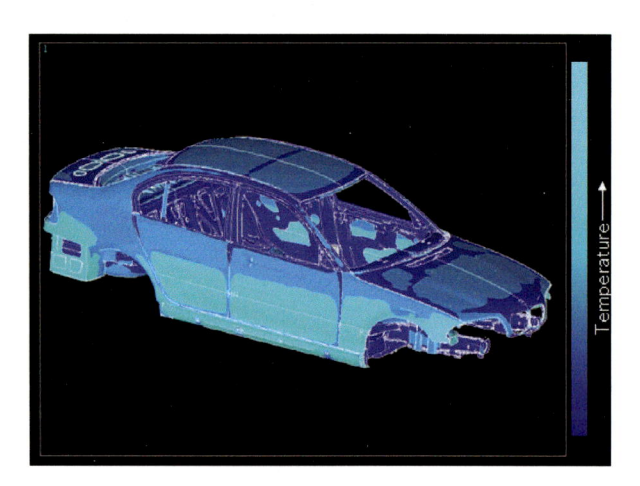

Figure 3-71 Temperature distribution in a car body during passage through the electrophoretic dip coating oven [5]

Following pre-treatment the car bodies run through the electrophoretic dip coating bath where anti-corrosive coating is applied. They next pass through the drying oven where coating, adhesives and structural foams are cured.

Figure 3-71 shows that the different temperature distribution in the body needs to be taken into consideration. The maximum object temperatures vary as the body passes through the oven. Lower temperatures are found in particular at joints in the frame.

For this reason the foams already react after 10 minutes at a temperature of 165 °C although the temperature in the electrophoretic dip coating oven is 195 °C.

The following reactions take place successively in the hollow sections:

1. Drying of electrophoretic dip coating

2. Curing of electrophoretic dip coating

3. Expansion of foam into the hollow sections

4. Adhesion of the foam to the electrophoretic dip coating

3.6.5 Outlook

Structural foams have potential for development in body frame design. A reduction in hygroscopicity of the materials could in the future simplify storage and handling of the components during the production process. These materials could then be used more efficiently in damp areas of the vehicle. Moisture can influence currently used material systems by considerably reducing expansion and compressive strength.

Present-day structural foam systems have rigidities up to 1100 MPa. Due to the further development of the foams with regard to higher moduli of elasticity, it is possible to increase component functionality with regard to their reinforcing effect on the vehicle.

Until now structural foam components have been installed in the vehicle in such a way that they are not bonded directly to the painted outer skin – for esthetic reasons. Before such components can be mounted over large areas of the outer skin, foam systems are required with linear expansion values close to those of steel or aluminum. New applications for structural foams will be accessed as long as there is close collaboration between material producers, processors and users.

References

[1] Brödner, S.: Bewertung des Einsatzspektrums von Strukturschäumen im Karosseriebau mittels dynamisch mechanischer Methoden [Evaluation of the application range for structural foams in body frame design by means of dynamic mechanical methods], thesis, Munich 2005

[2] BMW Group standard GS 97036 'Dynamic mechanical analysis (DMA) of polymers', Munich, 2002

[3] BMW Group standard GS 97061 'Structural foams – requirements and tests', Munich 2005

[4] Otte, W.: Laboratory report BMW Group structural foam, research report, Dingolfing, 1999

[5] Wimmer, W.: Faszination Lackiererei, Presentation, Dingolfing 2003

[6] Deckmann, H.: Two in One – A modern universal flexometer and a dynamic mechanic thermal spectrometer: heat build-up and viscoelastic data with one test, company publication by Gabo Qualimeter GmbH, Ahlden 1999

3.7 Body-in-White turns Black

MICHAEL BECHTOLD

3.7.1 Introduction

In the mid-fifties, three letters assumed a mythical status: SLR. They described a Mercedes-Benz racing sports car, which set new standards for sophisticated sedans cars. With stunning looks, as well as innovative performance and handling technology, the F1 partners Mercedes-Benz and McLaren have once again demonstrated their expertise in the development, design and production of high-performance sports cars.

In terms of body-frame and safety technology, the new Mercedes-Benz SLR McLaren fully justifies its reputation as the leading innovator among modern sedans. High-tech materials derived from the aviation industry are entering into automotive series production.

For some years now, fiber-reinforced polymers (FRP) have proven their worth in Formula 1 and the aerospace industry. The defined targets and objectives with regard to body stiffness, passive safety and lightweight design led to the large-scale specification of carbon-fiber reinforced plastics (CFRP) for the Mercedes-Benz SLR McLaren.

The partnership between DaimlerChrysler and McLaren stemming from Formula 1 was a crucial element in this decision. On the one hand, it enabled the partners to exploit their CFRP experience in vehicle design from Formula 1 and the McLaren F1 road car and, on the other hand, to make use of their automotive know-how originating from volume production. DaimlerChrysler's own CFRP expertise from the fields of process and materials engineering as well as research were also major factors in the decision-making process.

Components manufactured from carbon fibers weigh up to 55 per cent less than comparable steel parts, despite identical strength and stiffness values, and they are some 25 per cent lighter than aluminum components – making CFRP the material of choice for the production of high-performance automobiles. In addition, this cutting-edge lightweight material is characterized by excellent weight-specific energy absorption. The relevant values for CFRP are four to five times higher than those of metallic materials.

With the exception of the front crash structure, McLaren in Woking, England, bore responsibility for developing the complete vehicle. A significant proportion

Figure 3-72 Mercedes-Benz SLR McLaren

Figure 3-73 Body-in-white of the Mercedes-Benz SLR McLaren

Figure 3-74 Body in white without panels

of FRP body elements were manufactured by McLaren and completed with purchased parts to form the entire body-in-white. The overall vehicle assembly was carried out in Woking by McLaren. Support from Daimler-Chrysler was primarily provided in vehicle testing and in the implementation of the required process quality with regard to the manufacturing and assembly of the components.

This Formula 1 synergy on the one hand and volume production expertise on the other hand meant that, when it came to selecting the manufacturing technologies, the degree of automation would receive as much attention as quality and reproducibility. The exclusive use of classic CFRP technologies from F1 and aerospace was therefore out of the question. Although the Mercedes-Benz SLR McLaren represents a series application in an automotive niche, good commercial sense, not just technical feasibility was at the heart of the selection process. This distinguishes the solution discussed here from many previous ones.

3.7.2 FRP Manufacturing Technologies in the Mercedes-Benz SLR Mclaren

Besides the raw material costs for carbon fibers, that are still rather steep, a major hurdle for the use of FRP with continuous fiber reinforcement is the lack of optimized process and plant technology which would allow component production at a reasonable cost. In their component production, McLaren and DaimlerChrysler used technological innovations for the automation of FRP manufacturing processes in series production – some of them for the first time. This explains the use of a whole range of different technologies in one vehicle for the production of components from fiber-reinforced polymers (FRP) such as:

- Prepreg autoclave technology
- Resin Infusion (RI)
- Resin Film Infusion (RFI)
- Resin Transfer Molding (RTM)
- 3D textile preform technologies (braiding, knitting, sewing, tufting)
- Semi-automated manufacturing of preforms from multi-axial fabrics
- Low Density Sheet Molding Compound (LD-SMC)
- Advanced SMC

In this presentation, space constraints make it impossible to fully discuss all details of the FRP manufacturing processes used for the Mercedes-Benz SLR McLaren. The following chapter aims to present a selection of striking developments in the most important FRP processing techniques, the further development of SMC processes and preform technology in combination with resin injection methods.

3.7.3 Prepreg Technologies (Classical and RFI)

The classic prepreg method is the method familiar from aeronautics. In this process, the component is built up sequentially in accordance with its ply structure using woven or fabric mats (impregnated with resin) and then cured in an autoclave. This method is used, for example, for the extremely complex baggage cell.

The RFI method is a further development of the classical prepreg woven mats. Semi-finished fabrics or woven materials are used, whereby resin film layers are positioned between dry fiber mats. The most significant advantage of this method is that a high component quality – such as low pinhole incidence – can be achieved by oven curing alone, with no need to use autoclaves.

Figure 3-75 Male tooling approach

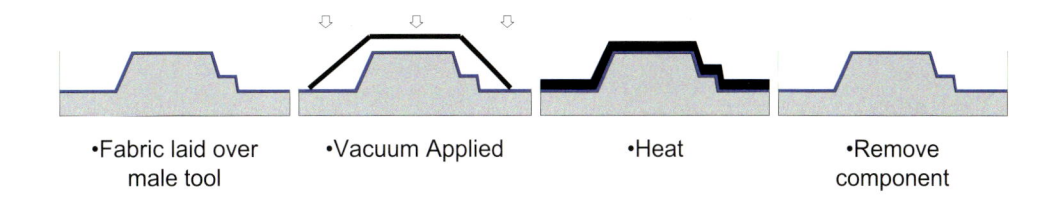

- Fabric laid over male tool
- Vacuum Applied
- Heat
- Remove component

3.7.4 Preform Technologies

In case of preform techniques, a distinction is drawn between sequential and direct preform manufacturing.

Sequential preform technology is used in most preform processes for the Mercedes-Benz SLR McLaren. The individual fabric and non-woven plies of the component are built up layer by layer on the basis of ply design plans.

Figure 3-76 Preforming bench

This preform process is used for the roof frame structure, the spider roof (refer to the section RTM process) and for all outer skin components manufactured using the RI process.

Semi-Automated Sequential Preforming

In order to be able to achieve the high quality requirements for the complex component geometries, McLaren uses a semi-automated process.

During this process, the defined ply structure of the component is placed on the molds positioned on preform benches. A binder was applied to the reinforcement materials already during production. Then a frame with a membrane is placed over the preform mold and a vacuum is applied, lending the preform its desired final contour which is stabilized by applying heat to activate the binder so as to allow the preform to be handled safely.

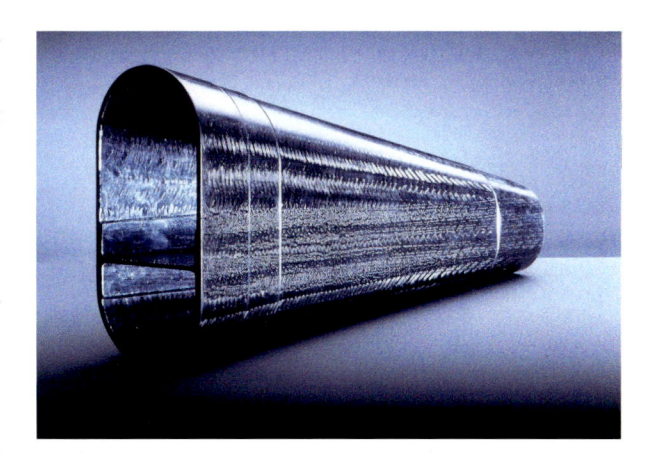

Figure 3-77 Front crash structure tube (FCS tube)

Among the innovations of the Mercedes-Benz SLR McLaren are the longitudinal members of the front crash structure, which are also referred to as FCS tubes. Apart from utilizing the extremely high weight-specific energy absorption potential of carbon structures, the main aim of development DaimlerChrysler at was to achieve as high a degree of automation as possible during the production of the component.

At the beginning of the development, the prototype parts were manufactured using "classical aeronautical technology"; i.e., manual lamination processes and autoclave technology. For the component geometries, which appear straightforward only from the outside, it took about one working day (curing overnight) to produce one component. This was not a major problem during the component design and trial phase, as the manufacturing method selected demonstrated the required flexibility.

Figure 3-78 Braiding machine

But once the geometry of the FCS tube had been finalized, preform production of the 3D textile processes was changed from the sequential to the direct method.

The core process is braiding on a circular braiding machine. A sewn multiaxial material stack is placed between two braiding cores manufactured from cellular plastics. The two ends are then flipped over to form a double "T" profile. This braiding core is then clamped in the circular braiding system and braided. The braided fabric with the core is then clamped in the tufting system, and the side flanks are tufted by a robot. During the so-called tufting process, a needle inserts yarn loops into the braiding core normal to the braid. For the component manufacturing process proper, the braiding cores are removed, the preform is inserted into an RTM mold core, the RTM mold is closed and the resin is injected.

Despite the highly complex geometry of the FCS longitudinal member, cycle times of 12 minutes are achieved for the braid on a fully automated computer-controlled circular braiding machine.

Resin Injection Processes RTM and RI

Besides the FCS tube already discussed, the so-called spider roof is produced using the RTM process. This is probably the most complex CFRP body-in-white module ever manufactured using an RTM production process. The spider roof includes the frame structures of the windshield, roof, A-pillar, C-pillar, door cutout and rear window in one CFRP module.

Once the individual RTM mold halves of the spider roof have been covered with ply stacks (sequential preform process), the cavities are filled with foam. In this process, EPP (expanding polypropylene) foam particles are blown into the cavity under high pressure. This ensures that a stable inner foam core is formed which can stabilize the ply stacks in the RTM mold. The resin can then be injected using the customary RTM method.

The RI process is used for all outer skin components. In contrast to the RTM process, resin is not injected into the mold under high pressure, but slowly flows into the mold cavity under the influence of the vacuum created. Once the preform has been completely impregnated, the preform is cured in the oven while maintaining the vacuum.

Figure 3-79 Spider roof

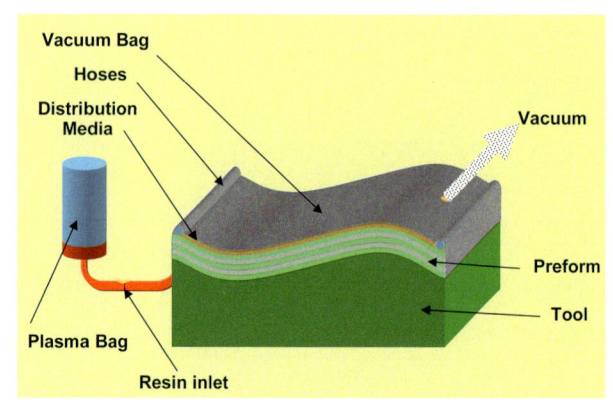

Figure 3-80 RI concept

Sheet Molding Compound (SMC)

SMC has been in use for a long time in the automotive industry. Compared to aluminum, which has established itself more and more in automotive engineering, classic SMC suffers from significant disadvantages with regard to weight. In order to be able to make use of the advantages of SMC, such as

- high temperature resistance
- manufacturing of complex component geometries with results close to the final contour
- lower thermal expansion coefficient than aluminum and greater design freedom

for the Mercedes-Benz SLR McLaren, despite the weight issue, two SMC lightweight versions were used. Apart from low density SMC at the front side panel, Advanced SMC is used for the first time in series production for the rear scuttle panel.

For low density SMC (LD-SMC), part of the calcium carbonate filler is replaced by glass microspheres. This results in a reduction of the material density by 20% to 1.5 kg/m^3. The proportion of microspheres was not increased further, despite further potential for reducing density, in order to ensure the required surface quality. For non-visible semi-structural applications, a weight reduction of up to 30% compared to Standard SMC could be achieved by increasing the glass microsphere content.

As LD-SMC is based on the Standard SMC formulation, the side panel did not require new developments in plant technology, process parameters and mold technology.

The manufacturing technologies for the production of carbon parts discussed so far are a first step towards somewhat higher unit numbers. Still, there is a long way to go until the volumes typical of the automotive industry can be produced. SMC technology has the potential for closing this gap. So why not use carbon fibers to replace the glass fibers in SMC? It should be taken into account, however, that this version, also known as Carbon SMC, results in reduced properties and a significant deterioration of the lightweight design potential, since the carbon fibers are chopped. This is a disadvantage that cannot be easily accepted, given the high material cost of carbon. If carbon is to be used, then the full weight-saving benefits should be enjoyed. Advanced SMC can provide a solution for small automotive series.

With regard to this process, too, the Mercedes-Benz SLR McLaren represents a platform for innovations. Although Advanced SMC is conceived for significantly higher volumes than required here, a semi-structural part, the so-called rear scuttle panel, is produced using Advanced SMC – the first production part to be manufactured using this technology. In this technology, unidirectional carbon fibers are processed using the compression molding technique.

Figure 3-81 LD-SMC microstructure (100x)

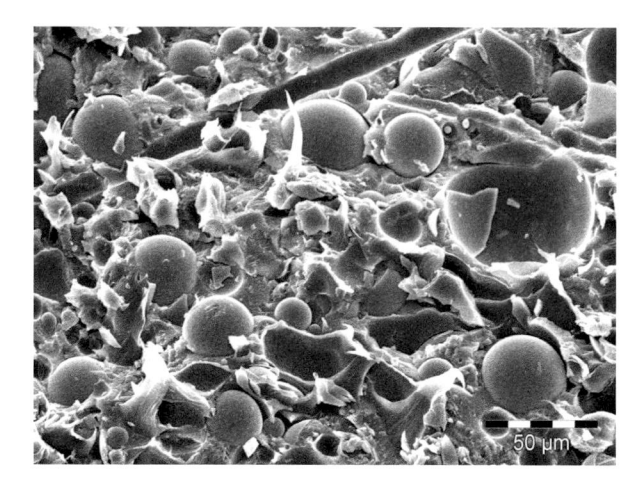

Figure 3-82 LD-SMC microstructure (500x)

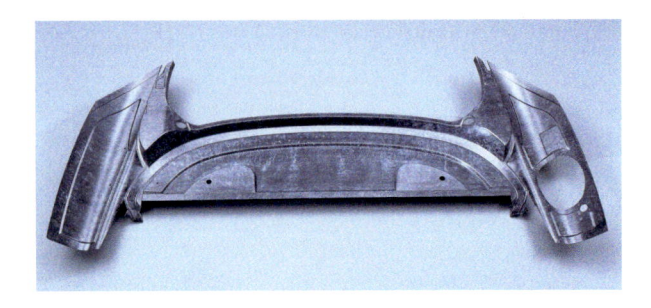

Figure 3-83 Advanced SMC rear scuttle panel

The Advanced SMC process already presented during last year's VDI meeting can be broken down into intermediate product, preform and component production. Thin carbon fiber tapes are placed on two carrier sheets impregnated with resin. The intermediate tapes are then wound onto rolls. In accordance with the ply structure previously specified for the rear scuttle panel, a varying number of intermediate product layers with the different fiber orientations required are combined into a preform. The flat preform covers approx. 85% of the subsequent component area.

Due to the low volume requirement for the Mercedes-Benz SLR McLaren, preform production has not been automated.

During component manufacturing proper, the preform is compressed in a heated compression mold at a temperature typical of the SMC process for a cycle time of approx. 3 minutes and cured. Breakthroughs and boreholes are then cut using a computer-controlled water-jet cutting system.

3.7.5 Summary

The use of FRP structures (in particular when using carbon) in small automotive production volumes is at present still hampered by two main obstacles. The first one is the very high raw material price, while the other one is the available production technology. Although the Mercedes-Benz SLR McLaren is a series application within an automotive niche, the project did not just focus on what was technically feasible. Quality, reproducibility and automation in production were important criteria in the selection of the FRP processing technology. McLaren and DaimlerChrysler are therefore using new technological developments in FRP manufacturing processes for the Mercedes-Benz SLR McLaren – some of which have never been used before in series production processes.

Innovations such as textile and semi-automated preform technologies, Advanced SMC, LD-SMC or RTM offer the technical and commercial potential required to lift CFRP applications from their niche to the next higher volume range of the small automotive series. To what extent FRP technology, which is still in its infancy, can be successfully scaled up to small series production will largely depend on the innovative will and power of OEMs and the supplying industry. The Mercedes-Benz SLR McLaren is a first step in this direction.

3.7.6 Acknowledgements

Due to the high number of development partners and suppliers involved in the development of materials, production processes and plants and in the component production, it is not possible to thank those individuals by name in this contribution. We would nevertheless like to express the gratitude of DaimlerChrysler and McLaren to all those companies and associates who were involved in the vehicle project and with their work made significant contributions to the realization of an extraordinary sports car.

3.8 Innovative Manufacturing Processes Adapted for the Production of Structural Modules

Philippe Dumazet

3.8.1 Introduction

In-line compounding technologies are very promising for several reasons. First, they permit each part to obtain a good ratio between performance and price on the basis of a compound tailored and adapted to each specific application. The flexibility of these technologies leads moreover to a very promising future based on new technical compounds. These compounds will be based on new matrix (PA, PET, etc.) combined with new fibers (natural, carbon, etc.) or fillers (talc, etc.).

3.8.2 The Evolution of Production Technologies for Structural Parts

For twenty years, production technologies for structural parts have led to a lot of innovations. The first answer was based on compression molding technology with a thermoset material (SMC) (see Figure 3-84).

The first evolution permitted thermoplastic material to be used for compression molding (GMT) leading thereby to reductions in density and cycle time. Faurecia then decided to develop a new technology called XRE [1] that leads to a cost reduction by in-line compounding of the raw materials (PP and glass fibers). In order to answer function integration requirements positively, Faurecia developed the XRI technology [2]. Indeed, this XRI technology enables combining the advantages of in-line compounding (cost, flexibility) with the advantages of injection molding (function integration, no extra work). The standardization of this technology was decisive for expanding this area of applications. Therefore, Faurecia and Krauss-Maffei decided to work together.

On a general point of view, the main target of the in-line compounding approach is to avoid the use of semi-finished products such as GMT and long-glass-fiber granules (see Figure 3-85).

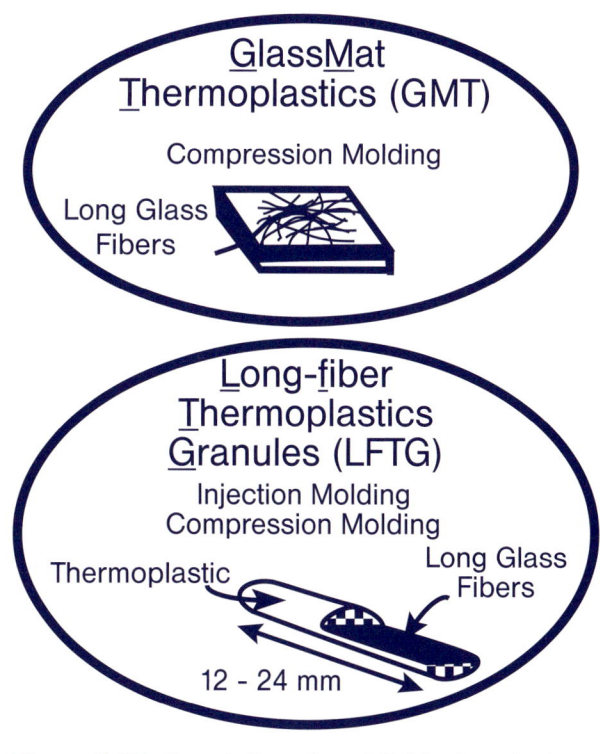

Figure 3-85 Description of semi-finished products

Figure 3-84 Evolution of production technologies for structural parts

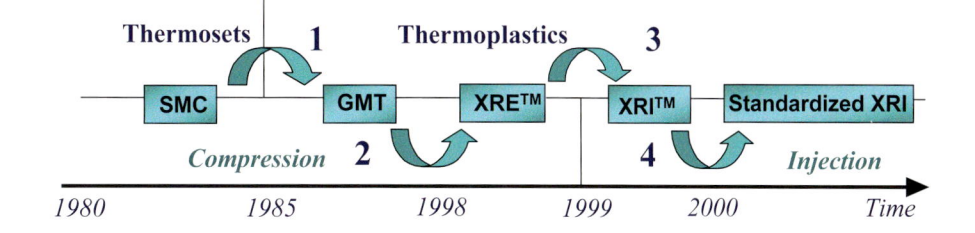

3.8.2.1 XRE Technology

The principle of the XRE technology patented by Faurecia is described in Figure 3-86.

On the basis of a single step process, this technology permits raw materials to be compounded and the part to be produced by compression molding. Since 1998, this technology has been used in industrial volume production of the Peugeot 206 front-end (see Figure 3-87).

3.8.2.2 XRI Technology

When structural modules are involved, one of the main requirements of the automotive industry that must be taken into account is their functional integration. In this area, injection molding technology, of course, plays a major role. Faurecia decided, therefore, to combine the advantages of in-line compounding of the XRE (cost, flexibility) with the advantages of injection molding technology. XRI technology is described in Figure 3-88.

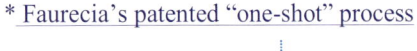

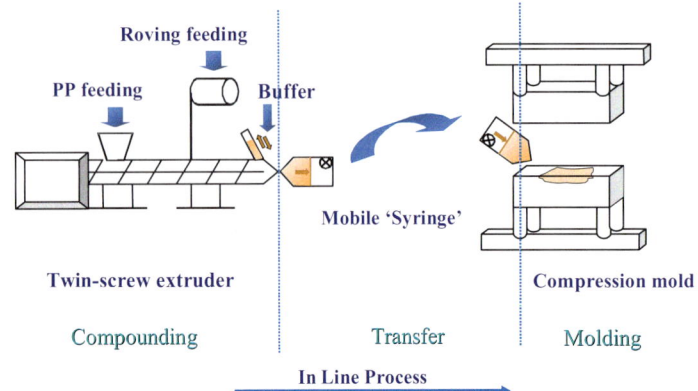

Figure 3-86 Diagram of XRE technology

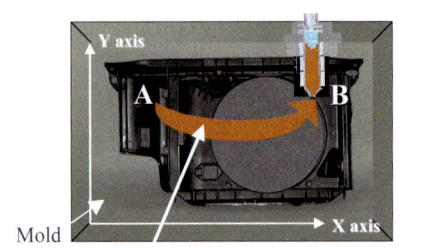

Figure 3-87 Front end panel of the Peugeot 206 (XRE technology)

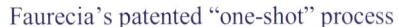

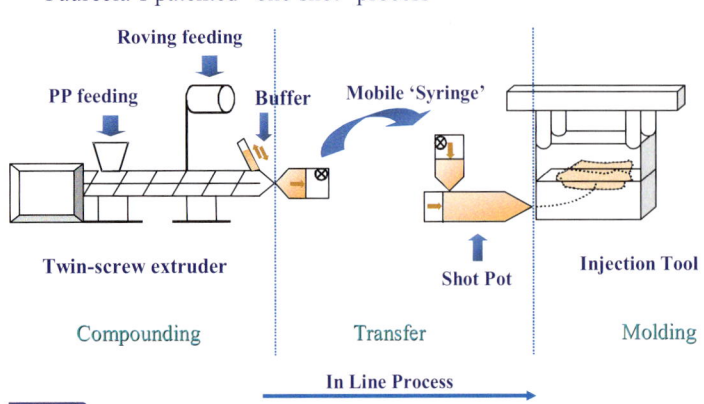

Figure 3-88 Diagram of XRI technology

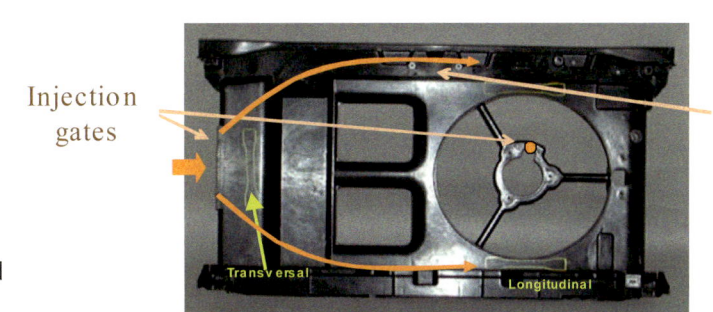

3.8 Innovative Manufacturing Processes Adapted for the Production of Structural Modules

Figure 3-89 Presentation of the Peugeot 307 front-end (XRI technology)

This technology is currently used on an industrial basis to produce the Peugeot 307 front-end (2000 parts/day) (see Figure 3-89).

In order to expand this innovative and promising technology, it was then paramount to standardize it. Faurecia and Krauss-Maffei decided therefore to work together. The result is the XRI-IMC concept illustrated in Figure 3-90.

The main principle remains the same: combination of a continuous extrusion system with discontinuous injection molding. This technology is currently used to produce front-end-modules for several vehicles (Citroen C2 and C3, Peugeot 1007 and 207, Audi A3). The transfer of the compound must be gentle enough to maintain good glass fiber length in the part. This is essential to obtain good mechanical properties (see Figure 3-91).

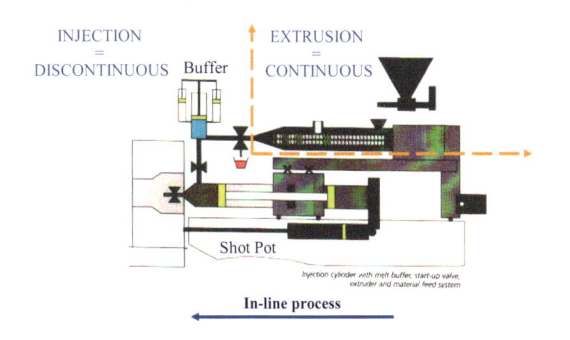

Figure 3-90 XRI-IMC concept

3.8.3 Developments of New Applications

On this basis, we are now investigating several approaches.

The first one consists in expanding the field of applications of the PP glass fiber solution, for which Faurecia has performed tests on an IP carrier. The first tests involved monitoring the composite properties and analyzing specimens cut from the full-scale component (see Figure 3-92).

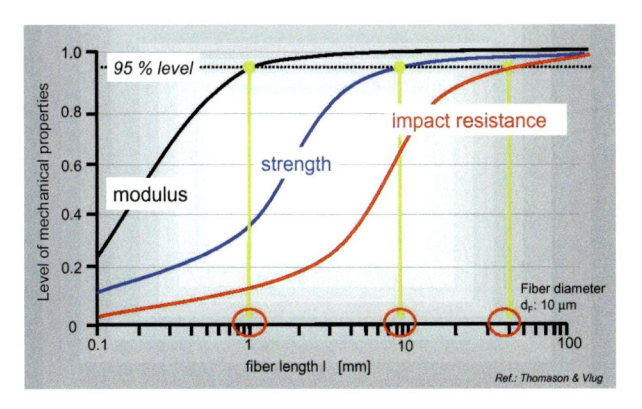

Figure 3-91 Evolution of the mechanical properties of a composite as a function of the glass fiber length

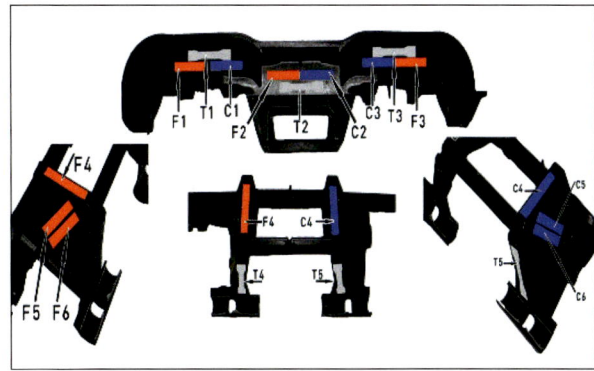

Figure 3-92 Specimen analysis based on a full-scale component (IP carrier)

- **Fullscale component profile**

  after falling dart impact (60Kg, 3m/s)

	PPGFL copolymere	I.P. XRI formulation
Energy absorbed (J)	268	296
Maximal effort (N)	13000	12100
Acceleration max (g)	22.4	24
Intrusion (mm)	35.5	38

PPLGF structure after impact

Copolymer structure after impact

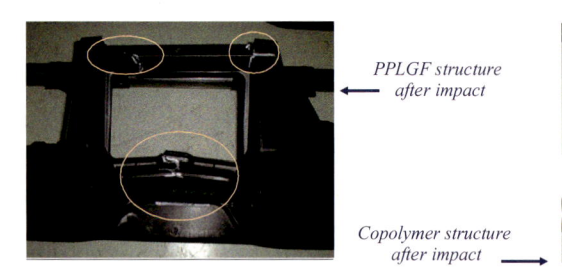

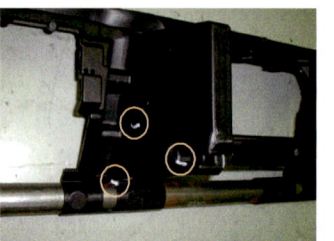

Figure 3-93 Falling dart impact test on IP carrier (comparison of long-glass-fiber granules and XRI solutions)

In order to compare the properties of LGF 40% – PP granules and XRI-IMC technology, tests were conducted in parallel on the same IP carrier presented above (Figure 3-92).

The results of a falling dart impact test (60 kg, 3 m/s) showed that component behavior based on the XRI-IMC technology is at least comparable to the alternative based on the LGF-PP solution (see Figure 3-93).

3.8.4 Future Potential of a New Process and Machine Technology

Besides the already discussed fiber reinforcement by long-glass fibers, a single-stage reactive compounding and injection molding process was developed in cooperation between Krauss-Maffei Kunststofftechnik and the Institute for Plastics Technology at the University of Stuttgart [3]. In this process, natural-fiber-reinforced PP composites (long fibers) and their molding are performed in an injection molding process (see Figure 3-94).

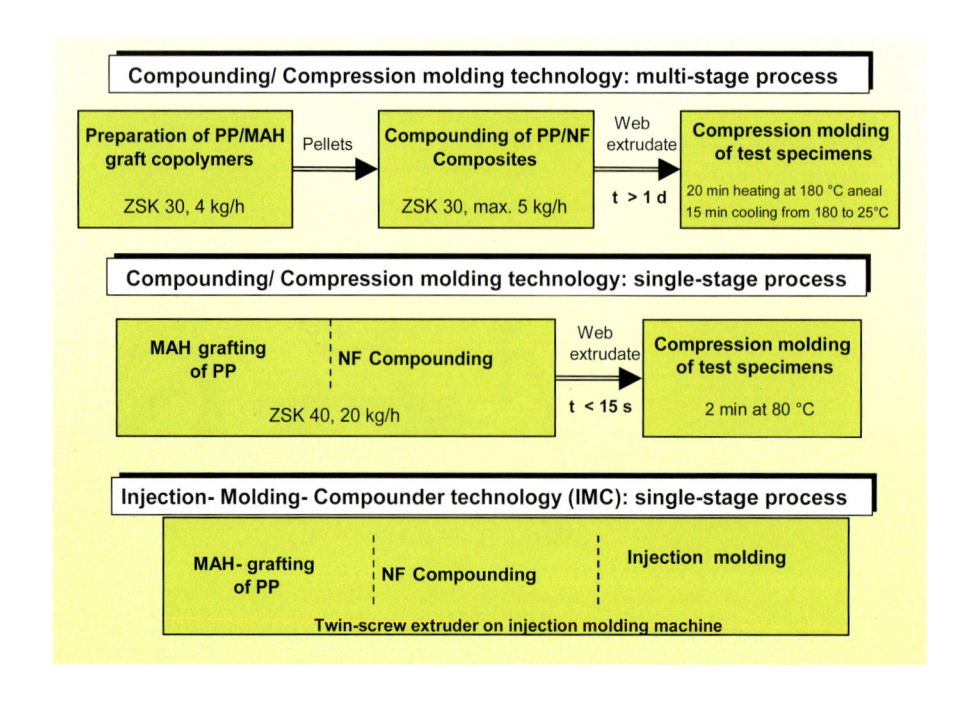

Figure 3-94 Generating NF-composite parts in a single-stage process [3]

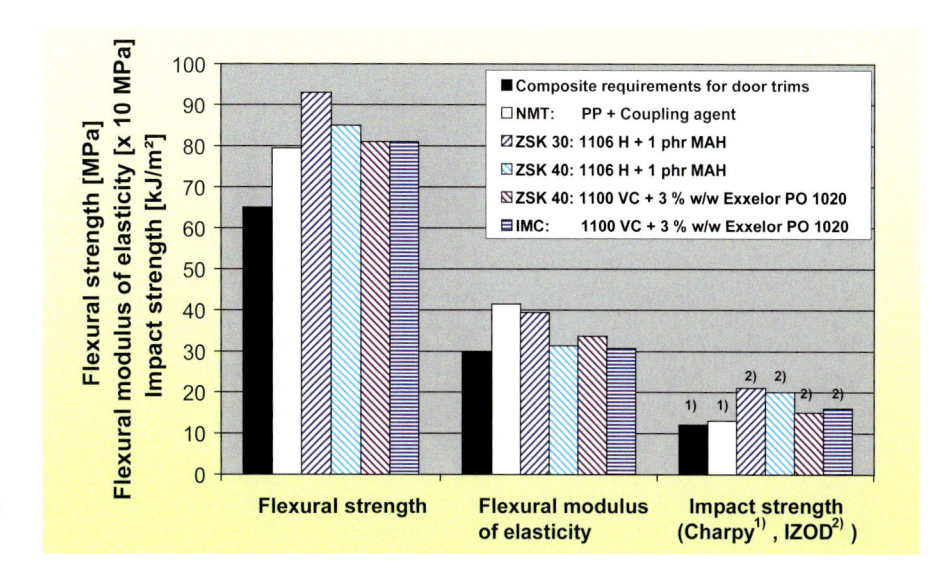

Figure 3-95 Mechanical properties of newly generated PP/flax fiber composites (30% w/w flax) [3]

The problem of the smell was solved by producing natural-fiber reinforced parts in "first heat" using a direct process. Taking flax and polypropylene as the basic materials, matrix functionalizing with the bonding agent molecule MAH (maleic anhydrid) was developed together with subsequent integration of the flax fibers in the co-rotating twin-screw compounding unit (a part of the injection molding machine). The subsequent injection molding process can follow in the same heat because of the new machine concept.

The aim was to thoroughly fulfill all relevant criteria of the automotive industry for such composites (see Figure 3-95).

In order to demonstrate the potential of this processing technology in practice, a pilot plant for the production of exterior parts was developed for another project – the K-show 2001 – in partnership with industry companies (see Figure 3-96) [4].

The most important aspects of this development were (see Figures 3-97 and 3-98):

1. the production of a long-fiber composite with a matrix made of technical thermoplastics and glass-fiber rovings,

2. the back-injection of a painted film in Class A quality,

Figure 3-96 Boot cover (fictitious part, edges not cut)

3. the preparation of the film by thermoforming as well as trimming by laser technology and

4. the creation of controlled conditions (clean room) in the important processing steps to fulfill the requirements on surface quality.

The specific potential of this machine technology is being further investigated in extensive projects with the industry and universities. The aims are, apart from the integration of reinforcing fibers, also its filling and/or functionalizing. Examples for this are a high filler content for noise reduction or generation of conductivity of high-performance polymers, but also the magnetizing of plastics by including high concentrations of iron.

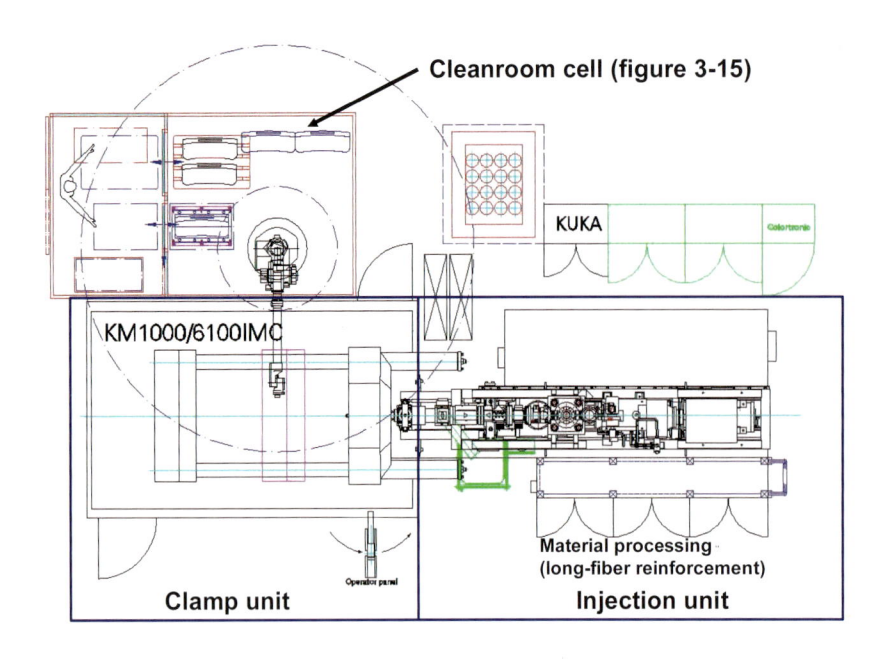

Figure 3-97 Plant layout

Figure 3-98 Clean room cell with handling device

In the future, polymer-blend technology will be an interesting field of application. For cost reasons, but especially in order to gain a time advantage in sample and pilot plant production, product-specific formulations of the compound can be created together with the raw material supplier directly on the injection molding machine. In this way, several steps are avoided for the composition and production of the compound. This time-saving advantage will play an increasingly important role in future.

Based on the new plasticizing and/or processing concept on the injection molding machine, new applications are feasible that were impossible to achieve with the injection molding process in the past. E.g., new materials characteristics can be created by reactive blending with co-continuous thermally unstable phase structures and then molded, which would otherwise be impossible to achieve.

In general, this process technology is of advantage where thermally sensitive material components have to be integrated. The process steps "compounding" and "injection molding" – in one heat – are directly in-line and offer decisive advantages [5].

A company that uses this process in the future will act simultaneously as a quasi compounder and injection molding processor. It might even be responsible for the quality of the material in some cases. In order to not fully transfer this responsibility to the processor, very close co-operation should be striven for with the raw materials industry. One idea would be for the raw materials supplier to offer certain formulations as so-called additive packages that can which are then filled into the injection molding machine (IMC) and/or mixed into the melt. A new approach may be required for this procedure.

Quality co-ordination should ultimately remain in the hands of the raw materials supplier and the part producer. To achieve this, a procedure must be defined in the future that is acceptable for all parties and that enables quality control and documentation directly at the point of production of the part (at the machine).

The degrees of freedom offered by this process technology should mainly serve to increase flexibility at the production site as well as to process difficult, thermally sensitive material components and to create new materials characteristics. Economic aspects will be an additional moving force behind this technology.

References

[1] Dumazet, P., Buron, M. P., Vallet, Y.: An innovative processing answer for low cost and high performance polymer composites, SAE 2001 World Congress, Detroit, USA, March 5–7, 2001

[2] Dumazet, P., Buron, M. P.: XRE et XRI: Procédés des transformation économiques adaptés à la réalization de pièces composites structurelles Allègement en Carroserie Automobile, Chalon sur Soane, 12. and 13.06.02

[3] Ruch, J., Fritz, H.-T. et al.: Innovative Direktverarbeitung von Naturfasern, *Kunststoffe* 92 (2002)2, p. 28–34

[4] Jensen, R., Mitzler, J. u. a.: Synergie schafft neue Technologie, *Kunststoffe* 91 (2001) 10, p. 96–102

[5] Jensen, R.: Synergien intelligent genutzt – IMC-Spritzgießcompounder erhöht Wertschöpfung, *Kunststoffe* 91 (2001) 9, p. 40–45

3.9 Innovations in Injection Molding for the Automobile of the Future

GEORG STEINBICHLER

3.9.1 Introduction

Plastics will continue to leave their mark on the conception and appearance of new automobile generations. The availability of plastics with the required property spectra, as well as a technically and economically interesting choice of combinable process technologies will play a leading role in realizing development targets for improving looks, comfort, safety and lightweight design, as well as quality and cost-cutting. Competition among the best materials for automobile manufacturing will further encourage innovations in plastic processing and in injection molding in particular.

This paper will attempt not only to describe the existing potential of new process technologies or technology combinations for applications in the automobile, but also to present some ideas for implementing future solutions. This will take the form of a brief tour of the automobile, starting with the interior, proceeding via exterior paneling and finishing with individual functional elements (Figure 3-99). Particular attention will be devoted to techniques for manufacturing plastic composite and hybrid designs for property-optimized large-area components. Material composite designs mean that cost-effective soft-touch surfaces, flexibly changeable design components, low-stress glazing

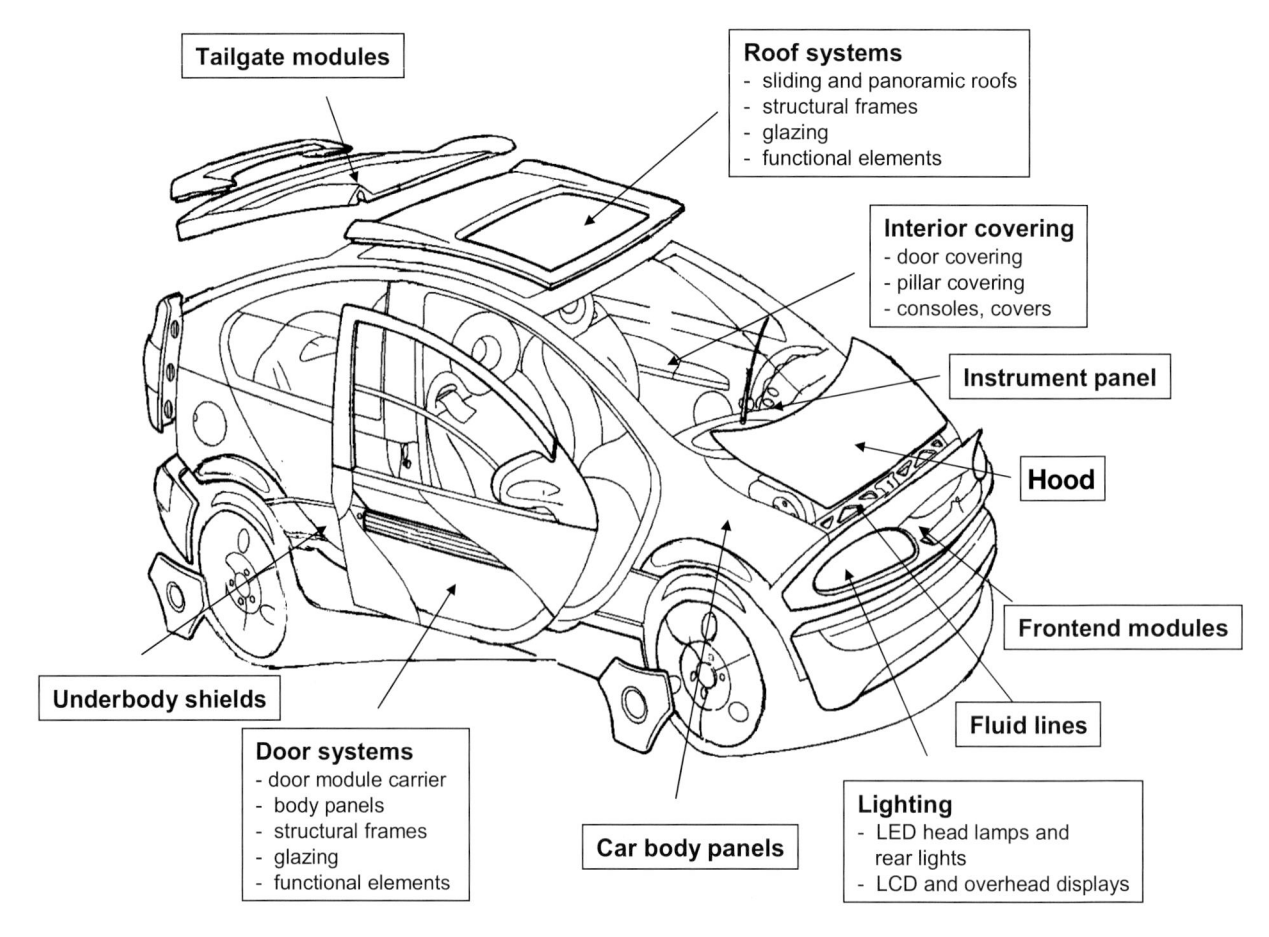

Figure 3-99 Systems, modules and functional parts with growth potential for plastics made possible by the innovative exploitation of new process technologies

components and also skin elements without sink marks for tailgates and doors, for example, can be manufactured in lightweight design. In addition, the possibilities of integrating further functions such as antennas will be discussed. Companies doing engineering work for the automotive industry in these areas of application will thus be able, even in the future, to strengthen their position, expand further and in many cases perhaps even create markets with innovative solutions. The important things here are to concentrate on technological core competences system among the parties involved that is clearly oriented towards benefiting the customer [1].

3.9.2 Innovative Areas for New Process Technologies

3.9.2.1 Door Systems and Modules

Due to new freedoms in esthetic design, the trend in developments of door systems is moving increasingly towards module carriers made of plastic [2]. In addition to its function as carrier for power windows, door locks, loudspeakers, motor and electronics, the module often also acts as water shield (Figure 3-100). Associated advantages are a reduction in assembly time, weight savings of up to 25% compared with sheet steel designs, improved sound insulation and also possibilities for integrating visible areas whereby parts of the cover panel could be omitted (Figure 3-101) [3]. Due to the required collision properties, plastic module carriers are made of long-fiber reinforced polypropylene. In order to prevent a great deal of fiber damage during the forming process, the method used is injection-compression molding with a good distribution of melt by the cascade technique via large needle-valve hot-runner nozzles and moveable cores in the mold to make cutouts (for loudspeakers, for example).

Particular flexibility in the manufacturing process is offered by the use of PP continuous-strand concentrates with a glass content of 75% by weight, which, depending on the application in question, are diluted with unreinforced PP via a gravimetric feed unit, this being carried out directly at the screw plasticizing unit. Door-module carriers with especially high flexural rigidity and low weight can be manufactured using a new precision air technique and chemical or physical blowing agents. The sandwich construction with integral foam structure (compact outer skin and foamed core) can be influenced via process control.

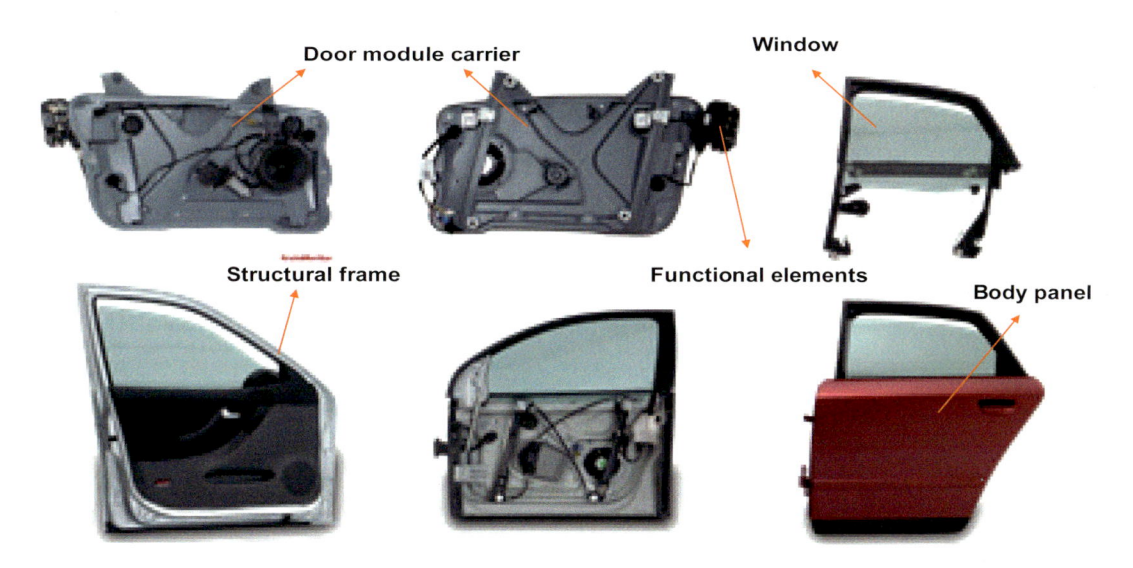

Figure 3-100 Door system as multiple-material composite part with high functional integration consisting of HIP plastic door module (highly-integrated plastic) with glass motion module (GMM) by ArvinMeritor, which replaces the top half of the door and consists of frame, window channels, glass, drive and center-piece reinforcement [2]

Figure 3-101 Door module concept with visible areas integrated into the module carrier, manufactured by the Polytec Group [3]

It is also possible to carry out partial modification of properties via foaming, or by reinforcement with fabric or continuous strands and by forming reinforcement profiles by the fluid injection technique. Until now, door interior panels have been decorated by in-mold laminating and back-molding using textiles, expanded films and so on. Here the component is shaped and the composite fabricated with the surface decoration material in a single-step process without subsequent adhesive laminating. In certain special cases, simultaneous fabrication of the composite with different decorative materials on a component is even possible without the use of adhesive. Single-material systems – that is, carrier and decorative material from a single group of materials, such as polypropylene or thermoplastic polyester – is being discussed with respect to recycling. New possibilities are also being opened up by foaming-on or injecting-on thin soft-touch layers made of polyurethane or thermoplastic elastomers while simultaneously creating surface-texture structures on door trim panels.

Cost-effective lightweight door designs are feasible, for example, when shell and structural-shape designs are combined intelligently. Here, semi-finished structural shapes close to the finished component – so-called "tailored tubes" – and made of steel, for example, can be combined in hybrid designs with shell-like plastic components for use as exterior paneling [4]. The new and further development of low-heat joining techniques is taking on considerable importance as regards hybrid constructions of this kind [5]. Design studies have also been made where transparent plastic elements are used in the door. However, there is currently no question of replacing glass in the moveable side window.

3.9.2.2 Front-End Module

These days, plastics are used for the form and function of the module carrier while sheet steel is used for its crash strength. Until now, plastic-metal hybrid solutions of this kind have been created by mechanical joining methods such as rivets or encapsulating by insertion in an injection mold. One example of a new developments is an injection molding made of long-fiber reinforced polypropylene bonded to an electrophoretically coated metal reinforcement [6]. Here, in contrast to conventional procedures, closed box-sections are used (Figure 3-102).

Figure 3-102 Bonded hybrid carrier for the front end module of the VW Polo consisting of PP-LGF injection molding with bonded-in metal box-section reinforcement. The new hybrid component is 25% (1.5 kg) lighter than its molded-in predecessor [6]

Using the box section and a continuous bonded joint produces higher rigidity combined with lower weight. Adhesive can still be applied and the parts joined in the automated injection-molding production cell. When a special adhesive is used, the material composite can be fabricated without pre-treating the PP plastic surface, such as flame treatment, plasma treatment or application of a primer coat. The bonded version has three further advantages. Risk of damage to electrophoretically coated metal parts is lower, since the metal sheets do not need to be placed in the injection mold; sealing with close-toleranced mold inserts is not required for encapsulation; and warpage adjustment by preventing shrinking is not necessary.

A new highly-integrated front-end module based on a Bodyflex concept offers a seamless flexible vehicle front end and higher design freedom than previous modules [7]. This concept is based on a functional expansion of the headlight outer lens cover which also serves as a bumper overlap and on the use of flexible seals between the outer lens cover and the headlight housing (Figure 3-103).

The large-area outer lens covers made of polycarbonate are only transparent where this is necessary for the light and electronics functions located behind them. The rest is painted in the body color. This eliminates the need for the usual joints between headlight and bumper. These two exterior parts, in conjunction with the center section, form the entire bumper overlap. Possible candidates for future realization of this Bodyflex concept include multi-component composite injecting molding and in-mold lamination of multilayer films with a scratchproof coating.

An important element in the front-end module providing improved pedestrian protection is the injection-molded pedestrian protection cross-member made of polypropylene with high energy absorption potential. In a head-on collision with a pedestrian, its positioning underneath the bumper causes the pedestrian's body to rotate and slide over the hood of the vehicle instead of being run over. A second load path for absorbing collision energy is created by means of integrated crash boxes made of EPP foam.

Figure 3-103 New highly-integrated front-end module manufactured by HBPO with polycarbonate headlamp outer lens cover which, in minimal installation space, simultaneously acts as a single-part bumper overlap and together with the polypropylene pedestrian-safety cross-member meets statutory requirements relating to pedestrian protection [7].
(1: Headlight outer lens cover;
2: Headlight housing; 3: Flexible seal;
4: Pedestrian-protection cross member;
5: Cooling module support; 6: Cross-member)

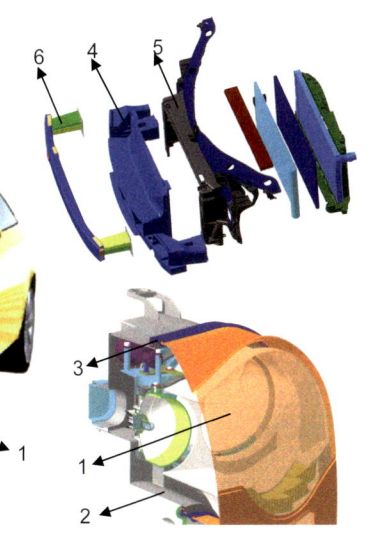

Compared with the simple roll-over protection components made of metal that were previously standard fittings in vehicles, the polypropylene plastic design results in savings of 25% in weight and 40% in assembly time. The microstructure foaming technique described in Section 3 might also be of interest for the cost-effective fabrication of the pedestrian protection cross-member. The cross-member consists of a weight-optimized dog-bone steel profile or is manufactured from long-fiber reinforced polypropylene by a fiber-friendly injection-compression or deposit compression molding process. Due to denser ribbing at the points of mounting on the vehicle frame, the PP+LGF cross-member can absorb energy up to an impact at 15 kph. Potential for tailoring the properties of cross-members in future designs is offered by the fluid injection technique for producing integrated hollow sections or by fiber mats and fiber tapes inserted into the injection or injection-compression mold by a robot system.

3.9.2.3 Hoods and Tailgates

Concepts for modular hoods and tailgates were presented at the last IAA 2005 in Frankfurt. To save on cost and weight, thermosets and thermoplastics should in the future be combined to produce components (Figure 3-104). An internal structure made of SMC is glued to an outer skin made of talc-filled polypropylene [8]. This hybrid tailgate, which is painted off-line, saves up to 50% by weight in comparison with conventional steel structures. It also means a reduction in investment costs and offers a high degree of design freedom [9]. Cost-effective lightweight designs are also possible when shell and structural-shape types of construction are combined intelligently. Here, semi-finished structural shapes close to the finished component – so-called "tailored tubes" – and made of steel, for example, can be combined in hybrid designs by gluing on shell-like plastic parts to form the vehicle outer skin. An economically efficient and low-warpage method of manufacturing large-area outer body parts is injection-compression molding with the cascade technique and parallelism control.

Integration of Antenna Systems into Plastic Body Components

Considering the constantly increasing number of one-way or two-way communication services available in modern vehicles, such as analogue and digital radio and television, telephone, satellite navigation, mobile telephony and the coming telematic functions, not to mention remote central locking or keyless entry and "keyless go", all using different receiving frequencies, there is a danger that automobiles will turn into a "forest of antennas" with a major effect on the design.

Figure 3-104 Hybrid tailgate and hood consisting of an SMC internal structure and a glued-on outer skin made of talc-filled polypropylene

This will be hard to avoid with conventional technology, since in the moving vehicle all information coming from outside or exchanged with the outside will have to be transferred by wireless communication paths – in other words, antennas.

The solution is antenna integration. Integrated antenna systems have their antenna structures, including the corresponding electronics, hidden inside or beneath the outer skin of the vehicle. This does not affect the appearance of the vehicle. In principle, antennas can be integrated into any non-metallic material used in the vehicle. It should, however, be noted that all integrated antenna systems have a higher interaction with the vehicle body. This means that the shape of the vehicle has considerable influence on antenna behavior which in turn means that antenna systems must be tailored to the individual model of automobile.

Antennas or the antenna structures of integrated receiving systems can, with the aid of foil wiring sets, be mounted directly on body parts made of plastic, such as windows, roofs, tailgates, wings, bumpers and so on (Figure 3-105). Here the antenna structures are first printed onto a special film, after which the in-mold method is used to integrate the pre-assembled film into the body component during the molding process without the need for a subsequent assembly operation [10].

The flat foil wiring set is precisely positioned in the mold by a robot system and held in the mold by mechanical means, such as pins, a vacuum or an electrostatic charge. When the melt flows into the cavity, the foil wiring set may be shifted or underwashed if no suitable flow pattern preparation and selection of the injection points (number and position) have been carried out. To prevent the thermoplastic backing film being washed away by the melting process, one process control method has proven itself which only subjects the film to the briefest and lowest possible temperature especially at the injection point. Particularly advantageous here are the low melt and mold temperatures as well as the short injection and holding-pressure times. The risk of film damage is also reduced by use of the injection-compression process if the melt is injected

Figure 3-105 Tailgate component with integrated antenna structures, made by film insert molding [10]

into a mold slit which is opened in addition over the thickness of the molding. This minimizes the shear stresses occurring at the film surface when the melt flows in and reduces the pressure requirements for the filling process.

Instead of visually bothersome sunlight blockers, films can be used which are transparent in their normal state. These films contain electrochromatic substances which turn dark when an electrical current is applied at the touch of a button.

3.9.2.4 Roof Modules with Large-Area Plastic Glazing

The roofs of our automobiles are currently being reinvented. Automobile manufacturers are taking an increasing interest in plastic roof modules pre-assembled completely off-line and supplied on a just-in-time basis to the assembly line. The major advantage in this is seen to be the fact that the vehicle interior remains accessible from above, thus simplifying assembly operations considerably. The roof systems are designed as a top-load system (in other words, to be installed from above) and are delivered ready to install. On the production line they are then glued onto an adapter frame welded into the body. This kind of system makes it a simpler matter to build further module versions with individual design features. This approach also makes aftermarket versions requiring little conversion work conceivable.

New panorama roof systems impressively demonstrate the advantages – in addition to weight savings – that plastics offer in this area of application. They open up enormous freedoms in aesthetic design and allow

Figure 3-106 Panorama roofs as a new design element with additional functional integration

designers and engineers to integrate a large number of functions in a single component. What all convertible drivers enjoy as a matter of course can now be extended to people driving sedans: an unobstructed view of the sky. With the new panorama roofs, the classic sliding sunroof is redefined. All the same, the open roofs of today share only the basic principle with the old sliding roof. Frequently we have large areas of glazing, thus not only providing ventilation but also functioning as an important design element and, thanks to all the incoming light, creating a new feeling of spaciousness. Roof boxes integrated into the roof system can provide the vehicle occupants with much more stowage space (Figure 3-106).

In the implementation of these new panorama roof systems, a strong trend towards replacing glass with polycarbonate can be seen. Freedom of design in particular and also the potential for weight reduction in the large glazed areas are decisive arguments here. An economically efficient and low-stress method of manufacturing the large-area glazing components is injection-compression molding. With the sandwich-molding technique, first a transparent polycarbonate window is produced and, in the second shot, additional function elements are molded onto it, such as rigid reinforcement frames with an integral foam structure, slider and guide rails, elements for mounting various components, and mounts for seals or sealing profiles.

Figure 3-107 Polycarbonate segmented roof of the Mercedes-Benz B-Class

As a result of this integration of functions, the overall system costs for assemblies with PC windows are in some cases less than with competing solutions.

This was the first time polycarbonate was used as a glazing material in transparent tailgate panels and fixed side windows. Currently the panorama roof of the Smart ForFour and the segmented roof of the Mercedes-Benz A- and B-Classes are made of polycarbonate (Figure 3-107). The PC segmented roof is about 3.5 kg lighter than a purely steel solution. This example shows the lightweight potential of plastic solutions in roof modules.

Figure 3-108 Mini concept car: "cargo roof" with integrated picnic table

The inside and outside surfaces of the PC windows are given surface hardening which consists of an adhesion-promoting primer and an abrasion-resistant top coat – based in most cases on polysiloxane. Ultraviolet absorbers are incorporated in this system to ensure the weathering resistance of the component. Pre-cured polycarbonate films are also being tested with an ultraviolet-curing clear lacquer coating. Full curing of the paint does not occur until the thermoforming and back-injection molding of the component. For very high abrasion resistance requirements – such as is the case with rear windows – an additional coating with a hardness close to that of glass can be applied by means of a special plasma technology. In combination with this system, complete solutions with integrated defroster and antenna module have also already been developed [11]. PC types with infra-red protection properties currently in development block a large part of thermal radiation from the sun, thereby preventing the vehicle interior from heating up so much. Solar cells or light-regulating liquid-crystal films can be integrated into these large glazed surfaces [12].

As an alternative to polyurethane systems, long-fiber, fiber mat or mineral-filled thermoplastics are also possibilities for the manufacture of the carrier frame as also for the outer skin of a roof module. When carrier materials of this kind are used, a high-quality class A finish is very difficult to achieve on account of this reinforcement material. For this reason, films with a high surface quality are used as weathering-resistant top-layer material or as a substrate for the paint coating [13]. By back-foaming these films by the thermoplastic structural foam molding process using chemical or physical blowing agents and followed by precision opening of the mold, sandwich designs can be produced with integral foam structure with adjustable compact outer-layer thickness and foamed low-density core. The composite with the surface decorative fabric on the inside can be fabricated in the same production step. With these production possibilities and the freedom of design afforded by plastics, new "cargo roofs" with additional functions can also be developed (Figure 3-108).

3.9.2.5 Vehicle Interior

The vehicle interior of today and tomorrow is caught between a large number of conflicting demands. Harmony and coherence, emotions, comfort, safety and functionality must all meld into an embracing whole while simultaneously preserving brand identity. Growth possibilities are forecast for the field of interior trim which are more than twice as great as the expected rise in automobile production [14]. Factors driving such growth include the increasing importance of the interior as a differentiation feature and expansion in the high-end sector. To help meet these growing demands on the vehicle interior with the required cost-efficiency, plastics are becoming increasingly important in material strategies as are the various injection molding methods. The production of

Figure 3-109 In-mold laminated door inserts and cockpit trim with metallic-effect surface

high-quality, appealing, soft-touch surfaces as single-material systems on door trim panels and instrument panels is currently being tried out with a combined method using composite and thermoplastic structural foam molding. The single-step production process for making the carrier and applying the soft-touch surface in an automated production cell brings economic advantages over the reaction injection molding methods in use today.

Until now, light in the vehicle interior has come from small hidden lamps or light-emitting diodes. With a new 'smart surface' technology which combines the phenomenon of electroluminescence with in-mold film laminating, flat three-dimensional plastic components can be made that light up in color over their entire surface when an A.C. voltage is applied [15]. Electroluminescence depends on the property of certain crystals to light up in an electric field – a process that is virtually the reverse of the solar cell. In the future, light will perform not only functional but also esthetic tasks in vehicles and in the form of ambient lighting will make orientation in the passenger compartment easier and create a cozier atmosphere.

In-mold film lamination (also known as IMD or in-mold decoration), similar to other low-pressure methods in the field of injection molding (back-injection molding with cascade hot-runner technology, back-compression molding, and back-compression molding after melt application by means of a multiaxial traveling injection unit), has been used successfully for years to make decorated three-dimensional plastic moldings cost-effectively even in small production runs (Figure 3-109).

With the smart-surface technology, a transparent polycarbonate film is screen-printed with a multi-layered electroluminescence system. This consists of a transparent electrode and a counter-electrode between which there is a dielectric layer – that is, a non-conductive layer. This contains special inorganic crystals which act as light emitters. Following molding, an inverter is applied to the rear of the component. This converts the D.C. voltage supplied by the car's electrical system into an a. c. voltage. In the switched-on state the surface of the component is illuminated, with hardly any energy being consumed. The light intensity can be controlled as a function of the voltage applied. Since this is a "cold" light source, no heat is emitted. This molding needs no maintenance and thus has an extremely long service life. Depending on the crystals used, blue, green, orange or white light can be generated, which is in addition glare-free. Another thing which makes this technology so interesting is that it has a great potential for modularization and automation. In the future, application of the converter is also to be integrated into the production process at the injection-molding machine. The transparent film of the smart surface molding which is visible

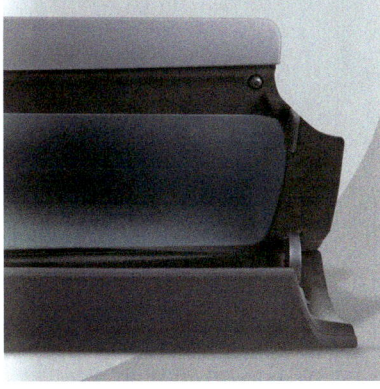

Figure 3-110 Smart-surface technology for creating new ambient lighting by in-mold laminating of electroluminescence films [16]

to the outside can be printed with a transparent or a translucent decoration, depending on design requirements. In addition the component can be covered with textile flocks (through-lighting effect). This is, for example, of interest in the case of parts of the headliner, glove compartment and stowage compartments (Figure 3-110).

It is conceivable that the film surface could be painted with translucent or transparent "soft feel" paints – for example, in the case of parts of the instrument panel or handles in the door area. Very appealing from the design point of view are metalized but translucent parts which during the day have a sporty metallic look but at night are illuminated. Examples of application here would be handles, decorative trim strips and badges.

3.9.2.6 Functional Parts in the Engine Compartment and for Lighting

Lights and Headlights with LED Light Sources

There is hardly a topic which is found more often in the studies and show cars at the automotive fairs of recent years than headlights and lights equipped by the designers with light-emitting diodes (LEDs) (Figure 3-111). In their innovations, virtually all automobile manufacturers are putting their money on the new faces that light-emitting diodes make possible.

Desires for differentiation on the basis of styling are thus helped by additional positive properties of the small semiconductor light sources. Semiconductors are not subject to wear and in comparison with the average life of the vehicle can last for just about forever. As a side effect, this opens up new possibilities for designing front modules and packaging the engine compartment, since servicing access flaps and lamp replacement are no longer required [17]. The desired light distribution can be achieved using reflectors, reflecting systems with lenses or supplementary lenses with efficiencies varying between 65 and 87%.

Transparent plastics and injection-molding methods for manufacturing optical components, such as injection-compression molding and molding nanostructures with variothermal cavity surface temperature control, are being developed further for these applications as well.

Figure 3-111 LED light sources are taking over rear lights and headlights

Figure 3-112 Dipstick tube made of PA 66 and manufactured by the water injection process for BMW

Media-Carrying Lines for the Engine Compartment

A large number of curved media lines, in some cases integrated into other function parts, are found in the automobile, especially in the engine compartment. Conventional processes used for manufacturing this kind of media line from metal or plastics in most cases require several production steps. A economically attractive alternative for the manufacture of a plastic media line of this kind is the fluid injection technique (Figure 3-112).

3.9.3 Current Process Innovations in Injection Molding

3.9.3.1 Injection-Compression Molding

The essential difference between injection molding and injection-compression molding is the way force is transmitted from the injection-molding machine to the melt in the mold. Whereas in injection molding, force is applied to the molding only via flow through the gates, pressure in injection-compression molding is also applied via the surface of the molding (Figure 3-113). In injection molding, to transmit energy to a cavity distant from the gate, dissipation energy must be applied all along the entire flow path. In injection-compression molding, on the other hand, energy is transmitted directly from the cavity wall to the melt. This form of force transmission accesses new degrees

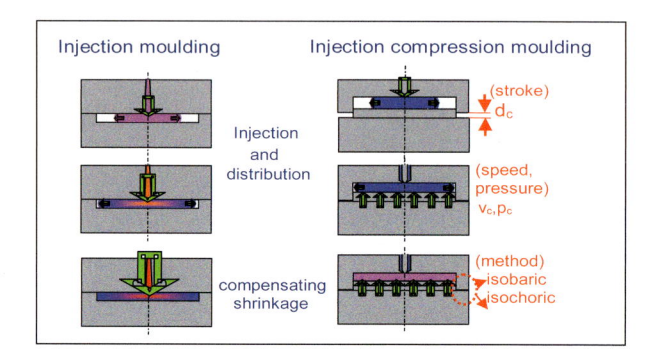

Figure 3-113 The main difference between injection molding and injection-compression molding is in how pressure is transmitted to the melt. This results in new process parameters for injection-compression molding.

of freedom for the molding process. Basic shaping is actively controlled not only by the fixed mold half of the machine, but also by the moving mold half and moving cores. The new degrees of freedom in process control can be expressed as additional process parameters, such as compression stroke, compression pressure and compression speed. Speed- and pressure-based closed-loop control strategies or combinations of both can be implemented for the embossing inserts as with the reciprocating screw.

A large number of useful variants are conceivable for the interplay of movements of the screw and inserts. A fundamental classification of injection-compression processes can be made on the basis of whether

3.9 Innovations in Injection Molding for the Automobile of the Future

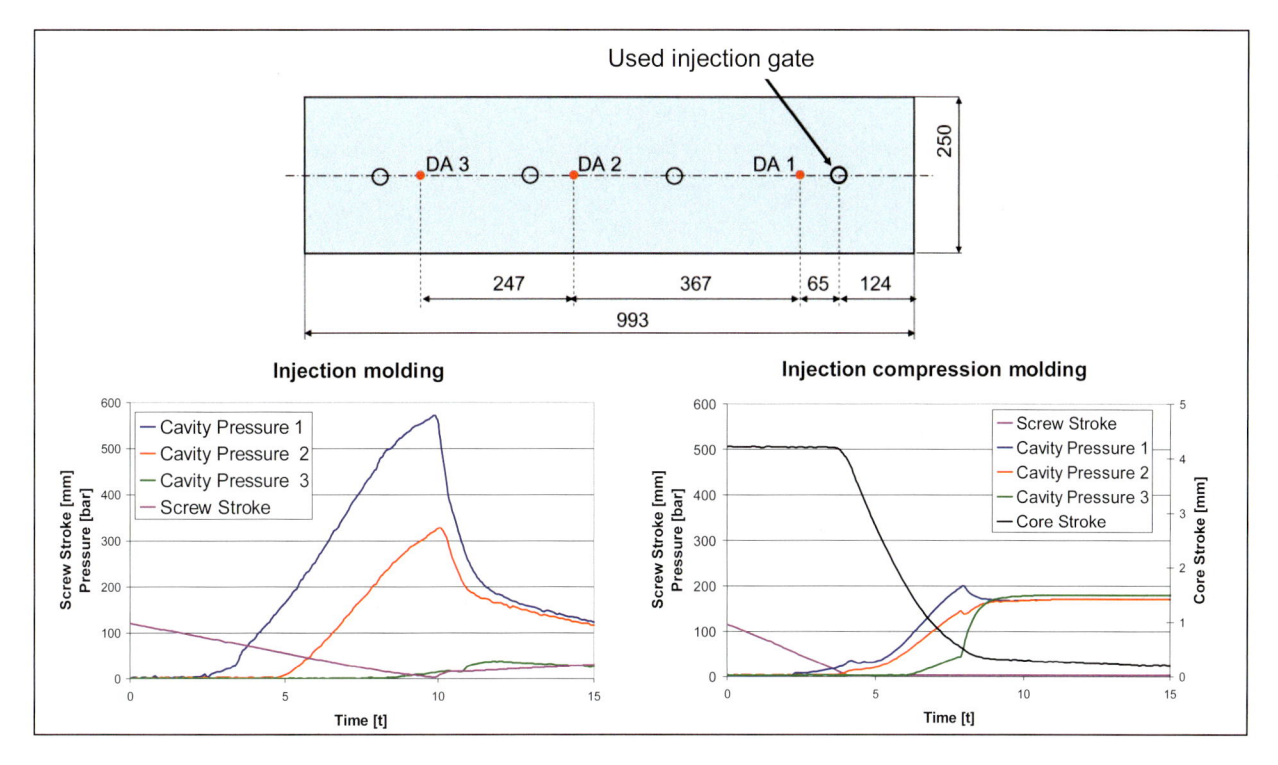

Figure 3-114 Comparison between injection molding and injection-compression molding in the production of a slab-shaped component made of PC Makrolon2607. Both the maximum pressures and the pressure gradients in the mold are considerably lower in injection-compression molding than in injection molding [18].

- the cavity is filled volumetrically before the compression stroke commences, or
- the compression stroke is started before completion of volumetric filling.

While the first type of technique is primarily used for the even application of follow-up pressure and the molding of structures, the second type is used in particular to enable parts with a high flow path/wall thickness ratio to be filled with little stress or even simply to be completely filled at all.

The details of designing molds for injection-compression molding are intimately involved with process control, but we cannot go into them in any detail. All that needs to be said is that even in mold making, there are different approaches to realizing the compression method, such as vertical flash faces, embossing frames or embossing stamps. Designing compression molds for parts without a flat or a symmetrical shape represents a special challenge here.

The positive effect of injection-compression molding on pressure distribution in a part typical of large-area applications is shown in Figure 3-114. A test slab was made 1000 mm long, 250 mm wide and 4 mm thick. The slab was gated via a hot runner at a point in its upper part.

In Figure 3-114, the graph on the lower left shows the cavity pressures measured with injection molding. The curve for pressure near the gate rises to just under 600 bar. The pressure differences along the flow path measured several hundred bar during the injection phase and did not fall below 100 bar even in the follow-up pressure phase. The graph on the lower right in Figure 3-114 shows the pressure curves for injection-compression molding of the same article. Here, the injection-compression process was controlled in such a way that the entire plastic melt was injected into the cavity which had been enlarged by 4 mm in the direction of closing. Next, the melt was pressed into the cavity by means of a parallelism controlled compression movement. At the 4 mm compression stroke selected it was possible to keep the pressure below 200 bar, even in the area close to the gate. The pressure gradients along the flow path were correspondingly low and it was possible to compensate for them almost entirely during the follow-up pressure phase.

Injection-compression molding as a manufacturing process is thus a very promising candidate for the manufacture of both transparent and also opaque parts of the automobile outer skin, particularly in combination with the processes dealt with below – long-glass-fiber injection molding, multi-component injection molding and sandwich injection molding.

The multi-component polycarbonate glazing part shown in Figure 3-115 for a panorama roof measures 1100 by 680 mm, weighs 4.8 kg and was injection-compression molded on an injection-molding machine with a clamping force of only 1500 tons. On the basis of calculations and practical experience with other parts, making the same item by the classic injection molding process would have required a machine with more than twice this clamping force. This means that sunroofs with areas of 1.5 m^2 and more would therefore not be producible on the injection-molding machines which are the rule today. The homogeneous input of force in injection-compression molding also gives considerable advantages as regards residual stress inside the components, simplifies the molding of surface details, and makes it easier to prevent sink marks.

For reasons of appearance, glazed parts of this kind can only be gated from the edge area outwards. However, this one-sided location of injection points results in asymmetrical mold opening forces not only during filling but also during the compression and follow-up pressure phases up to the point where the cavity has been completed filled and pressure equalization occurs. Particularly with large components, these asymmetrical mold opening forces in conjunction with the long flow distances can result in enormous torque

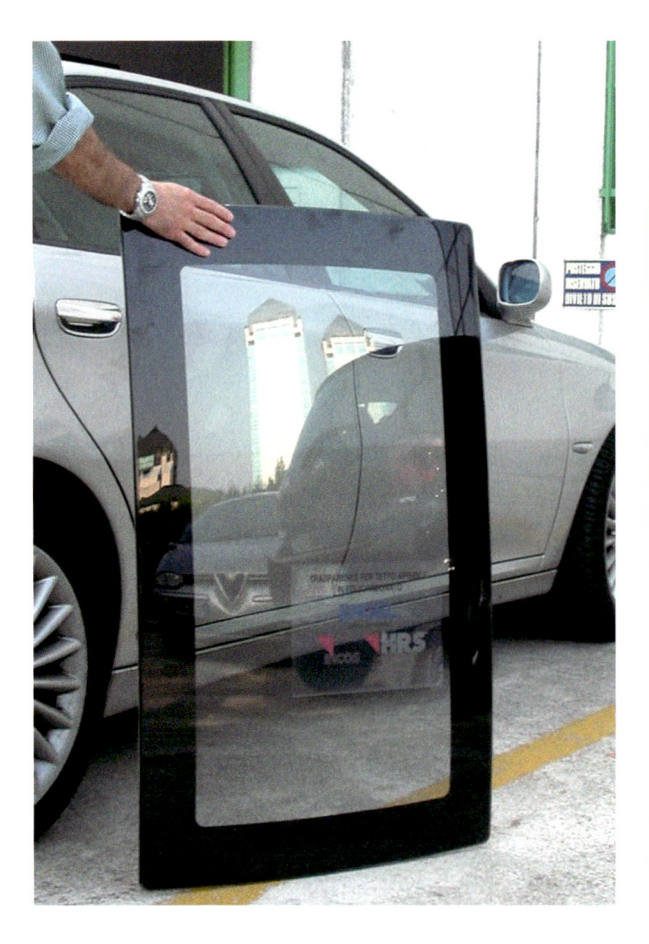

Figure 3-115 Glazing part made of PC for a panorama roof manufactured by injection-compression molding and sandwich molding [18]

values acting on the mold halves and platens. These torques are typically too high to be counteracted with use of mechanical guides of acceptable size. There are, however, machines equipped with embossing systems capable of precisely compensating asymmetrical forces in injection-compression molding.

The ability of injection-compression molding to apply an even follow-up pressure so as to minimize local sink marks and to deliver a particularly high surface structure reproduction quality is useful not only for manufacturing large-area class A surfaces but also, among other things, for fabricating optical elements of plastic, such as may well be used in the future in front and rear lights, sensors and display systems in the vehicle.

3.9.3.2 Processing Long-Fiber Reinforced Thermoplastics

Long-fiber reinforced thermoplastics (LFT) is the term for a group of products falling between the familiar group of short-fiber-reinforced thermoplastics with an initial fiber length of around 0.3 mm and the group of glass-mat-filled composites (GMT). LFTs have an initial fiber length of up to 50 mm and a reinforcement content of 15–60% by weight. The fiber most commonly used here is glass fiber, but there are other solutions, such as carbon fiber or natural fiber. In the majority of cases the matrix is polypropylene or polyamide. The rod-shaped pellets are made mostly by a pultrusion process similar to cable sheathing. In addition there are still mixtures of cut fibers and plastic or rovings which are fed directly into the melt (in-line compounding, direct LFT). Even so-called long-fiber concentrates – rod pellets with a glass-fiber content up to 70% by weight – are on the market. These can be cut with conventional non-reinforced granules in the injection-molding machine by means of a gravimetric feed unit to obtain the glass-fiber content required for the application in question.

This method offers the following advantages:

- Flexible, application-specific setting of the glass content without high storage requirements
- Reduced feedstock costs

The disadvantages are:

- A premixing station is required
- The processor has increased responsibility for the material

Figure 3-116 shows an injection-compression production cell with a gravimetric feed unit for mixing the thermoplastic long-fiber concentrate with recycled material and virgin material so as to obtain the required fiber content.

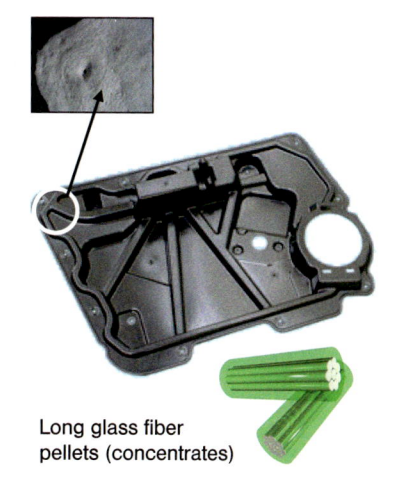

Figure 3-116 Production cell with a gravimetric feed unit for fiber-friendly processing of a polypropylene long-fiber concentrate for a door module support

In contrast to direct processing (dLFT) or even to the in-line method whereby fibers (as rovings or chopped fibers) are not added until the plastic has melted, processing of the LFT rod pellets can take place on near-standard production installations for injection molding or extrusion-compression molding. The mechanical shearing required for melting and homogenization should be as low as possible. This can be achieved by optimizing screw geometry and by appropriate process control measures. Preheating the rod pellets helps obtain the necessary increase in dwell time but without directly affecting the cycle time. Alternative molding versions, such as injection-compression molding, can, thanks to lower product pressures, get the best out of the material LFT.

3.9.3.3 Processing of Composite Materials

When component requirement profiles cannot be met (or not met optimally) by the property profiles of individual materials, composite materials and material composites will take on special importance. Due to the fact that (1) plastics have moderate melt temperatures compared with metal and ceramic materials, (2) homogenization is an integral part of the injection molding process, and (3) the molding process is highly flexible, injection-molding facilities can be used for making composites with tailored property profiles and represent an interesting basis for innovative combinations of materials.

Multi-component or sandwich molding is today already one of the key technologies in efficient plastics processing for manufacturing smaller and mid-sized components. Large-area glazed parts, body parts or instrument panels with esthetic or functional surfaces, as well as the fixed structures supporting them, cannot be produced economically until new machine technology in conjunction with a swivel-platen system can do so in a single-step process (Figure 3-117). Here the center section of the mold holding one molding component is rotated by 90° or 180° to allow another material component to be molded on or a metal composite carrier or fiber laminate to be inserted for decorative material. In most cases, the articles are molded by special methods such as injection-compression molding or structural foam molding.

Figure 3-117 Injection-molding machine with swivel-platen system for manufacturing multi-component glazing items. In both stations injection-compression molding takes place with parallelism control

In the case of in-mold laminating of large films for roof modules or for ultraviolet and scratch protection films for glazing, the handling and fixing of the preformed films in the hot mold is a demanding challenge. If film requires gentle treatment in the component molding process, structural foam molding is a suitable method to use.

3.9.3.4 Incorporation of Partial Reinforcements

In many cases, local stress varies a great deal in the component. This raises the justified question as to whether the complete component should be oversized or whether it makes more sense to incorporate reinforcements where they are really needed. One example is hybrid technology where a mechanically interlocking composite with sheet-metal insets is produced or the use of fabrics, continuous filament tapes or reinforcing mats which are immersed and coated (Figure 3-118).

The use of plastic-based inserts (glass-cloth-reinforced semi-finished thermoplastic products) rather than sheet metal can result in further weight savings and corrosion suppression. Also advantageous here is the adhesion between the thermoplastic components of the composite which results in a considerable increase in torsional rigidity. The individual alignment of the fabric in the insert offers additional potential for optimization. In some cases, the semi-finished product can be shaped directly in the injection mold after preheating.

The use of continuous filaments appears also to be of interest. They can be laid in the mold, positioned to suit the particular stress state in the molding. Components with tailored properties can be produced by an exact positioning of strand, tape or reticular fiber composite inserts. For example, a loop may be laid around a load application point and the force flow thus transferred on direct loading paths. One challenge is to incorporate and fix the reinforcement elements and to achieve the greatest possible embedding in the injected plastic.

3.9.3.5 Fluid Injection

While the gas internal pressure technique can already be counted among standard methods for producing hollow items by injection molding, the water injection technique is still in the growth phase. The best quality parts can be made using the "blow-out" technique. Here a main mold cavity is first completely filled with plastic, a subsidiary cavity is then opened by withdrawal of a split, and the still hot plastic from the main cavity is pressed out of the core of the main cavity and into the subsidiary cavity by water. Here the water is injected into the interior of the part by an injector at pressures ranging between about 30 to 100 bar. Depending on the plastic used and on the part geometry, pressing out may take between a few tenths of a second and several seconds. The in-flowing water cools the component from within. This bilateral cooling can cause shrinkage cavities to form in the component wall and must be prevented by a suitable selection of

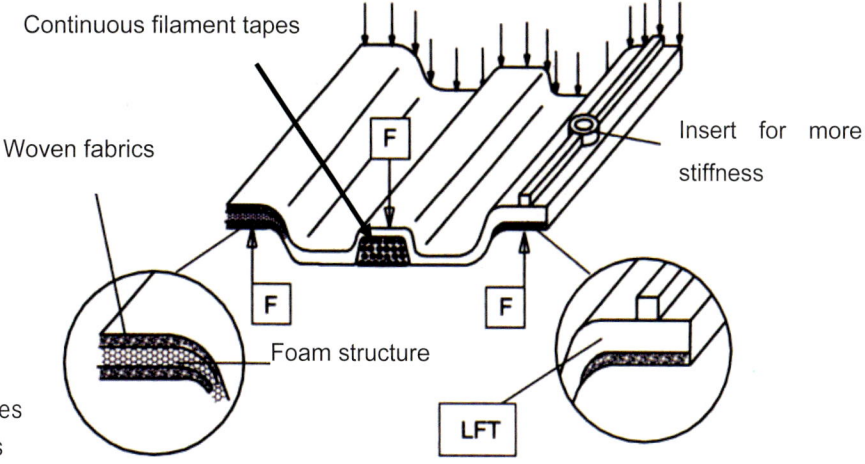

Figure 3-118 Tailoring properties by incorporating reinforcements

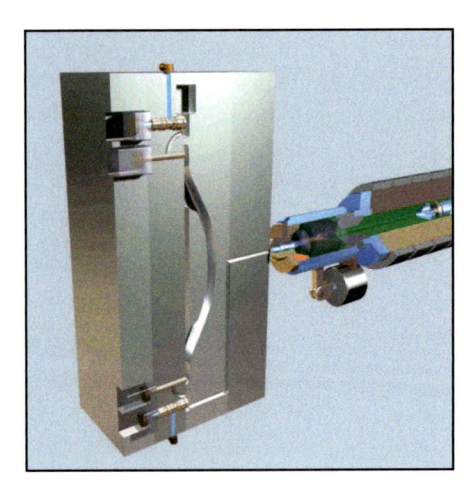

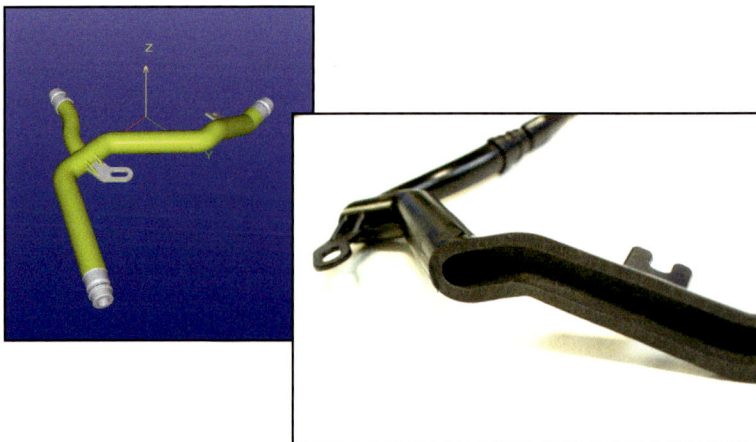

Figure 3-119 Production of media lines made of PA66 by water injection molding

parameters and the use of materials especially suitable for the water injection technique. Since the injected water heats up during and after injection, if rapid removal of large amounts of heat from the component is required after ejection of the core material, a second injector may be used to pump water into the cavity and this be extracted via the first injector.

In contrast to the gas internal pressure method, the additional cooling and the thinner wall thicknesses possible can reduce cycle times by as much as 70%, depending on the material and geometry.

The automotive parts most frequently made by the water injection method are pipe elements in the engine compartment such as media lines (Figure 3-119).

Future applications for the water injection technique in the automobile could be the production of hollow sections for reinforcing structural components or the integration of media lines into housings and trim panels. It is even possible to integrate inserts and to apply sandwich molding in conjunction with the water internal pressure technique. The proliferation of this method is supported by the availability of integrated systems in which the water injection module is controlled by the injection-molding machine and in which appropriate diagnostics of the water injection process are applied in quality monitoring and documentation [20].

3.9.3.6 Foaming with Physical and Chemical Blowing Agents

The injection molding of foamed parts has become more important recently.

The reasons for this are to be found not only in the better molding properties obtainable in comparison with conventional injection molding. These properties include:

- Minimization of warpage
- Prevention of sink marks
- Savings in weight and materials, and also

Improvements in or simplifications of process control due to

- The reduction in melt viscosity
- The possibility of reducing melt temperature and thereby reducing cooling time
- Evenly distributed build-up of "internal" follow-up pressure, and
- The possible reduction in clamping force and the associated economically efficient use of smaller injection-molding machines.

Basically, chemical and physical foaming methods are used, but the principle behind making foamed thermoplastics is similar. A gas is released in the thermoplastic melt and the subsequent pressure drop induces a thermodynamic instability that initiates the formation of foam. The difference between chemical and physical foaming lies in how the gas is brought into the melt. With chemical foaming, the gas is produced by the decomposition reaction of the blowing agent which occurs just below the processing temperature of the thermoplastic. With physical foaming, the gas is metered into the melt at high pressure.

The advantages of chemical foam lie in its practical simplicity. A foamed thermoplastic can be produced by adding blowing agent to the granules in, for example, the form of a master batch. The injection-molding machine must be equipped with a shut-off nozzle and a screw positioning control device so that, once the dispensing process is completed, a defined melt pressure is maintained in the space in front of the screw that will prevent premature foaming of the melt. The blowing agent system must be tailored to the plastic to be processed. The gas content and thus the blowing pressure required to finally fill the cavity are both limited. The foaming pressures achievable with chemical blowing agents are less than 30 bar. Low wall thicknesses or complex structures limit the range of application of this method, since complete filling of the mold is no longer guaranteed due to the low pressure.

Also to be taken into consideration are the negative influence on properties from blowing agent decomposition products, as well as occasional corrosion problems in the mold. Certain thermoplastics, such as high-temperature plastics or special thermoplastic elastomers cannot be foamed using chemical blowing agents.

Here we find the great advantages of physical foam. Every type of plastic can be foamed without the need for special additives and using just one medium which is also still inert (N_2 or CO_2). In addition, particularly small gas bubbles (cells) result in the finished component such as cannot be obtained with conventional thermoplastic structural foam molding using chemical blowing agents. The considerably higher internal gas pressure means that even thin-walled components can be foamed.

Two versions of physical foaming are available:

- Gas input directly into the cylinder during metering (the MuCell® process, for example)
- Gas input via a special mixing nozzle during the injection process (the Optifoam, for example).

The MuCell® technique in particular has recently become more important. Physical foaming by the MuCell® method originated in developments at the Massachusetts Institute of Technology (MIT) in Boston and by Trexel Inc. in Woburn, Massachusetts (MuCell® license required).

The MuCell® method has now been used successfully for processing not only standard plastics such as PP, PE-HD and PS, but also engineering plastics such as PC, PA and PC/ABS blends – with and without fillers or reinforcement materials – and also thermoplastic elastomers. Potential MuCell® components include virtually all parts with high requirements for dimensional stability but which, due to the irregular (swirled) surface structure typical of MuCell®, are not primarily required to function as visible parts. Even with thin-walled structures, reduced warpage tendency can be combined with weight savings and clamping-force minimization. Applications in the automotive sector include, for example, PP-T components for air-conditioning units and even textiles and films in the case of in-mold laminating (Figure 3-120).

In the case of technical components, a weight reduction of 5–15% can be achieved with the MuCell® process – depending on flow distance and component thickness. If markedly higher material savings are required, it is possible to combine MuCell® with the so-called "precise backstroke" (sometimes also called "breathing" or "negative compression"). Here, the volumetrically filled mold cavity is deliberately enlarged to the desired component thickness (Figure 3-121). This can be done simply, for example, by resetting or even completely removing the clamping pressure at the machine. Even large-area components made by this method have a very high flexural rigidity and a comparatively good surface (although top visual component quality cannot be obtained by this process).

Figure 3-120 Applications for physically foamed components are found in air-guidance parts in air-conditioning units, in film back-injected instrument panel faces and in mountings for fan motors (made by sandwich molding)

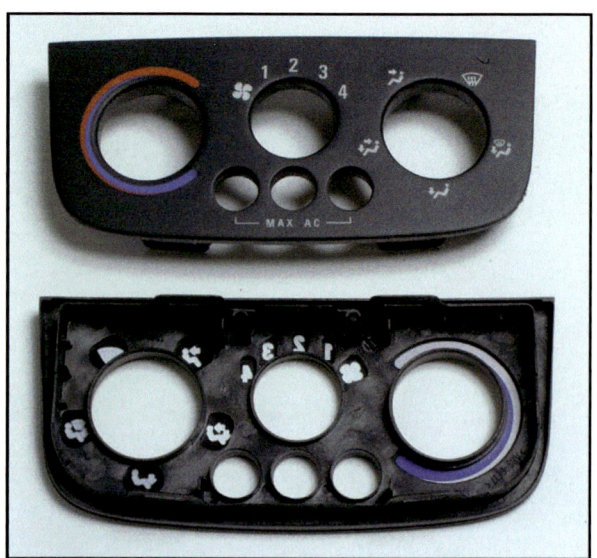

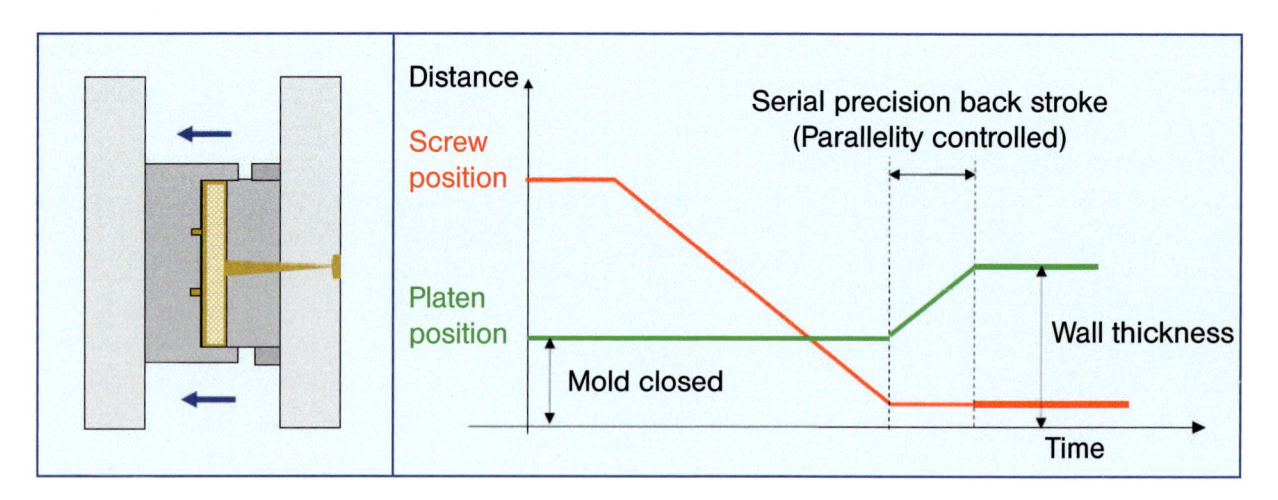

Figure 3-121 Molds suitable for the precision backstroke technique are those equipped with shearing edges (diagram on the left) or with moveable frames. Right: After volumetric filling of the cavity, it should ideally be opened under parallelism control.

Potential applications in the automobile for the MuCell® precision backstroke technique include door modules or instrument panel supports in which not only low weight but also dimensional stability and flexural rigidity are particularly important. The wall thicknesses obtainable with this process are in individual cases three or four times thicker than the initial wall thickness – foam moldings made of PE-HD and PP with final wall thicknesses of 10 to 12 mm have already been produced (Figure 3-122). Here it has proven advantageous to use a device to control parallelism in the platen. This prevents the platen from becoming skewed due to process irregularities (such as uneven foaming).

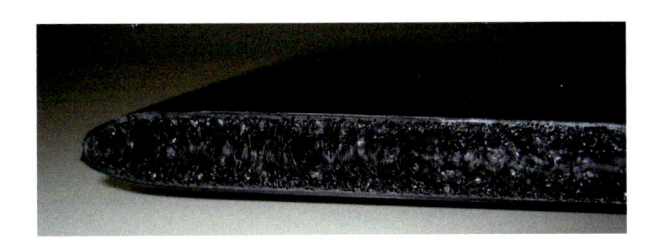

3.9 Innovations in Injection Molding for the Automobile of the Future

Figure 3-122 Foamed PP slab made by the MuCell® precision backstroke method: the melt is injected with a starting wall thickness (corresponding to the thickness of conventional molding) of 3 mm while the final thickness of the slab is 10 mm

In most cases foamed moldings have a visually unsatisfactory surface quality and are thus not normally used as visible components. The formation of the typical swirl patterning on the surface is due to the blowing fluid escaping at the melt front as the melt is injected. Due to shearing at the cavity wall, the melt skin on the bubbles tears and the escaping gas wanders between the flowing melt and the cooled mold surface, leaving the typical swirl pattern on the surface of the molding.

To prevent premature escape of gas at the melt front, the pressure before the melt front must be greater than the saturation pressure of the gas in the melt – this is achieved by having a gas counterpressure in the cavity.

To achieve this, the gastight mold is filled with nitrogen or dried air before the melt is injected (Figure 3-123). With an effective gas counterpressure of up to 80 bar the expansion of the blowing gas during injection can be prevented – the result is a smooth surface for the molding which is molded in the core region [21].

In the MuCell® foaming of polycarbonate (PC), the gas counterpressure technique creates a smooth transparent outer layer. The thickness of this layer can be set to match requirements by varying the process parameters. For example, by reducing the mean density by 5.5% it was possible to bring the surface roughness down from $Rz = 23.11$ μm to $Rz = 0.85$ μm (Figure 3-124).

The gas counterpressure method does however have its limits. Very high surface qualities are not possible unless there is a weight reduction of up to 10%, depending on polymer type. In addition, the mold technology is very complex – perfect sealing is required in the first phase and defined venting in the second. The actual process window turned out to be very narrow.

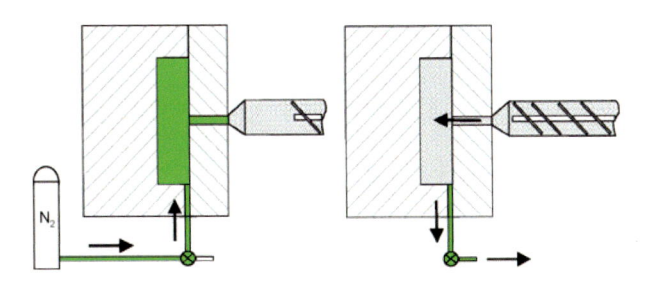

Figure 3-123 Process sequence with the gas counterpressure process. Gas pressure is created in the empty mold cavity (left) – melt is injected against the gas pillow (right) with the gas pressure in the cavity being regulated as a function of the injection distance with a rapid drop in pressure at the end of mold filling.

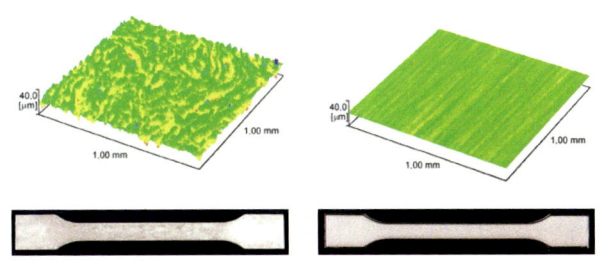

Figure 3-124 Influence of the gas counterpressure technique on the surface roughness of polycarbonate with a density reduction of 5.5%.
Left: MuCell® standard specimen, made without using gas counterpressure: $Rz = 23.11$ μm.
Right: MuCell® standard specimen, made using gas counterpressure: $Rz = 0.85$ μm.
The surface roughness was determined by laser-optical scanning with a 3D surface measurement system manufactured by UBM Messtechnik GmbH of Ettlingen, Germany.

References

[1] Stadler, R.: Innovationen – Motor der deutschen Automobilindustrie [Innovations: engine driving the German automotive industry]. *Kunststoffe* 11 (2005) p. 90

[2] Anon.: Light Vehicle Systems, company publication, ArvinMeritor, Detroit, 2005

[3] Anon.: Konzept Türmodul [Door module concept]. brochure, Polytec Group, Hörsching, Austria 2005

[4] Flehmig, T., Flöth, T., Hoffmann, O., Patberg, L.: Kostengünstige Profile für wirtschaftlichen Leichtbau [Inexpensive profiles for economic lightweight construction]. *Automobiltechnische Zeitschrift ATZ* 09/2005 Vol. 107 pp. 766–771

[5] Büdgam, S., Freitag, V., Hahn, O., Ruther, M.: Fügesystemoptimierung für Mischbau-weisen im Karosseriebau [Joint-system optimization for hybrid body designs]. *Automobiltechnische Zeitschrift ATZ* 12/2004 Vol. 106 pp. 1132–1141

[6] Droste, A., Janssen, R., Naughton, P., Röttger, J.: Kunststoff-Metall-Hybridlösungen für Frontend-Modulträger [Plastic-metal hybrid solutions for front-end module carriers]. *Automobiltechnische Zeitschrift ATZ* 12/2005 Vol. 107 pp. 1114–1117

[7] Opperbeck, G., Hassdenteufel, K., Krasenbrink, C., Cheron, H.: Hochintegriertes Frontendmodul [Highly integrated front-end module]. *Automobiltechnische Zeitschrift ATZ* 01/2006 Vol. 108 pp. 10–17

[8] Schlott, S.: Innovationsschaufenster der Autobranche [The automotive sector's shop window for innovations]. *Kunststoffe* 11/2005, pp. 96–101

[9] Anon.: Plastic Omnium press kit for IAA 2005 in Frankfurt

[10] Anon.: Integration von Antennensystemen in Karosserieteile aus Kunststoff [Intergration of antenna systems in plastic body components]. Company publication, Hirschmann Car Communication GmbH, Neckartenzlingen, Germany

[11] Krause, V.: Quo vadis Verscheibung aus Polycarbonat? [Future trends in polycarbonate glazing] *Kunststoffe* 11/2005, pp. 114–115

[12] Dudenhöffer, F.: Dachsysteme – Schub durch Innovation [Roof systems: growth due to innovation]. *Kunststoffe* 3/2005, pp. 96–98

[13] Gestermann, S., Krause, V., Zöllner, O.: Polycarbonat und seine Blends für Karosseriebauteile [Polycarbonate and its blends for body parts]. *Automobiltechnische Zeitschrift ATZ* 11/2005 Vol. 107 pp. 1010–1016

[14] Schlott, S.: Innovationsschaufenster der Autobranche – Trends auf der IAA 2005 [Innovation show case of the automotive industry: trends at IAA 2005]. *Kunststoffe* 11/2005, pp. 96–101

[15] Rothbarth, F.: Ein neues Beleuchtungskonzept für den Autoinnenraum [A new lighting concept for the automobile interior]. press release, BayerMaterialScience, Leverkusen Feb. 2005

[16] Anon.: Konzept Handschuhkasten mit Elektrolumineszenz-Folie [Concept for glove compartment with electroluminescence film]. company brochure, Polytec Group, Hörsching, Austria, 2005

[17] Hamm, M., Ackermann, R.: LED im Scheinwerfer – Revolution in Design und Technik [LED in the headlight: revolution in design and technology]. *Automobiltechnische Zeitschrift ATZ* 11/2005 Vol. 107, pp. 970–977

[18] Steinbichler, G., Gießauf. J., Pitscheneder, W.: Große Autoscheiben aus der Spritzgießmaschine [Large automobile windows from the injection-molding machine]. *Kunststoffe* 10/2004, pp. 164–170

[19] Koch, T., Schürmann, H.: Spritzgussteile lokal verstärken [Local reinforcement of injection moldings]. *Kunststoffe* 01/2006,

[20] Steinbichler, G., Egger, P.: Dem Experimentierstatus entwachsen – Wasserinjektion serientauglich [Matured beyond experimental stage: water injection suitable for full-scale production applications]. *Kunststoffe* 9/2004, pp. 196–202

[21] Egger, P., Fischer, M., Kirschling, H., Bledzki, A. K.: Serienfeste Vielseitigkeit beim MuCell-Spritzgießen [Versatility for full-scale production with MuCell injection molding]. *Kunststoffe* 12/2005

4 Modelling and Rapid Prototyping

4.1 Innovative Solutions for Prototype-Making in Automobile Manufacturing – Prototype Component Requirements and Practical Examples

Hartmut Albers

In passenger car development, prototype components are not only an important success factor but also a cost factor. This is why Mercedes-Benz goes to a great deal of trouble to match prototype components and the corresponding tools as optimally as possible to the requirements of the development process. This objective cannot be achieved without innovative suppliers. A number of examples of the requirements for prototype components in the development department will be described in some detail as well as the corresponding projects with prototype suppliers.

4.1.1 Passenger Car Development

Prior to Pleasure ...

The automotive industry is faced with major challenges throughout the world. Global overcapacities, vicious discounting battles in the USA (which in some cases have already reached Europe) and increasingly powerful competitors from the Far East are just a few of the problems which the automobile manufacturers have to deal with.

Naturally the DaimlerChrysler group is just as unable to isolate itself from these general conditions as any of its brands. The successful restructuring of the Chrysler Group in the USA – Chrysler is the only one of the "Big Three" which has growth in its market shares and an increasing profit on sales – has, however, shown that even in an environment as strongly affected by competition as the car market, there are still ways of ensuring survival: attractive products and optimum processes.

This is also true of the Mercedes-Benz brand. Innovative products such as the elegant new CLS class coupes demonstrate that Mercedes is not only coping successfully with the market challenges, but also regularly acting as the benchmark in the industry. The new flagship of the brand, the new Mercedes-Benz S class, in particular underlines this claim with the large number of innovations it incorporates.

... Comes the (Development) Pain

But, as we have already stated, attractive products only deliver a part of the advantages necessary in a competitive environment. Optimized processes are also necessary in order to be able not only to manufacture attractive products at attractive costs, but to sell them as well. In addition to production and marketing, processes within the development phase are of decisive importance.

Figure 4-1 The new S class from Mercedes-Benz (W221 series)

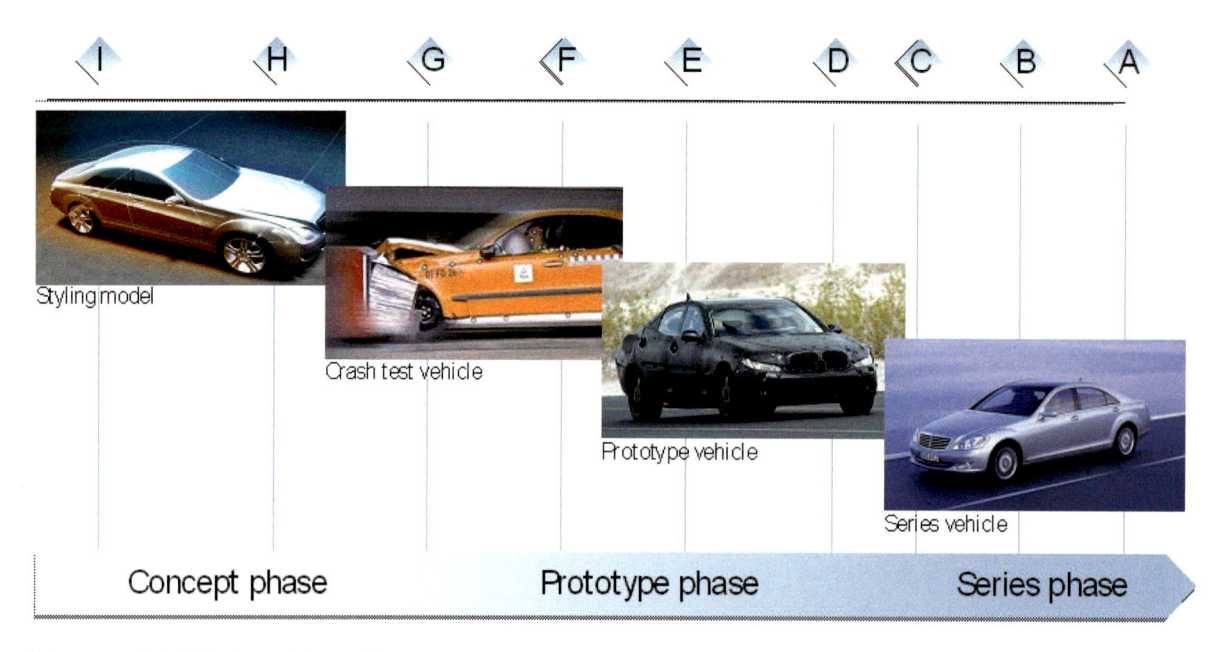

Figure 4-2 MDS plan with quality gates

New car models are developed at Mercedes-Benz on the basis of the **M**ercedes-Benz **D**evelopment **S**ystem (MDS). MDS defines the sequence of the individual development steps with regard to both the time scale and the contents: from the concept model right up to the first customer-ready vehicle. So-called "quality gates" form milestones where the progress of the development work can be checked against predefined content and quality criteria.

In addition to quality gate A (Job #1), i.e., start of production, quality gate E represents one of the most important milestones in the development project.

Quality gate E stands for "provision of test-worthy parts and components" – in other words, the test components. Immediately following quality gate E, the first development vehicles (DVs) of the new series are put together. Unlike the so-called structural validation vehicles (SVVs) from preceding process steps, the development vehicle is largely identical with the subsequent standard production vehicle both in equipment and appearance. Since the findings from trials and testing can however frequently result in modifications to individual parts or even to entire assemblies, production molds in this phase are used only in exceptional cases. As a rule, the test components are produced using specially made prototype molds.

Testing of the whole vehicle commences with the completion of the first development vehicles. Since these tests take place not only on special testing grounds throughout the world but also on the road system at home, it is this period that produces the various unauthorized press photographs of the new vehicles in disguise, known in Germany as 'Erlkönige', roughly equivalent to 'ghost riders'.

Figure 4-3 A disguised development vehicle (Erlkönig or ghost rider)

So as not to reveal specific design details too much ahead of time, test drives on public land or roads always use disguised vehicles.

4.1.2 Prototype Requirements

Prototype Requirements in the Development Phase

The success of current products should not blind us to the fact that there must be repeated checks and reviews to ascertain whether the current processes still correspond to the possible optimum. Every single process – from research to monitoring the after-sales market – has to be measured repeatedly against the competition and optimized.

This requirement, of course, also applies to vehicle development, especially as the decisions made here affect not only the automaker's own cost situation but also down the line from the production stage as far as sales and service aspects.

In addition to other factors, the costs for prototype components and tooling make up one of the big items in the development budget. Mercedes-Benz invests between 100 and 200 million Euros per year on the usual assembly parts alone (without E/E), depending on the model cycle and the development stages (Figure 4-4).

This huge figure makes it very clear that considerable importance must attach to an optimum design for this process: the availability of prototype components under competitive conditions is one of the success factors in passenger car development.

Two aspects here show clearly that in many cases completely different optimization approaches must be adopted for prototypes from what is subsequently the case in full-scale production.

The number of prototype components usually lies between only 20 and 500. A prototype mold is usually only used for between three months and a year.

Requirements for Prototype Components

The essential design criterion for a prototype component is its "testworthiness". In other words, the design must ensure that all objectives can be achieved in the component testing that follows. In addition to these direct component properties, which we will deal with in a little more detail later with the aid of an example, it is nevertheless also necessary in most cases to take into account a number of other constraints in designing the process.

For example, if a new kind of process is to be used in subsequent full-scale production of the component, it will often make good sense to fully reproduce this process even in the prototype phase. Late and expensive surprises during series production can in this way under certain circumstances be prevented, thereby justifying higher costs during the development stage. On the other hand, it does not generally make sense to build expensive prototype molds for components whose design has not yet been completed.

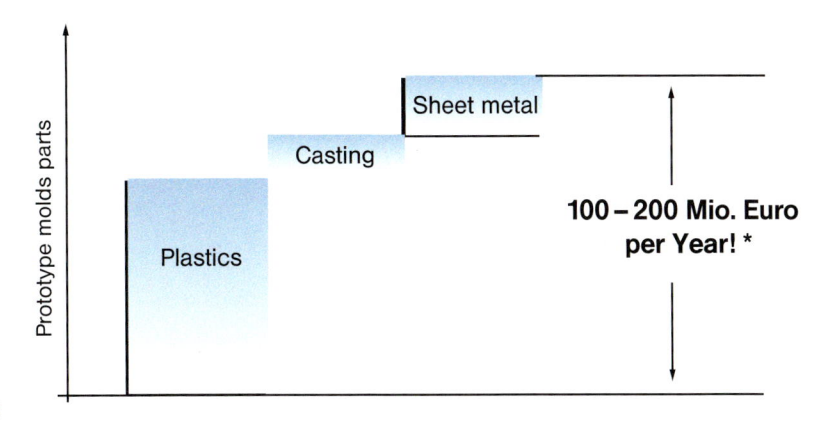

Figure 4-4 Prototype costs for assembly parts

*) Assembly parts, without BIW, E/E and powertrain

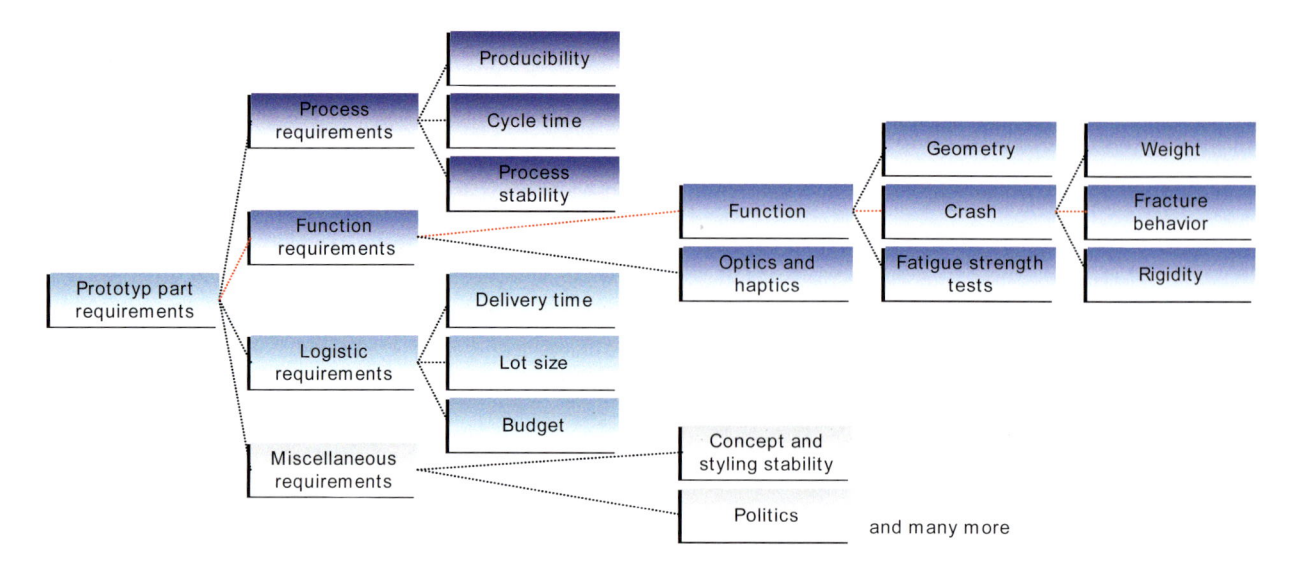

Figure 4-5 Some of the requirements for a prototype component

Example: Air Outlets in the Cockpit

The properties of a prototype component must ensure that all planned test objectives can be achieved. The obvious way of achieving this goal is to provide prototype components which are absolutely identical to the full-scale production components. Unfortunately this is not only the lengthiest way, but in most cases also the most expensive one and for this reason unfeasible in the light of cost efficiency requirements in the development stage (cf. Figure 4-7).

One possible solution to this design problem consists of examining the various test objectives and then deriving the required component properties in accordance with them.

Based on three "utilization scenarios" for a prototype component, we shall describe how this approach works taking a cockpit air outlet as our example:

1. Prototype as design model:
   Prototype components with various possible designs and surfaces are frequently required for design decisions. These are one-off pieces whose visual appearance should be as close as possible to that of the subsequent full-scale production component. To enable fast decision processes, the parts must be available in a very short time: mechanical functionality is not required in most cases.
   Prototype molds seldom need to be constructed for these model components, since the use of the original materials during this period would hardly result in any additional information; in most cases SLA models are prepared.

2. Prototype for interior climate testing:
   Some of the requirements for the ventilation system – for example, rapid windshield de-icing – have to be met early on in the development process, since under certain circumstances this may result in modifications to the positioning of the air outlets (design!).
   These tests call for a very small number of prototype components whose blades can be adjusted in the same way as blades on the production component. In general no visual requirements are made. In most cases it is not necessary to make prototype molds for these components since the temperature behavior of the usual materials is well known. Normally SLS components or polyurethane castings made from silicone molds are built for these tests.

Figure 4-6 Example of an air outlet

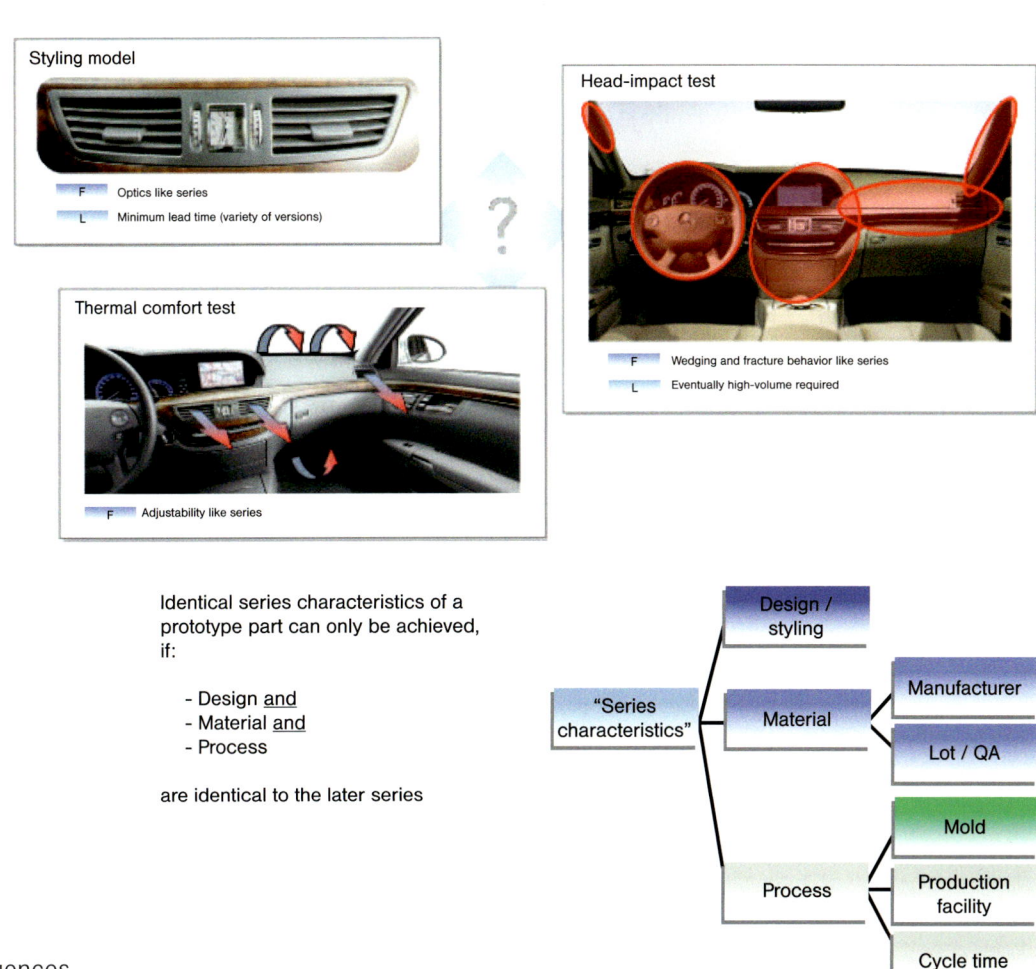

Figure 4-7 Influences on component properties

3. Prototype for head impact investigations in the crash test:
   One of the important properties of Mercedes-Benz cars is that in the event of a crash the loads on the vehicle occupants fall (considerably) below the maximum loads specified by law. This development objective is secured in a series of crash tests with the entire vehicle or on test stands. Before useful and informative results can be obtained from these investigations, the prototype components used must, as regards blockage formation and fracture characteristics, behave in precisely the same way as the later full-scale production components. A prototype tool is generally made for these components, since this objective can in most cases only be achieved by using the original material and manufacturing process, and since even relatively large quantities of the component are frequently required.

Prototype Part Identical with Production Part?

Particularly with completely new designs of components there are frequently uncertainties as regards the requirements which will have to apply to the prototype parts. To forestall possible risks (in full-scale production), in such cases a prototype mold that is close to or even identical with the final production mold is demanded not only of the designer responsible for the component but also of the system supplier.

When the decision is made regarding the design of the mold, it must be kept in mind that not only the mold but also a number of further parameters determine whether the component is identical to the full-scale production version. This means that under certain circumstances the test results for a component may become worthless if the design of the part is changed

– for reasons of appearance, for example – after the conclusion of testing. Production conditions or the material itself can have an equally great influence.

For these reasons, a cost-benefit analysis carried out during the prototype stage when the tooling and processes are being designed will usually result in a prototype component whose properties are all comparable with the full-scale production part.

Prototype Mold As Cost Driver

As has already been shown, the prototype mold represents only one item of the prototype costs, but if, for example, existing production systems and equipment can be used, it may be a very important one.

For this reason the design of the prototype molds may be of decisive importance to cost efficiency during the development phase. As Figure 4-8 shows, prototype costs increase markedly the closer the mold gets to the full-scale production version in terms of degree of automation, configuration and heat control. Unfortunately, the "closeness to full-scale production" of the prototype component does not usually increase so clearly.

Prototype Molds in Use

The corner stone is laid during the development phase for a new series to meet the (quite rightly) high expectations of the Mercedes customer right from the first vehicle of the series coming off the production line.

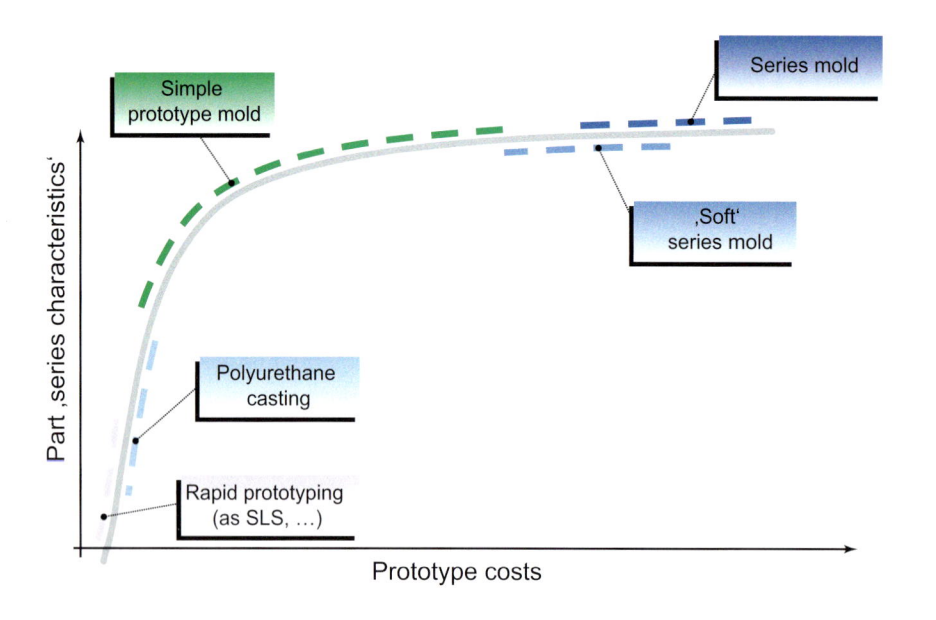

Figure 4-8 The prototype mold as cost driver

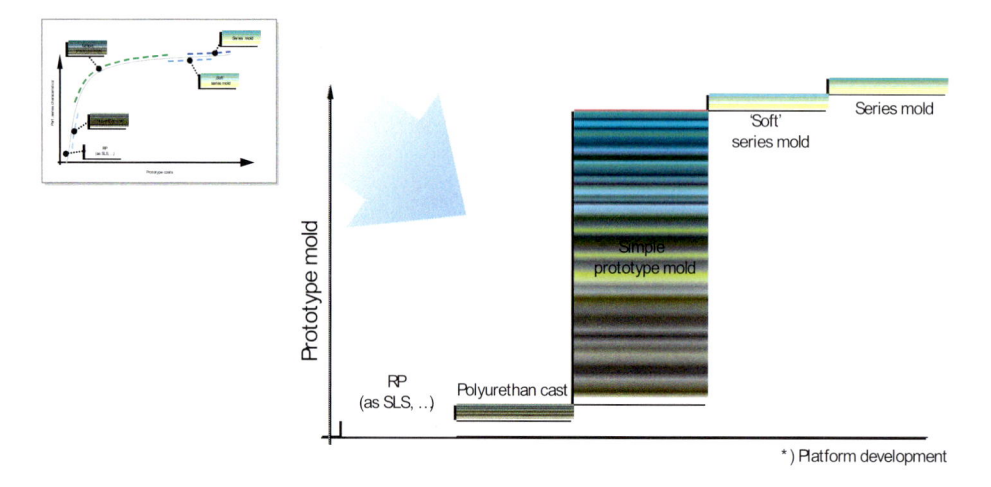

Figure 4-9 Prototype molds in new development series

In this regard the quality of the prototype component is a major factor. This is why more than 95% of plastic components in the development vehicle are made of the same material as the production vehicle and are manufactured by the same processes as the production vehicle. In most cases the testworthy component properties can already be achieved using simple prototype molds.

4.1.3 Selection of Suppliers

Collaboration with Prototype Suppliers

DaimlerChrysler does not operate its own production facilities for plastic components. These components are sourced from system suppliers. In the case of system suppliers, collaboration starts as early as the product creation phase. The supplier, working with a product brief, develops a component and subsequently supplies it to the production line. System suppliers are thus also responsible for providing prototype parts but usually obtain them, or their individual components, from their own (sub-) subcontractors in turn.

In a number of cases, such as when a system partner has not yet been decided on, the development department even procures prototype components directly from suitable prototype manufacturers.

Selection of the best prototype supplier is one of the most important (coordinating) tasks in this process. It forms the foundation for factors which will firm up later, such as scheduled time, quality and costs – not only in the first-off piece to specification but also in the event of modifications.

Selection Criteria

Due to both the regular occurrence of modifications and also a development process which in most cases has not yet completely concluded, prototype components and mold making are counted among so-called "products with a high clarification requirement" – even more so than their full-scale production counterparts. In addition to the criteria we have already mentioned,

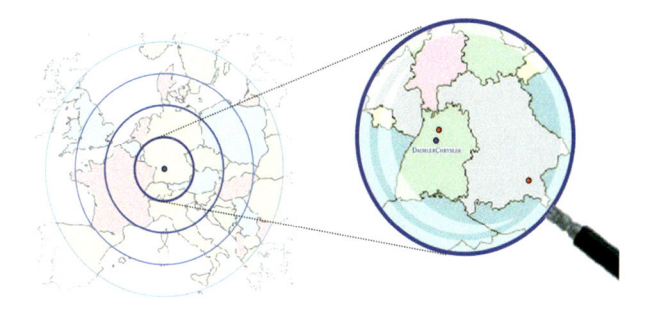

Figure 4-10 Regional proximity to the customer as a location advantage

geographical and communicative proximity (language, and so on) will therefore also be important factors in the selection of suppliers.

A poll conducted among a number of system suppliers revealed that preference is given to subcontractors who can be reached by car within two to three hours.

Nor can there be any doubt that it is no accident that the suppliers for the projects which we will be describing can also be reached from Stuttgart within this period of time.

4.1.4 Examples

Although geographical closeness to the customer is very important, it is obviously not in itself sufficient qualification for surviving under the very fierce competition in today's prototype market. Successful prototype suppliers are in addition usually characterized by employing innovative and customer-oriented technical concepts and/or processes.

The two examples which we will now go on to present should in each case throw some more light on one of these areas and thus describe some possibilities as to how very competitive prototype solutions can be implemented in a location such as Germany.

Testing of 'Class-A' Surfaces

The bumper fascia is one of the largest plastic components of a vehicle and one of the most important for

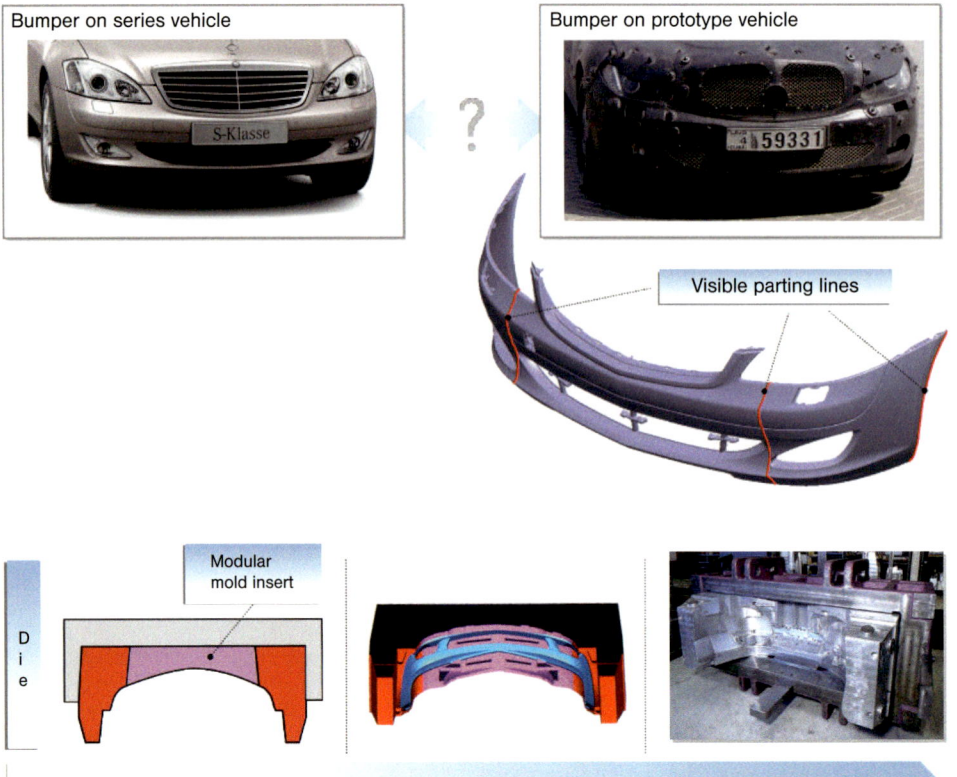

Figure 4-11
Bumper fascia: production version and prototype

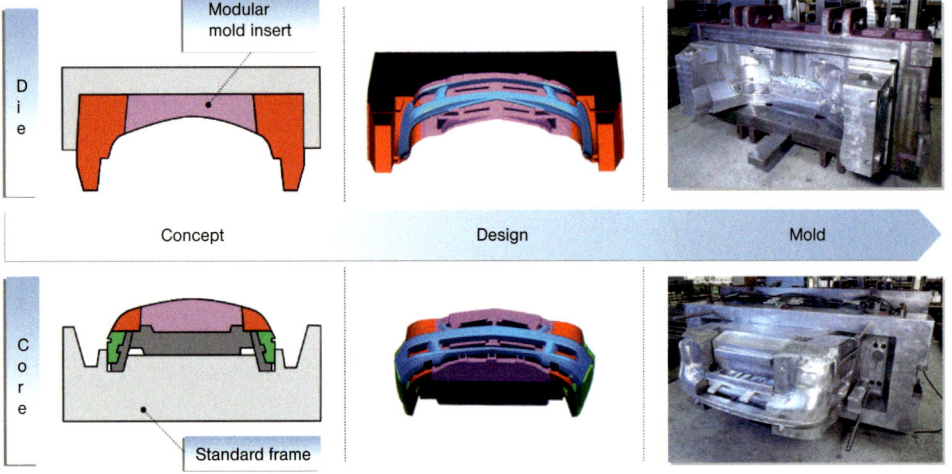

Figure 4-12
Bumper fascia mold in framework design

its esthetic design. There is no doubt that on account of their size and the class A requirements, the injection molds required for manufacturing these fascias can be regarded as belonging to the top rank of mold making.

However, during the development phase other criteria frequently have greater importance in the design of the prototype components. One very important requirement for modern cars is pedestrian protection, especially in case of accidents. Here the bumper system fulfills a very important function.

In order to ensure that the required pedestrian-protection properties are secured at an early stage, bumper fascias made of the original material are absolutely essential. But since the fascias have to be covered over as a disguise for the usual test drives, a class A surface or an expensive paint job for these parts is not usually necessary.

Framework Construction

In order to prevent visible mold parting lines on the exposed surface of the bumper fascia, the cavity side of the mold is normally made in a single piece. In the wheel housing area, the parting line is in addition routed along the inner side of the component thus making a demolding robot necessary for removing the component from the mold by force.

Dispensing with class A surface means that prototype molds can be made very rapidly and inexpensively on the basis of a framework design.

Here, instead of the solid one-piece cavity and core halves, a reusable mold framework is used which among other things already includes the guide elements and the ejector system. This framework accommodates mold inserts produced to customer-specifications and ensures the mechanical stability of the prototype mold. Since mold parting lines are permitted on the visible side of the prototype components, the mold inserts can be made on a modular basis from 'small' separate plates. In cases where visible parting lines are not desired, the mold inserts can of course be made in one piece as well. However modular design has the advantage not only that it uses very small quantities of material, but in particular also that the individual parts can be produced on several small machines simultaneously, thus saving time.

The modular structure was proposed for the framework-type injection mold during the course of a technical discussion with the prototype supplier and subsequently implemented in a pilot project. This example shows very clearly that the regular exchanges regarding component requirements on the one hand and technical possibilities on the other may be very advantageous to all parties involved.

Prototype Components Are often "Time-Critical" Components as Well

The MDS system lays down a clear time sequence for the course of a development project. This means that the individual steps can be tuned to each other and prepared. The same also applies to the structure of the development vehicles and of the prototype components required for them.

Unfortunately it is not such a simple matter to plan requirements during the subsequent testing. Findings relating to packaging and constructability or styling decisions may often make modifications to component design necessary in the same way as do negative results from the function tests. In such a case the prototype molds will then as a rule need to be modified to provide updated prototype components. In many cases it may even be necessary to redo the mold from scratch.

If this is not to endanger the time schedule for test drives or testing stands, the deadline for such items will mostly be "yesterday" ... but the data will, of course, have to be data from "tomorrow".

Some prototype suppliers for this reason specialize in this extremely fast production of prototype tools and components.

These companies are usually characterized by highly optimized processes and clearly defined standards, such as very flexible standard mold systems.

All of the parts shown in Figure 4-13 come from one of these suppliers. The delivery lead times they achieve here are oriented to meet requirements in the development process and in some cases even a shortening of the lead time would have been possible!

What is particularly interesting is that the continuous optimization of the processes in this company has not only enabled benchmark delivery times, but has also created a basis for benchmarking costs (and prices).

Standard Mold Unit

The standard injection mold unit developed and optimized within the last few years by the manufacturer of the parts shown in Figure 4-13 includes not only the machine connection surfaces and the installation surfaces, but also regularly required mold elements, such as the ejector system. The standard mold unit is available in a number of different standard sizes – matching the injection-molding machines that are available – and is used for virtually all prototype molds.

The system used here differs from other known standard units only in that it has support strips in the intermediate plate. In contrast to the frequently used cavity-mold-like arrangement of the ejector holes and the resultant restrictions in the positioning of ejector pins on the component, here the ejectors can be located virtually anywhere on the mold plate. Mechanical support of the mold plate is taken over by the support strips that can be positioned in unused areas around the ejector pins.

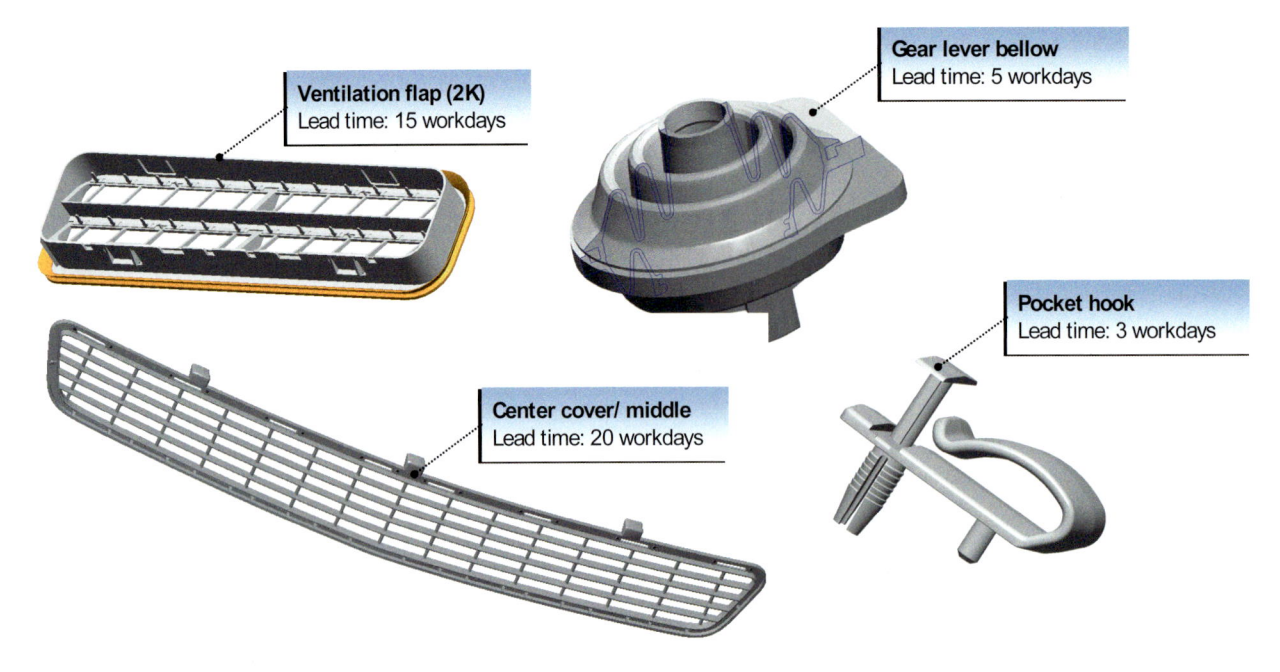

Figure 4-13 Some time-critical prototype components

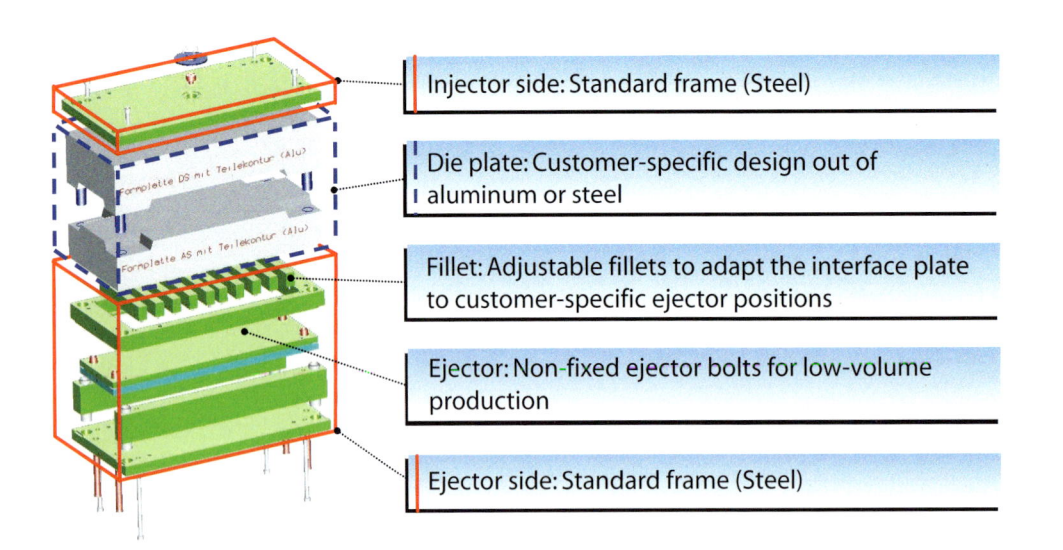

Figure 4-14 Flexible standard unit for prototype molds

The mold plates are usually made of aluminum. Here the focus is clearly on the actual cavity for the component; its external appearance attracts considerably less attention than is otherwise the norm in moldmaking.

An Outstanding Example of CIP Philosophy

Although the standard mold unit we have been talking about offers a large number of very good solutions for the rapid and cost-optimized making of prototype injection molds, it is certainly not the key to the success of the company. This key is rather to be looked for and found in the rigorous implementation of a simple – and for this very reason a very interesting – company philosophy. In the light of this, the standard unit is simply a natural outcome of the philosophy.

Some extracts from the process description and the company philosophy – in conjunction with our own experiences from collaboration – will throw light on the approach we have selected:

"... we work by the principle: what's inside, not the packaging, must appeal to the customer"
in each case, only those mold components which really are customer-specific are built from scratch and charged for accordingly the focus when building prototype molds is clearly on the cavity; the external shape is kept very simple

"... a costing calculation program that grows over the course of years (200–400 recalculations per year) makes it easier to prepare a quotation on short notice"
quotations can normally be prepared within a few hours, even within minutes if necessary, and are chargeable

"... productivity increases are passed on 1 : 1 to the customer"
learning effects are passed on in the form of cost reductions in subsequent orders

"... every order is maintained and planned in detail on the intranet ... the quantity of information is kept as small as possible ... everything necessary is included but a large quantity of standards is possible"

"... at least once per week, often even on a daily basis, an analysis is made where faults and bottlenecks have occurred"

"... employees can use the intranet to check the status of a project; this means time-consuming meetings can be avoided"

"... no plans or drawings are required, everything has been (and will be) standardized as much as possible".

The continuous improvement of business processes via, for example, the introduction and further development of company standards, and a clear focus on customer requirements in product design made it possible to build up a very competitive company.

Although these principles do not seem new or particularly complicated, implementing them calls for a high degree of discipline and follow-through in everyday company activities – this applies to both employees and managers.

This example does, however, show that this work can be very rewarding.

4.1.5 Summary

The motto propounded by Dr. Dieter Zetsche when he became managing director of the Mercedes Car Group applies in particular to the prototype process in product development: Achieve more with less!

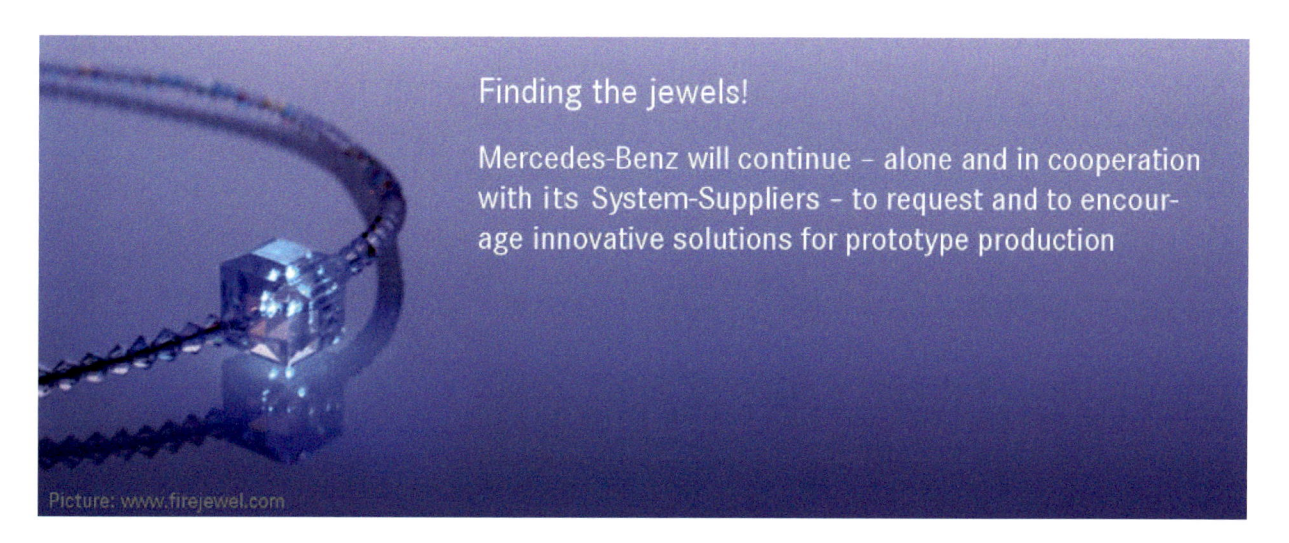

Finding the jewels!

Mercedes-Benz will continue – alone and in cooperation with its System-Suppliers – to request and to encourage innovative solutions for prototype production

Picture: www.firejewel.com

Figure 4-15 Find the jewels

Summary and Outlook

The on-time availability of high-quality prototype components under competitive conditions is one of the success factors in passenger car development. For this reason, the development department at Mercedes-Benz is just as intensively involved in this area as are its suppliers (and competitors).

Within the context of development, often very heterogeneous demands are made on prototype components. Optimum matching of the component to these requirements is therefore absolutely necessary for both technical and economic reasons.

Direct communication between the prototype manufacturer and the development department – who are the end users of the parts – has proved to be very helpful, even when procurement in most cases ultimately takes place via intermediate system suppliers.

The examples we have described clearly demonstrate the potential of such direct communication and collaboration. The location-related advantage which can arise in this connection from geographical (and linguistic) proximity to the customer should not be underestimated. In addition, the above examples reveal additional ways of manufacturing competitive prototypes – including locating in Germany.

4.2 Rapid Prototyping of Plastic Components for Efficient Vehicle Development

WOLFGANG GEISLER

Shortening development times require that design and development be supported by the availability of rapidly produced components for the most varied purposes. The prototype shop at Volkswagen meets this requirement by the use of rapid prototyping (RP). The strategic importance of RP is described and compared with virtual methods. An overview is provided as to which procedures are available and which of them are used in our department. In focusing on plastic components, we explain the purposes for which components made using RP can already be used today. This is backed up with examples.

Rapid prototyping (RP) and rapid tooling (RT) for rapid representation of parts.

The task of predevelopment in the prototype shop is to push these lines of advance ahead and to put the corresponding technologies and methods (processes) to efficient use in prototype assembly.

This means that rapid prototyping is a strategic element which in our view makes an positive contribution to the improvement not only of the development process (Figure 4-16) but also the product (Figure 4-17).

4.2.1 Introduction

The following contribution may be subdivided into two main parts. In the first part an attempt is made to describe the importance, basic principles, possibilities and development trends of rapid prototyping. In the second part, we look at the practical side, its use in Volkswagen's prototype shop, although our description will be limited to rapid prototyping of plastic components.

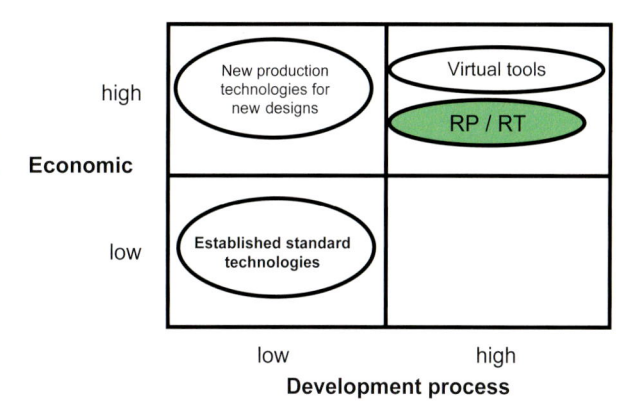

Figure 4-16 Strategic importance of RP for the development process

4.2.2 The Importance of Rapid Prototyping

Volkswagen's prototype shop (VB) in Wolfsburg constructs items for trials und testing, ranging from individual parts to vehicle prototypes, everything in fact needed to get a new vehicle on the road. In performing these tasks, the prototype shop must, by shortening its construction times, make its own contribution to shortening product development times.

This objective is pursued with two main aims: rigorous and logical use of virtual tools as a means of verifying constructability and the construction process.

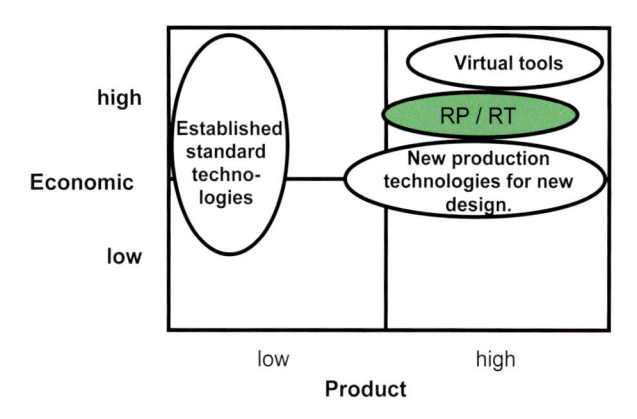

Figure 4-17 Strategic importance of RP for the product

Do the virtual tools assessed in Figures 4-16 and 4-17 in fact supplement or compete with RP? The answer is: both.

By relocating development loops into the virtual world – in other words, replacing the development loop "design – prototype part – test" by the loop "design – mapping onto an FEM network – simulation" – they can be speeded up considerably and costs lowered. However, test results and thus real components continue to be required as was previously the case in order to be able to compare simulation programs with reality (preparation of virtual development loops), to supply input data for calculations, and later on in retrospect to confirm simulation results.

Due to the innate limits (which still exist) of virtual tools (for example, virtual parts are not fastened and do not fall off) there are different purposes for which VR and RP are used within the prototype shop.

- Virtual tools are used for adjusting and tuning the producability and installability of designed parts
- Rapid prototyping is used for coordination and defining improvisations.

Furthermore, the prototype shop as a service provider manufactures

- RP viewing models for the design departments
- RP communications models for development co-ordinating and adjusting in the design and with suppliers,
- RP functional models (usually with limited testing capability for the testing departments).

What needs to be determined are the similarities and differences in the methods between working with RP and with virtual tools:

- In both cases these auxiliary resources can be included at an early stage in the development process.

There are differences in the sequence of the coordination process:

- In team sessions with virtual tools, the group is led by its chairman through the problem areas to be found with a particular vehicle. The group has a shared and identical view (forced on them by virtual tools) of the problem situation. The meeting calls for a disciplined attitude and for emotions to be kept in check.

- In the team sessions with RP parts the group stands in a circle around the vehicle. Each person can select his own particular angle for viewing the installation problem. The group members can in the truest sense of the word "grasp" the problem and work off their emotions by trying things out hands-on (what will work and what won't).

What this all shows is that rapid prototyping has its own autonomous place as a strategic element which it will maintain despite the orientation of product development processes towards development loops in the virtual world.

4.2.3 Rapid Prototyping – a Definition

The term 'rapid prototyping' is generally applied today to all methods of developing the first samples at a fast speed. This may also include, for example, procedures for the rapid development of software prototypes.

In particular, however, generative procedures for making parts from 3D CAD data are referred to as rapid prototyping on account of the little time required to make them.

As a representative of all methods we will provide a brief description of selective laser sintering (Figure 4-18).

The model is built up layer by layer on the build platform. Here, by means of a mirror galvanometer, the laser beam is directed at all points on a layer which belong to the work piece. At these points the laser melts the powder-like build material. Once a layer has been fixed, the build platform is lowered by one layer thickness and a new thin powder layer applied by the coating mechanism. The procedure is repeated until the entire component is completed. Typical build materials are nylon, wax, polycarbonate, polystyrene, plastic-coated metal powder or even uncoated metal powder.

Currently there are about 20 different approaches to RP. Figure 4-19 attempts to show the connections between the most important procedures, technologies and materials.

Some of the materials that can be used for building the models include liquid, powder or solid materials, and most frequently plastics. The technologies used for curing or melting the material include ultraviolet light, laser light and now also the electron beam. Particularly for purely viewing models, techniques similar to the inkjet printer tend to be favored today since they can even be used in the office. For al good overview of methods and vendors, see the internet site www.rp-net.de from which Figure 4-18 has also been taken.

Since all generative methods run automatically, the work pieces must be available in 3D CAD data. These are then converted into a special data format (STL data) which is understood by all RP machines. Here it is necessary with some methods to apply additional support structures to enable overhangs to be generated.

As the result of the generative process, a work piece model is obtained that is built up from bonded layers. Depending on the particular process which was used, the individual layers may be visible as steps on the work piece surface, either clearly or almost imperceptibly. The surface will need reworking to a greater or lesser extent depending on what surface quality and "steppedness" is required or acceptable.

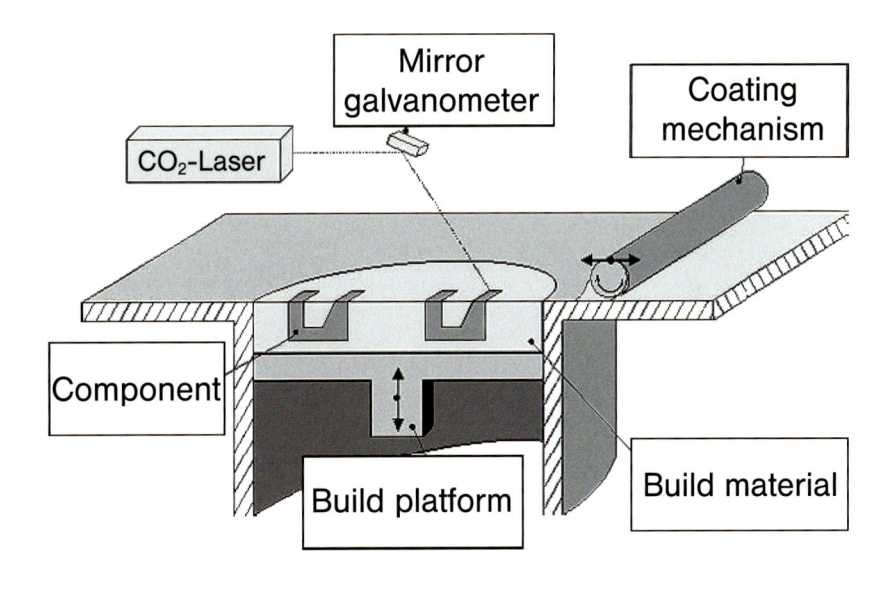

Figure 4-18 Principle behind selective laser sintering (SLS)

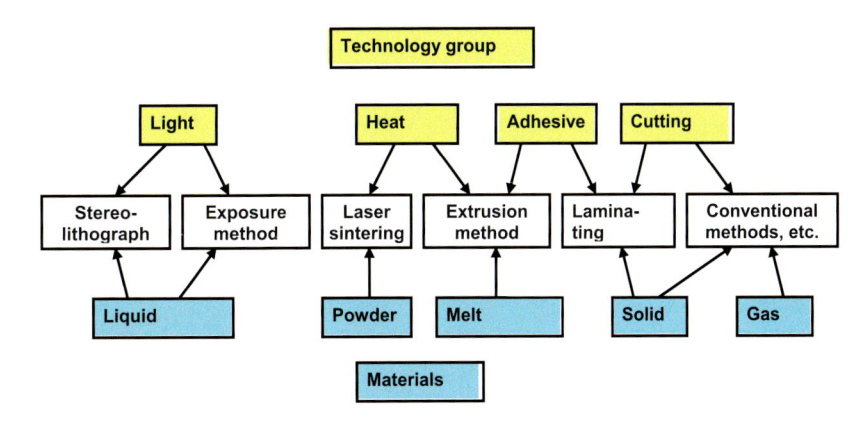

Figure 4-19 Rapid prototyping methods, relation between technologies and materials

4.2.4 Rapid Prototyping (RP), Rapid Tooling (RT), Rapid Manufacturing (RM): Some Differences

Since the generative RP model is not usually created from the material used for series production of the work piece, it will have only limited capability for trials or testing. To generate parts in the final material at a quality close to that of series production, the RP process is used for rapidly creating an injection mold, for example (rapid tooling). Since rapid tooling is dealt with in another contribution, at this point we will only be concerned with explaining the differences which influence whether RP or RT is selected. On the one hand, these are the quantity of items which can be produced over time (Figure 4-20) and also the curve of cost plotted against quantity (Figure 4-21).

These diagrams show that rapid prototyping has advantages over RT at low quantities and when parts are needed immediately.

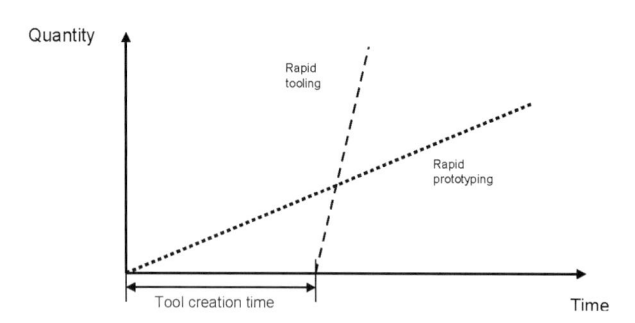

Figure 4-20 Relative time required by RP and RT for producing parts

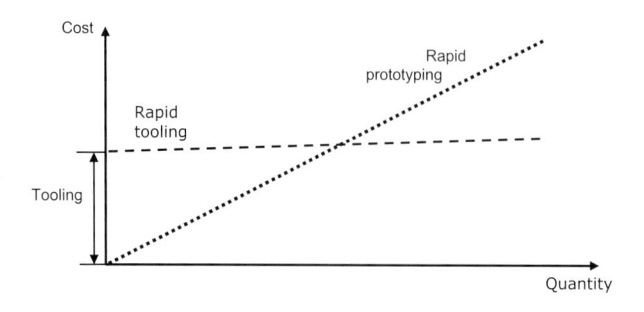

Figure 4-21 Cost plotted against quantity with RP and RT

The term rapid manufacturing is used when generative methods are used to create the end product. This is economically possible and reasonable when, for example, individually adapted parts with a complex geometry are required. A manufacturer of hearing aids uses RP, for example, to manufacturer the plastic housing. Since even highly stressable parts can be made by direct metal laser sintering, an example of one area of application is turbine blades. Since in automobile manufacturing even small series mean unit quantities greater than 10,000 per year, rapid manufacturing is appropriate for RT, but not for RP.

4.2.5 The Prototype Assembly Department as Internal RP Service Provider

In addition to production for its own purposes, the prototype shop (VB), as has already been mentioned, offers rapid prototyping as a service for the entire development division. As a way of conveying information about RP and its possibilities to the customer, the prototype assembly department makes an extensive range of relevant material available on the intranet. This includes a choice of selectors which the customer can use to find the right method and material for his particular problem. But naturally the pointers on the intranet all lead to the prototype assembly experts who can give the customer the best advice about solving his problem. Despite all scientific advances the selectors are not yet capable of functioning simply for the customer on one hand nor on the other of converting the complex description of his problem into a clear proposal as to which method should be used. To be able to assess one's own performance and that of the various methods, a number of measurements are necessary. The following process steps are defined for this:

- Receipt of order and preparation of the CAD data record
- Fabrication of the models by means of generative manufacturing processes
- Reworking of the models
- Possible casting of the models
- Use of the model by the customer.

4.2 Rapid Prototyping of Plastic Components for Efficient Vehicle Development

Figure 4-22 uses the example of the VW emblem to show how data were prepared for evaluation. Figure 4-23 shows the processing times for different components. The surprising conclusion drawn from this is that the speed achieved by using RP is in part lost again by the customer. In order to monitor performance, it will also be necessary to determine both the fabrication times and costs as a function of the quantity. For reasons of commercial confidentiality we are unfortunately unable to reveal these results here.

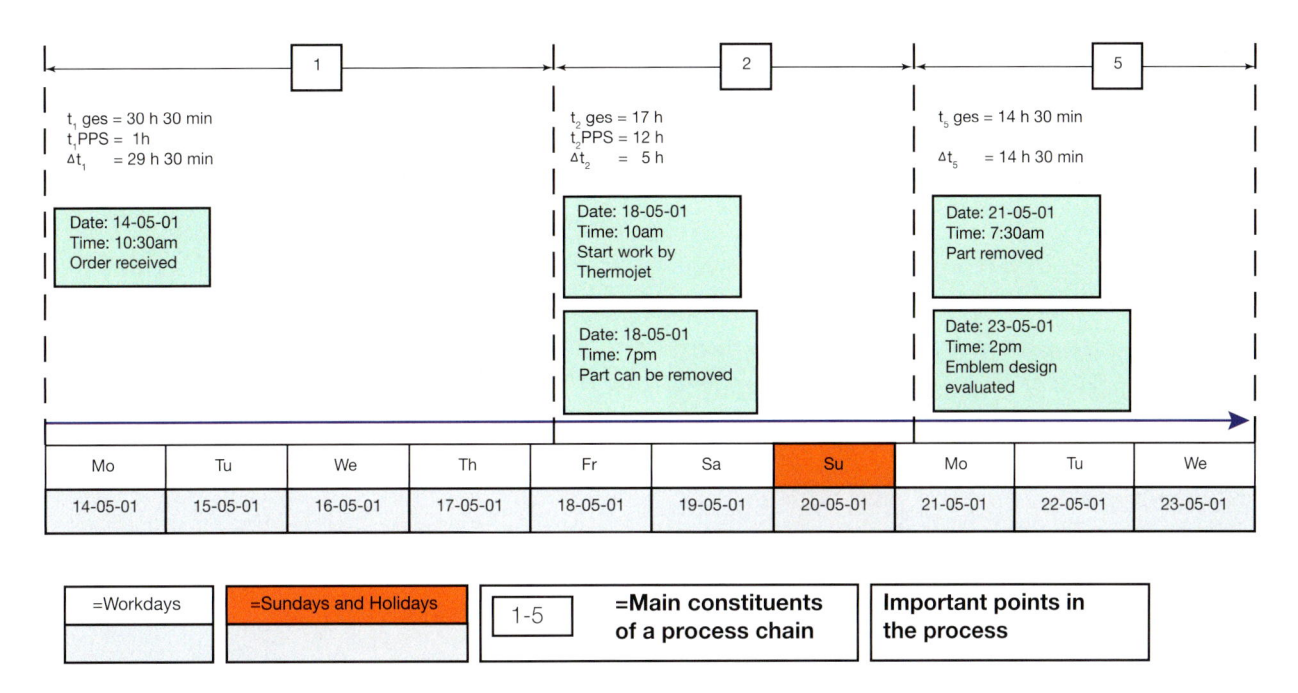

Section 1
- t_1 ges = 30 h 30 min
- t_1 PPS = 1 h
- Δt_1 = 29 h 30 min
- Date: 14-05-01, Time: 10:30am, Order received

Section 2
- t_2 ges = 17 h
- t_2 PPS = 12 h
- Δt_2 = 5 h
- Date: 18-05-01, Time: 10am, Start work by Thermojet
- Date: 18-05-01, Time: 7pm, Part can be removed

Section 5
- t_5 ges = 14 h 30 min
- Δt_5 = 14 h 30 min
- Date: 21-05-01, Time: 7:30am, Part removed
- Date: 23-05-01, Time: 2pm, Emblem design evaluated

Mo	Tu	We	Th	Fr	Sa	Su	Mo	Tu	We
14-05-01	15-05-01	16-05-01	17-05-01	18-05-01	19-05-01	20-05-01	21-05-01	22-05-01	23-05-01

Legend: =Workdays; =Sundays and Holidays; 1-5 =Main constituents of a process chain; Important points in the process

Figure 4-22 Measurement of speed for RP component VW emblem

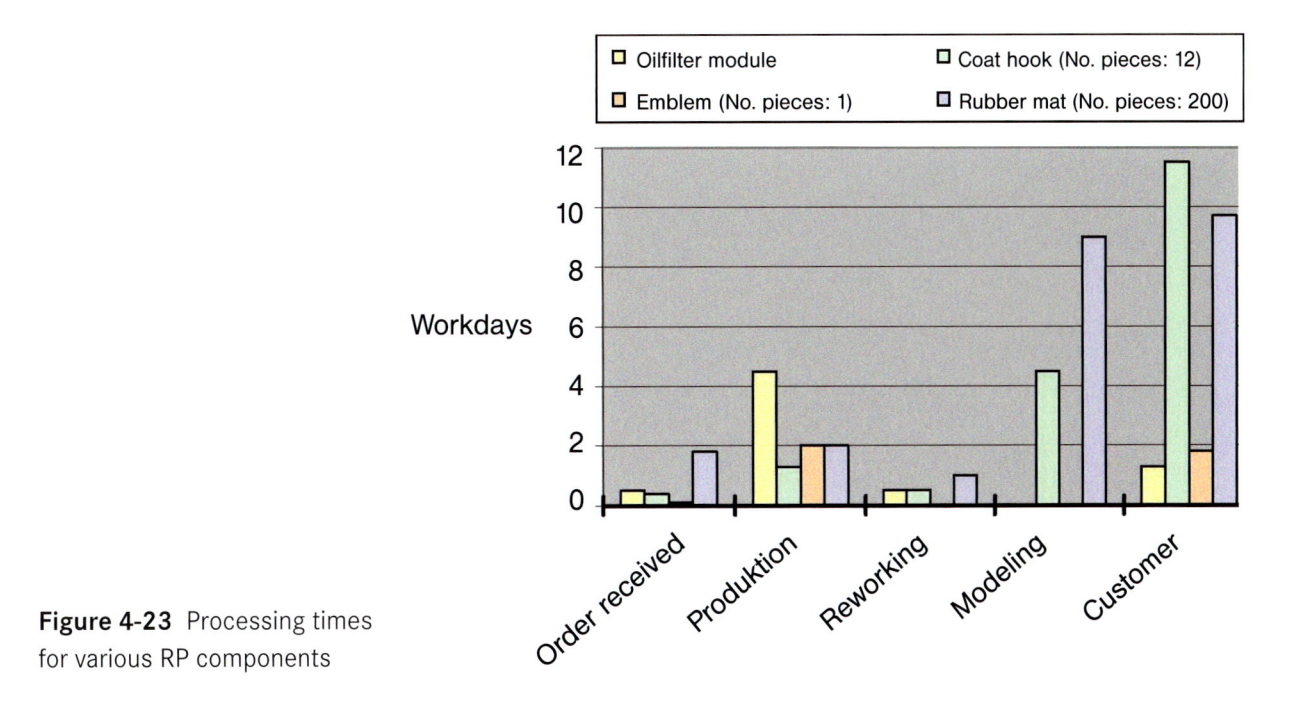

Figure 4-23 Processing times for various RP components

(Bar chart: Workdays vs. stages — Order received, Produktion, Reworking, Modeling, Customer — for Oilfilter module, Emblem (No. pieces: 1), Coat hook (No. pieces: 12), Rubber mat (No. pieces: 200))

Figure 4-24 EOS P360 laser powder sintering unit

Figure 4-25 Components made of polyamide

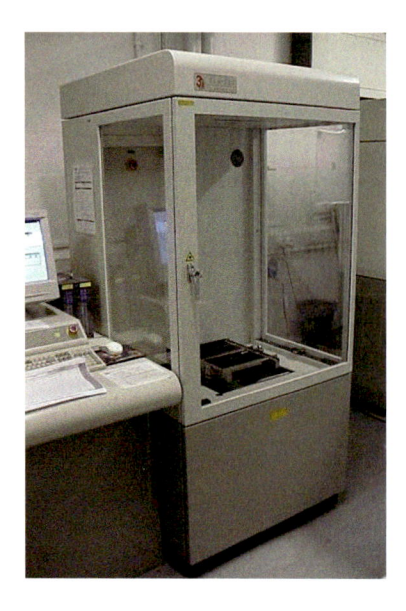

Figure 4-26 SLA350 stereolithography unit

Not until one has gathered experience using his own equipment and installations is it possible to give the customer the best advice or to estimate the performance of external service providers. For this reason, the prototype assembly department uses its own systems although extensive use is made of outside RP suppliers to cover the overall demand. To avoid losing any time in the procurement process in the individual case, general contracts are concluded within which services with a defined scope can be rapidly ordered.

4.2.6 Systems and Applications in the Prototype Assembly Department

A brief sketch is now provided of the systems and associated applications.

The *EOS P360 laser powder sintering unit* (SLS) (Figure 4-24) is used for manufacturing parts from polyamide powder (Figure 4-25) which are to a limited extent capable of undergoing trials and tests.

These parts are used for investigating installation space in, for example, the engine compartment or for carrying out a limited check of function. Examples include sill strips, A, B, C pillars, radiator protection grilles, vent grilles, air outlets, steering column panels, wheel housing liners.

SLS parts are becoming more and more important since the material properties of the powders used have improved constantly over the years and are still being developed further.

The *SLA350 stereolithography unit* (Figure 4-26) is used for making sample pieces from epoxy resin – in general, these are not suitable for testing purposes. For this reason, the unit is normally used for our process chain *epoxy-resin sample piece – silicone mold – polyurethane part (cast)* (Figure 4-27) when a quantity of 10–30 is required or the economic efficiency of other methods is poorer. Typical applications are trim panels, switch covers, air outlets, seat covers, headlights. Further areas of application include flow models, water cores and optical parts for the design

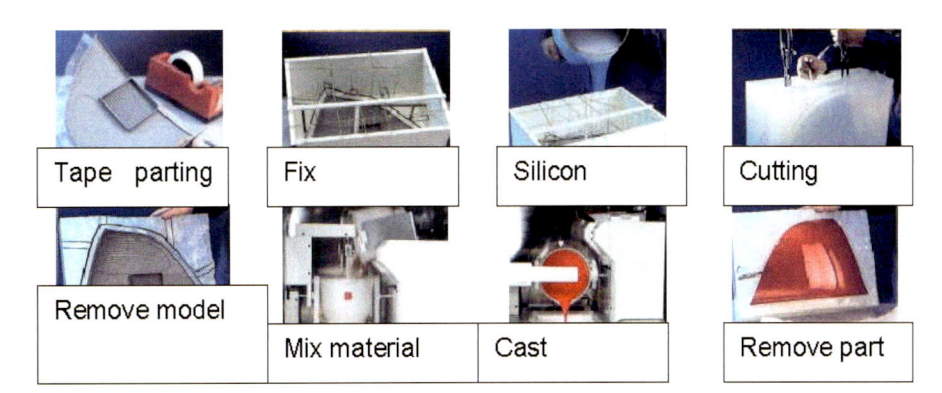

4.2 Rapid Prototyping of Plastic Components for Efficient Vehicle Development

Figure 4-27 Process chain for a polyurethane part

Tape parting | Fix | Silicon | Cutting
Remove model | Mix material | Cast | Remove part

Figure 4-28 Thermojet 3D printer

Figure 4-29 Z8 10 3D plaster printer

department. These parts are very accurate and can be very easily machined or further processed.

The *Thermojet 3D printer* (Figure 4-28) is used for making design models and manual samples from wax. These are not suitable for testing purposes. This method is characterized by good surface quality since it combines very high local resolution with small layer thickness (0.025 mm). Examples include rim sections at a scale of 1 : 18, automobile models at a scale of 1 : 20, or interior parts.

The Z810 3D plaster printer (Figure 4-29) is primarily used for making higher-volume models (up to $600 \times 500 \times 400$) for design purposes (Figure 4-30).

Figure 4-30 Semi-finished volume model for installation checks

For installation checks, the models are infiltrated with polyurethane or epoxy resin to make them more robust. Applications include transmission housings, automobile models at 1 : 10 scale, rim models, water tanks. The method is very fast (design space in 20 hours) and is the cheapest RP method as far as the material used is concerned.

4.2.7 Outlook

Both machines and procedures have reached a stage of relative maturity. As far as automotive engineering is concerned, larger design spaces would be desirable. We know of 24 different equipment manufacturers with a total of around 8000 RP systems in operation worldwide. The methods are for the most part protected by patens and allow little room for the development of special systems and materials. We expect the greatest developments in the materials sector. The trend will be towards materials with properties similar to the materials used in series production, up to and including flame resistant materials (for example, for aircraft engineering) or the use of series-production materials. In order to be able to participate in these developments, the prototype shop is collaborating with the Rapid Prototyping Alliance of the Fraunhofer Society. Detailed information about this may be found on the internet at www.RapidPrototyping.fhg.de.

In automobile manufacturing, even small series still have high unit quantities and short cycle times as compared with other industries. For this reason, for the automotive industry rapid manufacturing is currently neither economically efficient nor can it offer the required performance. In small-scale series, rapid tooling is worth considering as a way of producing small-series molds quickly.

References

[1] Rapid Prototyping and Tooling, VDI-Berichte 1686, 2002, ISBN 3-18-091686-9
[2] Gebhardt: Rapid Prototyping, Hanser Verlag, 2002, ISBN 3-446-21242-6

4.3 Calculating the Fatigue Life of Short-Glass-Fiber-Reinforced Thermoplastics

Martin Brune

4.3.1 Introduction

Due to their low weight combined with high strength and rigidity, plastic composites have already in some cases usurped the place of metallic materials in the field of structural components. Nevertheless plastic composites have so far not been used to the extent one would have expected from the quality of their material properties. The reason for this may basically be found in the high costs involved in manufacturing components from these materials. The cost aspect will have decisive importance particularly when components are concerned that are required in large numbers combined with a correspondingly short production time. This is one reason why this group of materials is underrepresented in the automotive industry, despite having great advantages over competing materials for lightweight design.

One way of reducing manufacturing costs is to use the injection molding process to produce these components. With this production method, components can be produced rapidly at a relatively low unit cost. However, there is one disadvantage to this method that will have to be accepted, i.e., such injection-molded chopped-strand-reinforced thermoplastics have a shorter fatigue life under dynamic operational loading than long-fiber- or continuous-strand-reinforced materials. Before inexpensive chopped-strand-reinforced thermoplastics can qualify for further applications in vehicle design, the behavior of the material under cyclic loading will have to be investigated and understood in detail. In the following we will attempt to describe the characteristics of this material on the basis of an examination of the relevant literature and our own research. Armed with this knowledge, we have developed one possible way of calculating the fatigue life of components made of chopped-strand-reinforced thermoplastics [1–3]. We will illustrate our approach with a number of examples of such calculations.

4.3.2 Material Behavior under Cyclic Loading

As part of our work in developing a procedure for calculating the estimated fatigue life of thermoplastic materials we made use of existing knowledge gained in the field of metallic materials. Our intention was to take the fatigue life hypotheses developed in particular for steel and to transfer them, after making suitable modifications, to the group of materials we are concerned with here. The first step in developing a new fatigue life prediction method is to analyze material behavior under cyclic loading so as then to be able to formulate the corresponding hypotheses in subsequent steps. To do so, not only are comprehensive experimental investigations carried out on material samples but studies from the relevant literature are also evaluated. The samples used for our own tests consist of the material copolyamide (CPA) with glass fibers 0.4 mm long and 10 µm thick. This material was made available to us by EMS-Chemie. Table 4-1 shows the variation in samples for the test program [1].

The samples are 150 mm long, 10 mm wide and 2 mm thick. A sketch of the shape of the samples is shown in Figure 4-33. With these samples, tests were carried out at constant stress amplitudes with fatigue tensile stress at a stress ratio of $R = 0$ and with cyclic tensile/compressive stress at a stress ratio of $R = -1$. The stress levels were normalized to the tensile strength of the materials is use and appear in Figure 4-31 (left), taking fatigue stresses, by way of example, as a Wöhler curve. This represents the results in the case of a 50%

Table 4-1 Fiber orientation and weight content in the material samples used

Fiber orientation [deg]	0	45	90
Fiber weight [%]	30 and 50	50	30 and 50

probability of occurrence while Eq. (4.1) describes the so-called Wöhler master curve.

$$\frac{S_a}{S_t} = g + h \log N \qquad (4.1)$$

The variables g and h in Eq. 4.1 describe the position and slope of the straight lines and are a function of the stress ratio R [1]. The high degree of conformity between the Wöhler master curve and the plotted test points led to the idea of using Eq. 4.1 with stress ratios ranging from static tension ($S_m / R_m = 1$ (Figure 4-31, right)) to $R = -1$ as a basis for calculating the Wöhler curves for thermoplastic materials.

The mean stress susceptibility of the material used is shown in the *Haigh-Goodman* diagram (Figure 4-31, right). Once again, this is a visualization normalized to the tensile strength and determined with the aid of the Wöhler master curves for the stress ratios being investigated. Figure 4-31 (right) shows an example of samples with a 0° fiber orientation and consisting of 30% fibers by weight.

In order to determine the applicability of the Wöhler master curve for different thermoplastic materials, a thorough examination of the literature [4–16] was undertaken. All available test results were plotted in a Wöhler diagram in each case for the same fiber materials and with the same stress ratio of $R = 0.1$.

Figure 4-32 shows these diagrams for carbon-fiber- and glass-fiber-reinforced thermoplastics. In addition the Wöhler master curve is drawn in as per Eq. 4.1 in each case. The high conformity observable here between calculated Wöhler curves and test results obtained from the relevant literature corroborates the validity of the procedure we have described.

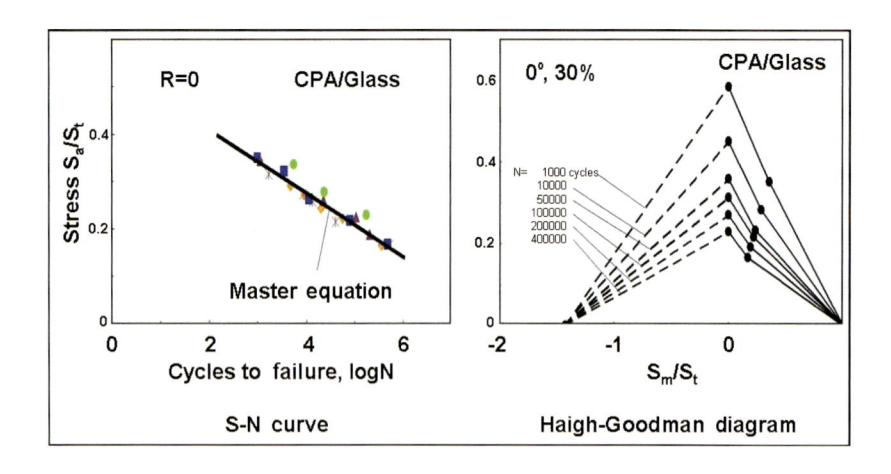

Figure 4-31 Wöhler tests and mean stress susceptibility for specimens made of short-glass-fiber-reinforced polyamide

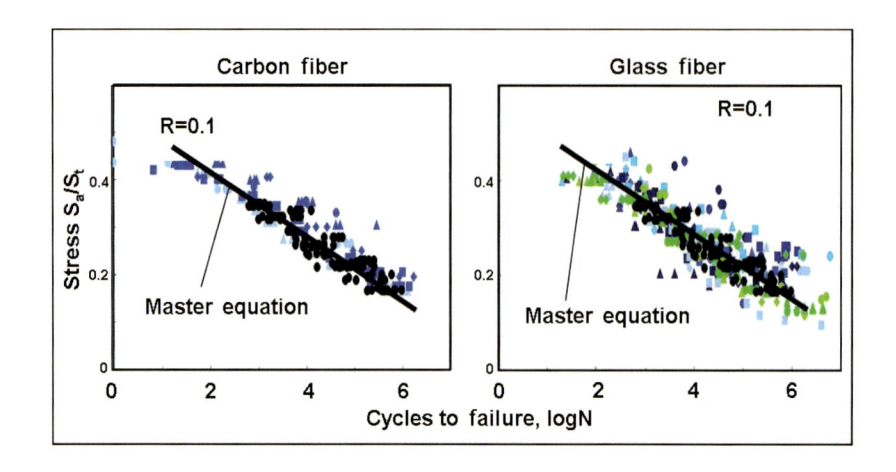

Figure 4-32 Wöhler tests obtained from the literature compared with the calculated Wöhler master curve

The next step in the analysis is investigation with respect to damage accumulation [2]. In the case of metallic materials, the use of linear damage accumulation according to *Miner* is very widespread. Whether application to thermoplastic materials is permissible should be revealed by tests with the material samples already described. These tests were carried out with different load-time sequences (block program and random) until failure of the test pieces. The *Miner* rule as per Eq. 4.2 can now be used for calculating fatigue life:

$$\sum_j \left(\frac{n_j}{N_j}\right)^b = 1 \tag{4.2}$$

The variable quantity is the exponent b which is now set in the provisional calculation with the Wöhler master curve on the load capacity side such that the best possible agreement with the test results is achieved. This is averaged over all test results at $b = 0.92$. Figure 4-33 presents a comparison of the results calculated with this provisional formulation as well as the test results. A high level of conformity may be seen even for random stress loading.

4.3.3 Formulation for Calculated Fatigue Life Estimation

The findings obtained regarding material behavior under cyclic stress are now input into a general formulation for calculating the estimated fatigue life of components made of short-fiber-reinforced thermoplastics. Development of such a formulation took the following steps [3]:

- Calculation of fiber distribution and orientation in the component.

- Calculation of the stress distribution effective in the component.

- Estimation of the component's fatigue life on the basis of the *Miner* calculation taking mean stress as per the *Haigh-Goodman diagram* into consideration.

The main steps in the calculation will now be dealt with in more detail.

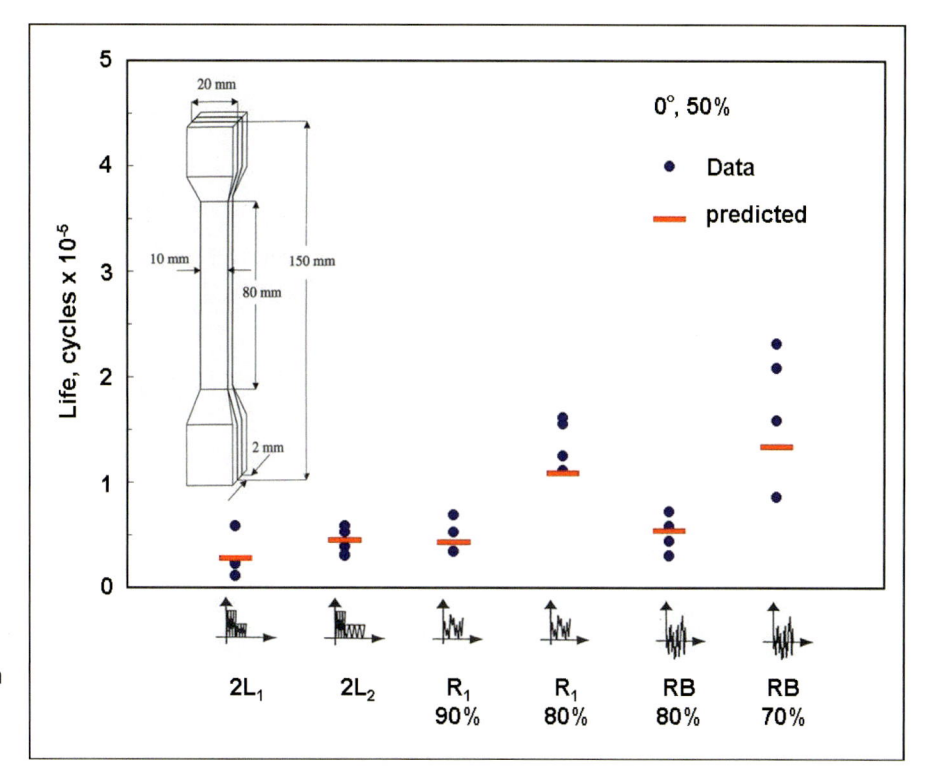

Figure 4-33 Comparison between the new calculation formulation and the test results

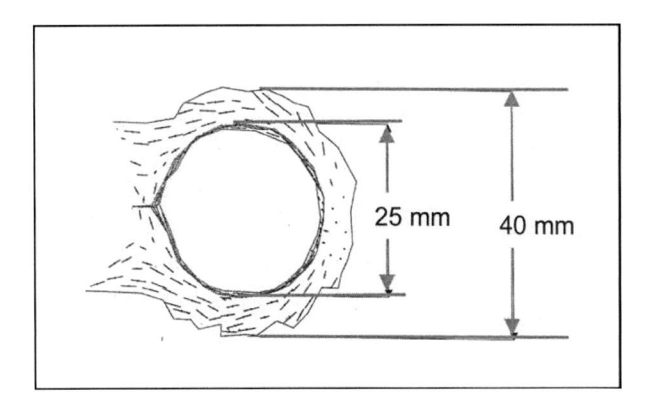

Figure 4-34 Calculated fiber orientation in a gear-shift linkage component using the C-Mold application [17]

4.3.3.1 Fiber Distribution and Orientation

In the early stages of the vehicle design process, no components are as yet available in physical form, meaning that fiber orientation and distribution cannot be measured directly. During this phase it is therefore necessary to determine these quantities by theoretical calculation. Investigations are carried out using the software programs C-Mold [17] and Moldflow [18].

Figure 4-34 shows an example of computer-calculated fiber orientation.

4.3.3.2 Calculation of the Stress Distribution Effective in the Component

The next module to be tackled in our journey towards calculating fatigue life is stress analysis for the component under consideration. Proceeding in three steps appears advisable here.

1. Determining fiber distribution and orientation in the component.
   The procedure has already been described.

2. Stress calculation with static unit loading cases.
   The normal stresses are calculated and the smallest and the greatest stresses identified. The nodes in the finite element network are recorded with the minimum and maximum normal stresses.

3. Stress calculation with stress amplitudes varying over time.
   Dynamic stress in the form of a load-time function with stress amplitudes of various sizes is classified with the aid of the rain flow counting procedure [19]. All stress amplitudes together with information as to amplitude size, mean stress and frequency of occurrence are stored in the rain flow matrix. Next, the normal stresses $\tilde{\sigma}^{max}$ and $\tilde{\sigma}^{min}$ are calculated for each matrix entry as a function of F^{max} and F^{min} for every node in the finite element network.

$$\tilde{\sigma}_{max} = \tilde{\sigma}^0 \frac{F_{max}}{F^0} \qquad \tilde{\sigma}_{min} = \tilde{\sigma}^0 \frac{F_{min}}{F^0} \qquad (4.3)$$

In Eq. 4.3, $\tilde{\sigma}^0$ is the normal stress with the static unit load F^0 under tensile or compressive stress.

Component stresses are generally calculated using the finite element method. In the case of very long load-time signals, correspondingly large amounts of computing time will be required. This problem can be avoided by assuming the material behaves linear-elastically. If this assumption is proceeded with, it is recommended that the static unit loading cases be determined first.

An advantage here is that these calculations only need to be carried out once; not only that, but commercial software tools very commonly encountered in industry are used. These calculation programs must be capable of dealing with anisotropic material behavior, such as is, for example, caused by differences in fiber orientation. For this purpose we used a software application brand-named ABAQUS [20].

Following calculation of the unit loading cases for static loading, stress calculation was carried out under dynamic loading by linking the corresponding factors in the stress amplitudes to the unit loads. In the case of stress on a component involving several stress input variables, each load is first considered individually. Once loading from all stress input variables has been calculated, the corresponding stress tensors are overlaid.

4.3.3.3 Calculated Fatigue Life Estimation

The fatigue life calculation method described above is now applied using the Wöhler master curve and the modified *Miner* calculation with the exponent $b = 0.92$. The main steps in the calculation are as follows.

1. Calculation of the angle between fiber orientation and maximum/minimum normal stress
2. Calculation of stress ratio R and stress amplitude $\tilde{\sigma}_a$

$$R = \frac{\tilde{\sigma}^{min}}{\tilde{\sigma}^{max}} \qquad \tilde{\sigma}_a = \frac{\tilde{\sigma}^{max} - \tilde{\sigma}^{min}}{2} \qquad (4.4)$$

3. Application of the values calculated using Eq. 4.4 for evaluating mean stress susceptibility as per the *Haigh-Goodman diagram*.
4. Application of the modified *Miner* calculation with the sustainable stress weighted for the mean stress.

The damage thus calculated is linear accumulated, vibration cycle by vibration cycle, in accordance with Eq. 4.2. By definition the part will fail when the accumulated total damage reaches a value of 1.

4.3.4 Examples of Calculation

The modified accumulation of damage for short-fiber-reinforced thermoplastic materials shown by us was developed with the aid of test results from the literature and also from our own tests. The basis for our hypotheses is the corresponding test results for sample materials. In order to be able to judge the accuracy of the method or its applicability to real components, the results of grading and endurance strength tests carried out on two automobile components were compared with the results of calculations. The components in question are, first, a drive component from the gearshift linkage (Figure 4-35) and, second, a control arm from the running gear (Figure 4-39). Both components are made of polyamide with 50% short-glass-fiber by weight (length: 0.4 mm, diameter: 10 μm) and manufactured by injection molding. These components, made by EMS-Chemie in Switzerland, are not currently used in series production but have the status of test samples for feasibility studies.

Figure 4-35 Geometry and load application for the gearshift linkage component

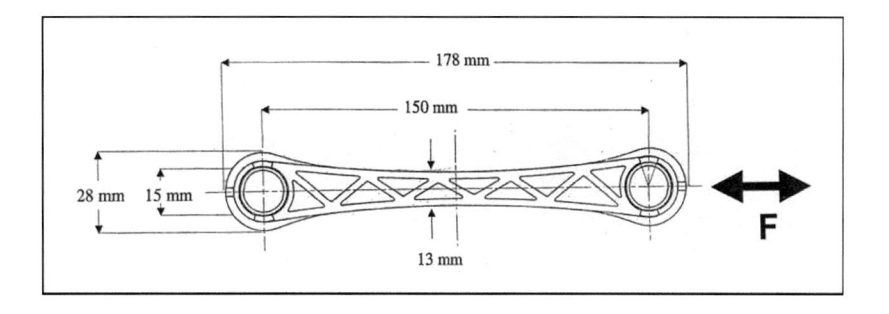

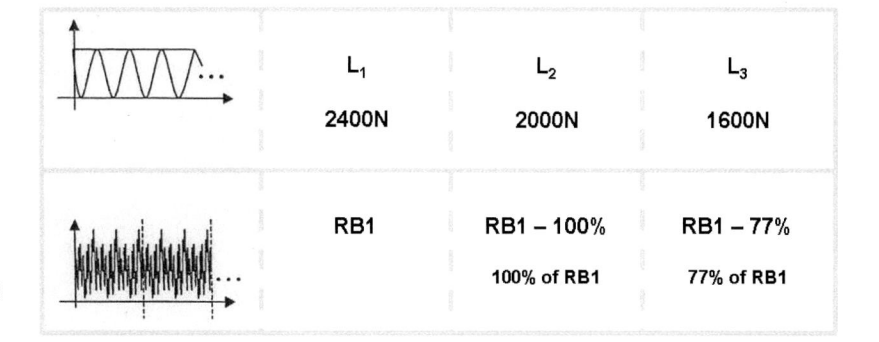

Figure 4-36 Constant amplitude loads and random load sequences for the test program

	L_1	L_2	L_3
	2400N	2000N	1600N
	RB1	RB1 – 100%	RB1 – 77%
		100% of RB1	77% of RB1

4.3.4.1 Gearshift Linkage Component

The geometry of this part of a gearshift linkage is shown in Figure 4-35. For the purposes of the test, metal bearings free of play were set into the component eyes. One component eye was clamped immovably in the servo-hydraulic test rig with load being introduced axially via the other eye.

Constant amplitude (Wöhler) and fatigue strength tests were both performed on this component. The load horizontals of the unit load tests and the random load/time functions for the fatigue strength tests are shown in Figure 4-36. Comparison of the test results with the results of the calculation is shown in Figure 4-37.

The method previously described is used for making the corresponding calculation for this drive component. Accordingly, the application C-Mold [17] is used for determining fiber orientation in the plane (Figure 4-34). The stress relating to static unit load in the component is determined using the ABAQUS program [20]. In all calculations it is assumed that all fibers lie in the same plane, as appears in Figure 4-34.

As regards the finite element calculation, it is taken as a condition that the fibers in an *individual* element have the same orientation and that every element is treated orthotropically. The bearings set in the eyes are assumed to be rigid for purposes of calculation. A fatigue life estimate calculation is made for all load sequences shown in Figure 4-36. Results of these calculations are shown in Figure 4-37.

Since the fatigue life of a component is dependent not only on the stress it is subjected to, but also on fiber orientation, for example, the damage occurring during cyclic loading need not necessarily be found at the location experiencing the highest amount of stress. In the case of the gearshift linkage component, the C-Mold program [17] calculates a fiber orientation of 0° at point A in Figure 4-38. This value is used for the fatigue life forecast. In order to check the plausibility of this calculated fiber orientation, the stresses were determined for uniaxial static loading at point A. The maximum normal tensile stress theoretically reaches the tensile strength of the material (fiber orientation 0°) at a force of 4241 N. In the case of quasistatic tensile testing, the component fails when a force of 4690 N

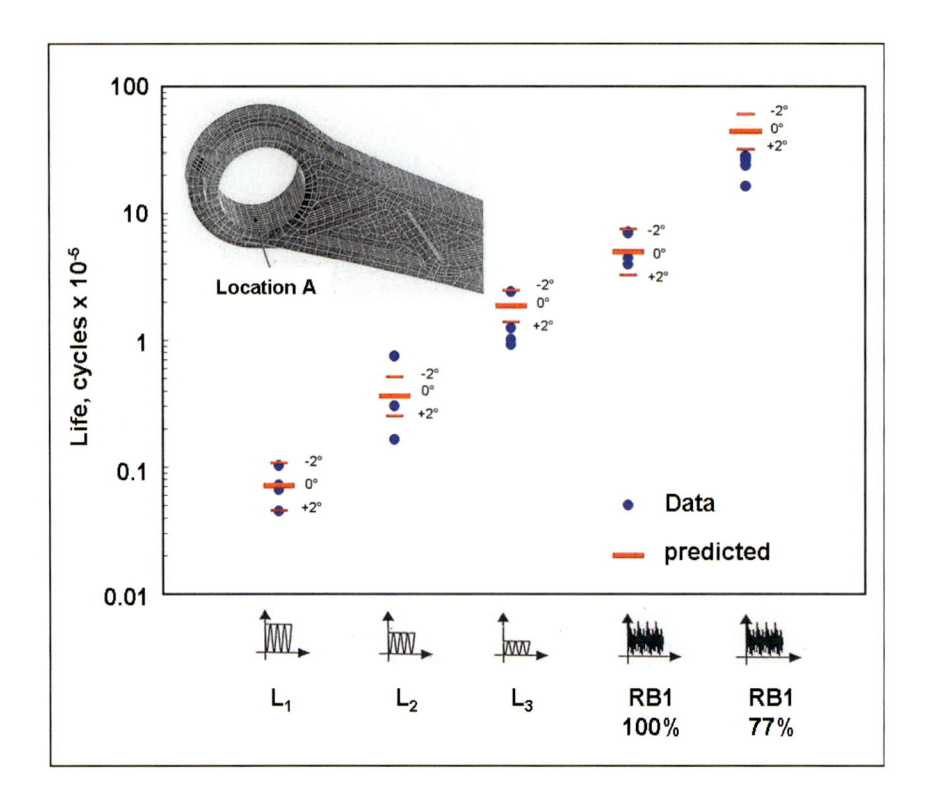

Figure 4-37 Comparison of test results with results of calculations

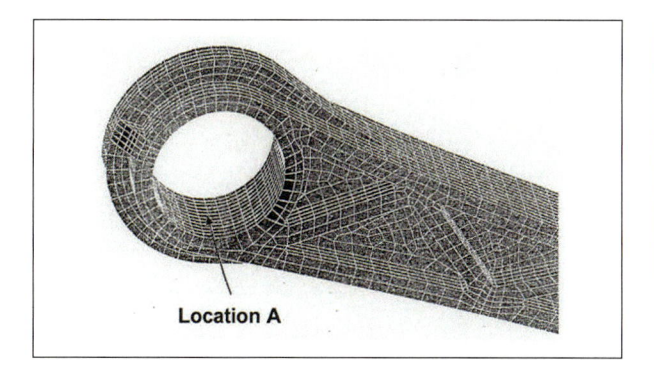

Figure 4-38 Finite element networking of the gearshift linkage component

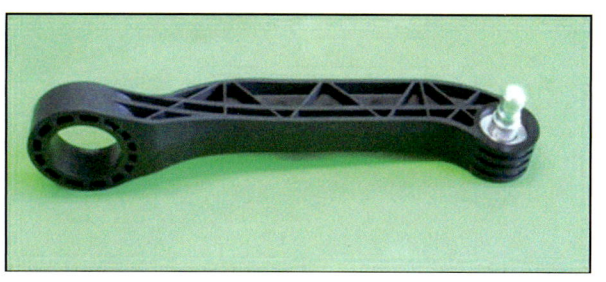

Figure 4-39 Track control arm made of glass-fiber-reinforced polyamide with aluminum bearing (photograph: Georg Fischer)

is applied. The difference between the force measured in the test and the calculated force that would result in component failure is thus 9.5%. The order of magnitude of the detected error is justification for stating that the fiber orientation calculated for point A on the component is plausible.

As can be seen from Figure 4-37, the results of the calculated fatigue life of the gearshift linkage component fall approximately within the scatter range of the test results, something which supports the validity of the calculation method developed here. All the same, caution is advised in applying the method generally, since changes in the parameters in the calculation may have a very marked effect on the results. An example of this is the influence of fiber orientation. In the present case, 0° calculated orientation was used. As a way of assessing the influence of this factor, fatigue life calculations were also made on the basis of fiber orientations of −2° and +2°. Figure 4-37 demonstrates that even small changes in the angle of orientation have a clearly visible influence on the calculation results.

4.3.4.2 Track Control Arm

The track control arm is shown in Figure 4-39. In a way similar to the testing of the gearshift linkage component, here the corresponding bearings were fitted into the force transfer eyes of the control arm with one end clamped firmly in the test rig and load was applied axially via the other end. The load cycles used for the tests and for the calculation are shown in Figure 4-40. The results are presented in Figure 4-41.

Control arm fatigue life is calculated the same as described above for the gearshift linkage component. In this case, fiber distribution and orientation are calculated by the Moldflow [18] program. The ABAQUS [20] program is used for determining the stresses in the component for two static unit load cases, tension and compression. In our calculations for the control arm, the same assumptions were made as in the case described above. The results of fatigue life calculations are compared with the test results in Figure 4-41.

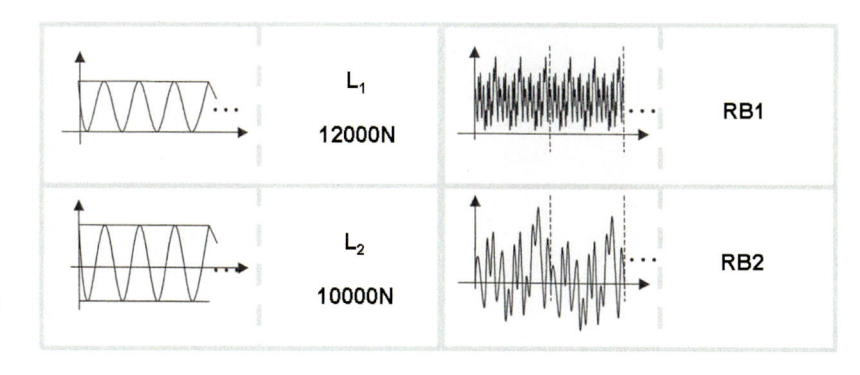

Figure 4-40 Constant amplitude loads and random load sequences for the test program

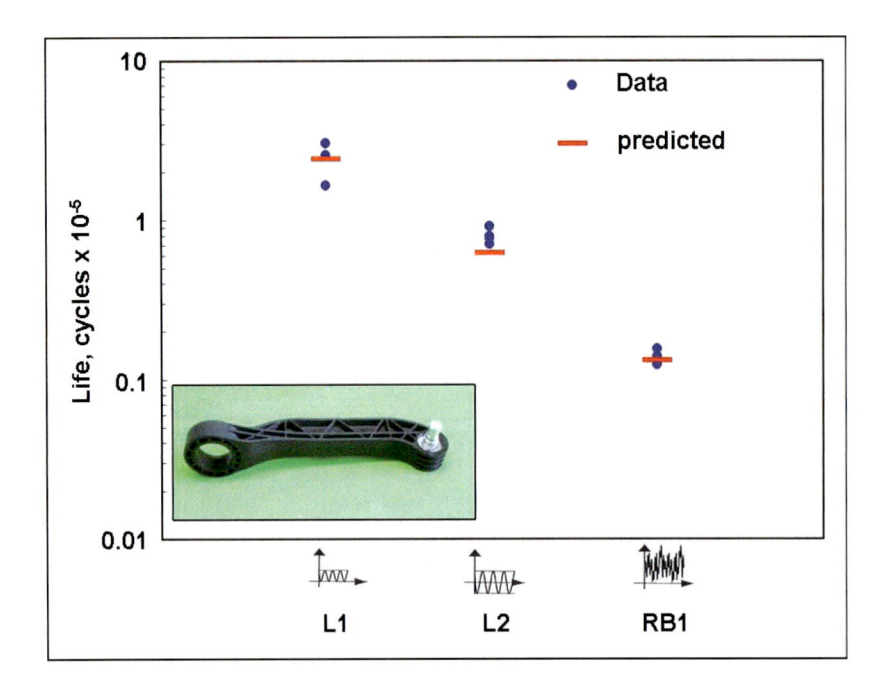

Figure 4-41 Comparison of the test results with the results of calculations

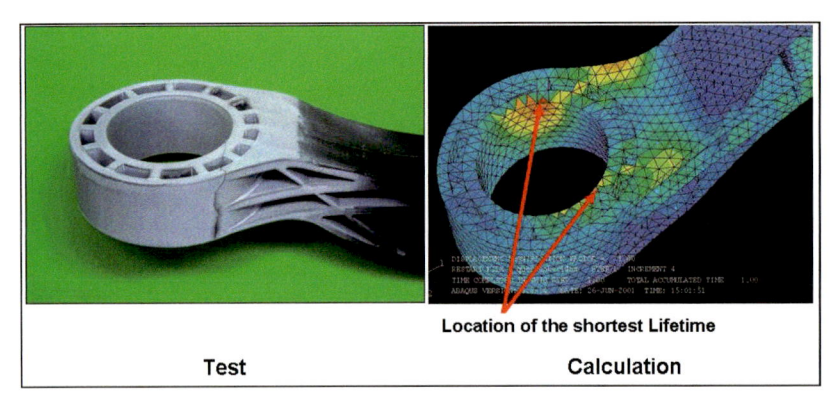

Figure 4-42 Failure location with fatigue tensile stress (photograph: Georg Fischer)

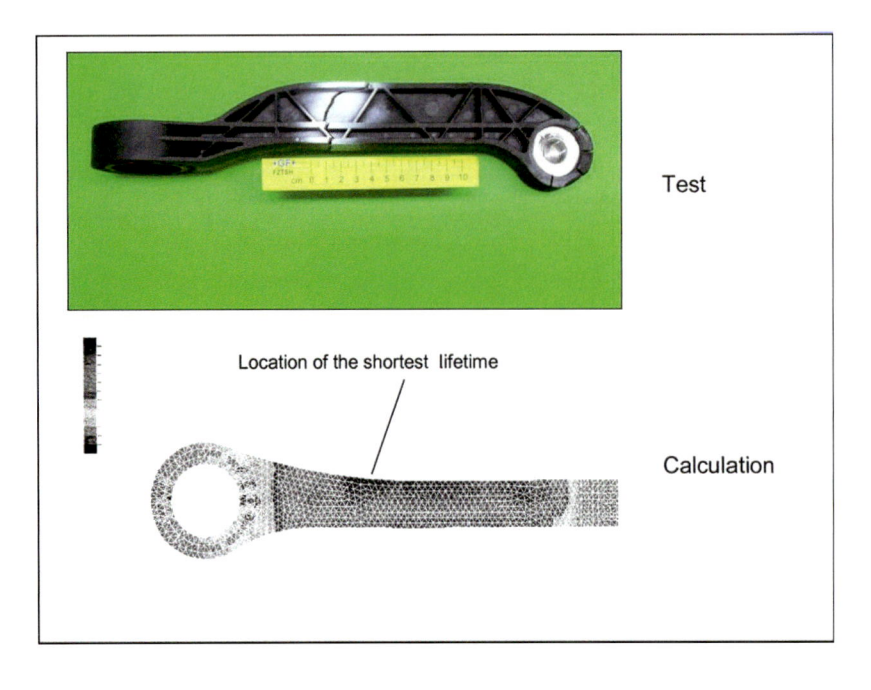

Figure 4-43 Failure location with cyclic tensile/compressive stress (photograph: Georg Fischer)

The tests demonstrate that the control arm fails at the upper eye under static and fatigue tensile stress (Figure 4-42) and under repeated tension/compression loading in the middle of the component where the stress exhibits a maximum/minimum (Figure 4-43).

At both failure locations a fiber orientation of 0° was calculated using the Moldflow [18] program. This value is also used for determining the theoretical fatigue life of the component. To check the plausibility of the calculations, a comparison was also made here between the calculated tensile strength and that measured in the test. Under static uniaxial tension, the maximum normal stress was calculated to reach the component's tensile strength at a tensile load of 33 152 N. In the quasistatic tension test, the control arm failed at a tensile load of 26 940 N. This deviation of approx. 20% is still acceptable and depends among other things on the fineness of the finite element network.

A comparison of the test results for grading and for random stress with the results of fatigue life calculations once again shows good conformity (Figure 4-42 and 4-43). One exception here is the investigation regarding random load case RB2 (Figure 4-10). Here calculation predicts damage will occur in the middle of the component, but in the test, the component failed once at the top eye and once at the bottom eye under approximately the same number of load cycles. One explanation of this is that the *Haigh-Goodman* diagram has not yet been adequately validated for the stress ratios occurring here (compression range). It is planned to apply the method on further components made of similar materials in order to secure the basis and further development of the method and to increase the amount of empirical data available.

4.3.5 Summary and Outlook

We described our work in developing a procedure for calculating fatigue life prognoses for components made of short-glass-fiber-reinforced thermoplastic materials. The necessary hypotheses could be formulated with the aid of our own tests of sample material and by reference to the relevant literature, with verification then carried out on the actual components. These components are test samples made of thermoplastic material not yet used in series production. The hypotheses and procedures used are already established in the field of metallic materials. They were modified in accordance with the material behavior analyzed. Mention can be made by name of the linear damage accumulation according to *Miner* which was adapted for use with fiber-reinforced thermoplastics. The examples of calculations we presented demonstrate astoundingly good agreement with the test results and permit us to predict optimistically that this procedure will also be applicable for thermoplastics with carbon-fiber reinforcement with the proviso that further verification work will also be necessary to corroborate the validity of the method or to develop it further. It will not only be a case of expanding the range of materials considered, but also of taking environmental influences on the material into account. Here, to take the example of material aging, solutions will have to be found with regard to simulation and incorporation into fatigue life calculations.

By applying, continuously improving and expanding the method we will succeed in building a store of empirical data and transferring this type of theoretical fatigue life calculation into productive use in the product creation process.

References

[1] Zago, A., Springer, G. S.: Constant amplitude fatigue of short glass- and reinforced-reinforced thermoplastics, *Journal of Reinforced Plastics and Composites* Vol. 20 (2001) 7 pp. 564–595

[2] Zago, A., Springer, G. S., Quaresimin, M.: Cumulative damage of short-glass-reinforced-reinforced thermoplastics, *Journal of Reinforced Plastics and Composites* Vol. 20 (2001) 7 pp. 596–605

[3] Zago, A., Springer, G. S.: Fatigue lives of short reinforced-reinforced thermoplastic parts, *Journal of Reinforced Plastics and Composites* Vol. 20 (2001) 7 pp. 606–620

[4] Mandell, J. F., Huang, D. D., McGarry, F. J.: Fatigue of glass- and carbon-reinforced-reinforced engineering thermoplastics, 35th Annual Technical Conference of Reinforced Plastics/Composites Industry – SPI (1980) pp. 1–11

[5] Adkins, D. W., Kander, R. G.: Fatigue performance of glass-reinforced thermoplastics, 4th Annual Conference on Advanced Composites (1988) pp. 437–445

[6] Jinen, E.: Accumulated strain in low-cycle fatigue of short carbon-fiber-reinforced nylon 6, *Journal of Material Science* (1986) pp. 435–443

[7] Horst, J. J.: Influence of fiber orientation on fatigue of short glass-fiber-reinforced thermoplastics, Delft University dissertation 1992

[8] Jia, N., Kagan, V. A.: Effects of time and temperature on the tension-tension fatigue behavior of short-reinforced-reinforced polyamides, *Polymer Composites* Vol. 19 (1998) pp. 408–414

[9] Hamada, M., Hiragushi, K., Tomari, K.: Effect of surface treatment on fatigue properties of glass-reinforced-reinforced injection-molded parts, 53rd Annual Technical Conference – Society of Plastics Engineers (1995) pp. 3981–3985

[10] Mandell, J. F., McGarry, F. J., Huang, D. D., Li, C. G.: Some effects of matrix and interface properties on the fatigue of short reinforced-reinforced thermoplastics, *Polymer Composites* Vol. 4 (1983) pp. 32–39

[11] Mandell, J. F., McGarry, F. J., Li, C.: Fatigue crack growth and lifetime trends in injection-molded reinforced thermoplastics, ASTM-STP 873 (1985) pp. 36–50

[12] Malzahn, J. C., Schultz, J. M.: Tension-tension and compression-compression fatigue behavior of an injection-molded short-glass-fiber/poly(ethylene terephthalate) composite, *Composites Science and Technology* Vol. 17 (1986) pp. 253–289

[13] Barré, S., Benzeggagh, M. L.: On the use of acoustic emission to investigate damage mechanisms in glass-reinforced-reinforced polypropylene composites, *Science and Technology* Vol. 52 (1994) pp. 369–376

[14] Grove, D., Kim, H., Cooper, D., Ellis, C.: Longitudinal fatigue behavior of long- and short-glass-reinforced, injection-moldable polypropylene composites, 26th International SAMPE Technical Conference Vol. 26 (1994) pp. 281–295

[15] Silverman, E. D.: Material properties of unreinforced and short-glass-fiber- reinforced injection-molded PVC composites, 40th Annual Conference Reinforced Plastics/Composites Industry – SPI (1985) pp. 1–5

[16] Hoppel, C. P.: The effect of tension-tension fatigue on the mechanical behavior of short-reinforced-reinforced thermoplastics, 7th Technical Conference Of the American Society for Composites – ASC (1992) pp. 509–518

[17] C-Mold: C-Mold 99.1: Release notes, Advanced CAE Technology (1998)

[18] Moldflow Corporation: Moldflow Plastics Insight Manual, Release 2.0 (2000)

[19] Downing, S. D., Socie, D. F.: Simple rainflow counting algorithms, *International Journal of Fatigue* Vol. 4 (1982) pp. 31–40

[20] Hibbitt, Karlsson & Sorensen Inc.: ABAQUS Standard User's Manual, Version 6.2 Vols. 1, 2, 3 (2001)

4.4 Calculated Estimate of Service Life for Components Made of Short-glass-fiber-reinforced Plastics

Martin Brune

4.4.1 Introduction

Lightweight vehicle design is becoming more and more important due to the continuously tightening requirements on the automotive industry with regard to the reduction of CO_2 emissions and thus to reductions in fuel consumption. Involved in this is the growing use of plastics in automobile manufacturing, not only in the vehicle interior but increasingly as well for load-bearing structures in the drive-train and body. Mostly, it is fiber-reinforced plastics which are used here. Due to ever shorter development times, numerical analysis techniques methods in engineering design work have come to play a very great role in the vehicle development process in both the design and evaluation of these components with regard to strength. A special challenge to computer-aided component design is posed by the case of dynamic operational loading on components made of reinforced plastics. The so-called calculated life prediction estimate has become firmly established in the development process for metallic materials, but this is virgin territory as regards plastics.

In this paper, we describe a method by which operational strength design can be carried out in the future for plastic components as well with the aid of the service life calculation, even in the early stages of vehicle development.

The work presented here was carried out under the central coordination of the Polymer Competence Center in Leoben (PCCL) as part of a current project sponsored by the Austrian Federal Ministry of Traffic, Innovation and Technology (BMVIT), by the provinces of Styria and Upper Austria, and by the University of Leoben.

Bodies participating in the project included not only the BMW Group, but also the Engineering Center Steyr (MAGNA Powertrain ECS), the raw materials supplier EMS-GRIVORY, and also the Institute for Materials Science and Plastics Testing (IWPK) and the Chair of General Mechanical Engineering (AMB) of the University of Leoben.

4.4.2 Initial Situation and Objectives

Thanks to their low weight coupled with high strength and rigidity as well as their suitability for large-scale production by different manufacturing processes, fiber-reinforced plastics have the potential to replace metal materials as a structural component in a number of application cases. Before fiber-reinforced plastics can be used to any greater extent in automobile manufacturing, material behavior under cyclic loading must be understood and described comparably as with metals [1].

As part of the project, it is intended to develop and verify a service life calculation method based on the testing of specimens and components made of short-fiber-reinforced plastics. This method is then to be implemented in a commercially available software package, one which is in fact already in use today for service life calculations for metal components.

The tool to be developed should make it possible to dimension components based on the stresses they experience and taking into account material-specific properties and advantages. This should increase concept maturity in the early development stages and ensure that better use is made of potential for lightweight design.

It is important here, too, – particularly with this group of materials – that not only mechanical stresses are taken into account but also a large number of further influencing factors. This makes designing and dimensioning a complex task.

To take all influencing factors into account, such as

- Material anisotropy
  (acquiring of anisotropic material data)
- Influence of the material
  (type and nature of the fiber-matrix system)
- Multiaxial loading
- Mean stress
- Temperature
- Frequency-dependence
- Water absorption/ambient medium
- Creep behavior
- Sequence effects
- Production
- Ageing, and so on

would exceed the scope of the research project on account of the great outlay in time and testing it would involve. For this reason, initially only those parameters should be studied and described which appear to have the greatest influence on component fatigue, such as material anisotropy and various mean stresses.

Building on the findings reported in [2] and to be backed up in the present project, investigations will be carried out regarding the influence of temperature on component service life and other matters as an additional task in the second phase of the project.

4.4.3 Theoretical Considerations

4.4.3.1 Fiber-Reinforced Plastics

In the automotive industry, fiber-reinforced plastics are used above all for increasing the strength, rigidity and service life of components. Reinforcing fibers may be subdivided into natural and synthetic, organic and inorganic fibers whereby the advantages of synthetically produced fibers lie in the reproducibility of their properties (strength and rigidity) [3].

In the present case, the reinforcing fibers are made of glass with a length of 150 to 300 μm and are embedded in a thermoset or thermoplastic matrix.

Characteristic of fiber orientation with injection molding is the three-dimensional orientation of the fibers without any aggregation worth mentioning. The type of failure depends on fiber orientation.

Before fiber-reinforced plastic components can be designed appropriately for the material used, information is required on local fiber orientation (averaged direction of fibers) and distribution in the component. For this, not only experimental methods can be used, but also finite-elements programs for calculating fiber orientation and the degree of orientation of fiber-reinforced fill masses.

Following calculation of local fiber orientations, the resulting local material properties can be determined. In order to reduce their numerical complexity, properties are not determined for every single fiber orientation found in the component. Instead it is sufficient to average out the orientation (weighted mean value) and use this averaged orientation to assign the corresponding material properties to the individual cells (elements).

If the fibers are oriented in one plane, one parameter will be sufficient to describe fiber orientation. Here the degree of alignment (degree of identical orientation of the fibers) will lie between unidirectional and random arrangement. With respect to a reference direction (such as the direction of the gate, for example), the fiber orientation parameter f_p may take on values between −1 (all fibers oriented 90° to the reference direction) and +1 (all fibers arranged along the reference direction) [4] (Figure 4-44).

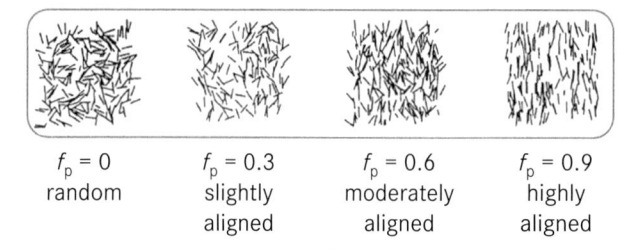

f_p = 0	f_p = 0.3	f_p = 0.6	f_p = 0.9
random	slightly aligned	moderately aligned	highly aligned

Figure 4-44 Fiber orientation parameter between stochastic arrangement (f_p = 0) and nearly unidirectional alignment (f_p = 0.9) [4]

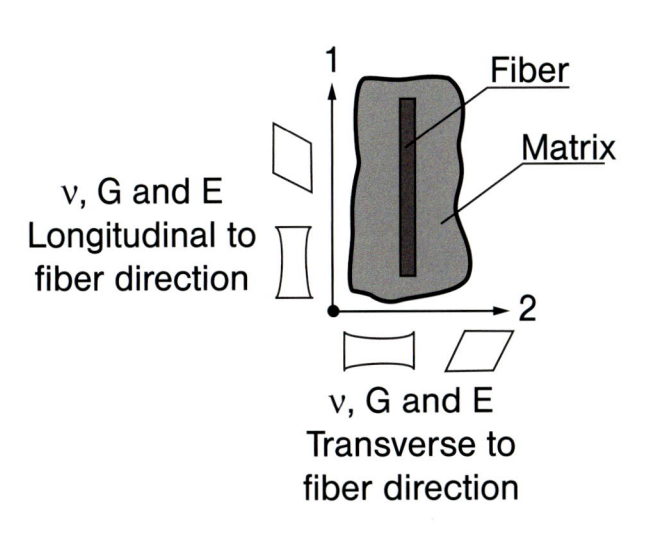

Figure 4-45 Material properties as a function of fiber orientation [5]

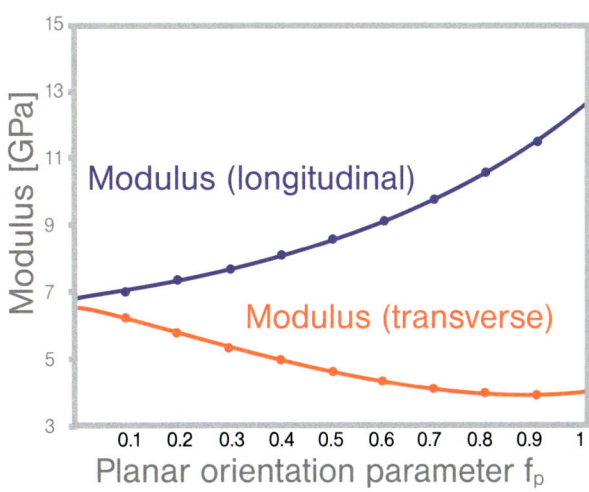

Figure 4-46 Modulus of elasticity and fiber orientation for nylon 66 (33% glass fiber, L/D = 25)

Once fiber orientation and distribution have been determined, the resulting material behavior must be described. The mechanical behavior of a material is described by means of the stress-strain relation. In the case of anisotropic or orthotropic materials – unlike isotropic materials – strain response differs with the direction of loading.

If only one single fiber-matrix element is considered, the material behavior will be identical in every plane normal to that fiber (Figure 4-45).

In other words, the elongation corresponding to the load will always be the same irrespective of the location of the load applied. Due to this transverse isotropic material behavior, only five material properties are required to describe the stress-strain relations. These are the modulus of elasticity in the direction of the fibers and perpendicular to it (E_1, E_2), the Poisson's ratios (ν_{12}, ν_{23}) and the shear modulus G_{12}.

Characteristic values for the unidirectional reinforced material are needed as a basis for calculating the local anisotropic material properties as a function of fiber orientation. However, obtaining unidirectional reinforced specimens in reality requires a great deal of work, which is why the required material properties are obtained via micromechanical methods (Halpin-Tsai, Tandon-Weng) [5 to 7] or FE analyses [8].

On the basis of the simulated averaged fiber orientation and distribution, with the aforementioned micromechanical methods, the material behavior for the unidirectional reinforced material can be determined solely from the mechanical material properties of the matrix and fiber materials. This allows properties of the material with different fiber orientations to be estimated later on by means of macromechanical approaches (mixing rules, for example). Thus a set of transversally isotropic material parameters can be calculated for each element, thereby providing all of the data required for finite element calculation [5].

The influence of fiber orientation on material anisotropy and, bound up with this, on locally varying mechanical properties can be seen in the example of the path taken by the curves for the modulus of elasticity plotted along and normal to the direction of flow for a layer of glass-fiber-reinforced nylon 66 [4]. Anisotropy rises with increases in the fiber length ratio (L/D), fiber content and increasing fiber orientation. In general, as fiber length increases, an increase in tensile strength and modulus of elasticity can be expected.

4.4.3.2 Stress and Load Capacity of the Material

Stress on the material due to fatigue and ageing should be quantified in order to dimension operationally stable fiber-reinforced plastic components (Figure 4-47).

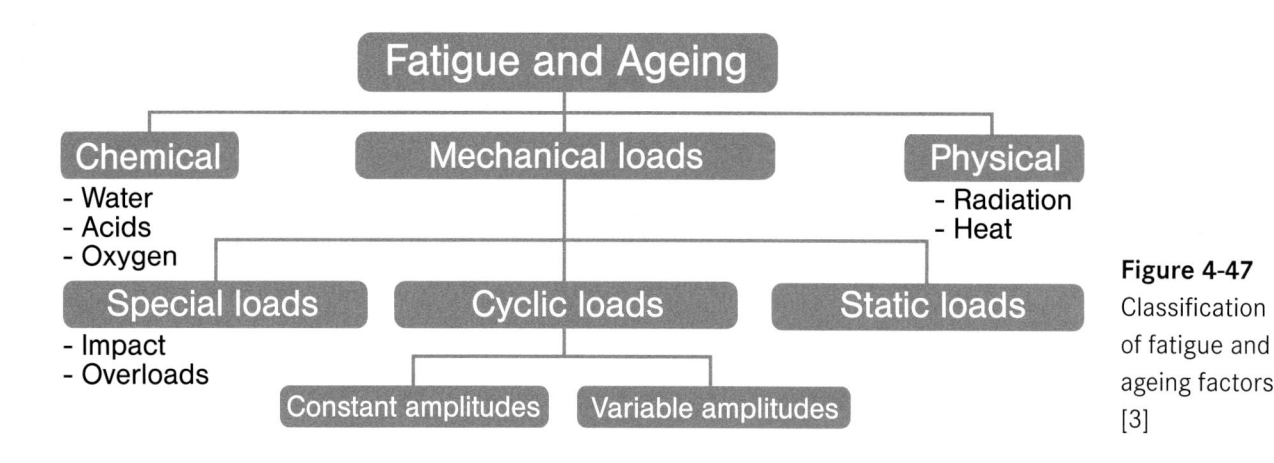

Figure 4-47 Classification of fatigue and ageing factors [3]

In order to limit the complexity of this, the first approach to estimating service life was based only on mechanical stresses. Other aspects, particularly such influencing factors as are dominant in plastics, such as temperature, will also be taken into consideration as the project proceeds.

Durable component design requires an evaluation of material fatigue due to stresses changing with time. Durability and static strength are two properties which for the most part are treated separately. Static strength describes the resistance which the material or component exerts against a single occurrence of stress, the maximum stresses and strains being of decisive importance here.

Durability, on the other hand, derives from the resistance to cyclical stress and is not markedly influenced by the absolute values of stress and strain amplitudes, but rather by the frequency and form of the stress-strain hystereses which occur. For this reason, static stress analysis cannot deliver any information on the durability of a component. Neither are special loadings, such as impact or overload, taken into account in life prediction, since an evaluation of these types of loading requires different models for describing material behavior and also a different material database (material properties at high strain rates, for example) [9].

Components relevant to system durability are often subject to non-proportional multiaxial loading. With orthotropic or anisotropic materials, strain response depends on the direction of force application, in contrast to isotropic materials, whose material behavior is independent of the direction of force application.

As a basic rule, with short-fiber-reinforced materials, multiaxial stress states occur in the component not only under multiaxial cyclic loading but, due to the anisotropic properties of the material, local multiaxial loads as well, even in cases of uniaxial external vibrational stress [5, 10].

4.4.3.3 Failure Mechanisms and Failure Criteria

With fiber-reinforced plastics, failure mechanisms can essentially be subdivided into fiber failure, inter-fiber failure (IFF) (Figure 4-48) and delamination. Frequently, fiber detachment from the matrix occurs initially, causing the maximum loading capability of the matrix to be exceeded, and microcracking commences. As the result of further cyclic loading, the detached areas become larger as do the microcracks. Finally these processes result in the complete detachment of the fibers from the matrix, so that the matrix material has to support virtually the entire load. The microcracks lengthen and become macrocracks which ultimately lead to component failure. Here the fibers are pulled out of the matrix (fiber tearing). Due to the shortness of the fibers, fiber failure occurs infrequently.

Due to the different loading capacities in different directions and the multiaxial loads that occur, it is a very complex matter to describe the failure or fracture behavior of fiber-reinforced materials.

4.4 Calculated Estimate of Service Life for Components Made of Short-glass-fiber-reinforced Plastics

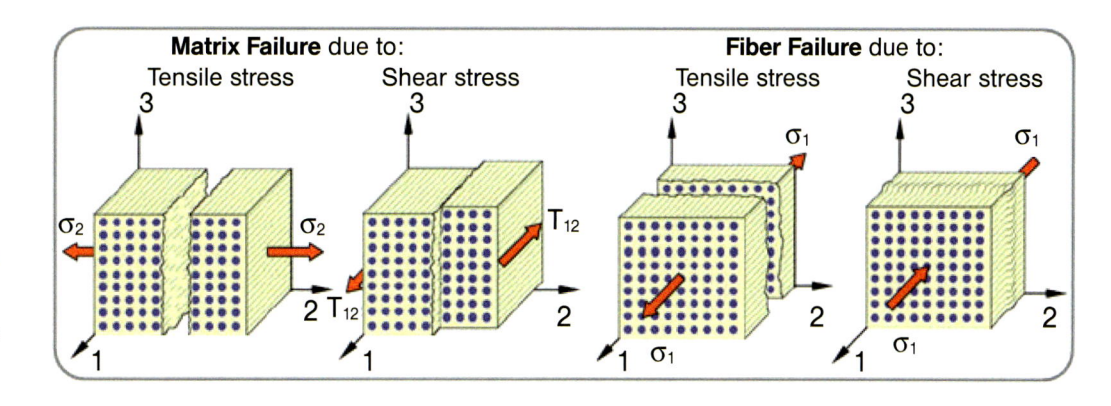

Figure 4-48 Failure mechanisms [11]

Figure 4-49 Failure criteria [3]

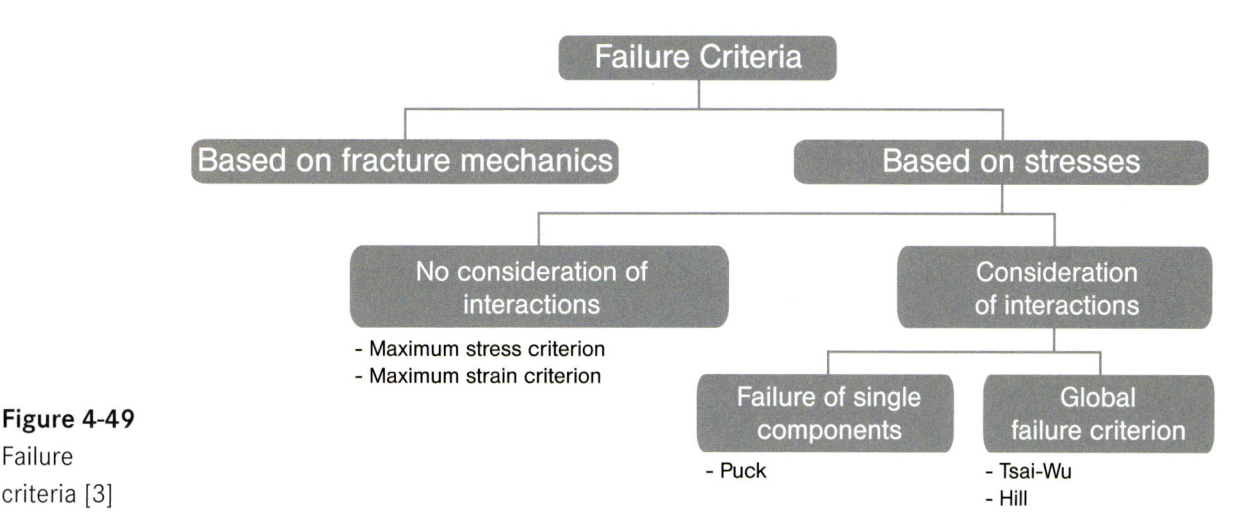

For cost-effective dimensioning, it is necessary to determine the material loads caused by a multiaxial stress field as precisely as possible. For this, equipotential conditions (failure potentials) are required with which these the local stresses can be converted into a fictive reference value by means of a suitable, material-related hypothesis. In most cases, a reference stress is calculated here which is set in relation to a material property (load capacity) preferably determined in a simple and, in most cases, uniaxial loading test. It is assumed here that there is an equivalent material stress exposure in both loading cases [10, 12].

The failure criteria or the failure hypothesis behind it can be based on physical reasons or be of a purely analytical nature (by adaptation to test results). With short-glass-fiber-reinforced plastics, it should be noted that it is not sufficient to identify the maximum loading. Instead, all of those stresses are critical which are maximum in comparison with the load capacity.

Here the load capacity of the material depends on the number of load cycles. The relationship of load to load-ability – so-called stress exposure – is thus a variable dependent on the number of vibrations. It therefore makes sense to determine the stress exposure for each number of load cycles for which the component is to be designed [10].

One possible classification of failure criteria is shown in Figure 4-49. The simplest criteria to apply are those of maximum stress and of maximum strain. With these criteria, only the stresses and strains occurring on the x, y and z axes are compared with the maximum permitted values for tension and compression. The interaction of stress and strain components in different spatial directions is ignored here. This is why it does not make sense to apply them in cases of multiaxial loading.

Criteria that take stress interaction into account are subdivided into those which take the fracture behavior of individual components of the composite into consideration (Puck, for example) and those which can only make global statements about the failure of the entire composite (Tsai-Wu, Hill, for example) [3].

Another kind of failure criteria is based on failure-mechanism approaches which assume that the material already contains crack-like flaws that enlarge due to the effect of external loads and finally result in failure.

4.4.4 Procedure for Estimating Service Life

Before the actual service life calculation can be carried out with the durability program FEMFAT [13], the database shown in Figure 4-50 must have been set up. The first step here is the filling simulation to determine fiber orientation and distribution from which the locally divergent material properties can be derived that are required as input information for finite element analysis.

On the basis of the load-time curve and the material characteristic functions (Wöhler curves) obtained from specimens, local component Wöhler curves are calculated taking further influencing factors into account. These local curves form the basis for the service life calculation.

We shall now go on to briefly sketch the procedure behind the calculated estimation of component service life with the FEMFAT program. With FEMFAT it is possible to carry out durability calculations on the basis of finite element stress results.

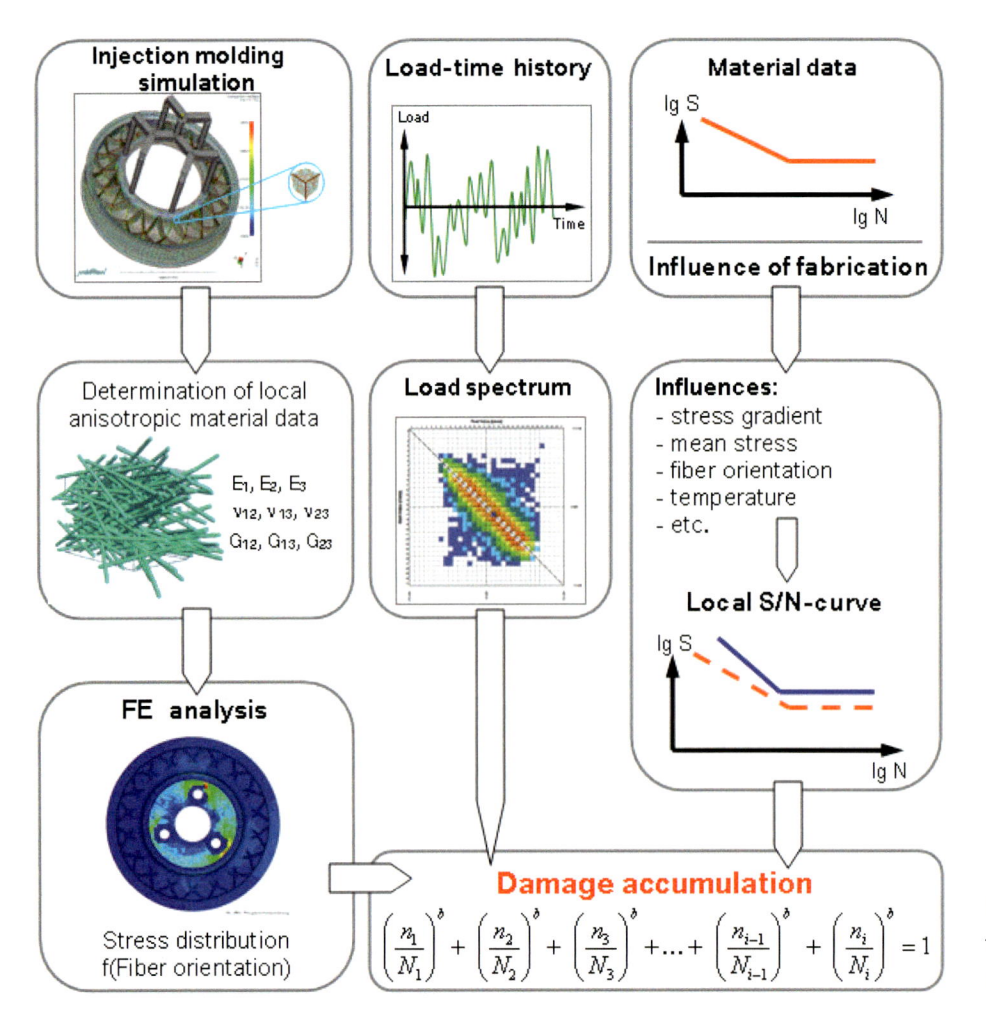

$$\left(\frac{n_1}{N_1}\right)^\delta + \left(\frac{n_2}{N_2}\right)^\delta + \left(\frac{n_3}{N_3}\right)^\delta + \ldots + \left(\frac{n_{i-1}}{N_{i-1}}\right)^\delta + \left(\frac{n_i}{N_i}\right)^\delta = 1$$

Figure 4-50 Procedure for calculating service life

4.4 Calculated Estimate of Service Life for Components Made of Short-glass-fiber-reinforced Plastics

In the case of multiaxially loaded structures in which loading cannot simply be subdivided into a mean stress component and an amplitude stress component, the method of the critical section plane is applied [14]. To be able to evaluate the interaction of several simultaneously acting loads, the stress information is superposed on every section plane (144 planes per finite element node) and converted into reference stresses. With the aid of local component Wöhler curves that depend among other things on the loading state, the corresponding instances of microdamage are then determined on the basis of the reference stresses. The section plane with the most serious microdamage ultimately has the most relevance for component damage. In this way the instances of damage are determined and linear accumulated load cycle by load cycle (linear damage accumulation according to Miner).

Life estimation is carried out using the influencing factors method. Here the course of the local component Wöhler curve is described by the parameters of fatigue strength ($\sigma_{D, B}$), vertex load cycle number ($N_{E, B}$) and angle of inclination (k_B). These characteristics must be determined under local conditions and are in each case dependent on various influencing factors and their interaction [13].

The main factors influencing the durability of the component are geometry, material, loads and environmental influences acting on it, as well as influences resulting from the fabrication process (Figure 4-51). The more precisely these factors can be described and incorporated in service life calculation, the better will also be the quality of the results of the life estimation.

The FEMFAT program is to be expanded by the following features to enable evaluation of plastic components:

- Interfaces for inputting orientation data; calculation of the resulting material data using conventional finite-element programs

- User interface for inputting the material data in the parallel-to-fiber direction

- Material generator for calculating material properties (fatigue strength, endurance limit, and so on) and interpolation of the material properties for other fiber orientations

- Creation of a material database for various (fiber-reinforced) polymers

- Non-linear damage accumulation

- Taking ageing effects into consideration.

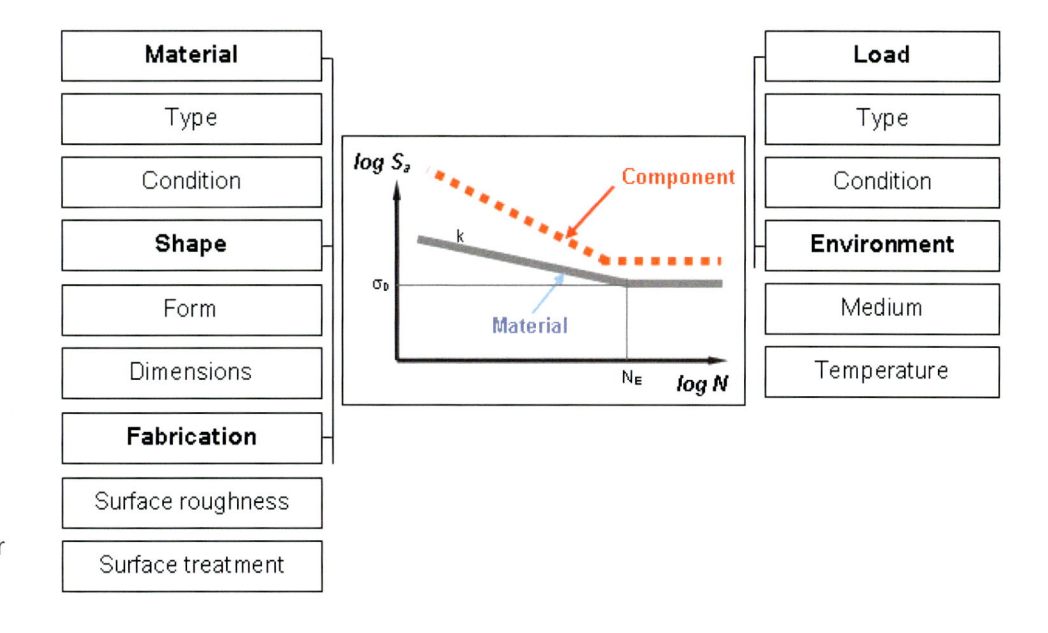

Figure 4-51 Influences on the shape of the component Wöhler curve [13]

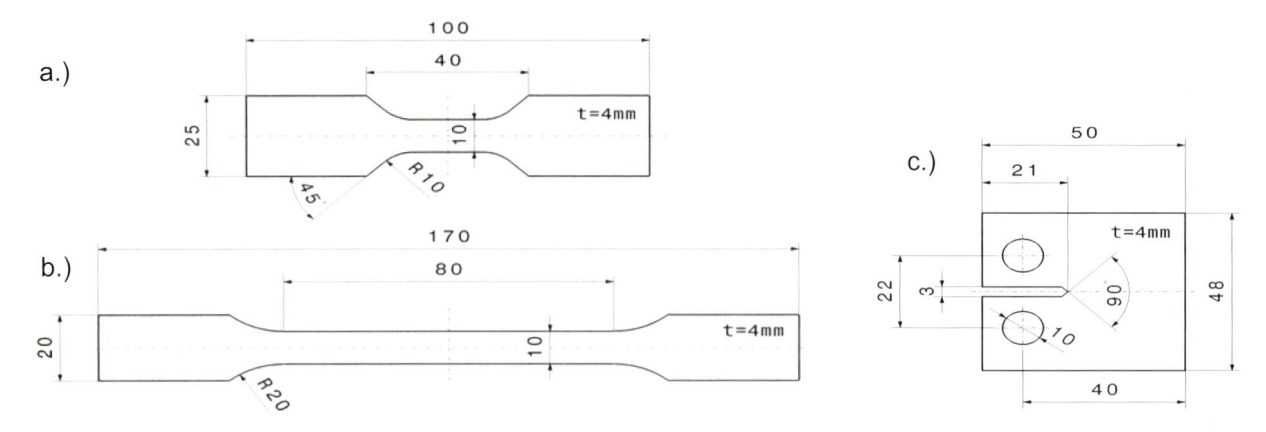

Figure 4-52 Specimen shapes [16]
(a) Short test specimen, (b) Multipurpose test specimen type A as per ISO 3167, (c) CT specimen

4.4.5 Experimental Investigations

4.4.5.1 Materials and Specimen Shapes

Various specimen shapes have been defined for use in obtaining the material data that form the basis for service life estimation (Figure 4-52). In order to characterize short-glass-fiber-reinforced thermoplastic and thermoset materials, a thermoplastic polyphthalamide (PA 6T/6I-GF40+MX25) and a thermoset phenol-formaldehyde resin (PF-GF30+MX30) were used respectively [15]. Depending on the manufacturing process used, there were differences in the morphology of the test specimens and these also were characterized. The standard test pieces made of each of the two materials are produced by injection molding. The short test pieces and CT specimens for fracture-related investigations are in each case cut from injection-molded thermoplastic or compression-molded thermoset slabs. To evaluate the influence of fiber orientation on the mechanical properties of the materials, specimens were cut from injection-molded thermoplastic slabs in the direction of gating and also at 90° to the direction of injection.

Figure 4-53 Servo hydraulic testing machine MTS 359 axial/torsional [16]

4.4.5.2 Test Methods and Testing Program

Monotonic Tension and Compression Tests

To determine the mechanical properties of both materials, a servo-hydraulic test machine manufactured by MTS Systems GmbH [16] was used (Figure 4-53). Monotonic tensile tests with controlled application of force in each case were performed on the standardized multipurpose test specimens and short test specimens under standardized conditions at 23 °C and 50% relative air humidity.

Due to the danger of buckling, the standard test pieces were not used in the compression tests. Strain rates were kept constant for the tests performed on both types of test pieces (different clamping distances). The test speed was 1 mm/min for the standard test specimens and 0.4 mm/min for the short test pieces. In each case at least three monotonic tests were carried out for each specimen shape.

Figure 4-54 Standard specimens made of PA 6T/6I-GF40+MX25 following Wöhler tests with $R = 0.1$

Cyclic Tests

With a view to characterizing fatigue behavior under cyclic loading, the integral reaction of the material to forced cyclic loading was investigated in Wöhler tests (based on Alstädt, 1987 and Buxbaum, 1992 in [15]). In other words, material fatigue strength was determined under sinusoidal loading at a constant stress amplitude.

The disadvantage of Wöhler testing of fiber-reinforced plastics is that the data obtained for fatigue strength are clearly dependent on the shape of the test piece. This circumstance has to be considered before applying such data to components of any shape whatsoever. In addition, the complexity involved in such testing makes it virtually impossible to differentiate between crack initiation and the crack growth phase, i.e., conclusions regarding the damage mechanisms determining failure can only be drawn with very great difficulty.

Multipurpose specimens (Figure 4-54) and short specimens were tested with stress ratios ($R = \sigma_u / \sigma_o$) of $R = 0.1$, $R = -1$ and $R = 10$ in order to determine Wöhler curves for material. In addition to this, a temperature sensor was added to the test rig in order to be able to measure specimen heating at the surface.

Cyclic tests were carried out at different frequencies to determine a test frequency at which hysteretic self-heating of the specimen during the tests could be kept low. From these preliminary tests, a suitable test frequency of 10 Hz was obtained at which the specimen would heat up by approximately 5–8 °C.

Figure 4-55 shows by way of example a comparison of Wöhler curves obtained on multipurpose test specimens of both materials.

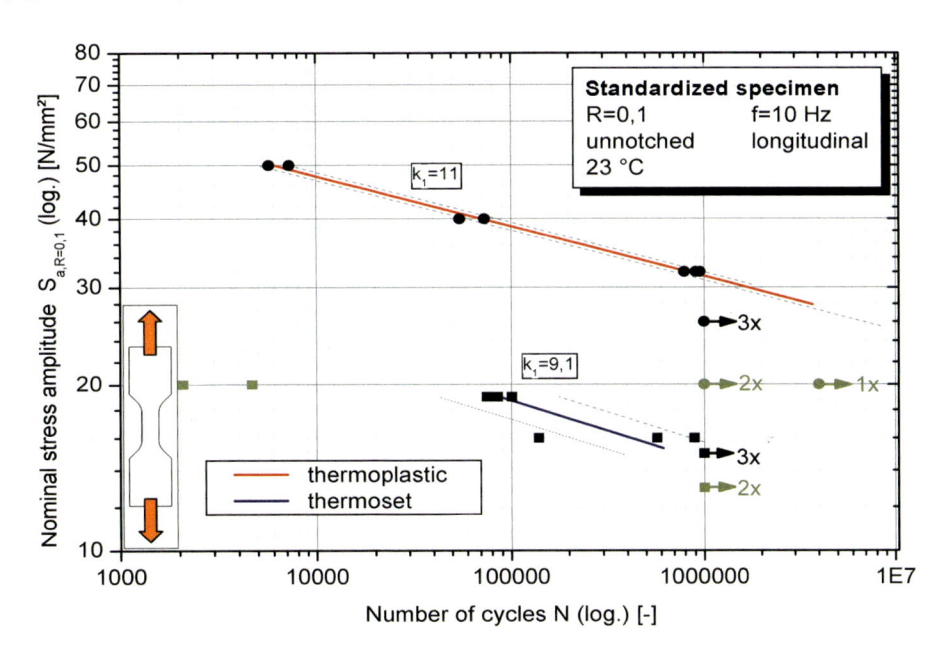

Figure 4-55 Comparison of Wöhler curves for standard specimens made of PA 6T/6I-GF40+MX25 and PF-GF30+MX30 with an R ratio of $R = 0.1$, 10 Hz and 23 °C [17]

Here a larger statistical scatter of test results emerged for the thermoset material at lower sustainable stress amplitudes. Furthermore, the fatigue endurance line for the thermoplastic has a slightly steeper angle than that of the thermoset material.

The more favorable fatigue characteristics of the specimens made of PA 6T/6I-GF40+MX25 are due, among other things, to the higher proportion of glass fibers as well as to better fiber-matrix adhesion.

In the compression range ($R = 10$), the Wöhler curves are flatter and in the case of the thermoplastic material, for example, the tolerable nominal stress amplitudes are roughly four times higher than under conditions of repeated tensile stress.

Fracture-Mechanical Investigations and Correlation With Wöhler Tests

The fracture-mechanical concept is based on the assumption that any component has flaws (cracks, inclusions, and so on) and describes the propagation of these flaws or crack growth resulting from external loads. The kinetics of fatigue crack growth under cyclic loading can here be described even in the case of plastics using the methods of linear-elastic fracture mechanics [15].

The example in Figure 4-56 shows two CT specimens used for preliminary fracture-mechanical investigations. Tests to determine fatigue crack growth were carried out on the testing machine shown in Figure 4-53 at 10 Hz under conditions of repeated tensile stress ($R = 0.1$). Here the initial crack was made using a precision diamond saw.

The areas of stable and unstable crack propagation are marked. In principle it is true to say that the stable propagation of the cracks takes place in both materials at right-angles to the direction of loading. The crack deflection in the unstable phase in the thermoplastic material can be explained by the higher degree of fiber orientation in the direction of the gate. Since the specimen was cut transversely to the gating direction of the injection-molded slab, the crack tends to deviate towards the direction of fiber orientation. This does not happen with the CT specimen made of PF-GF30+MX30, since these specimens were cut from compression-molded slabs which, due to the manufacturing process used, have no marked preferential direction of fibers. For both materials, the brief phase of stable crack propagation was followed by rapid, brittle failure of the specimens.

The use of linear-elastic fracture mechanics is a valuable supplement to the Wöhler tests. Fatigue strength, among other things, can be estimated from the data from the crack propagation testing of CT specimens. This is possibly one way of reducing the time-consuming tests ($N > 10^6$) for fatigue strength at low load amplitudes (Figure 4-57).

To make it possible to transfer the results of fracture-mechanical testing to Wöhler tests, fatigue crack growth has to be the major determinant in the total service life of both standard and short specimens, since the crack growth curve provides no information

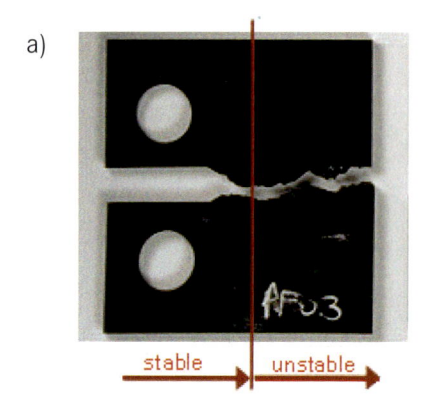

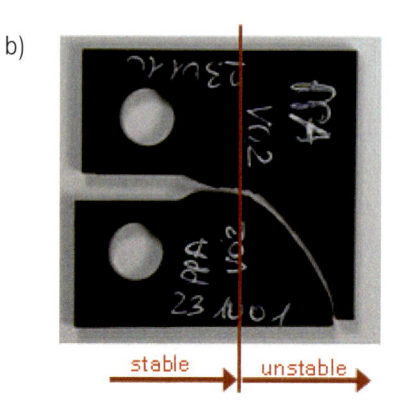

Figure 4-56
Preliminary tests for cyclic crack propagation (RT, 10 Hz, $R = 0.1$) [16]
(a) PF-GF30+MX30,
(b) PA 6T/6I-GF40+MX25

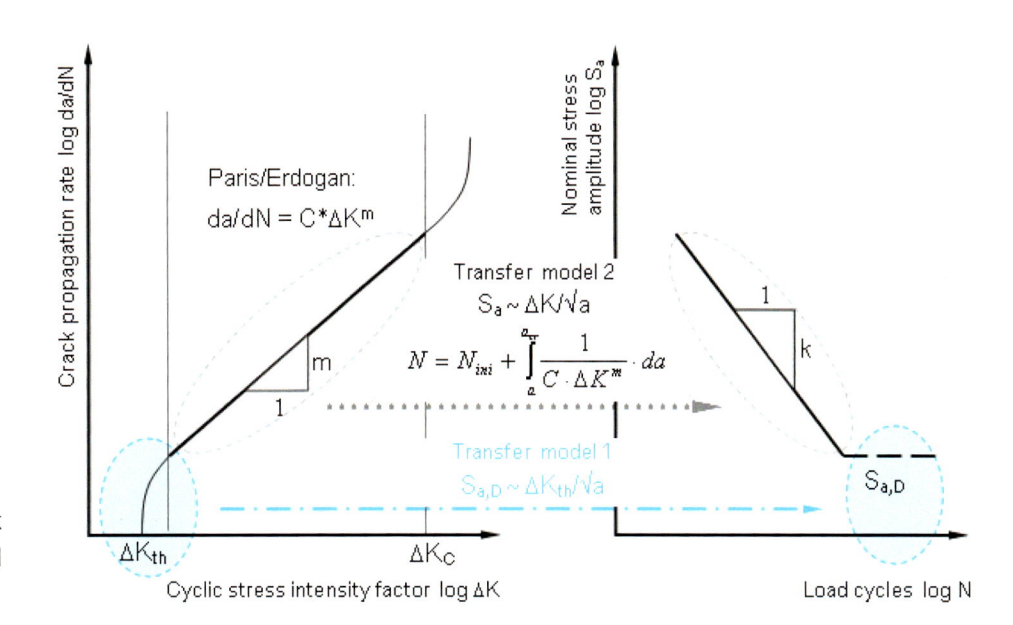

Figure 4-57 Transfer models between crack propagation curve and Wöhler curve [15]

whatsoever on crack initiation time which, on the other hand, is included in the Wöhler curves. Taking the necessary preconditions into account, it is assumed in the literature that the increase k in the fatigue endurance range of the Wöhler curve corresponds to the reciprocal of the exponent m of the Paris/Erdogan equation (Figure 4-57).

Using this relationship, models can be drawn up for transferring between the limit region of fatigue crack growth and the region of fatigue endurance of the Wöhler curve, as well as between the region of stable crack growth and the region of straight fatigue endurance lines in the Wöhler diagram. However, these theoretical models require further experimental validation.

4.4.6 Components and Component Tests

A number of suitable components were identified at the start of the project for use in developing material and damage models and for subsequent validation of the calculation results (Figure 4-58). Since characterization was required for both the thermoplastic and the thermoset materials, test components made of these two groups of materials were used.

The test piece used for the short-glass-fiber-reinforced thermoset material PF-GF-30+MX30 is a belt pulley

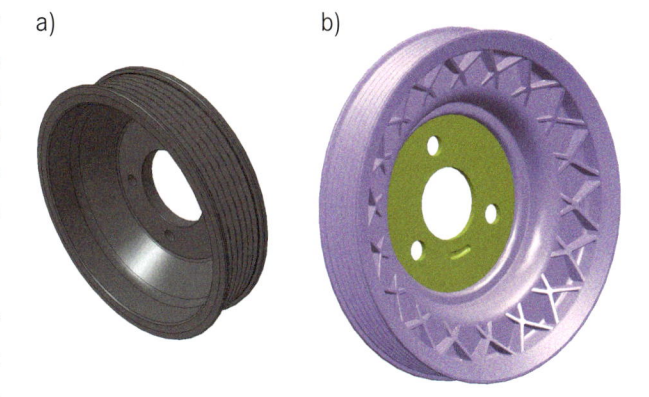

Figure 4-58 Selected test components
(a) Belt pulley made of PF-GF-30+MX30,
(b) Belt pulley made of PA 6T/6I-GF40+MX25 with molded-in steel insert

with a strongly offset shape (asymmetrical distribution of force). In the case of the thermoplastic material PA 6T/6I-GF40+MX25, a belt pulley was used which also had an offset connection of the belt contact surface with the hub, although the offset is not so marked as in the thermoset belt pulley. In addition, this pulley has a molded-in steel insert in the hub area.

Figure 4-59 shows stress images derived from the static strength calculations carried out for the selected test components. As expected, the highest reference stresses, disregarding the mounting holes, are found in the transitional radius of the offset.

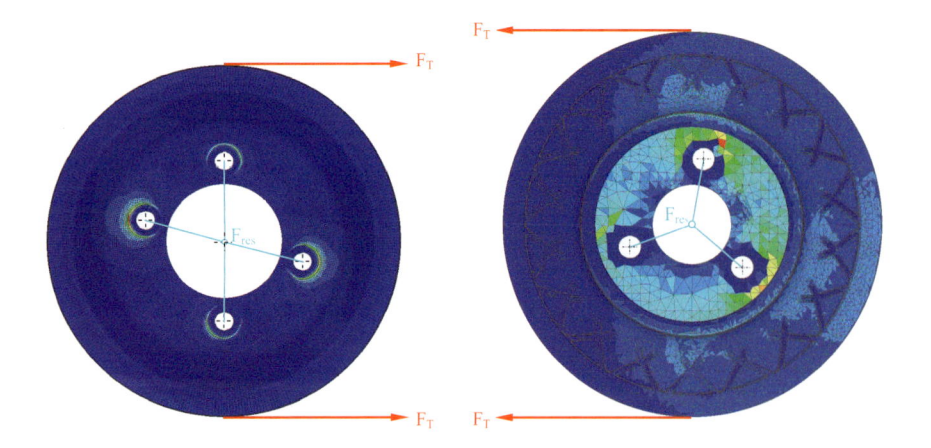

Figure 4-59 Stress images of thermoset and thermoplastic belt pulleys

During the course of the project, cyclic tests were carried out at a constant amplitude on these components in order to determine the component Wöhler curves. The results from these investigations are required on the one hand for comparing the data from specimen testing and also, as stated above, to coordinate the calculated load cycle number with the experimentally determined one. Furthermore, the fracture surfaces of failed belt pulleys are examined in order to deduce from their morphology what the decisive damaging mechanism would have been.

Figure 4-60 shows a test rig manufactured by Joma-Polytec GmbH on which service life testing of the thermoplastic belt pulley was carried out. The test rig has been set up so that two belt pulleys can be tested simultaneously. Here the pulleys are clamped and subject to constant loading via a V-belt wrapped 180° around the pulley circumference (without transfer of moment). The test rig also has a hot-air blower to simulate the higher ambient temperatures encountered in service.

4.4.7 Summary and Further Procedure

Composite materials have a great potential for future applications of lightweight design in the automotive industry. That this potential cannot yet be fully tapped is partially due to a lack of simulation tools for cyclic component design in the early phase of vehicle development. Advanced calculation tools for industrial application are required to enable inexpensive design of fiber-reinforced plastics to exploit their potential for lightweight construction and to open up new areas of application for these materials, something which has not as yet been possible due to insufficient information on cyclic loading and loadability.

This paper has attempted to present one possible procedure for the calculated estimation of service life for components made of short-glass-fiber-reinforced plastics subject to cyclic loading. The basic principle here rests on findings already published by Zago [1, 2]. The data and methods for service life calculation obtained during the current collaborative project of the companies mentioned at the beginning were implemented using the commercially available FEMFAT software package from MAGNA Powertrain/ECS.

Figure 4-60 Belt-pulley test rig manufactured by Joma-Polytec GmbH

Local component Wöhler curves are required in component design or service life calculation carried out with FEMFAT if damage is to be assessed correctly. Since no prototypes are as yet available in the early development stage, the designer turns to simulation in order to estimate local Wöhler curves on the basis of material Wöhler curves (determined using unnotched specimens) and different influencing factors.

Here, once again, caution is advised regarding transferability from specimen to component, since in most cases there is no uniaxial vibrational stress in the component directly comparable with the loading used to test specimens. Furthermore, additional above-mentioned influencing factors must be taken into account, such as temperature, humidity, notches and fiber content.

The long-term goal is to create a broad database similar to that which exists for metal materials which, in combination with suitable failure and damage accumulation hypotheses, would make it possible to perform service life estimations for a wide variety of (fiber-reinforced) polymers.

4.4.8 Acknowledgments

The research presented was carried out at the Polymer Competence Center Leoben GmbH as part of the K_{plus} competence center program of the Federal Ministry of Traffic, Innovation and Technology with the participation of the Chair of General Mechanical Engineering (AMB) and of the Institute for Materials Science and Plastics Testing (IWPK) at the University of Leoben, Austria, and also of the BMW Group, the Engineering Center Steyr (MAGNA Powertrain ECS) and EMS-GRIVORY and subsidized with funds from the Austrian government, the provinces of Styria and Upper Austria and from the University of Leoben. The authors should like to express special thanks to Prof. W. Eichlseder and Dr. I. Gódor (AMB), Prof. R. W. Lang and Dr. Z. Major (IWPK) for their important contributions in designing and monitoring the project and also to B. Unger (ECS) and T. Jeltsch (EMS-GRIVORY) for their expert contributions to this paper.

References

[1] Brune, M., Zago, A.: Lebensdauerberechnung für kurzglasfaserverstärkte Thermoplaste [Service life calculation for short-glass-fiber-reinforced thermoplastics], VDI Society of Plastics Engineering, Plastics in Automotive Engineering, conference Mannheim 13/14 March 2002, pp. 129–145

[2] Zago, A., Springer, G. S.: Life Prediction of Short Fiber Composites, Final Report to BMW AG, Department of Aeronautics and Astronautics, Stanford University, July 2000

[3] Sedlacik, G.: Beitrag zum Einsatz von unidirektional naturfaserverstärkten thermoplastischen Kunststoffen als Werkstoff für großflächige Strukturbauteile [On the use of unidirectional natural-fiber-reinforced thermoplastics as material for large-area structural components], Faculty of Mechanical Engineering, Technical University of Chemnitz, dissertation, 2003

[4] Tutorial on Polymer Composite Molding, http://islnotes.cps.msu.edu/trp/inj/mic_ornt.html, Intelligent System Lab, Department of Computer Science and Engineering, Michigan State University, 1999

[5] Immersive SIM Engineering GmbH, www. Immersive-sim.de, 2005

[6] Numerical Simulation of Composite Materials, www.mat.ethz.ch/about_us/material_world/, Department of Materials, ETH Zürich, 2005

[7] Haag, R.: Anisotrope Bauteilauslegung – Einfluss der Faserorientierung auf die strukturmechanische Auslegung [Anisotropic component design: influence of fiber orientation on structure-mechanical design], SimpaTec GmbH, ABAQUS Users Conference, Nuremberg, September 2005

[8] Lusti, H. R.: Computerunterstütztes Design von Verbundwerkstoffen mit Palmyra [Computer-aided design of composite materials with Palmyra], ETH Zentrum Zürich, CADFEM Users Meeting, Friedrichshafen, October 2002

[9] Köttgen, V. B., et al.: Optimierung im Hinblick auf Betriebsfestigkeitseigenschaften am Beispiel von Fahrwerkskomponenten [Optimization of engineering strength properties in car body components], Symposium on Simulation in Product and Process Development, Bremen, November 2003

[10] Bolender, K., Büter, A., Gerharz, J.: Entwicklung eines einfachen numerischen Bemessungswerkzeuges zur Bewertung mehraxial beanspruchter kurzfaserverstärkter Kunststoffe [Development of a simple numerical dimensioning tool for evaluating multiaxially loaded short-fiber-reinforced plastics], Fraunhofer Institute for Durability and System Reliability (LBF), Darmstadt, Congress on Intelligent Lightweight Systems, 2005

[11] Middendorf, P.: Composites – Materialmodellierung und Anwendungen im Flugzeugbau [Composites: material modeling and applications in aircraft construction], EADS Corporate Research Center Germany, 3rd LS-DYNA Users Forum, Bamberg, 2004

[12] Bardenheier, R.: Mechanisches Versagen von Polymerwerkstoffen, Anstrengungsbewertung mehrachsialer Spannungszustände [Mechanical failure of polymer materials and stress exposure factor evaluation of multiaxial stress states], Kunststoffe-Forschungsberichte, Vol. 8, Carl Hanser Verlag Munich and Vienna, 1982

[13] FEMFAT BASIC Theorie-Manual, Engineering Center Steyr GmbH & Co KG, Magna Powertrain

[14] FEMFAT MAX, Benutzerhandbuch [User's manual], Engineering Center Steyr GmbH & Co KG, Magna Powertrain

[15] Schreibmeier, M.: Ermüdungsverhalten von kurzglasfaserverstärkten Polymeren für Anwendungen im Automobilbau unter Zug- und Druckschwellbelastung [Fatigue behavior of short-glass-fibber-reinforced polymers used in automobile manufacturing under tensile and repeated compressive stress], Institute for Materials Science and Plastics Testing, University of Leoben, Polymer Competence Center Leoben, dissertation, July 2005

[16] Guster, Ch., Balika, W.: Ermüdungsgerechte Gestaltung von Kunststoffbauteilen in der Automobilindustrie durch Modellierung und Simulation [Fatigue-oriented design of plastic components in the automotive industry by modeling and simulation], Institute for General Mechanical Engineering, Institute for Materials Science and Plastics Testing, University of Leoben, Polymer Competence Center Leoben, Status Report WP-04, Domat/Ems, May 2005

[17] Guster, Ch., et al.: Fatigue Analysis of Polymer Components for Applications in the Engine Compartment, Polymer Competence Center Leoben GmbH, Department of Product Engineering, Chair of Mechanical Engineering, Institute of Materials Science and Plastics Testing, University of Leoben, 7th Austrian Polymer Conference, Graz, July 2005

4.5 Material Data for Imaging Thermoplastic Plastics in Crash Simulation

Helge Liebertz

4.5.1 Introduction

The development of automobiles is strongly affected by ever shorter development times. Even in the early phase of design, compliance with statutory provisions must be ensured such as those concerning passenger protection according to ECE R21.

Functionality is ensured by the use of calculation and simulation with virtual prototypes even before the actual physical prototypes are made.

At Volkswagen, the finite element calculation program PAM-CRASH is used for collision loading cases. The quality of the calculation results here depends very much on the quality of the material properties.

This paper will give an overview of the material properties required for crash calculations. It will describe the current state of imaging capabilities in material models and point out the wide variety of possibilities for error in the generation of these data.

4.5.2 Input Data for Crash Calculation

Problems in the analysis are frequently due to missing or erroneous material data. Many parameters that are helpful in strength calculations involving static loads cannot be used in calculating highly dynamic events with major deformation. Even today, adequate data are still not available.

The direct relationship between stress and strain under uniaxial loading and a constant strain rate is the input variable used in most FEM programs. For steel, these variables can be simply obtained from tensile tests (DIN EN 10 002), since they assume volume constancy and homogeneous deformation up to elongation before reduction of area.

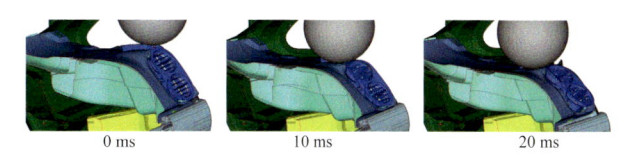

Figure 4-61 Simulation of head impact on air outlets in the instrument panel of the Golf Plus

In most cases, deformation is inhomogeneous with thermoplastics. Once yield stress is reached, necking occurs (Figure 4-62). Of the quantities measured in the tensile test, only the displacements over the entire specimen are considered (referred to here as global measurement) and force is expressed in relation to the initial cross-section.

These global displacements as determined in the test according to EN ISO 527 do not express the local relationship between stress and strain in the material.

How then can data be obtained in the form of true stress-strain curves for use in the calculation?

Most available data are based on point values obtained from quasistatic tests. Product data sheets only provide individual values such as the modulus of elasticity, yield stress and elongation at break.

It is not practice for manufacturers to publish actual stress-strain curves. Information systems, such as

Table 4-2 Mechanical properties from a data sheet

Mechanical properties	Value	Unit	Test Standard
Tensile modulus	2200	MPa	ISO 527-1/-2
Yield stress	52	MPa	ISO 527-1/-2
Yield strain	4.2	%	ISO 527-1/-2
Nominal strain at break	42	%	ISO 527-1/-2

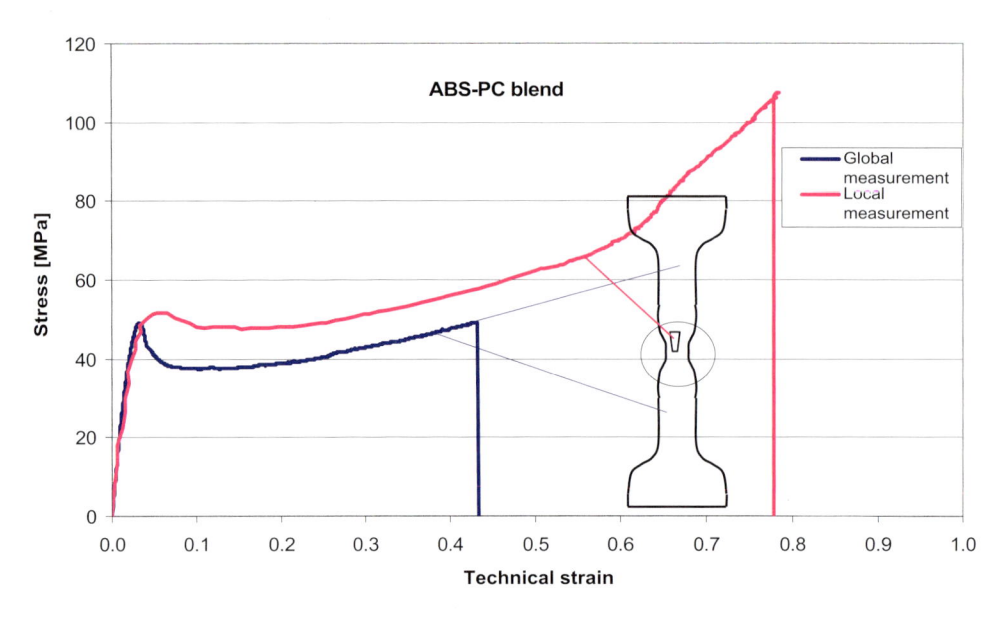

Figure 4-62 Comparison of the stress-strain curves from local and global measurement

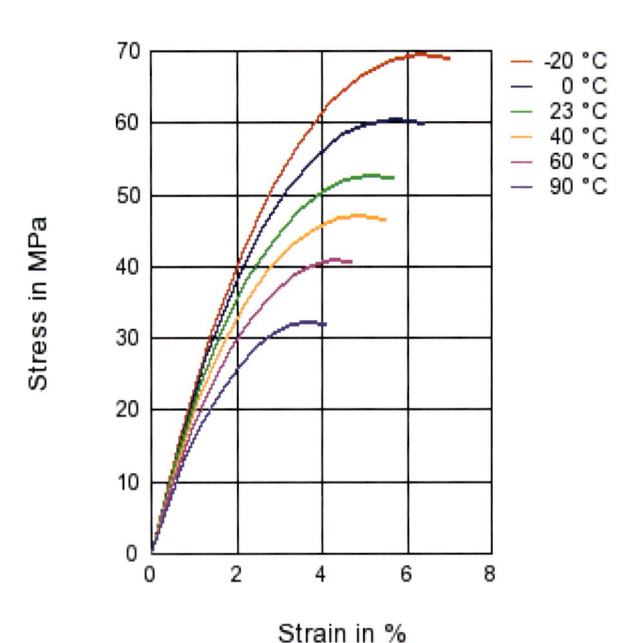

Figure 4-63 Example of stress-strain curves from the CAMPUS system

CAMPUS, for example, plot curves from tensile tests, but unfortunately only up to just beyond the yield point (Figure 4-63). Although the data for the complete nominal stress-strain curve exist, they are not presented.

An exact measurement of true stress and true strain in the necking zone can only be made using local measurement. The true figures for elongation can only be obtained from the current local elongation over the momentary cross-section at a particular location. The magnitude of elongation depends here on the three-dimensional resolution. It should lie in the same range as the mesh size of the analysis model. Simultaneous measurement of specimen width and local elongation for large elongations and high speeds works only if optical measurement systems are used. Due to their mass, mechanical sensing elements are not suitable for dynamic measurements. Strain gauges are only suitable for elongations of about 20% and thus cannot be used for very ductile materials [4]. Unfortunately optical measuring equipment is expensive and not generally available. For this reason another approach was taken in measuring the real stresses and real strains. Fictitious stress-strain curves are defined and then optimized until the entire course of the global force and displacement in the tensile test agrees with them. This procedure, referred to as reverse engineering, can deliver the right results [5]. It does, however, call for a certain amount of experience and the results are not always clear-cut. It should be noted here that the force-deformation characteristic is not the only optimization goal of the reverse engineering approach, but kinematic aspects as well – in other words the spreading of specimen necking must be correctly described. Deformation will thus be frequently limited in the analysis to a few elements in the loading direction, as can be seen on the right-hand side of Figure 4-64.

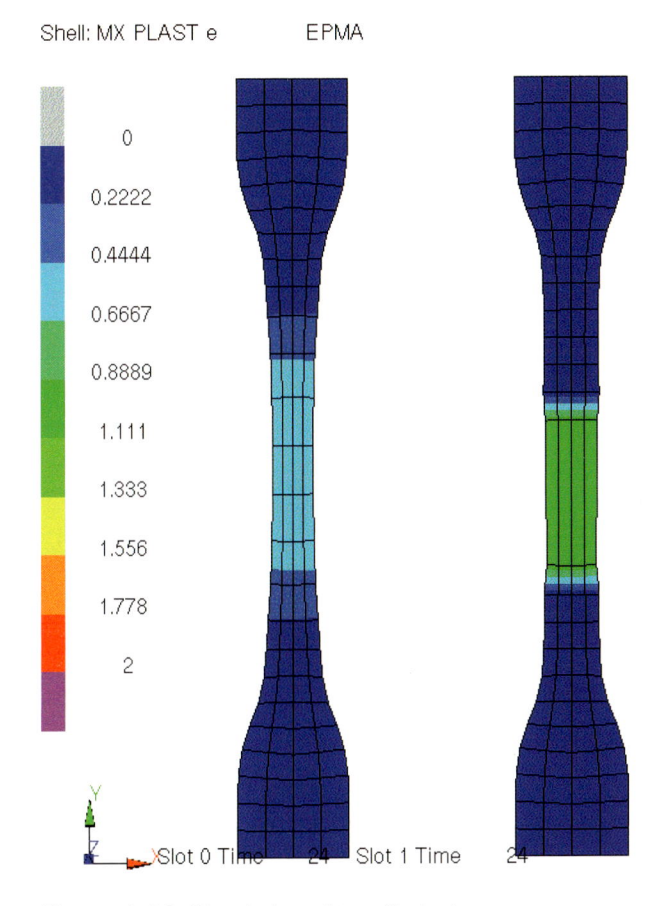

4.5 Material Data for Imaging Thermoplastic Plastics in Crash Simulation

Figure 4-64 Simulation of tensile test

4.5.3 Representation of Non-Linear Behavior in the Simulation

4.5.3.1 Time-Dependent Behavior

In commercial program systems for FEM analysis, several approaches are available for describing non-linear material behavior. In these, deformation under an applied force is described. One visually satisfactory form of these material models displays equivalent elements as springs, dampers and friction elements (Figure 4-65).

a) Elastic-plastic model   b) Viscoplastic model

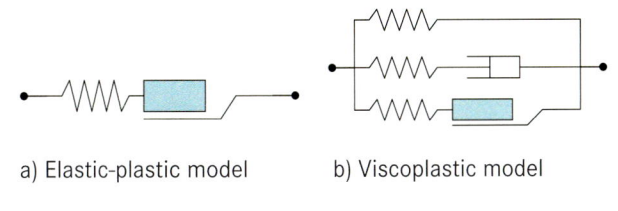

Figure 4-65 Equivalent model in the FEM program

Here the spring describes linear-elastic behavior and the friction element describes plastic behavior. In descriptions of non-linear deformation behavior, a distinction is drawn between behavior dependent on the yield rate and time-dependent behavior.

Behavior dependent on the yield rate can be described by the elastic-plastic model shown in Figure 4-65 a. Time-dependent behavior, such as creep and non-linear elasticity, can only be described by viscoplastic models (Figure 4-65 b).

Adjustment of a viscoplastic material model for large deformations and a wide speed range is not done, since it is too complex a matter to identify for each material the parameters for the coupled system consisting of springs and dampers. No procedure exists as yet for deriving generally valid parameters of viscoplastic material models directly from tests. There is a need here for research institutes and software manufacturers to come up with the corresponding solutions. In crash loading cases where impact-like loads are applied in a period of just a few milliseconds, these time-dependent variables play only a subordinate role if only accelerations and not the final state of deformation are considered.

4.5.3.2 Behavior Dependent on the Yield Rate

In their mechanical behavior, thermoplastics exhibit great dependence on speed. Taking an ABS-PC blend as an example, the change in yield stress is seen as a function of the strain rate (Figure 4-66).

This yield-rate-dependent behavior is a considerable influencing factor in crash analyses of plastics. Quasi-statically based tensile tests always deliver results for behavior that are too soft. The range of strain rates relevant in collisions extends from quasi-static loading up to about to 500/s. In determining the full range of behavior dependent on the yield rate, it has proven useful to carry out a logarithmic subdivision of the strain rates for test series into quasi-static, 2/s, 20/s and 200/s. The material models currently used contain the real stress-strain curves and extrapolate to current strain rates.

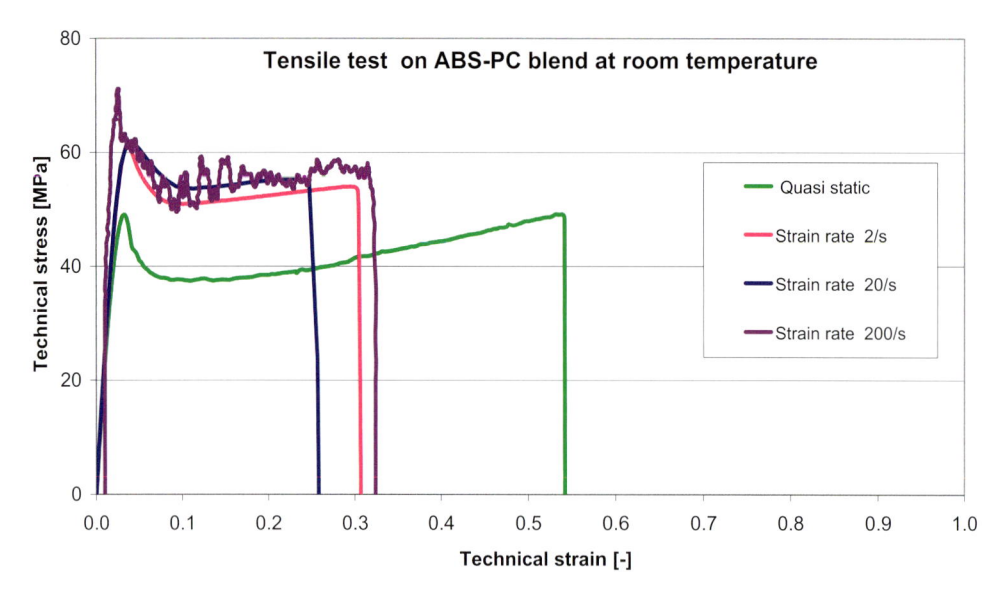

Figure 4-66 Influence of the strain rate on mechanical properties in the tensile test

4.5.3.3 Temperature-Dependent Behavior

The temperature dependence of mechanical parameters lies within the same order of magnitude as dependence on strain rates (Figure 4-67). Since the entire temperature range cannot be continuously tested, testing is restricted here to the extreme values plus a small safety margin. In addition to room temperature the temperatures of −35 °C and 85 °C are taken into consideration in testing.

Temperature dependence can also be described by analytic functions within the analysis program [1, 2].

These approaches work for very ductile materials, but the transferability of parameter approaches to a different plastic is limited, which means that here, too, speed- and temperature-dependence have to be measured for each material. Either there are only a few data sets for plastics investigated in such detail or they have not been made generally accessible.

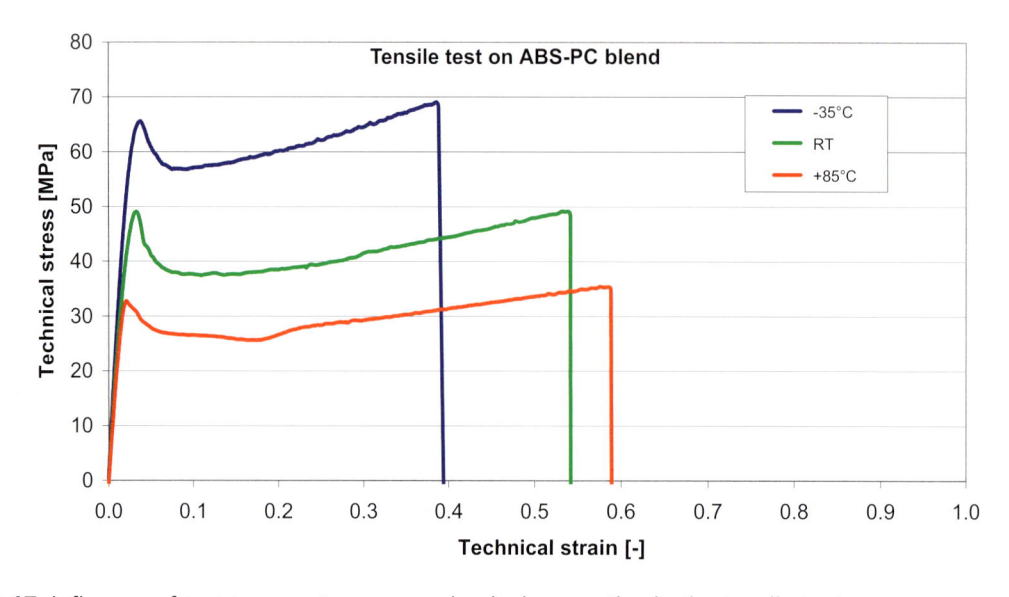

Figure 4-67 Influence of test temperature on mechanical properties in the tensile test

4.5.4 Fiber-Reinforced Thermoplastics

Fiber-filled plastics are finding application with the trend towards plastics which become ever stiffer but nevertheless remain tough. The fibers become oriented during the injection process. The proportion of oriented fibers is to a very great extent determined by the manufacturing process.

Samples taken from a component longitudinally or transversely to the direction of injection will behave very differently not only in the force level but also in elongation at break (Figure 4-68). It is not possible to tell from the data sheets how strong the anisotropy is or can be.

Thermoplastic fiber-filled plastics only receive their characteristic mechanical properties from the manufacturing process while orientation and crystallinity form first in the component. It therefore makes sense to define parameters for simulating such materials only on the basis of test samples made close to the start of series production. The mechanical properties of an injection-molded standard specimen and a sample cut from a plate resembling the component can exhibit greatly differing behavior. For this reason, samples for the analysis parameters of fiber-filled thermoplastics should only be taken under conditions close to production as regards the extruder, molds, flow paths and curing times. A directly injection-molded standard specimen 4 mm in thickness, such as is used for tensile tests with CAMPUS data, is not suitable for this purpose. Flat sheets with a typical wall thickness of 2 mm have proven suitable for making close-to-production samples for automotive applications. Samples with different fiber orientations can be taken from these sheets.

Fiber orientation in components can be determined with the aid of injection-molding simulation and these data transferred into the crash simulation [3]. This requires injection-molding simulations and material models for fiber orientation. These are not generally available in the concept phases in which the materials are frequently defined. The current solution is to use averaged curves and to apply isotropic material models for the simulation. The mean value curve is obtained from samples cut longitudinally and transversely to the fiber orientation; minimum and maximum values are taken as a safeguard.

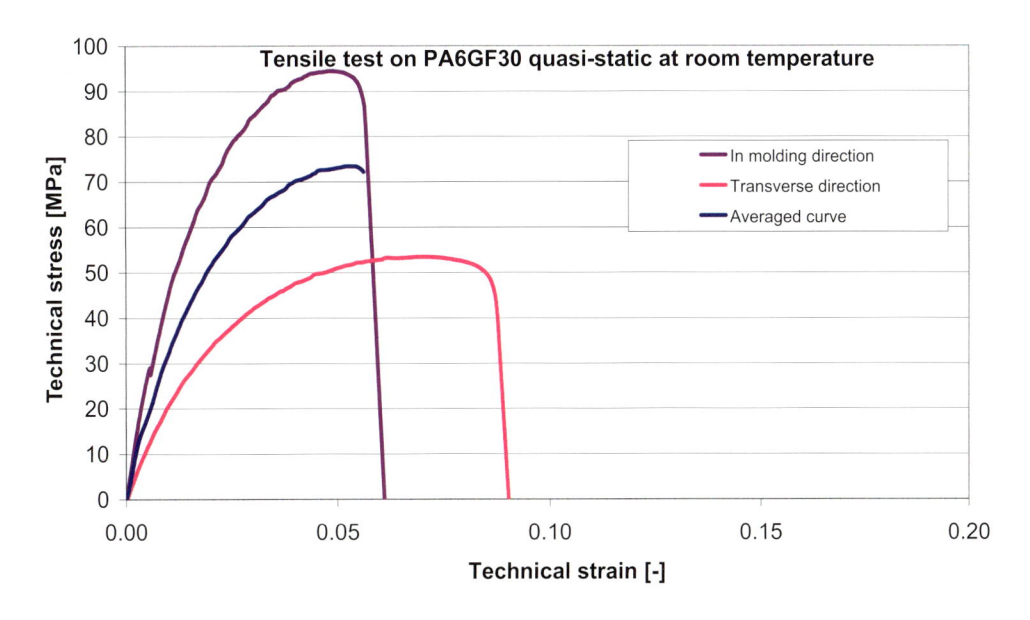

Figure 4-68 Influence of direction of injection on mechanical properties in the tensile test

4.5.5 Validation of Material Data and Material Models

The correct reproduction of material behavior in the material model is checked in complex models before use. The simplest way to validate material data is to simulate the tensile test curve (Figure 4-64) in real geometry. As a rule, the plasticity model used is not transferred to any load states other than uniaxial tension.

There are two reasons for this. Firstly, only very little information is available about material tests involving other kinds of load such as compression, shear, bending or biaxial loading. Secondly, material models currently do not provide any method for a simple input of such information without getting lost in complex parameter identifications. The only way to improve matters here is for material producers, developers and suppliers to collaborate with one another. Clear definitions are required as to which tests are needed for the simulation parameters and how they are to be performed. The basics are currently being worked out by working parties at FAT (the Automotive Engineering Research Association) [6].

4.5.6 Failure Analysis

A description of failure in the simulation of plastics that can support prognostic activity represents a major research challenge. Unlike metals, accumulated effective strain is not an adequate criterion for failure under tensile stress. The dependence of elongation at break on speed, temperature and orientation can be seen in the diagrams (Figures 4-66 to 4-68). Procedures for taking these influencing factors into account cannot be found in commercially available crash analysis programs. For this reason, failure can only be set for the specific loading case and after comparison with tests in simulations.

4.5.7 Outlook

It would be desirable for parameters suitable for use in simulations to be made more accessible. In the case of plastics whose mechanical properties vary according to the processing method, parameters need to be determined using specimens closely resembling the final component. The concept used by CAMPUS – that of making a straightforward comparison between the polymers – could be expanded. A joint recommendation as to what such specimens must look like and what other component parameters (such as the strength of flow lines) can also be tested, is a necessary and exciting goal for the future, one which can bring about a marked improvement in the quality of plastic component simulations.

References

[1] Zerilli F. J., Armstrong R. W.: Thermal activation based constitutive equation for polymers, *J. Phys.* IV France 10 (2000)
[2] Duffo, P., Monasse, B., Haudin, J. M., G.'Sell, C., Dahoun, A.: Rheology of polypropylene in solid state, *J. of Mat. Sc* (1995)
[3] Glaser, S., Wüst, A.: Der Einfluss des Herstellungsprozesses auf die mechanischen Eigenschaften thermoplastischer Bauteile, [The influence of the manufacturing process on the mechanical properties of thermoplastic components], LS-DYNA user conference 2005 Bamberg
[4] Michalke, W.: Kunststoffe simulationsgerecht charakterisieren, [Characterization of plastics for simulation purposes] *Kunststoffe Automotive* 1-2004
[5] Lutter, F., Münkers, M., Wanders, M.: Rechnen bis zum Versagen, [Analysis to the point of failure] *Kunststoffe* 1-2002
[6] FAT Arbeitskreis 27 Finite-Element-Anwendungen im Automobilbau, [Finite element applications in automotive engineering], UA Crash und Insassensimulation AG Kunststoffe

5 Joining

5.1 Bonded Joints in and on the Vehicle Structure

Ulrich Walther

5.1.1 Introduction

The development of new kinds of products suited to market requirements is a process which, when considered in its technological aspects, represents a linking of design, material and production concepts. Due to the increasing demand for thorough-going lightweight design and multifunctionality of the product, as well as special requirements regarding the esthetics of the component, the production concept has acquired a high degree of importance in the development process. In this connection, even the selection of the bonding technique is becoming increasingly important. One goal is the best possible use of materials in a "multi-materials mix" of different metallic materials. Another is in particular to combine metallic with non-metallic materials. This means that so-called "thermally poor" bonding and joining techniques, such as gluing and riveting, are becoming key techniques in the development of concepts using a combined-materials approach. Equally important are design concepts, such as concealed joints in visible areas, which are forcing more widespread use of bonding techniques in the automotive industry [1, 2].

Here, a number of adhesive applications are presented in different vehicle areas from a materials- and design-specific point of view by describing one example of the procedure used for selecting suitable adhesives for a component.

5.1.2 Assembly Bonding

Assembly bonded joints (Figure 5-1) are bondings made, not during production of the body shell, but during the subsequent assembly stage. At this time, the body is electrophoretically dip coated or has a top coat. The main difference from body shell bonding is the "corrosion-proof" substrate, enabling a much wider range of adhesives to be used than on an untreated body shell. The relaxation in temperature stability restrictions, by contrast with body shell adhesives that have to withstand temperatures in excess of 180 °C in the electrophoretic coating process, also means that a wider variety of adhesives can be used in the assembly process.

Adhesives currently in use include predominantly moisture-curing polyurethanes, two-pack polyurethanes and epoxy resins, as well as adhesive tapes for the various substrates and components. Nevertheless other types of adhesive are also used, though to a lesser extent, such as a wide variety of rubbers, methacrylic and cyanoacrylate systems, as well as MS polymers. The wide variety of materials used in an automobile (Figure 5-2) gives some idea of just how many material pairings there are and also the complexity of what has to be taken into account before assembly bondings can be implemented in the automotive industry [2].

In addition to gluing on separate body components, such as spoilers, decorative trim, emblems and so on, particular emphasis must also be given to the gluing-on of the windshield, rear and rear side windows (Figure 5-3). Compared with rubber-mounted glazing, gluing fixed vehicle glass satisfies special collision requirements, increases torsional rigidity of the body considerably and improves the aerodynamics and airtightness of the passenger compartment.

As early as 1976, AUDI became the first European automaker to use direct glazing installation by means of adhesive in series production [3]. There have never-

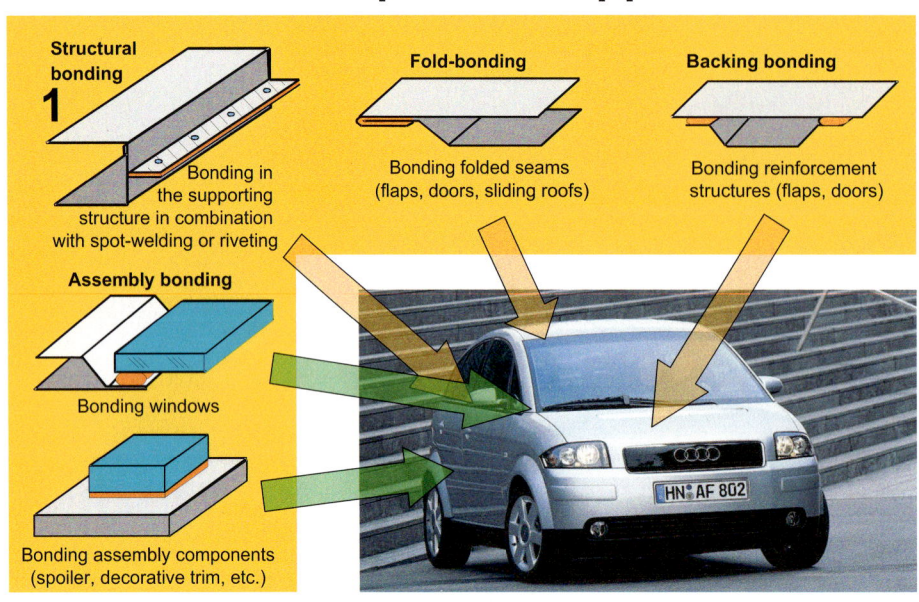

Bonded joints in and on the vehicle structure

Adhesive techniques and applications

Structural bonding
1
Bonding in the supporting structure in combination with spot-welding or riveting

Fold-bonding
Bonding folded seams (flaps, doors, sliding roofs)

Backing bonding
Bonding reinforcement structures (flaps, doors)

Assembly bonding
Bonding windows

Bonding assembly components (spoiler, decorative trim, etc.)

Figure 5-1 Overview of adhesive applications in the automobile [2]

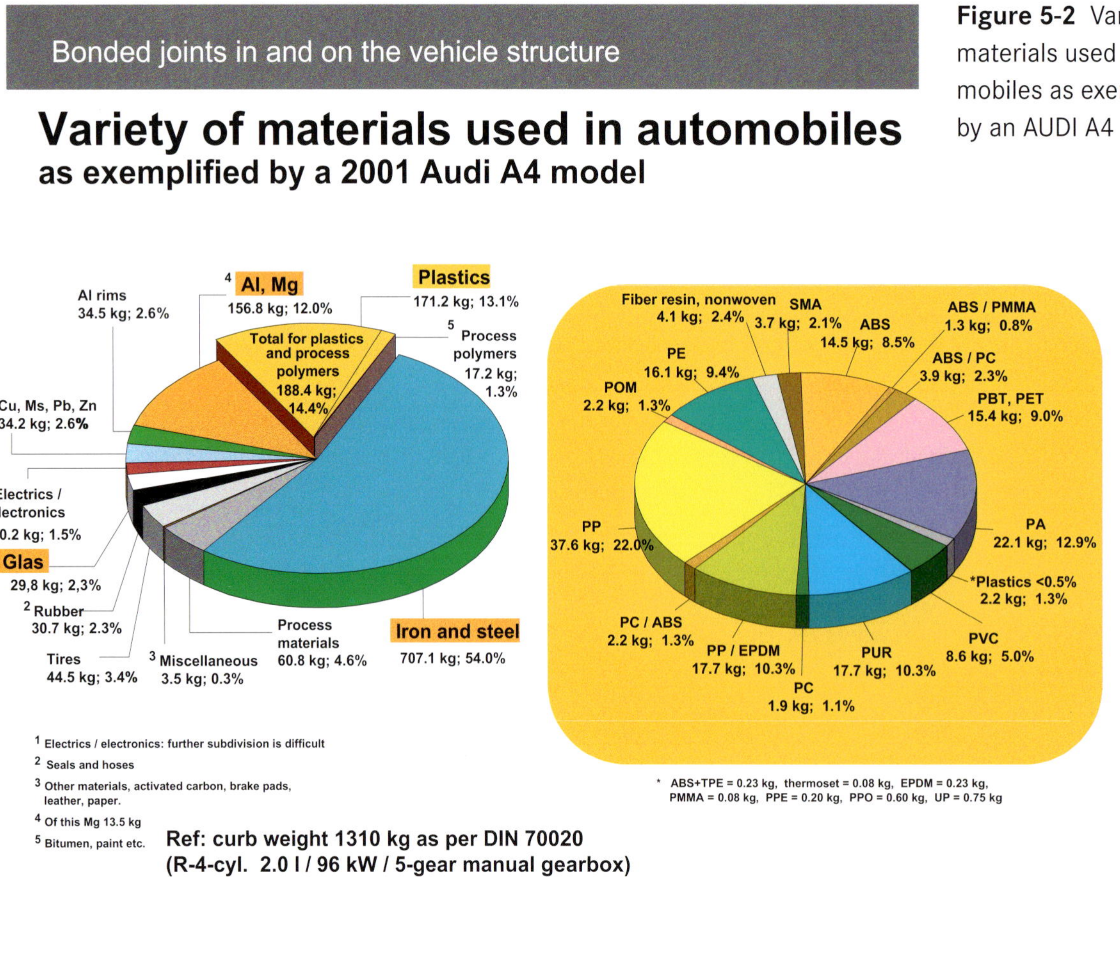

Bonded joints in and on the vehicle structure

Variety of materials used in automobiles
as exemplified by a 2001 Audi A4 model

Figure 5-2 Variety of materials used by automobiles as exemplified by an AUDI A4

Pie chart 1:
- Al rims 34.5 kg; 2.6%
- [4] Al, Mg 156.8 kg; 12.0%
- Plastics 171.2 kg; 13.1%
- Total for plastics and process polymers 188.4 kg; 14.4%
- [5] Process polymers 17.2 kg; 1.3%
- Cu, Ms, Pb, Zn 34.2 kg; 2.6%
- [1] Electrics / electronics 20.2 kg; 1.5%
- Glas 29.8 kg; 2.3%
- [2] Rubber 30.7 kg; 2.3%
- Tires 44.5 kg; 3.4%
- [3] Miscellaneous 3.5 kg; 0.3%
- Process materials 60.8 kg; 4.6%
- Iron and steel 707.1 kg; 54.0%

Pie chart 2:
- Fiber resin, nonwoven 4.1 kg; 2.4%
- SMA 3.7 kg; 2.1%
- ABS 14.5 kg; 8.5%
- ABS / PMMA 1.3 kg; 0.8%
- ABS / PC 3.9 kg; 2.3%
- PBT, PET 15.4 kg; 9.0%
- PE 16.1 kg; 9.4%
- POM 2.2 kg; 1.3%
- PP 37.6 kg; 22.0%
- PA 22.1 kg; 12.9%
- *Plastics <0.5% 2.2 kg; 1.3%
- PC / ABS 2.2 kg; 1.3%
- PP / EPDM 17.7 kg; 10.3%
- PUR 17.7 kg; 10.3%
- PVC 8.6 kg; 5.0%
- PC 1.9 kg; 1.1%

[1] Electrics / electronics: further subdivision is difficult
[2] Seals and hoses
[3] Other materials, activated carbon, brake pads, leather, paper.
[4] Of this Mg 13.5 kg
[5] Bitumen, paint etc.

Ref: curb weight 1310 kg as per DIN 70020
(R-4-cyl. 2.0 l / 96 kW / 5-gear manual gearbox)

* ABS+TPE = 0.23 kg, thermoset = 0.08 kg, EPDM = 0.23 kg, PMMA = 0.08 kg, PPE = 0.20 kg, PPO = 0.60 kg, UP = 0.75 kg

Assembly bonds

Figure 5-3 Assembly bonding in the automobile [2]

(Callout: windows, spoiler, decorative trim, emblems, door step trim, panels, headliner, protective moldings, spare wheel well ...)

theless also been further developments in this field, such as the improvement of the collision performance of the adhesives or the reduction in the fixing and curing times of autoglass adhesives.

5.1.3 Bonded Joints in and on the Body

The increasing requirements made on the automobile, particularly as regards exhaust emissions, energy consumption, active and passive safety, comfort, reliability and even driving pleasure, mean that innovative vehicle concepts are needed. Especially in the body sector, further developments are necessary regarding:

- Collision characteristics
- Dynamic flexural and torsional rigidity of the first and second order, and also
- Weight reductions

if the requirements mentioned above are to be satisfied. New material and process developments here form the basis for lightweight-oriented design principles which permit direct weight reduction in the vehicle as an overall system or make it possible to compensate additional weight required by measures to improve both comfort and safety. However, the weight savings achieved, for example, by using higher-strength steel materials and/or hybrid designs, cannot be maximized unless the specific material properties are retained at the bonded interfaces where forces are transmitted. Against this background, adhesive bonding as a technique still has great potential for improving body properties without adding a lot of weight or for enabling new types of design by using different materials [4]. Bonded joints used in the body may be divided into the following three categories:

- Folded bonds
- Backing bonds
- Bonds in the structure.

Folded bonds – also known as folded seam bonds – are found in the vehicle in the outer parts of wheel housings and in add-on parts such as doors and flaps. This type of bond has been state-of-the-art for many years in the automotive industry. All the same, there have been additional advances in this type of bond to be found primarily in the field of process technology. In response to increasing customer pressure for individualization, more and more niche products and derivates are being offered. The consequence of this is that production equipment must be made more flexible and even more compact. What this means for folded bonds is that, in addition to conventional processes, new techniques, such as "folding without fold bed", induction heating and innovative application techniques also need to be investigated. Developments in fold bonding thus primarily involve improvements in the process capability of the adhesives. With regard to their adhesive mechanical properties, the adhesives used for fold bonds have to be classed under structural adhesives. Their shear strength lies between 200 and 1000 MPa.

Backing bonds have also been used regularly in the bodywork for many years. These are bonded joints that create a connection between thin outer body panels and inner reinforcement profiles. The primary objective of this bond is to increase the rigidity of the component. For this reason, adhesives are used whose shear strength is as high as possible But on the other hand, with this kind of connection, there is also the requirement that the connecting points are not visible on the esthetically demanding exterior surfaces of the vehicle.

Telltale markings can arise from reaction-related shrinkage of the adhesive during the curing process as it shifts from its liquid or pasty state to its solid state. If the adhesive has to cure under the non-homogeneous temperatures experienced by the individual components during coating processes, so-called thermally induced relative movements may take place and be frozen by the adhesive curing, thus possibly leading to sinkholes or other markings on the outer skin. An effective way of preventing such manifestations is to use relatively soft adhesives. The selection of suitable backing adhesives thus requires a compromise between fewer markings and greater rigidifying effect. Tests and FEM calculations indicate that, with many add-on parts, component rigidity increases markedly but only up to an adhesive modulus of about 10 MPa. For this reason, adhesives with a shear modulus of approx. 0.5 to 10 MPa are currently used for backing bonds in the automotive sector.

The term structural bonding is frequently used in a generalized manner, not only for strength- and rigidity-related connections in the body structure, but also for fold-bonded joints. "Pure structural bonding" is not currently used in bodywork applications, unless to a very limited extent. In most cases, gluing is used for backing up conventional bonding techniques such as spot welding, punch riveting, clinching or screwing.

A wider range of additionally glued flanges in the body structure could exert a considerable positive influence on the properties of the total system, i.e., the vehicle. For example, the service strength of body components can be considerably improved by additional gluing. Accordingly, comparative tests involving spot-welded and spot-weld-glued bodies have indicated strength-relevant improvements of between 20% and 300% [5]. Damage symptoms change here in such a way that, with spot-welded bodies, 90% of cracks start at the welding points, whereas with glued and combination-bonded bodies failure occurs outside the glued areas. This means that, due to the shift of stresses, additional gluing does not necessarily result in improvements in the service strength of the system as a whole [5]. Sudden changes in rigidity arise between combined bonded and exclusively spot-welded flange areas; these can be the initiators of damage in adjoining areas. For this reason, areas where adhesive is to be used need to be very carefully analyzed and defined.

A new generation of structural adhesives currently frequently referred to as collision-optimized adhesives, offers advantages over systems previously available with regard to loadability under dynamic collision-like loading and also under dynamic cyclic loading. This in turn opens up new opportunities for increasing the rigidity of the body (Figures 5-4 and 5-5).

The rigidity of a body can also be considerably increased by combined bonding techniques. Depending on the body concept, additional adhesive bonding can bring improvements of 15% and more with regard to static and dynamic torsional and flexural strengths, and do so without adding any significant weight (Figure 5-6).

5.1 Bonded Joints in and on the Vehicle Structure

Bonded joints in and on the vehicle structure

Mechanical properties of bonds: dynamic-cyclic

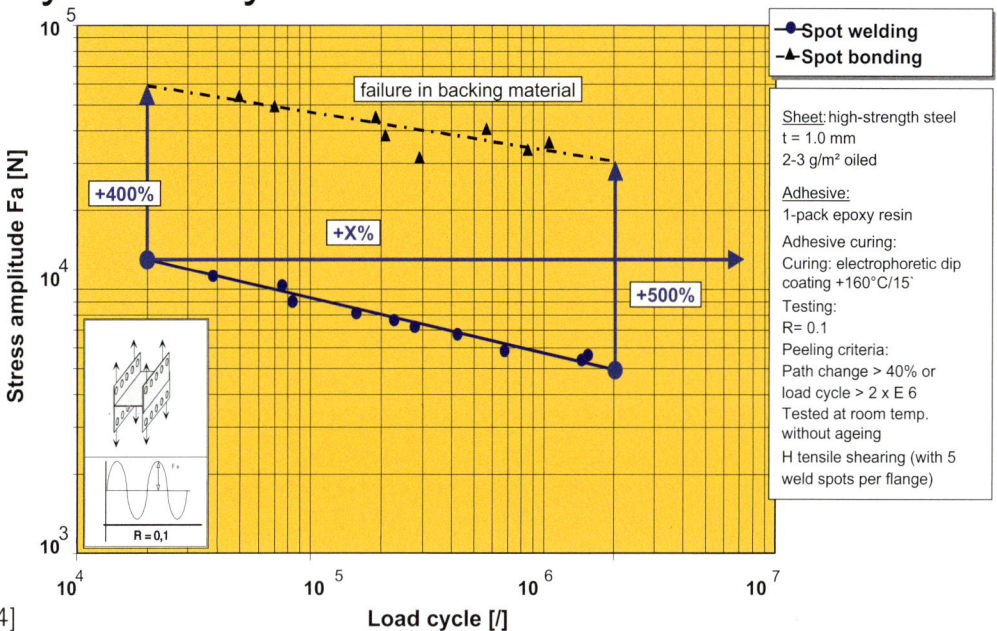

Sheet: high-strength steel
t = 1.0 mm
2-3 g/m² oiled

Adhesive:
1-pack epoxy resin

Adhesive curing:
Curing: electrophoretic dip coating +160°C/15`

Testing:
R= 0.1
Peeling criteria:
Path change > 40% or load cycle > 2 x E 6
Tested at room temp. without ageing
H tensile shearing (with 5 weld spots per flange)

Figure 5-4
Comparison of the fatigue strength of spot welding and spot-weld bonding [4]

Mechanical properties of bonds: dynamic - collision

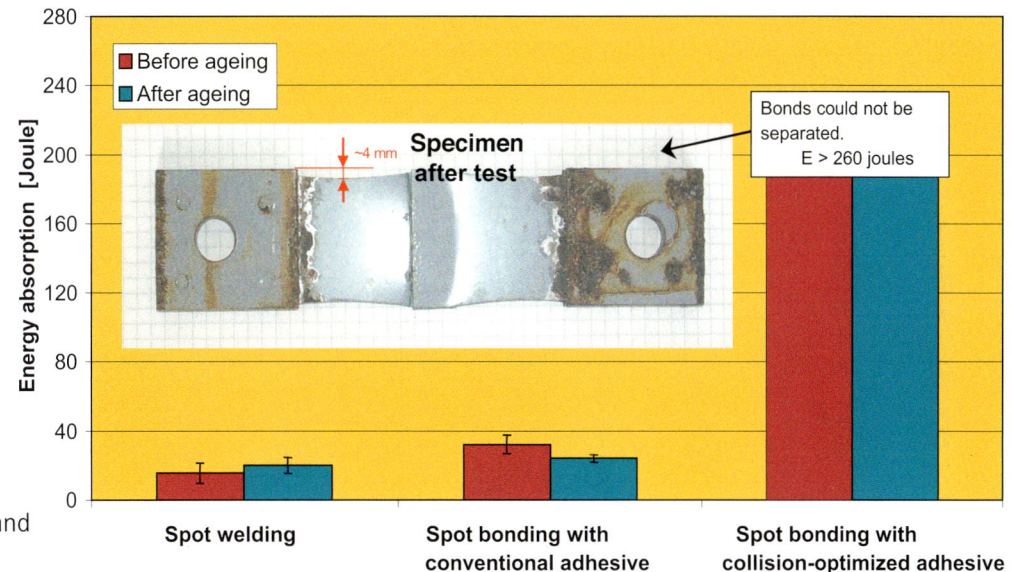

Bonds could not be separated.
E > 260 joules

Figure 5-5
Comparison of collision strength with spot welding and spot-weld bonding

Torsional stiffness - Bod-in-White

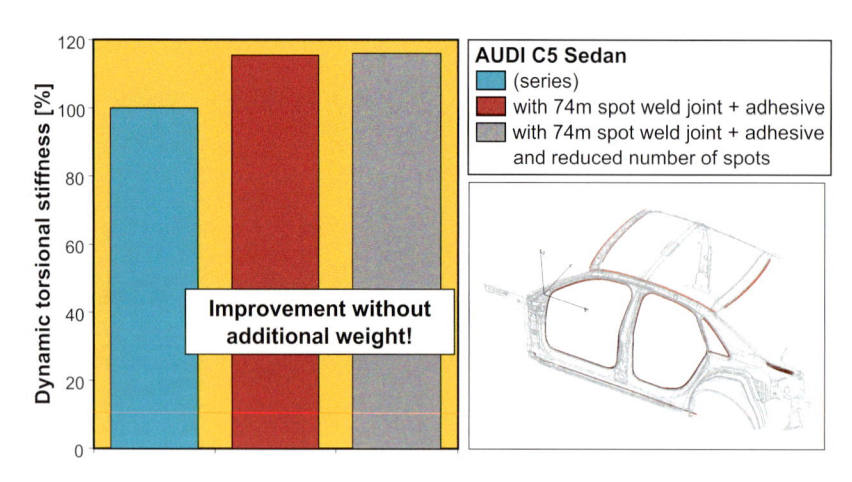

Figure 5-6 Altering dynamic body rigidity by additional bonding

A considerably more rigid body structure offers better ways of optimizing the total vehicle system with regard to vibration comfort, handling dynamics and running-gear design.

Additional adhesive bonding has a positive effect regarding corrosion warranties against rusting-through, which has been considerably expanded by all automobile manufacturers, by hindering crevice corrosion in joint flanges. Automatic application of adhesive together with the corresponding process control measures should be designed to ensure the sealing effect of the adhesive by applying an adequate amount of adhesive. The "combined bonding" technique offers above all a relatively inexpensive method for locally matching the properties of the structure to modified requirements. The improvements resulting from additional adhesive bonding may also make it possible to reduce the scope of the elementary joints (spot welding, punch riveting, etc.). This in turn, under certain circumstances, can compensate additional costs incurred by the use of additional adhesive.

Besides the adhesive bonding applications in the above-mentioned three categories, adhesives will also be used in the future for bonding exterior components to the body.

In the development of new vehicles, styling concepts are being increasingly presented which require shaping of the materials in a manner either originally or subsequently unfeasible. Such styling concepts can then be implemented by attaching the outer skin to the body. However, outer skin attachments to a class A surface quality will not be feasible with existing bonding techniques unless considerable compromises are accepted with regard to reworking or additional panels or decorative trim are employed. However, for styling-related reasons, the latter approach is not acceptable in most cases.

Outer skin bonds created by laser welding, laser soldering or adhesive bonding can be given an adequate surface quality, for example, by using adhesive to apply a subsequent finish. The first series application of this method is the three-component SMC tailgate of the AUDI A4 convertible (Figure 5-7).

5.1.4 Testing, Development, Dimensioning

Before the properties of bonds can evaluated or compared with other joining methods, it is necessary in the early stage of development to use suitable specimens and specimens shaped similar to the component that can reproduce, as accurately as possible, the full range of loading cases that a vehicle is subjected to in service.

This must be done if useful data relating to the entire life cycle of a vehicle are to be obtained as early as the concept phase (Figure 5-8).

5.1 Bonded Joints in and on the Vehicle Structure

Outer-skin bonds with adhesive finish

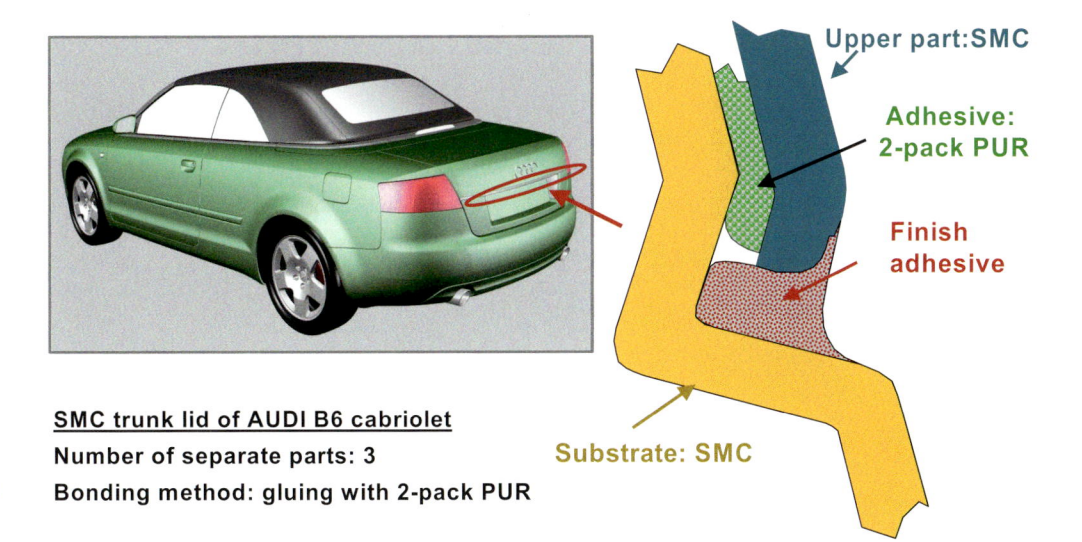

Upper part: SMC
Adhesive: 2-pack PUR
Finish adhesive
Substrate: SMC

Figure 5-7 Outer skin bonding with adhesive finish on an SMC convertible trunk lid

<u>SMC trunk lid of AUDI B6 cabriolet</u>
Number of separate parts: 3
Bonding method: gluing with 2-pack PUR

Bonded joints in and on the vehicle structure

Types of stress experienced by cars

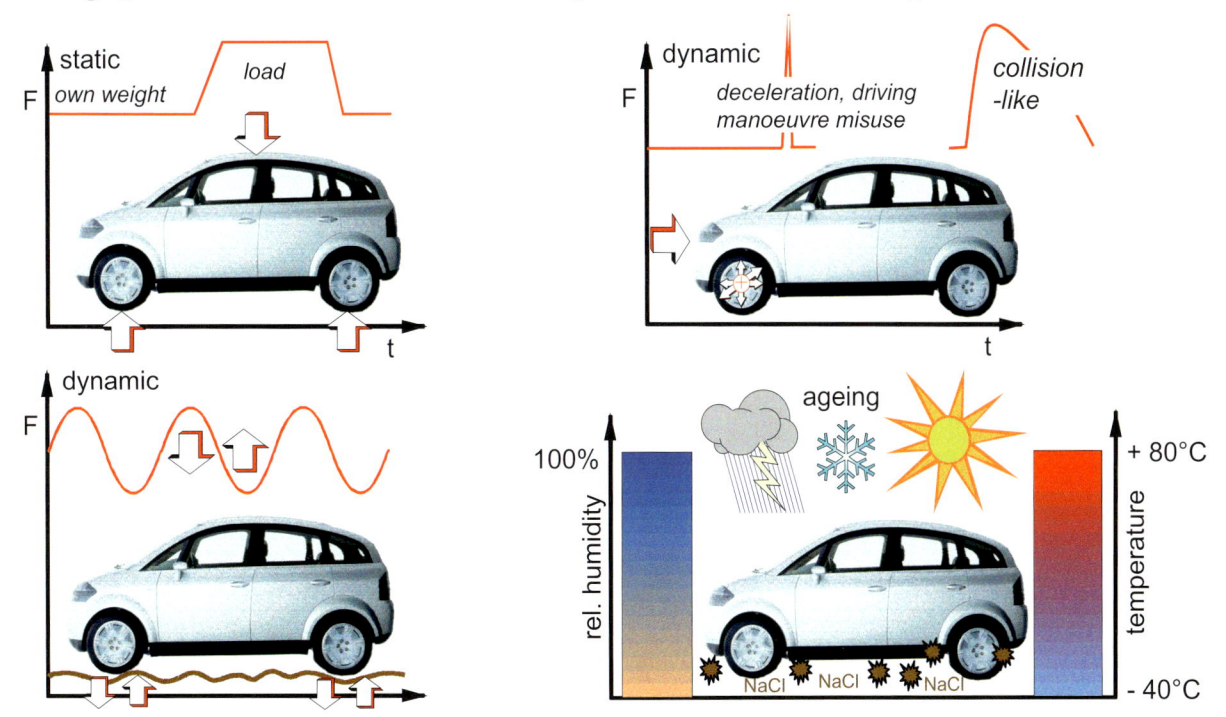

static — own weight, load

dynamic — deceleration, driving manoeuvre misuse, collision-like

dynamic

ageing — rel. humidity, NaCl, temperature (+80°C / −40°C), 100%

Figure 5-8 Types of load experienced by cars

These stresses are not only static and dynamic-cyclic but also shock-like forces. Long-term durability in particular – i.e., securing vehicle properties throughout its entire service life – calls for suitable ageing or corrosion tests and also the ability to transfer the results from artificial ageing to the bonding used in real components [1]. Iterative test comparison with the real-world equivalent is available at AUDI due to series-production use of adhesive bonds with worldwide lifespan analyses (AUDI 100 fixed glazing more than 25 years; AUDI 80 spot-weld adhesive bonding with steel bodies more than 10 years; and AUDI A8 adhesive bonding of aluminum almost 10 years) [2].

Of decisive importance for the appropriate selection of suitable bonding techniques is well-founded information on material and joint properties. These form the basis for the development of numerical joint or bond simulations which can be regarded as an essential requirement for further optimizing development time by means of virtual development methods in the product creation process. But even as early as the concept phase it is important not to ignore the basic principles behind the design of an adhesive bond.

5.1.5 Summary and Outlook

New materials and design methods provide the basis for developing attractive lightweight products with high innovatory content. The mixed approach using lightweight materials, such as very high strength steels, aluminum, magnesium, fiber-reinforced plastics and so on, offers considerable potential here, but one which can only be exploited if suitable bonding techniques are available.

Even with conventional sheet steel construction, it will be possible in the future to implement the objective of lightweight design even more consistently. The automotive industry will use high-strength or very high-strength steel materials more frequently for body components in order to enable further decreases in sheet thickness and thereby more weight reductions. Of course, one necessary condition for this is that the good strength properties of these materials can be retained and exploited in the area where they attach to other structural components. As has been shown, bonding techniques and combined methods offer great potential for the desired high level of exploitation of materials in this area as well.

As a consequence of the development tendencies described, the scope and extent of adhesive bonding in automobile design will increase further. However, when compared with conventional mechanical and thermal joining techniques, the technique of adhesive bonding will continue to be regarded as a new-fangled method and frequently regarded skeptically.

References

[1] Haldenwanger, H.-G., Walther, U.: Entwicklung wärmearmer Fügetechniken für die Mischbauweise [Development of thermally poor bonding techniques for mixed-material construction], 7th Paderborn Symposium in Joining Technology, Paderborn 2000

[2] Walther, U.: Kleben in der Automobilindustrie – Werkstoff- und konstruktionsgerechter Klebstoffeinsatz [Adhesive bonding in the automotive industry – using adhesive appropriate to material and design], Conference in bonding techniques, design and engineering, Ludwigsburg 2001

[3] Beerman, H. J., Haldenwanger, H.-G., Herrmann, M.: Einkleben von Fensterscheiben von Fahrzeugen mittels pumpbarem Klebdichtungsband [Attaching automotive glazing by pumpable adhesive sealing tape], ATZ, 81 (1979) 11 pp. 587–596

[4] Haldenwanger, H.-G., Korte, M., Schmid, G., Walther, U.: Funktionsverbesserung der Karosserie durch kombinierte Fügetechniken [Functional improvement of the body by combined bonding techniques], Hybrid bonds in automobile manufacturing, Bad Nauheim 2001

[5] Lechner, H.: Punktschweißkleben im Vergleich zu anderen Fügeverfahren aus der Sicht der Betriebsfestigkeit [Spot-weld adhesive bonding compared with other joining methods from point of view of fatigue strength], VDI Reports, No. 883, VDI-Verlag Düsseldorf 1991

5.2 Joining Techniques Used for Integrating Fiber Composite Plastic Components in Metallic Body Structures

GÜNTER H. DEINZER

When lightweight design concepts are transferred from motor sports to standard production vehicles, this reveals again and again the limits on using fiber composite plastic components: component costs, inexpensive processability within a given production process, adjustment of collision and strength properties to required limit values, and joining techniques suitable for assembling them together with metallic structures in mass production. This situation calls for specific properties to be taken into consideration as early as the concept phase and has serious repercussions affecting even the structure of the fabric.

Figure 5-9 Audi R 10 with TDI technology in CRP monocoque design

Joining technology is the key to implementing these innovative ideas economically and suitably for their application. Both mechanical joining and adhesive bonding offer technical solutions for implementing very different materials combinations in mass production. A feasibility assessment must take into account not only the problems of electrochemical corrosion but also problems arising from the differences in thermal expansion between metals and plastics.

5.2.1 Introduction

Fiber-reinforced plastics (FRP) have long been tried and tested in motor racing on account of their potential for lightweight strength-optimized design of the body structure and are state-of-the-art in many racing classes, such as Le Mans standard cars. Even the new Audi R10 has a monocoque made of carbon-fiber-reinforced plastic (CRP) (Figure 5-9).

When concepts of this kind are transferred to standard production vehicles, this always reveals the limits on using fiber composite plastic components: component costs, inexpensive processability within a given production process, adjustment of collision and strength properties to required limit values, and joining techniques suitable for assembling them together with metallic structures in mass production.

However, since there will be an increased push in coming years towards lightweight design as the result of the general energy and environmental politics situation, as well as further increasing demands relating to driving performance and driving dynamics, it is absolutely essential that automobile manufacturers tackle the issue of practical integration of fiber-reinforced plastic components into the body structure. One important link in this is joining technology, which can after all exert considerable influence on the properties of the resulting mixed-design body [1].

5.2.2 Lightweight Design Using Body Structures

The term hybrid is not uniformly applied to automotive bodies in the literature. In the case of a 100% steel body shell, it means the use of different types of steel ranging from soft deep-drawing steel to case-hardened, extremely high-strength heat-treatment steel. If a certain proportion of lightweight materials such as aluminum is also used in a steel body, this body concept is also referred to as hybrid. Finally, an aluminum body containing sheet steel components in the appropriate areas is also called a hybrid body.

Here the aim of body design will always be to achieve the best results under the given general constraints arising from body rigidity, vehicle safety, fatigue strength, weight and costs by using the right material in the right place. Naturally, the manufacturing method will also play a decisive role here, too.

In addition to body concepts based on the use of particular materials, categorization can be based on different structural and/or design concepts. In the case of standard production vehicles we distinguish between the so-called steel shell design (monocoque) and designs that make intensive use of structural shapes. Virtually all steel body shells use the monocoque approach, but a number of studies (ULSAB-AVC, ATLAS, VW) do exist which indicate that, even with this material, advantages may accrue from structures shaped by internal high pressure (hydroforming) [2, 3, 4].

When the lightweight material aluminum is used, the implementation of a space-frame concept (for example, ASF® = Audi space frame) has proved advantageous as this allows best use to be made of manufacturing possibilities by using sheets, extruded profiles and highly integrative cast components [5, 6].

By combining an aluminum space frame with flat FRP components, more recent body concepts for niche vehicles are now attempting to exploit the remaining weight-saving potential and to smooth the way for FRP components in (large) volume production.

5.2.3 Joining Techniques for Hybrid Aluminum-FRP Designs

Weight savings from the use of lightweight materials, such as aluminum or fiber-reinforced composite plastics or the corresponding hybrid designs, cannot be regarded as optimally achieved unless specific material properties are retained at the joints. The long-term strength of these joints is a basic requirement for securing vehicle properties over the entire life of the vehicle and must also be secured when leaner, less expensive manufacturing processes are used.

Methods of making metal/plastic connections include mechanical joining, gluing and also the corresponding combined technologies.

5.2.3.1 Process Strategies for Joining

The selection both of adhesives and the mechanical fastening elements is subject to the given process conditions in the manufacturing sequence. In many cases this sequence cannot be designed with the flexibility that would be desirable from the point of view of individual part function or an optimum balance of joint properties.

For this reason even the capability of the joining method for integration in an existing manufacturing sequence is an aspect with decisive influence on the development of joining techniques. Figure 5-10 shows the various ways in which the joining process can be integrated into the manufacturing sequence for a vehicle body [7].

Besides a possible lack of heat resistance in fiber composite components, a problem emerges from the fact that, with aluminum and fiber-reinforced plastics, two lightweight materials with very different coefficients of thermal expansion are being joined together. Depending on how the joining process is integrated in the manufacturing sequence for the vehicle body, following the joining operation this part will pass through processes involving different temperature loads. The

Online process	Inline process	Offline process
Joining process in the body shop	Corrosion protection (E-coat, 190°C / 30 min)	Corrosion protection (E-coat, 190°C / 30 min)
Corrosion protection (E-coat, 190°C / 30 min)	Joining process in the paint shop	Paint application (160°C / 15 min)
Paint application (160°C / 15 min)	Paint application (160°C / 15 min)	Joining process in the assembly line
1C adhesives	1C / 2C adhesives	2C- and Assembly adhesives

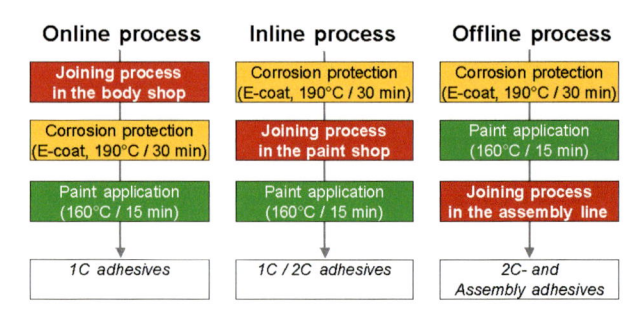

Figure 5-10 Possible ways of integrating the joining process in the manufacturing sequence

consequence of different expansion behaviors, and possibly of different heating and cooling gradients in the joined parts, is that the connections can be subject to high levels of mechanical loading. This can result in damage and, in the worst case, to immediate destruction of the joints even during the production process.

In on-line processes, one-pack heat-curing bodywork adhesives are normally used. The drying ovens of the electrophoretic dip coating process – at approx. 190 °C/30 min – are used for curing these adhesives. During passage through the oven, relative movements of the joint members occur. These result from the different thermal expansion properties and become, as it were, frozen due to hardening of the adhesive. The frequent result is internal stresses and component deformations once the body has cooled down (Figure 5-3).

An alternative to this is the in-line process in which the bond is made in most cases using cold-curing adhesive systems applied after electrophoretic coating. Since relative displacement of the joint members is prevented, the layers of adhesive, which have cured at room temperature, are subjected to mechanical loading in the drying ovens, as are the other hybrid types of joint in the car body. The result can be damage to and deformation of the components. In contrast to mechanically jointed connections, the mechanical and physical properties of polymer adhesive layers also vary with temperature.

In the off-line process, joining takes place during assembly – that is, after passage through all drying ovens. The only adhesives that can be used here are cold-curing systems.

It is obvious that the positioning of the joining process also affects corrosion protection of the joints when combined with application of electrophoretic coating [8, 9].

5.2.3.2 Mechanical Joining

In DIN 8593, mechanical joining methods belong to the group "Joining by forming processes". This classification groups together the cold manufacturing processes that form a joint without any melting of the joint materials and where the joint is generally held together by mechanical interlocking combined with frictional interlocking. They can be subdivided into techniques which do and do not use an auxiliary joining element.

One mechanical joining technique which does not use an auxiliary joining element and is widely used in automobile manufacturing is clinching. Here predominantly single-step non-cutting clinching techniques are used that demand a sufficient level of plasticity in the parts being joined. For this reason, they cannot be considered for joining fiber composites to metals. For this type of materials combinations, special clinching techniques have been developed, such as TOX Variopunkt or Eckold Confix. Figure 5-12 shows a Variopunkt clinching joint of steel and CRP. A hole is first punched through the CRP component, then the steel is pushed into the hole and shaped by means of a punch. Use of the corresponding female die causes a radial flow in the steel, producing an undercut.

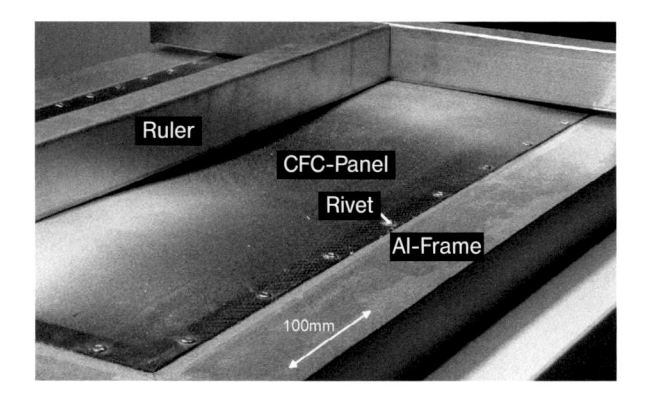

Figure 5-11 Deformation following the drying of electrophoretic dip coating on an aluminum/CRP specimen similar to an actual component that has been pop-riveted and glued using body adhesive BM 1494 [8]

The preliminary hole-cutting operation, the difficulty of automating the process due to the need to locate the hole, and the relatively low strength values of the joint currently prevent this technique from becoming more popular in vehicle design.

Currently the most widespread mechanical joining technique in vehicle body manufacture is self-pierce riveting with a semi-hollow punch rivet. This technique is especially popular with aluminum and hybrid design bodies. The body of the current Audi A8, for example, uses about 2400 punch rivets. In punch riveting with a semi-hollow rivet, the joint is made in one step without any preliminary hole-cutting required. Here the rivet is punched through the rivet-side component of the joint and, as it penetrates into the joint component on the side of the female die, it spreads at its base. This is achieved with the help of the female die, thus creating an undercut.

With this technique, only the material on the female die side has to have sufficient plasticity. This is why semi-hollow punch riveting is suitable for joining FRP components to aluminum or steel structures, provided the FRP material is on the punch side and the ratio between the wall thicknesses of the two joined components permits a marked undercut in the metal component on the female die side (Figure 5-5). The properties of an FRP/metal punch-riveted joint under mechanical loading can be improved by using large-head or flat-head rivets instead of the usual flush rivets employed with purely metal structures.

If the joint location is only accessible on one side, such as is the case, for example, with space-frame structures with closed aluminum extruded profiles, joining techniques such as clinching or punch-riveting cannot be used since they require access from both sides. Blind riveting on the other hand is a joining technique requiring access on only one side. Blind rivets can also be used with hard-to-form materials. Their greatest disadvantage is the required preliminary hole-cutting operation that currently limits application of this technique in automated production in automobile manufacturing.

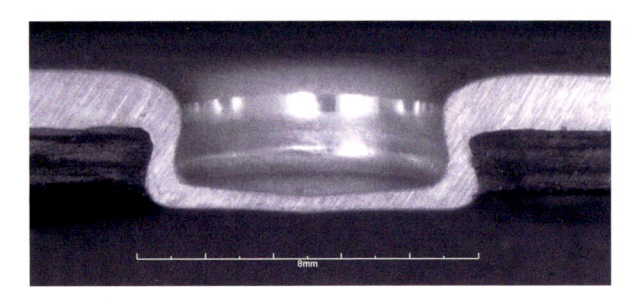

Figure 5-12 Clinching joint (Variopunkt) between steel (St07; 1.2 mm) and CRP (1.5 mm)

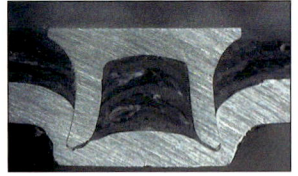

Figure 5-13 Punch-riveted joints of steel (ZStE; 1.5 mm) and CRP (2.0 mm) with a flush rivet (left) and a large-head rivet (right)

With blind-riveted metal/CRP joints, care should be taken to limit the pressure on the face of the hole when the rivet is inserted; otherwise the laminate can be damaged. What is preferable is for the joint to be made by the clamping force between the rivet head and the folded hollow section. If possible, the CRP component should be on the same side as the rivet head and the rivet head has to be large enough (Figure 5-14).

The creep behavior of matrix plastics can, however, have a negative effect on strength properties of the blind-rivet joint, especially on stiffness and failure behavior, particularly in the case of loading due to high temperatures.

Where access is from one side only, hole- and thread-cutting screws are also suitable. Continuous-hole-forming screws (FDS® system) are currently becoming more popular in body design at Audi for the automated

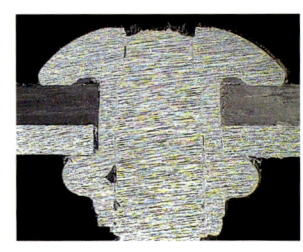

Figure 5-14 Blind-riveted joints between steel (ZStE; 1.5 mm) and CRP (2.0 mm) – both joining directions

joining of steel and aluminum components, as well as in hybrid designs of steel with aluminum. As a rule, first a hole is made in the part to be clamped. Since the through-hole must be formed in the part on the screw entry side and the thread cut by the screw in a single operation, with metal-CRP joints the metal must be on the screw entry side (Figure 5-15).

Various threaded functional elements are found in vehicle structures for fastening add-on and assembly components. These could be blind rivet nuts and bolts, standard rivet nuts and insertion pins. If metal components are replaced by CRP structures, ways must be found for fastening them. In the case of high loading, appropriately shaped threaded sleeves must be laminated in during the component fabrication process. Blind-riveted functional elements on the other hand are fitted after completion of fabrication of the polymer component and are considerably cheaper. The transferable forces are smaller but nevertheless adequate for most assembly parts.

Figure 5-16 shows an example of a closed stainless-steel blind rivet nut with a small head countersunk in CRP. If the design permits, a blind rivet nut with a countersunk head is preferable since this reduces the surface pressure on the laminate if there is high clamping force.

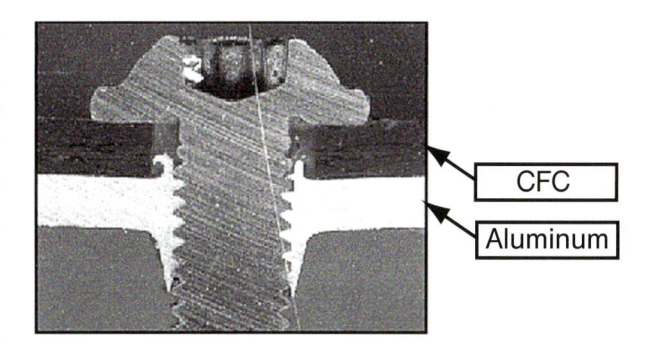

Figure 5-15 FDS system – CRP/aluminum screwed joints

Figure 5-16 Blind rivet nut made of rust-resistant steel in CRP

Corrosion

When fiber-reinforced polymers are used as material for the body, considerable bimetallic corrosion problems will occur if, for example, joining or cutting processes expose the carbon fibers (which behave like a noble metal) and come into contact with other metal materials in the body. Even the auxiliary joining element can play an important role here. In many cases punch rivets, blind rivets, screws and functional elements are made of carbon steel and given metallic coatings to allow them to be used in steel and aluminum structures. In the case of CRP components in areas highly subject to corrosion, this protection is often inadequate and corrosion phenomena occur at auxiliary joining elements or functional elements (Figure 5-17).

Figure 5-17 Corroded blind rivet nut made of carbon steel (coated) in CRP

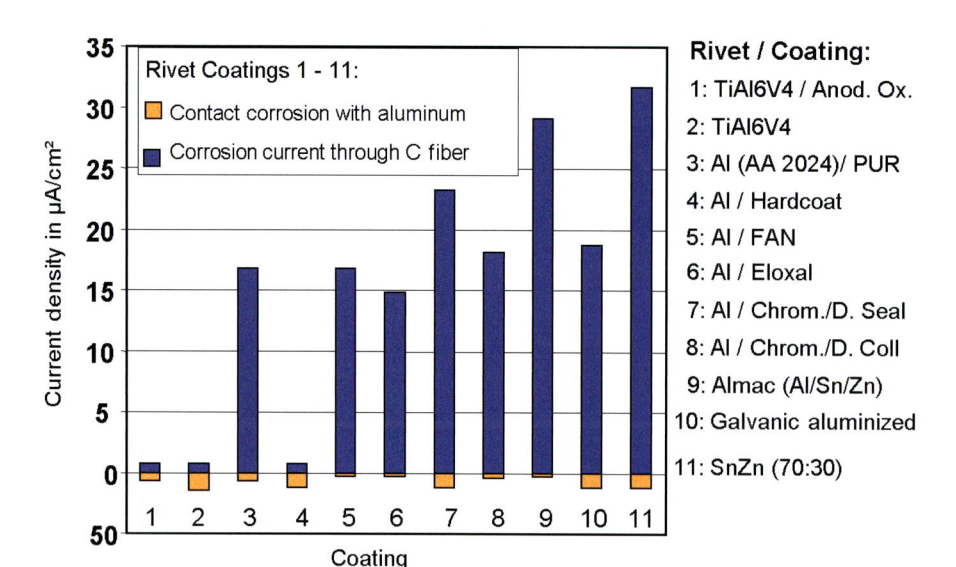

Figure 5-18 Bimetallic corrosion currents in selected rivet coatings with aluminum (AA 6016)

Rivet Coatings 1 - 11:
- Contact corrosion with aluminum
- Corrosion current through C fiber

Rivet / Coating:
1: TiAl6V4 / Anod. Ox.
2: TiAl6V4
3: Al (AA 2024)/ PUR
4: Al / Hardcoat
5: Al / FAN
6: Al / Eloxal
7: Al / Chrom./D. Seal
8: Al / Chrom./D. Coll
9: Almac (Al/Sn/Zn)
10: Galvanic aluminized
11: SnZn (70:30)

Figure 5-19 Separation of CRP and the steel blind rivet by means of a plastic sleeve

Even the punch rivets used in volume production of the Audi A8 and which have been given a ternary composite coating of aluminum, tin and zinc (commercial name: ALMAC) cannot be considered for joining aluminum and CRP (Figure 5-18: coating no. 9). For this combination of materials, special requirements must apply to the materials of the auxiliary joining element and functional elements and to their coatings for it to be possible to guarantee the durability of the joints.

Particularly when CRP components are to be fitted to the vehicle structure after electrophoretic dip coating and no paint has been applied to the auxiliary joining elements, further corrosion protection measures must be taken [10]. Riveted joints can, for example, be sealed with adhesive or sealants, but here it must be borne in mind that that adhesives do not provide complete basic protection against electrochemical corrosion. For example, the addition of conductive fillers such as carbon black lowers system conductivity. This can lead to corrosion phenomena and thus to a loss of adhesion due to corrosive undermining of the adhesive [7]. Additional organic coatings or sleeves are conceivable as an insulating barrier layer (Figure 5-19).

The potential for corrosion can also be reduced by selecting the right material. Auxiliary joining parts made of titanium alloy offer the best corrosion resistance for CRP applications (Figure 5-10), but due to their cost, production and processing, they are only suitable in exceptional cases for automobile manufacturing. Applications do, however, exist in aircraft design. One alternative is the use of connecting elements made of rust-resistant steel. Blind rivets and blind rivet nuts made of this type of material are state-of-the-art and available in a wide variety of designs. They provide considerably better corrosion protection than elements made of carbon steel with the corresponding coatings.

Defects occurring during production must also be excluded. If the preliminary hole in the CRP is poorly

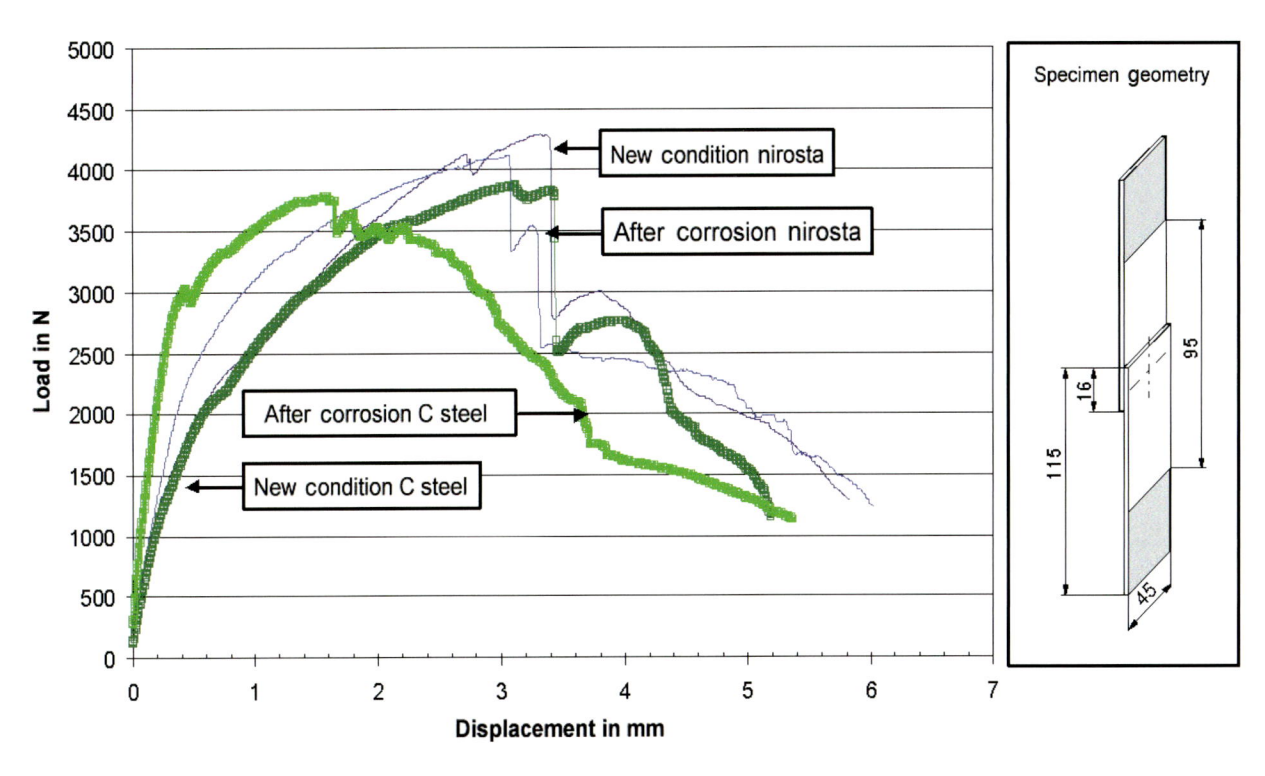

Figure 5-20 Force-deformation curve of blind rivet joints with and without corrosion loading

drilled this can aid the formation of corrosion. Care should be taken when drilling that neither the material is overheated nor that delamination occurs nor fibers become exposed. In addition, crevice corrosion can also occur if the auxiliary joining elements or functional elements have screw head contact on one side.

No negative influence of corrosion on strength properties could be detected in the quasi-static loading range (Figure 5-20).

Despite what are sometimes clearly corrosion products, no negative influence on shear tension forces occurs when either a coated carbon-steel rivet or a blind rivet made of rust-resistant steel is used. The rigidity of joints following exposure to corrosion in the INKA test (Ingolstadt corrosion and ageing test) actually increases with respect to its initial state, as can be seen from the steeper rise in the shear tension force. This is a known phenomenon with mechanical joints and can be explained by an increase in the clamping effect caused by the creation of corrosion products with enlarged volumes.

5.2.3.3 Gluing

The search for a suitable joining strategy for hybrid aluminum/FRP designs must also lead to bonding techniques that make it possible to join practically all materials with positive bonding. The technique offers a large number of advantages for use in hybrid designs. Two that stand out are even stress distribution (compared with punctiform connections) and low impact of heat on the materials thanks to low process temperatures. The sealing and electrically insulating properties of adhesives are particularly favorable for the corrosion protection of hybrid components. As mentioned in Section 5.3.2.1, with all adhesives there is a lack of basic protection against bimetallic corrosion.

The disadvantages of bonding – such as the need to hold the component in place until the adhesive has cured, the inadequate heat resistance encountered with some production processes, and the low durability to peeling stress – can be compensated by combining the technique with mechanical joining methods.

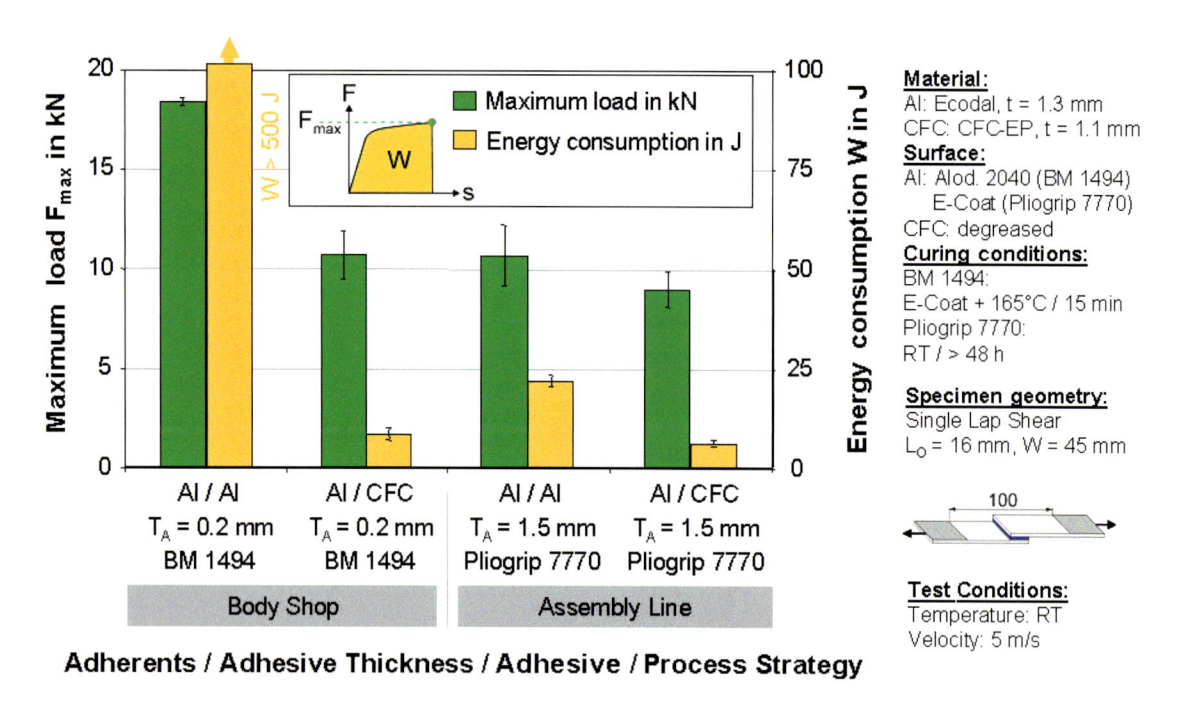

Figure 5-21 Strength properties of variously bonded aluminum/aluminum and aluminum/CRP combinations under impact stress

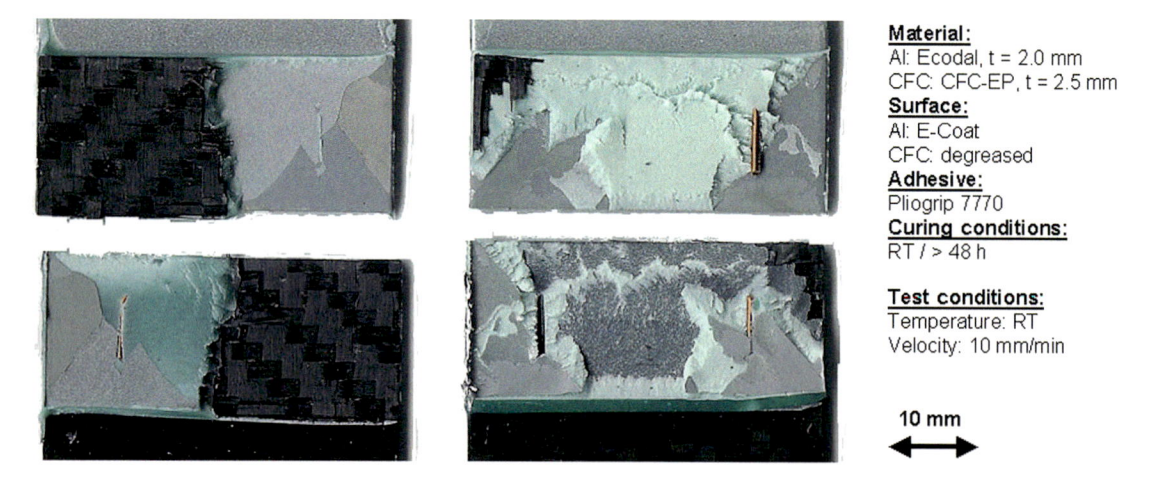

Figure 5-22 Fracture photomicrographs in aluminum/CRP combinations bonded with Pliogrip 7770

Figure 5-13 shows the maximum force and the energy absorption of variously bonded aluminum/aluminum and aluminum/CRP combinations at an impact stress rate of 5 m/s.

Shifting an on-line bonding process for aluminum/aluminum combinations to assembly (off-line process) led to a considerable reduction in strength or energy absorption of the joint. In this test, the bonding process took place on the electrophoretic dip coating line with Pliogrip 7770 – a two-component polyurethane adhesive. With aluminum/CRP combinations, the differences in strength and energy absorption between on-line adhesive bonds and off-line adhesive bonds are not very serious. Due to a delamination of the CRP component in the overlapping part, the collision strength of modern, tough and resilient vehicle body adhesives (BM 1494, for instance) cannot be fully exploited.

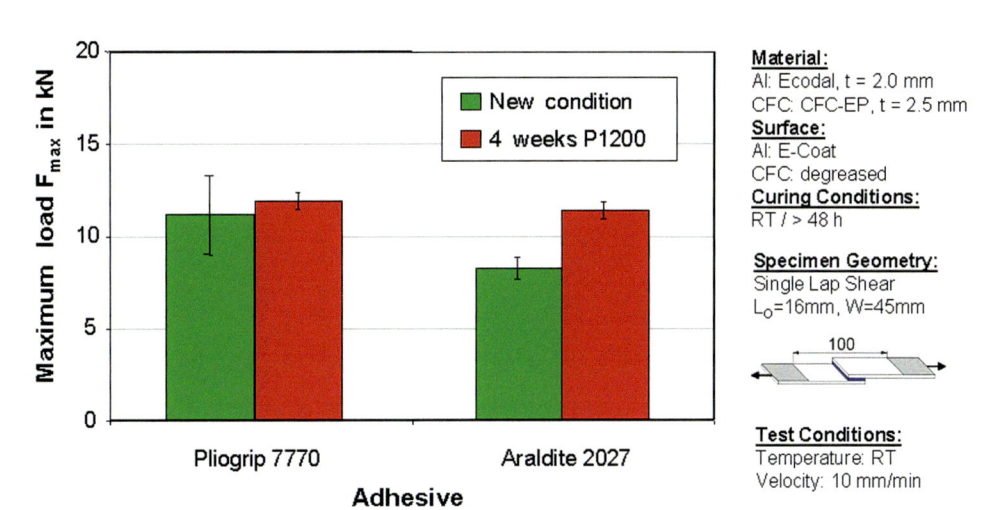

Figure 5-23
Influence of P1200 ageing on the strength of bonded aluminum/CRP combinations as a function of the adhesive

Material:
Al: Ecodal, t = 2.0 mm
CFC: CFC-EP, t = 2.5 mm
Surface:
Al: E-Coat
CFC: degreased
Curing Conditions:
RT / > 48 h
Specimen Geometry:
Single Lap Shear
L_0=16mm, W=45mm
Test Conditions:
Temperature: RT
Velocity: 10 mm/min

Even with the aluminum/CRP combinations bonded off-line, delaminations occurred in the CRP component in addition to cohesive adhesive layer failure and tearing of the electrophoretic coating (Figure 5-22).

Integration of fiber-reinforced plastic components in metal supporting structures by means of bonding techniques requires that the joint be specially designed. If the glueline is suitably designed (adherent surface, adhesive film thickness, load transfer) and the appropriate adhesive systems are selected (strength, elongation at break, shear modulus), a joint with a satisfactory material utilization factor can be achieved, even with the off-line process. Considering the high degree of rigidity to be achieved by the vehicle body, rigidity is just as important for the joint as are its strength properties under quasi-static, vibratory and impact loading. Here too it will be necessary to compensate the different thermal expansion coefficients of the components in the joint area (see Section 5.2.3.1).

Figure 5-23 shows the influence of a P1200 test involving storage under changing climatic conditions (12 h cycle at –40 °C to +80 °C/80% relative humidity) on the strength of bonded aluminum/CRP joints. A slight increase in strength can be detected. This is due firstly to greater hardening of the adhesive systems resulting from the higher temperature during changing-climate storage and secondly to elastification effects that produce better stress ratios in the bonded joint [11].

5.2.4 Study: Audi A2 with FRP Body Components

In the Technical Development department at Audi AG in Ingolstadt, studies were carried out using Audi A2 bodies or entire vehicles and focusing on the subject of lightweight design using FRP components (Figure 5-24). The aluminum space-frame is an obvious candidate for predevelopment work in CRP technology or joining techniques for subsequent full-scale production of ASF-CRP designs.

Roof and floor of the A2 body were replaced by CFK-EP prepreg parts. In the case of the front floor and bulkhead, fiber composite components with an aramid/carbon fiber hybrid fabric were used (ACFK). The greater tensile elongation at break value of the aramid fiber laminate and the resistance to crack propagation should, in comparison with a carbon fiber laminate, increase the energy-absorbing capacity of the components in this safety-related area.

All substituted components were glued into the supporting structure in an off-line process using Pliogrip 7770 – a two-component polyurethane adhesive. At room temperature this adhesive has a shear modulus of about 300 MPa and a bond strength of about 15 MPa. Blind rivets are used for holding the components until the adhesive has cured. The use of rivets with an aluminum sleeve prevents bimetallic corrosion occurring

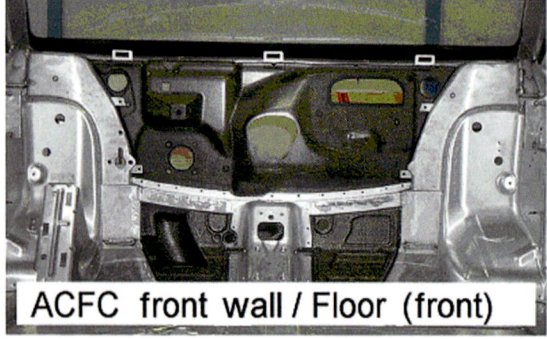

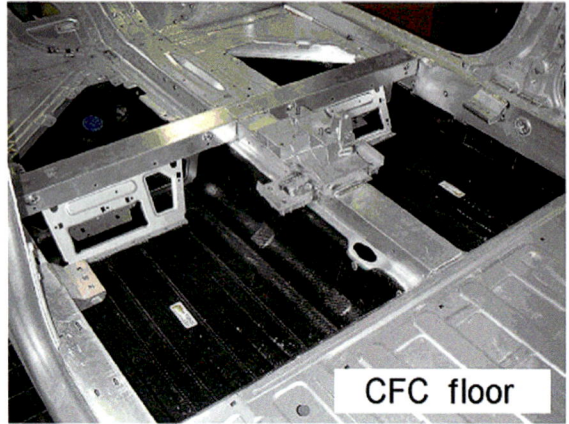

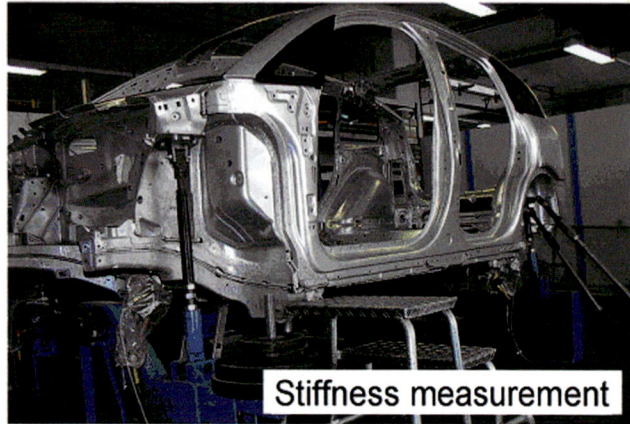

Figure 5-24 Audi A2 body with FRP components

with the aluminum space-frame. Electrical isolation with regard to the fibers exposed in the holes was secured by means of plastic sleeves which were also used at the same time for setting the adhesive film thickness of 1.0 mm.

The result of these lightweight design methods is a component weight saving of 26% over production parts coupled with a 6% increase in rigidity for the body [11, 12].

In addition to the tests conducted on the Audi A2 body, which included measurement of rigidity and weight, two more complete vehicles were built. One was subjected to a crash test while the other has been in service for more than two years at Audi AG as a function vehicle for testing customer-related suitability.

5.2.5 Summary and Outlook

The integration of FRP into metal body structures calls for its specific properties to be taken into consideration as early as the concept phases and can have serious repercussions affecting even the structure of the fabric. A feasibility assessment must take into account not only the problems of electrochemical corrosion but also problems arising from the different thermal expansion coefficients of metals and plastics. In addition, viable repair possibilities must be developed for all hybrid design concepts and it is absolutely essential that the requirements for dismantling and recycling also be satisfied.

In future vehicles, it will very much be a question of achieving new targets in driving performance and fuel consumption via lightweight design by making sensible use of high-performance materials while incurring only reasonable additional costs that the market will accept. Hybrid materials concepts that also involve using FRP in structural components will therefore have a major role to play in future vehicle models [9].

Joining is the key technology for implementing these innovative ideas economically and meeting the requirements made on it. Not only mechanical joining but also adhesive bonding represent possible technical methods for allowing a large number of materials to be combined in volume production. What would be desirable is adhesive systems that cure "on command" in a short period of time and deliver energy-elastic behavior with good strength values at temperature ranges above process temperatures in the manufacturing sequence for a car body. In order to optimize the use of material by means of high-strength bonded joints, delamination of the FRP component must be prevented at least in the joint area by appropriate material- or process-related measures. First approaches in the direct thermal joining of FRP to metals [13] exhibit potential and have to be followed up on, for example, by developing innovative joining elements made of plastic.

References

[1] Haldenwanger, H.-G., Schmid, G. u.a.: Leichtbau mit alternativen Werkstoffen und Verfahren; Ingenieur-Werkstoffe 7 (1998), Nr. 4 – November (Teil 1) Ingenieur-Werkstoffe 8 (1999), Nr. 2 – April (Teil 2)

[2] Grüneklee, A.: NSB – New Steel Body – Ein gewichtsoptimiertes und kostengünstiges Stahlleichtbaukonzept; 8. Europäische Karosserie-Leichtbaukonferenz "Zukünftige Stahlkarosserien in Hybridbauweise", Bad Nauheim, April 2004

[3] Rudolf, H., Jüttner, S., Dorn, L., Koppe, K.: Profilintensive Karosseriebauweise – Eine Herausforderung für die Fügetechnik; DVS-Berichte 237, Schweißen und Schneiden; Große Schweißtechnische Tagung 2005, Essen, September 2005

[4] Schulz, R., Freytag, R.: Neue Potenziale für den Karosseriebau in Stahl; 8. Europäische Karosserie-Leichtbaukonferenz "Zukünftige Stahlkarosserien in Hybridbauweise", Bad Nauheim, April 2004

[5] Haldenwanger, H.-G., Korte, M., Schmid, G., Walther, U.: Mischverbindungen im PKW-Karosseriebau; Tagung Dünnblechverarbeitung, "Fügen von Stahlwerkstoffen" München, March 2001

[6] Schmid, G.: Innovationen durch Fügetechnologien im Karosseriebau; 1. Fertigungstechnisches Kolloquium der TU Ilmenau, March 2005

[7] Haldenwanger, H.-G. Walther, U.: Entwicklung wärmearmer Fügetechniken für die Mischbauweise; 7. Paderborner Symposium Fügetechnik, Paderborn, 2000

[8] Kläger, O., Meschut, G., Wetter, H.: Multi-Material-Design – Thermische Beanspruchung in Lacktrocknungsprozessen; Adhäsion (2003), Nr. 4 – April (Teil 1), und Nr. 5 – May (Teil 2)

[9] Deinzer, G.: Werkstoff und Fertigungskonzepte für das Automobil der Zukunft; 11. Sächsische Fachtagung Umformtechnik, Feiberg, October 2004

[10] Reinhold, B., Brettmann, M.: Oberflächenschutz für Mischbau-Konstruktionen; 4. Werkstofftechnisches Kolloquium, Technische Universität Chemnitz, 2001

[11] Deinzer, G., Schmid, G., Wetter, H., Bangel, M.: Leichtbau durch hybride Karosseriestrukturen – Schlüsseltechnologie Fügen; 12. Paderborner Symposium Fügetechnik – Mechanisches Fügen und Kleben, Paderborn, November 2005

[12] Reim, H.: Internal Audi-Test report (unpublished)

[13] Velthuis, R., Schlarb, A. K.: Induktionsschweißen von faserverstärkten Kunststoffen für Automobilanwendungen; DVS-Berichte 237, Schweißen und Schneiden; Große Schweißtechnische Tagung 2005, Essen, September 2005

5.3 Bonding Automotive Bodies

Udo Buchholz

In small to medium size series vehicle, fiber composites are increasingly the materials of choice for exterior body panels. Adhesive bonding is rising in importance here, since it is often the only suitable joining technology.

Plastics currently make up 13 to 18% of a motor car. This is predicted to grow at an annual rate of 4.7% in the West European automotive industry. Plastics applications are distributed among the car interior – by far the greatest portion – followed by body exterior parts, engine compartment and electrical/electronics systems. New plastics applications are mainly to be expected in the body, that part of the vehicle where fiber composites are being increasingly used. This is driven by the trend towards increased diversification and niche vehicles. Visible body panels of fiber-reinforced plastics (FRP) are most economical on vehicles produced in small production runs, where they can replace steel or aluminum [1]. Fiber-reinforced plastics also offer relatively high strength and stiffness, combined with low weight. They are highly corrosion resistant and permit a high degree of freedom of design, which, if possible for metals, would be extremely expensive to achieve [2]. For example, the rear door of the Mercedes-Benz C class coupe owes its complex form to an innovative plastics solution using SMC (sheet molding compound) [3].

Fiber composites, such as SMC or GMT (glass-mat-reinforced thermoplastics) have become established for mass production, while high-performance fiber composites, such as CFRP (carbon-fiber reinforced plastics) with thermoset matrix are suitable for small-series vehicles or technology leaders [1]. Hence, Volkswagen's "1-liter car" has an outer skin made of CFRP [4]. Other examples of CFRP applications can be found in the Aston Martin Vanquish [5]. Numerous SMC applications illustrate the high status of fiber-reinforced composites for automotive production. Typical SMC applications include the rear cover of the Mercedes-Benz CL, the rear door of the Volvo V 70, the roof spoiler of the BMW X5 and the Audi A2 and the rear door and front mudguards of the Renault Vel Satis [6, 7].

As early as 1999, approx. 35% of European SMC consumption, or in absolute figures 66,000 t, was used in automotive applications. A large portion of these are so-called class A outer skin applications, the class that is expected to show the highest annual growth rates of over 6% [8]. Class A parts make the highest demands on the visual quality of the surfaces.

5.3.1 Bonding – Often the Only Feasible Joining Technology

For automotives with composite parts, the choice of joining methods is just as important for quality as the performance of the composite material itself. For FRP parts with a thermoset matrix, joining is limited to snap connections, rivets or screws and adhesive bonding. When conventional mechanical joining processes

Figure 5-25 In the Audi R8 Infineon, the aluminum engine radiator is bonded to the CFRP frame of the vehicle structure by a high-strength, extremely tough and resilient adhesive

5.3 Bonding Automotive Bodies

in composite applications reach their limits, bonding technology can contribute to optimally exploit the potential of composite materials and designs. Bonding offers numerous advantages over mechanical joining processes, including:

- No boreholes to weaken fibers and load-bearing cross-sections;
- No visible joining elements;
- Sealed joints;
- More uniform stress distribution;
- Superior fatigue strength, and
- The ability to compensate for dimensional tolerances.

Adhesive bonding can therefore play a crucial role in achieving the objectives that prompted the use of composite materials.

The adhesives most often used for the structural bonding of fiber-reinforced polymers in automotive engineering are generally based on epoxy, polyurethane or methacrylate-base formulations. These adhesive groups, as well as individual adhesives, exhibit widely different properties depending on the chemical base or the formulation actually used by the adhesive manufacturer. Typical mechanical properties of structural adhesives are shown in Figure 5-26.

Where FRP is used in automotive manufacture, bonding is often the only feasible joining technology. One example of this is the rear spoiler of a high-performance sports car (Figure 5-27) produced by CBS, an Italian supplier to the automobile industry, from several CFRP components. The CFRP components are bonded with a two-component toughened epoxy adhesive (Araldite 2015 manufactured by Huntsman Advanced Materials, Duxford, UK). The combination of high strength and high toughness of this adhesive permits bonds to be created that withstand high dynamic loads, even at the sports car's top speed of 320 km/h. The adhesive bond is cured at 90 °C in a heating device to ensure short production cycles, and, after one hour, the spoiler can be transferred to the downstream production stages – polishing, painting and final assembly on the vehicle rear.

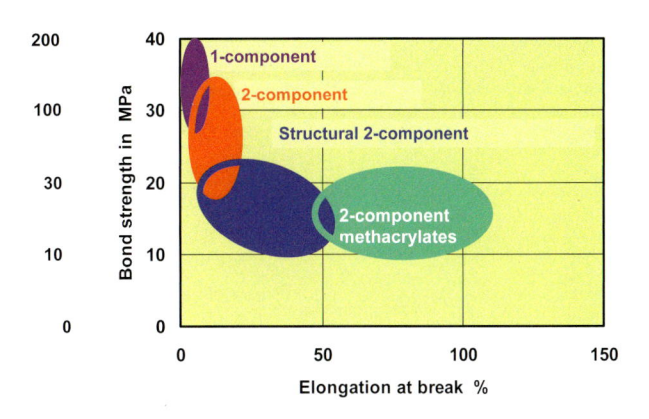

Figure 5-26 Adhesive strength: typical mechanical properties of structural adhesives at room temperature

Figure 5-27 Bonded CFRP spoiler of a high-performance sports car

In race car production, too, bonding of high-performance composites is an established joining technology. For years, the Audi R8 Infineon has been setting the pace in races at Le Mans. The different coefficients of thermal expansion of aluminum and CFRP made adhesive bonding the preferred method for joining the aluminum radiator for the 610 horsepower V8 bi-turbo engine of this racing car to the CFRP frame. For this bonding task, the adhesive must meet extreme requirements. It must retain high strength at the peak temperatures of 130 °C, and reliably withstand sudden mechanical loads and vibrations. It must also still have adequate toughness to compensate for the different thermal expansions of the joint counterparts throughout the operating temperature range. Audi Sport chose an extremely tough and resilient two-component epoxy adhesive of the latest generation (Epibond 1590 A/B manufactured by Huntsman Advanced Materials, UK), which stands out from other adhesives because of its outstanding test results. This "cold-curing", epoxy-base

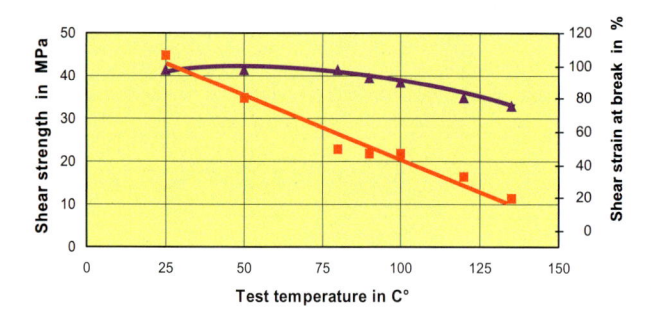

Figure 5-28 Heat resistance: shear strength and shear strain at break of Epibond 1590 A/B "cold-curing" two-component epoxy adhesive as a function of temperature

Figure 5-29 Luxury sports car with bonded body parts of fiber-reinforced plastics (model: Mangusta, manufacturer: Qvale Automotive Group S.R.l., Modena/Italy)

two-component paste adhesive still achieves strengths of 10 MPa at 130 °C, with a shear-strain at break of approx. 100%. (Figure 5-28). This shows that it is possible to develop an adhesive that offers an excellent combination of high strength and heat resistance, combined with high toughness.

On the Italian luxury sports car, Mangusta (Figure 5-29), components of the vehicle outer skin are made of fiber-reinforced plastics, as are the vehicle doors, produced with a double-shell design. The two shells of the vehicle doors are also bonded with a two-component epoxy adhesive (Araldite 2015). The main demands on these adhesive bonds, proven in numerous functional tests, are high strength, high heat and impact resistance and excellent sealing properties.

Class A components made of fiber-reinforced plastics for the vehicle outer skin are produced from SMC in large quantities for medium and large series. Bonding is the established joining technology for such SMC components. Examples of class A SMC adhesive bonds include the rear license plate panel for American and Japanese Audi 4 Avant versions, in which the outer SMC shell is bonded to the inner one (Figure 5-30). For the Fiat Multipla, the SMC headlight housings are bonded from the inside to the SMC upper headlight panel (Figure 5-31).

Figure 5-30 Rear license-plate panel of SMC bonded with two-component epoxy adhesive

Figure 5-31 SMC headlight housing of the Fiat Multiplia bonded with a primerless two-component polyurethane adhesive

5.3.2 New Adhesives for SMC Components Meet Maximum Requirements

The adhesive bonding of class A SMC components makes particularly high demands on the adhesives. Though in the past, special surface preparation, such as grinding, sandblasting or priming were still acceptable for bonding SMC, the trend now is clearly towards pretreatment-free, so-called primerless bonding of SMC. The SMC surfaces to be bonded are merely wiped dry with a clean cloth to ensure that surface contaminants such as dust are removed before bonding. Adhesives for primerless bonding of SMC save costs by eliminating additional production steps for surface preparation of SMC. Another substantial advantage of primerless bonding of class A SMC parts is the elimination of the risk of contaminating visible SMC surfaces with the primer that would otherwise be required, thus impairing surface quality. To avoid print-through of the bonded seam on the vehicle outer skin, adhesives for class A SMC applications should exhibit minimal cure-shrinkage (< 1%).

The automotive industry frequently demands of primerless adhesives that their bonded SMC joints fail with fiber tear in the SMC in the test, that is to say, the cohesive strength of the adhesive and the bonding of the adhesive layer to the SMC should be greater than the cohesive strength of the SMC material itself. The latest generation of primerless polyurethane (PU) adhesives for SMC only require curing at room temperature to meet these demands for SMC fiber failure (Figure 5-32). This can save considerably on investment for thermal curing equipment. Adhesives for primerless bonding of SMC are matched to conventional SMC grades containing internal release agents to assist in demolding. For good integration of the adhesive processes in the overall production, primerless PU adhesives with various curing rates are available. For large production runs, SMC bonding nevertheless often requires precuring at elevated temperatures to achieve shorter cycle times. In this case, adhesives that only require room-temperature curing for primerless bonding to SMC can produce high-quality bonds even if less than the entire length of the adhesive seam is heated in the precuring device.

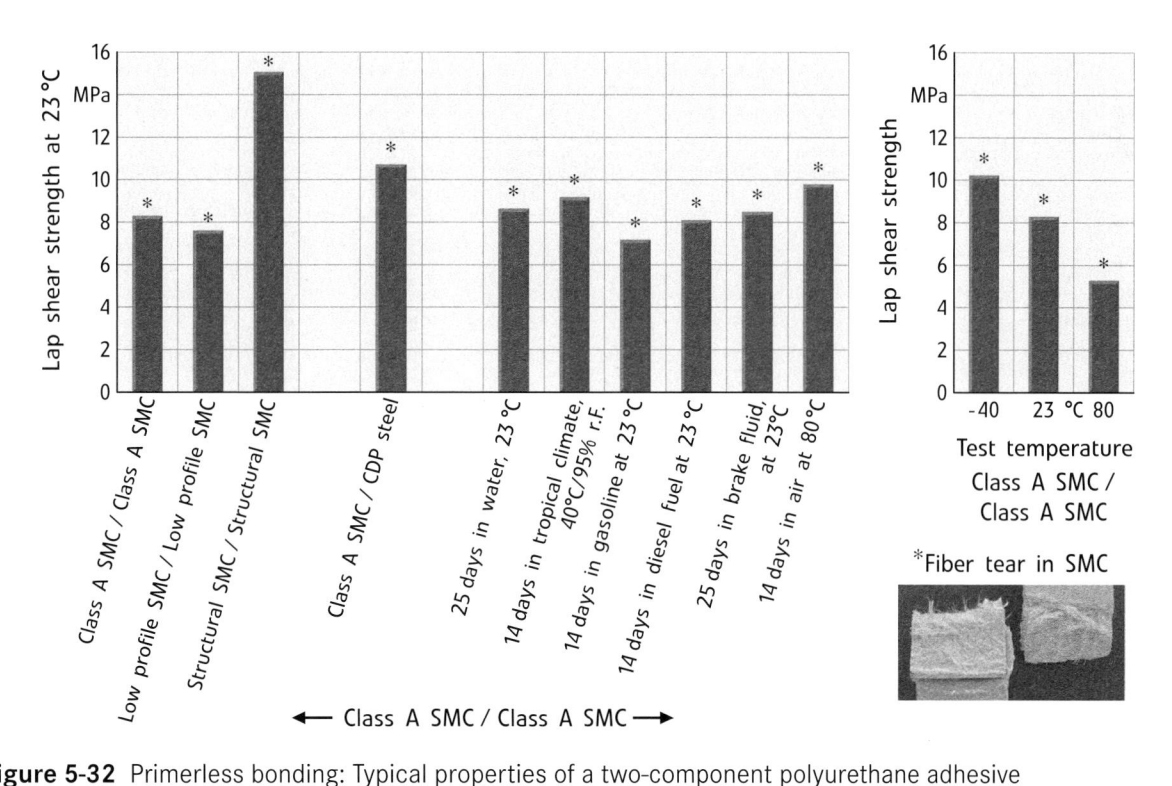

Figure 5-32 Primerless bonding: Typical properties of a two-component polyurethane adhesive (Araldite xB 4712 A/4713 B) for primerless bonding of SMC, preparation by dry wipe

The high adhesion capability of the latest generation of primerless SMC adhesives to electrophoretic dipcoated (CDP) steel has extended the range of applications of these adhesives (Figure 5-32).

Automotive body parts are subject to various environmental influences and in some cases to chemicals such as fuels or brake fluids throughout the lifetime of the vehicle. The adhesive bonds used here are also required to show long-term resistance to these environmental and chemical effects. This is verified for a particular adhesive/substrate combination in accelerated ageing tests (Figure 5-32).

Service temperatures for body applications in automotive engineering are in the general range from −40 to +80 °C. As with all polymeric materials, the properties of adhesives also change with temperature. Many users require that the loss in strength of adhesive bonds at high service temperatures should be as low as possible. Figure 5-32 shows what is possible with the latest primerless two-component polyurethanes for bonding SMC.

On the assumption that the adhesive adheres adequately to the SMC to achieve fiber tear in the SMC, the joint strengths of SMC adhesive bonds, in the event of failure, may be of variable magnitude depending on the mechanical properties of the adhesive (Figure 5-33). This is caused by the typical stress distribution in adhesive bonds. To exploit the inherent strength of SMC as efficiently as possible, it is therefore advantageous for users if adhesives are optimally tailored to SMC's mechanical properties by a well-balanced combination of strength and deformability. Such adhesives optimized for primerless bonding of SMC must have strength sufficient to surpass the inherent strength of SMC in the joint configuration. However, at the same time, they must exhibit relatively low moduli and high deformability to achieve maximum load transmission of the bonded SMC joint by minimizing stress concentrations (Figure 5-33).

Optimization of the mechanical properties of the adhesive in this way also benefits the dynamic and impact properties of SMC adhesive bonds.

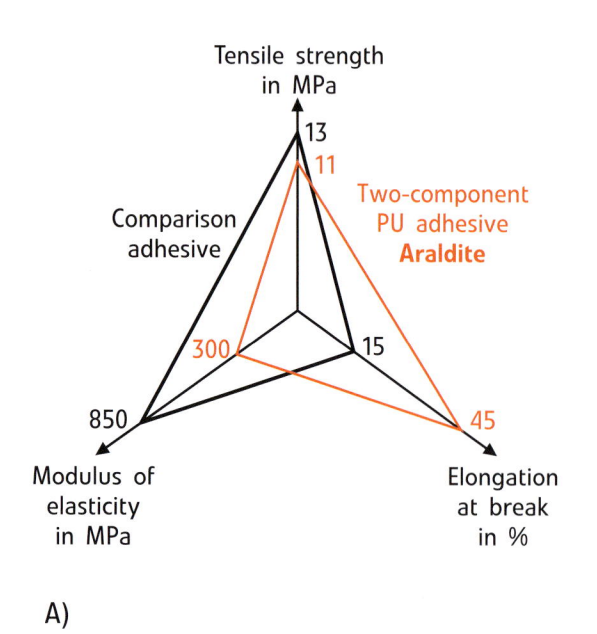

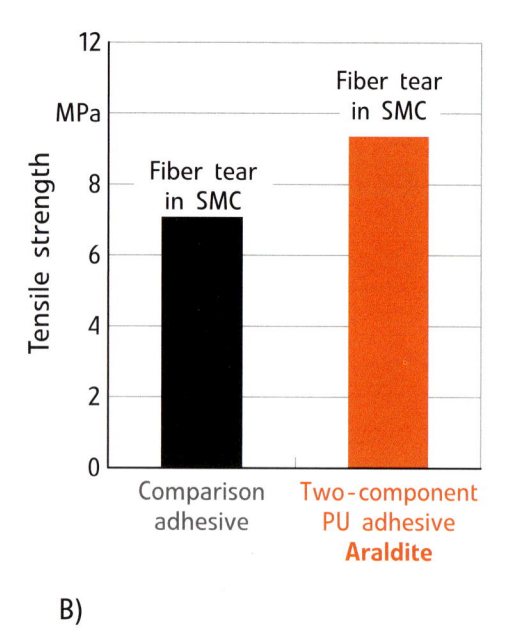

Figure 5-33 Optimum material exploitation: efficient utilization of the inherent strength of SMC (B) thanks to optimized adhesive properties (A)

5.3.3 Chemical Thixotropic Adhesives

The two individual components of two-component adhesives must be metered and homogeneously mixed in a specified ratio. For applications with low adhesive consumption, or where the adhesive must be applied in numerous areas of production, dual barrel cartridges with a static mixer nozzle offer a fast, clean and reliable method of adhesive application. For applications with large adhesive consumption, and where the adhesive is required centrally at a few places, it is advisable to apply two-component adhesives from drums with meter-mix-machines available from a large number of suppliers [10]. For adhesive application, the application nozzle can then be either manually guided or coupled to an industrial robot.

For FRP applications in automotive manufacture, the beads of applied adhesive must generally be sag resistant. This is achieved with thixotropic adhesives. In the case of physically thixotropic adhesives, the thixotropy of the adhesive mixture or adhesive beads is achieved by making both individual components of the adhesives thixotropic. In this case, drum pumps with follower plates are required for feeding both thixotropic adhesive components from the drums to the meter-mix machine. With chemically thixotropic adhesives, on the other hand, the two individual components are free flowing. A rapid thixotropic reaction in the mixer produces sag-resistant beads of chemically thixotropic adhesives. Since the two individual components of chemically thixotropic adhesives are free-flowing, they can be easily fed to the metering pumps of meter-mix machines through hoses from drums positioned at a higher level. This simple principle is called gravimetric feeding. Chemically thixotropic adhesives, unlike physically thixotropic adhesives, therefore offer users the option of saving substantially on investments for drum pumps with follower plates. Because of its simplicity, gravimetric feeding can therefore be regarded as maintenance-free and trouble-free.

5.3.4 Summary

Adhesive bonding can contribute to exploiting the full potential inherent in innovative fiber-reinforced plastics and designs in automotive manufacture. This has been proven by well-established applications of composite adhesive bonds. New generations of adhesives with excellent property profiles offer users in the automotive manufacturing industry new technical possibilities and potential for considerable savings for joining fiber-reinforced plastics.

References

[1] N. N. Polymere bauen Positionen aus *Automobil Entwicklung Kunststoffe* 3/2002, p. 54-58
[2] Potter, L.: An Introduction to Composite Products. Chapman & Hall, London 1997, p. 210-213
[3] Engelen, Pl, Hörsting, K.: Thermoplast und Duroplast vereint, *Automobil Produktion*, 3/2002, p. 74
[4] Imhof, T.: Die Formel Eins, *Auto Motor Sport*, 10/2002, p. 12-16
[5] Mortimer, J.: Aston Martin – Uses Composite Bodysides, *Auto Technology*, Vol. 1, August 2001
[6] Sommer, M.: Neueste Anwendungen von SMC/BMC im Automobilbereich, AVK-TV Tagung, Baden-Baden, September 2000
[7] N. N.: SMC/BMC im Automobilbau, *Kunststoffe* 91 (2002) 8, p. 72-73
[8] Icardi, G.: Market overview of industrial processes in composites: Future trends, AVK-TV Tagung, Baden-Baden, September 2000
[9] de Witt, F., Sauer, J.: Kaltkleben – Innovationssprung im Flugzeugbau, *Adhäsion Kleben & Dichten*, Part 1 in Vol. 7-8/2001, p. 10-12 and Part 2 in Vol. 9/2001, p. 34-37
[10] N. N.: Marktübersicht 2K-Dosieranlagen *Adhäsion Kleben & Dichten*, 10/2002, p. 29-33

5.4 Warm Reactive Bonding

Peter W. Merz

5.4.1 Headlight Bonding

Individual parts of an automobile headlight consisting of various plastics can be bonded economically using a reactive warm melt or warm fusion adhesive and a specially developed technology. The adhesive hardens quickly with environmental moisture and bonds well to plasma pre-treated plastics.

Automobile headlights made by the system supplier Hella KG Hueck & Co., Lippstadt, Germany, consist in principle of two parts: a transparent cover shield and a housing (Figure 5-34). The cover shield is usually a polycarbonate (PC) protected from scratches by a hard layer, e.g., a UV hardening system. The housing is generally polypropylene (PP) which when modified with fillers such as talcum powder can have high scratch resistance. Also polybutylene terephthalate (PBT) or PC are used for production of headlights.

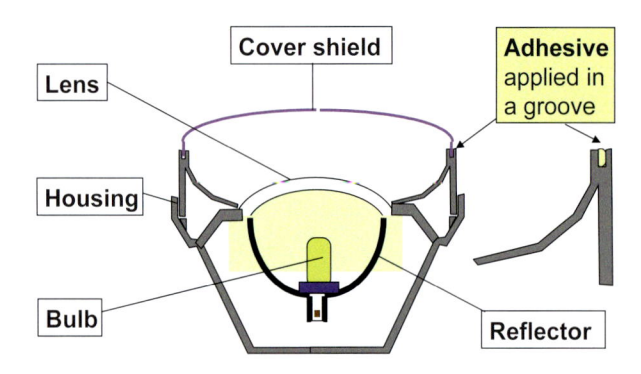

Figure 5-34 Design of a headlight
Automobile headlights consisting of a transparent cover shield and housing

5.4.2 Current Status of the Technology

Sealing materials or adhesives have been used for a long time in the production of headlights. So far, two technologies have been established: reactive and non-reactive adhesive systems.

- Reactive systems include
- Single-component adhesives based on silicone,
- Two-component adhesives based on polyurethane (PU) or silicone,
- Humidity-reactive fusion adhesives or hot melts (RHM) and
- Humidity-reactive fusion adhesives or warm melts (RWM).

Hot melt sealants based on butyl rubber are used as non-reactive systems.

To manufacture automobile headlights efficiently, a durable connection has to be achieved without mechanical fixation, such as clips. At the same time, a leakage test must be performed for quality control after bonding. Single-component adhesives based on silicone do not fulfill all the requirements of the intended market. These pasty adhesives are sticky and, in the uncured state, inclined to contamination, leading to problems with bonding as well as additional costs and delay in the leakage test. Moreover, additional adjustments are necessary.

Two-component adhesives also require clips to hold both parts together. The low initial viscosity after mixing excludes immediate leakage testing. Also the danger of contamination is clear. If reaction of the two-component systems is too slow, cross-linking time can be shortened in a heater, such as a hot-dip system. Such two-component systems therefore require a reaction time of 50 minutes prior to the leakage test. Thus at a cycle time of two to three headlights per minute, in the case of a mixture error, upwards of 150 defective headlights may be manufactured.

Two-component systems require complex delivery systems with pumps, dosing equipment and expensive static mixers. The static mixers must be cleaned, meaning waste and increased maintenance costs. On the positive side, a firm elastomer bond forms after

5.4 Warm Reactive Bonding

Figure 5-35
Plasma pretreatment – composition of the plasma

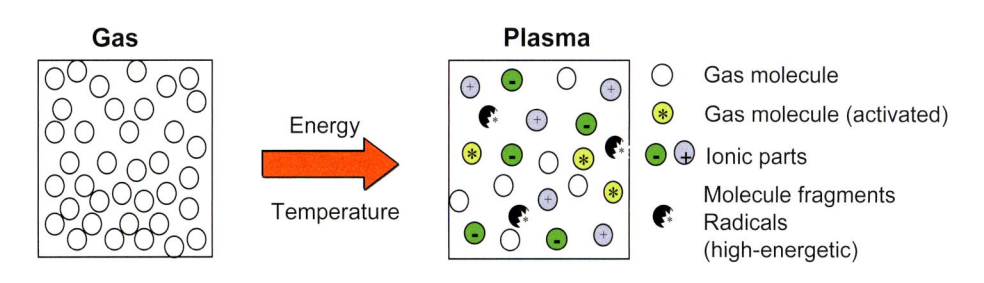

the cross-linking of the two-component adhesive. At the same time this cross-linking ensures higher resistance to heat generated by the headlight or the vehicle engine.

Reactive hot melts are processed at temperatures above 100 °C and are often brittle due to their poor adhesion as a result of their quick structure formation ("quick-fix"). Due to rapid crystallization, the "quick-fix" characteristic sometimes leads to poorer wetting of the substrate. At the same time, short furnace times and decreased stability limit processing flexibility.

Reactive warm melts are applied below 100 °C and are described in detail later. Non-reactive sealing materials are rubber-like hot melts applied at high temperatures (> 160 °C). Since they do not cross-link, they remain thermoplastic. In order to prevent resultant creep, mechanical fixation must be done at higher temperature. A great advantage is that a leakage test can be performed immediately after application, thus achieving 100% quality control.

5.4.3 Market Requirements and Wishes

Mass production of automobile headlights is a very fast process, making important demands on any adhesive plastic compound system:

- No relative displacement of the housing and the cover shield;
- Must be touchable, i.e., to prevent danger of contamination;
- The leakage test must be integrated in the process;
- Requires neither clips nor mechanical adjusting.

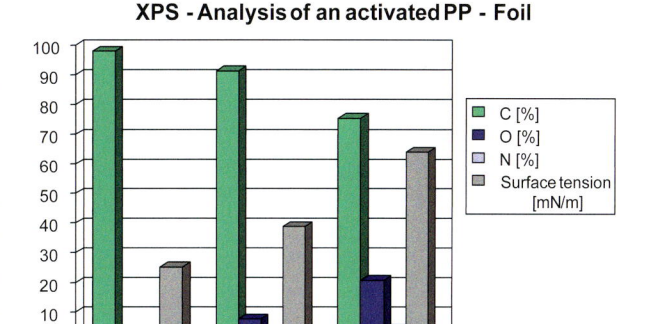

Figure 5-36 Surface pretreatment – XPS analysis of a PP foil

Of the adhesives and sealants used so far, the reactive warm melts (RWM) fulfill headlight manufacturers' requirements best.

5.4.4 Just-in-Time Production Feasible

Particularly with adhesion-weak plastics, the pre-operative method plays a role. It must be easy to apply and provide good resistance to ageing of the bond. The PlasmaTreat system fulfils these requirements. The principle of the plasma pre-treatment is shown in Figure 5-35 [1] where different excitation states are represented schematically. The PlasmaTreat process introduces active carboxylates, ether, ketones or other oxygen-containing groups at the plastic surface without the use of a primer. The resulting surface tension is increased to over 60 mN/m (Figure 5-36) [1].

In this method, the substrate is treated briefly with a potential-free flame (Figures 5-37 and 5-38). This obtains good wettability and the adhesive can be applied directly. The substrate keeps its optimal adhesion quality for more than seven days. Even with metal-containing materials, short-circuits cannot develop to damage the substrates, since the ionic parts of the plasma are separated out in the plasma jet nozzle.

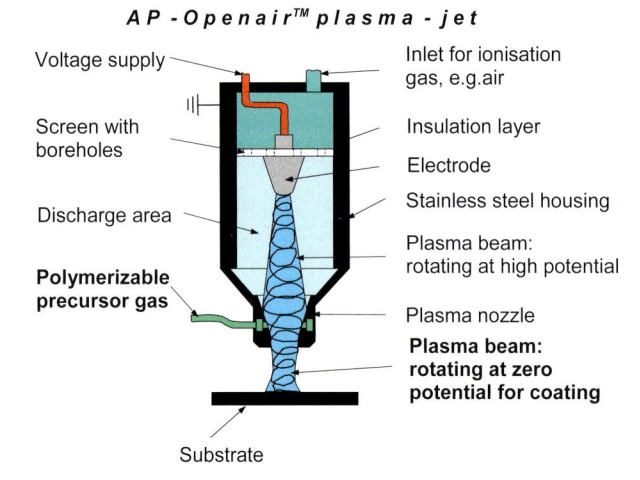

Figure 5-37 Functional principle of atmospheric pressure plasma

A newly developed reactive warm melt (RWM) designated Sikaflex-630 (manufacturer: Sika Technology, Zurich, Switzerland) combines the advantages of well-known technologies. Table 5-1 shows some technical characteristics of this warm-hardening adhesive that provides "instant-fix" handling strength. Sikaflex-630 hardens in environment moisture to an elastomer and develops good adhesion with plasma pre-treated polypropylene, untreated polycarbonate and hard-coated polycarbonate.

The advantages of this warm-melt adhesive include:

- Single component, i.e., no mixing,
- Instant handling strength for headlight adhesion,
- Touchable and non-sticking shortly after application,
- Loadable directly following leakage test under pressure (owing to rubber-like rheological behavior of the adhesive) and
- Only temporary mechanical fixation (clip) during leakage test.

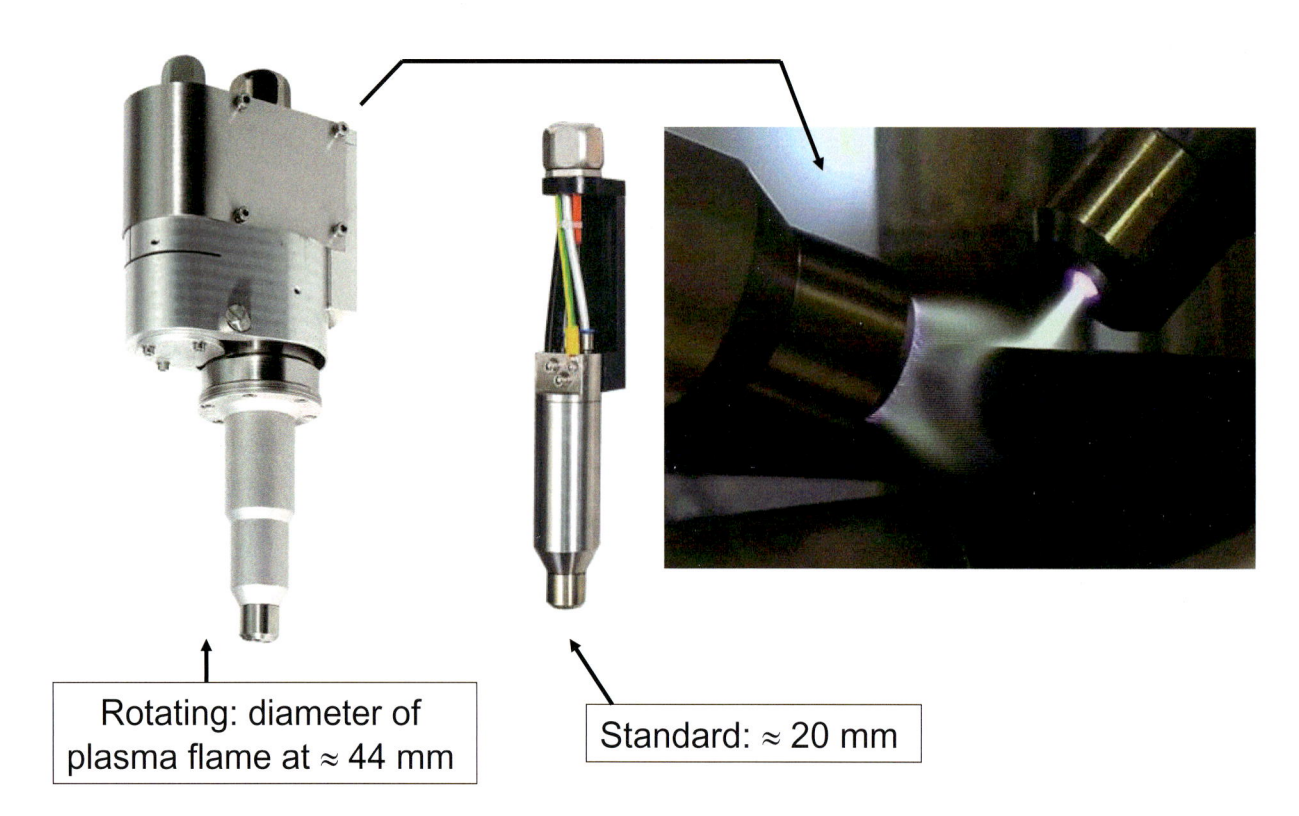

Figure 5-38 Plasma generator

5.4 Warm Reactive Bonding

Table 5-1 Reactive warm melt (RWM) Specifications of Sikaflex-630 warm-curing melt adhesive

Properties	Data
Application temperature:	95 °C
Non sagging:	3 mm bead okay
Fast physical hardening:	< 2 minutes
Tensile strength:	≈ 4.5 MPa
Elongation at break:	> 800%

The "instant" solidification of the Sikaflex-630 adhesive enables "just-in-time" production and packing of the headlights. Final cross-linking of the adhesive takes place during storage and transport. Further advantages result from

- Increased productivity (more headlights per minute),
- Compared with two-component adhesives, considerably higher throughput (fewer cleaning interruptions, short maintenance time and lower costs),
- Lower initial cost (e.g., no heater or hot-dip system) as well as
- Smaller number of rejects.

5.4.5 Influence on Handling Strength

Handling strength, an important factor in the leakage test, depends to a large extent on the reactive, warm-hardening adhesive. RWMs usually contain rubber-like thermoplastic components or a crystallizing resin. These are defined in Figures 5-39 and 5-40.

5.4.5.1 Instant-Fix RWM

"Instant-fix" RWM has a handling strength that permits leakage testing immediately after bonding the headlight assemblies. It consists of a rubber-like thermoplastic material where the polymer chain is comparatively long and has a molecular weight of more than 7000 g/mol. Between the chains there are intermolecular forces that permit plastic deformation.

Therefore simple assembly is feasible without the danger of cracking. The open time of this RWM is long, and at the same time it shows good wetting. This provides good adhesion without the danger of stresses building up in the adhesive.

5.4.5.2 Quick-Fix RWM

"Quick-fix" RWM is semi-crystalline and requires up to five minutes to develop the handling strength required for the leakage test. Since this product does not contain a rubber-like component, the molten adhesive takes time to crystallize. This procedure is comparable with that of candle wax, which also solidifies slowly. In contrast, the thermoplastic material described above consists of a liquid resin and a liquid chain-extender so that it behaves like chewing gum.

When the "quick-fix" RWM cools, the open time also ends immediately due to the high injection forces arising at that point. Cooling leads to material shrinkage due to crystallization, possibly leading to stress cracks in the adhesive as well as between adhesive and substrate.

"Instant-fix" handling strength
- Butyl rubber-like thermoplastic material
- Immediate leakage control
- Open time long: up to 30 minutes
  => good wettability and higher shapeability
- Non-sagging and cut-off string reduced

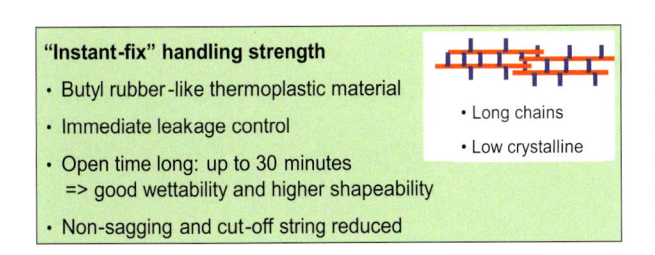

• Long chains
• Low crystalline

Figure 5-39 Adhesion characteristics
Characteristic properties of a reactive warm-hardening adhesive or warm melt with instant-fix initial strength

---

"Quick-fix" handling strength
- Crystalline binder (short chains)
- Up to 5 minutes for leakage control
- Open time short: below 5 minutes
  => limited wettability and less shapeablity
- Non-sag and cut-off string improved

• Short chains
• High crystalline

Figure 5-40 Initial strength
Characteristic properties of a reactive warm hardening adhesive or warm melt with quick-fix initial strength

5.4.5.3 Combined Handling Strength

The interdependence between the chain length of the thermoplastic material and crystallinity is shown in Figure 5-41. Here, number 1 stands for an "instant-fix" RWM1, number 2 for a "quick-fix" RWM2, number 3 for an "instant-fix" and "quick-fix" RWM3 and number 4 for a pasty (reactive) adhesive.

Figure 5-42 makes clear that the handling strength of RWM3 is higher than that of RWM1, Sikaflex-630. RWM3 has instant strength and simultaneous quick strength build-up. Its crystalline bonding agent increases the initial strength of the reactive thermoplastic (= RTP) after five minutes. RWM1 and RWM3 permit bonding with immediate leakage testing for headlights.

In Figure 5-43 the viscosities of the thermoplastic RWMs are plotted over the relevant temperature range. Their processing temperatures differ according to the type of RWM. Experience teaches that a viscosity range between 800 and 1200 Pa s is the limit for good pumpability. RWM1 has linear viscosity and requires 95 °C for processing by extrusion. The rubber-like thermoplastic causes this behavior and exhibits outstanding bonding characteristics. Good adhesion is achieved at ambient temperature. RWM2 contains a crystalline solid bonding agent without thermoplastic and, compared with RWM1, requires lower processing temperatures of approximately 50 °C. Its rapid solidification with cooling in five minutes to achieve handling strength is nevertheless too long to integrate a leakage test into the production process.

RWM3 is a combination of the rubber-like thermoplastic with the crystalline solid bonding agent. It can be processed at 80 °C. Above 60 °C, the viscosity curve is more or less identical to that of RWM1 and lower than that, approx. the same as that of the quick fixing warm melt RMW2. This combination results in a solidification behavior between RWM1 and RWM2.

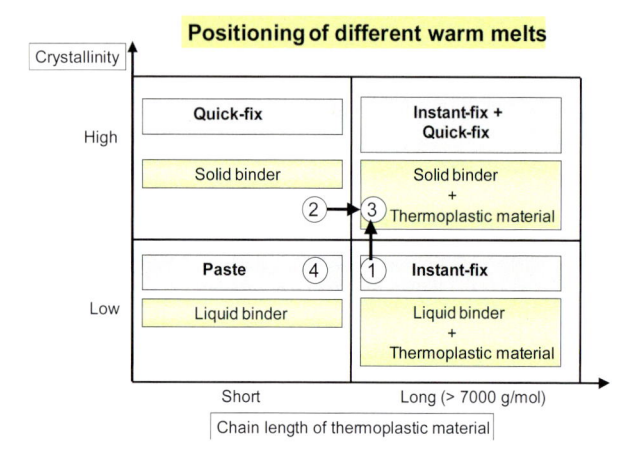

Figure 5-41 Positioning the warm melt – dependence on chain length of a thermoplastic material and crystallinity of a warm melt

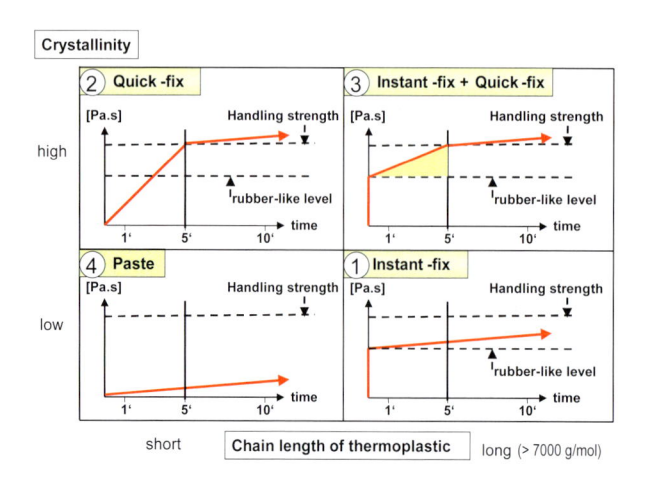

Figure 5-42 Handling strength – comparison of initial strength for various warm melts

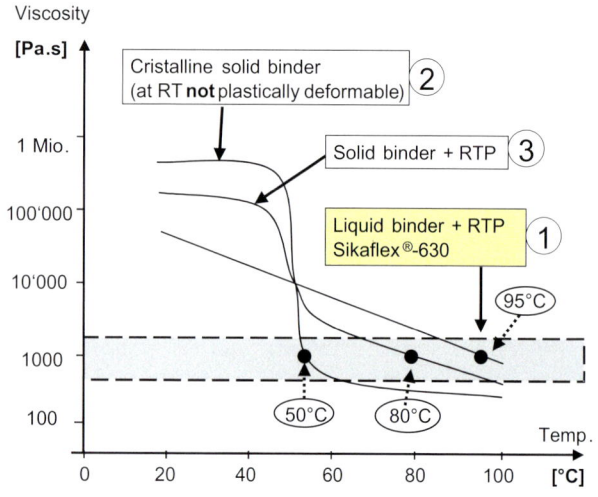

Figure 5-43 Comparative influence of viscosities on the processing properties of warm melts

5.4 Warm Reactive Bonding

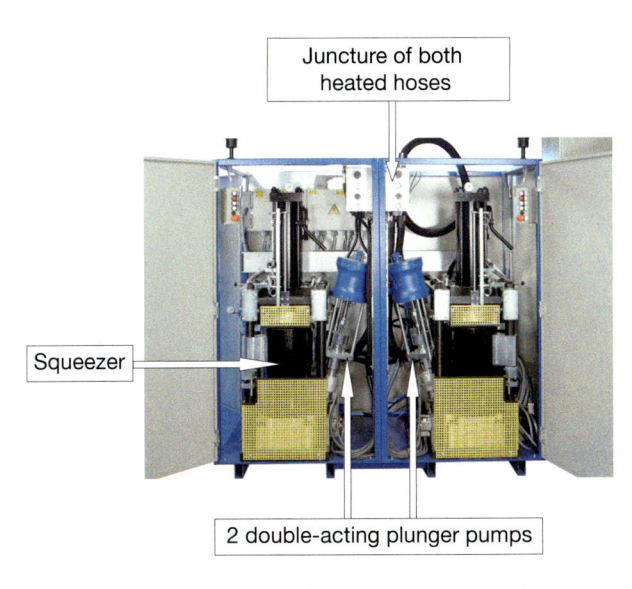

Figure 5-44 Adhesive is fed to a tandem application unit in headlight production

Figure 5-45 Headlight assembly line

5.4.6 Pump and Application System for Production

A discharge capacity of at least 350 g/min is required for producing headlights. Conventional pumping equipment is too weak for such discharge quantities, as the rubber-like behavior of Sikaflex-630 (RWM1) requires high processing pressures. New pumps had to be developed in order to achieve the required capacity. This new pump, RT Warm Melt 800, has up to 800 g/min discharge capacity depending on material viscosity [2]. The adhesive is contained in a composite foil bag in a hobbock 280 mm in diameter and 380 mm high. After the plastic bag is pressed into the squeeze pipe, a hydraulic cylinder presses the adhesive into the heating chamber and then into the feed pump.

The adhesive proceeds through heated hoses to the feed nozzle, where a dosing unit can also be installed. Material changes can be performed very easily. Almost 100% of the material is pressed out and empty bags can be disposed of (Figure 5-44). For reasons of thermal stability, 200 liter barrels cannot be used. Since the filling temperature of warm melts is the same as the temperature during production and application, heat builds up in the center of the barrel, leading to uncontrolled hardening.

5.4.7 Utilization in Series Production

This new system for assembling headlights is currently in use in series production of automobile headlights (Figure 5-45).

This system for bonding headlights is unique due to the special characteristics of its components:

- Reactive warm melt,
- Plasma pre-treatment and
- Pumping and feeding system.

This procedure is not limited to assembling headlights, but can also be used successfully wherever high requirements on bonding quality are combined with short cycle times.

References

[1] Documentation by PlasmaTreat
[2] Documentation by Reinhardt Technik

6 Case Studies – Design, Production, Performance

6.1 Structures and Body Panels

6.1.1 Film Technology in Automobile Manufacturing – a Comparison

Hans-Joachim Ludwig

The reduction of emissions, lightweight design, design freedom, functional integration, reduction of investment requirements, low system costs – these are some of the demands and wishes of automobile manufacturers and their suppliers. Short model life cycles and an increasing number of versions demand flexibility and modularization. Plastic body panels satisfy some of these requirements. Bumper and front-end systems, body side moldings, covers, rocker panels, roof panels, fenders and tailgates are in serial production.

Off-line painting by spray application is carried out on paint lines to create plastic parts in body color with class A surface (Figure 6-1). Today most of our paint systems are waterborne base coats.

Are there any other coating methods that satisfy the demands and wishes mentioned and can supplement or even replace off-line painting?

On-line painting in the body painting facilities of the automobile manufacturers saves the separate off-line process step, but also calls for materials with extreme heat resistance and only very few thermoplastics or thermosets satisfy this criterion. There are components are in serial production and the technical limits are known.

The use of films with corresponding surface quality is another possible way of avoiding the wet-coating of plastic components.

Figure 6-1 Products and processes

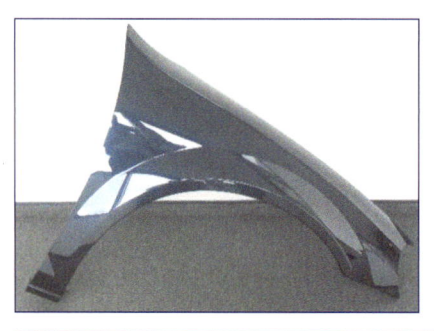

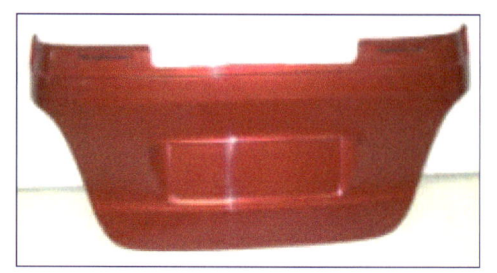

Figure 6-2 Development projects in mold film technology at Decoma

Surface properties
- Color/Color match
- Gloss
- Surface structure
- OEM specifications
    - Chemical resistance
    - Scratch resistance
    - Temperature/Climate tests
    - Weatherability

Physical/Mechanical properties
- Thermoformability
- Temperature resistance
- Adhesion to backing material
- Impact behavior

Economics
- Film price
- Piece price
- System costs

Figure 6-3 Requirements for IMD films for class A body parts

Structured films and films in body color, have been and are being tried out in development projects and are also in serial production (Figure 6-2).

Until now, however, the actual number of examples of applications has been limited due to the very high requirements for films to manufacture components in class A quality.

The films currently available cannot satisfy all requirements (Figure 6-3).

Weak points lie in the field of color matching, particularly in matching of components painted by different methods and also in the presentation of the total range of color of the vehicle lines.

In some aspects chemical resistance, scratch resistance and heat resistance do not meet requirements and specifications either.

6.1.1.1 Films Used for Body Components

Soliant and Avery Dennison use PVDF-based thermoplastics for the clear film (Figure 6-4).

Thermoplastic materials based on PVDF or acrylic are used for the color layers. In both cases, the dry paint film is laminated onto a backing film.

Examples of applications in body color are found on rocker panels and body side moldings of American car models.

6.1 Structures and Body Panels

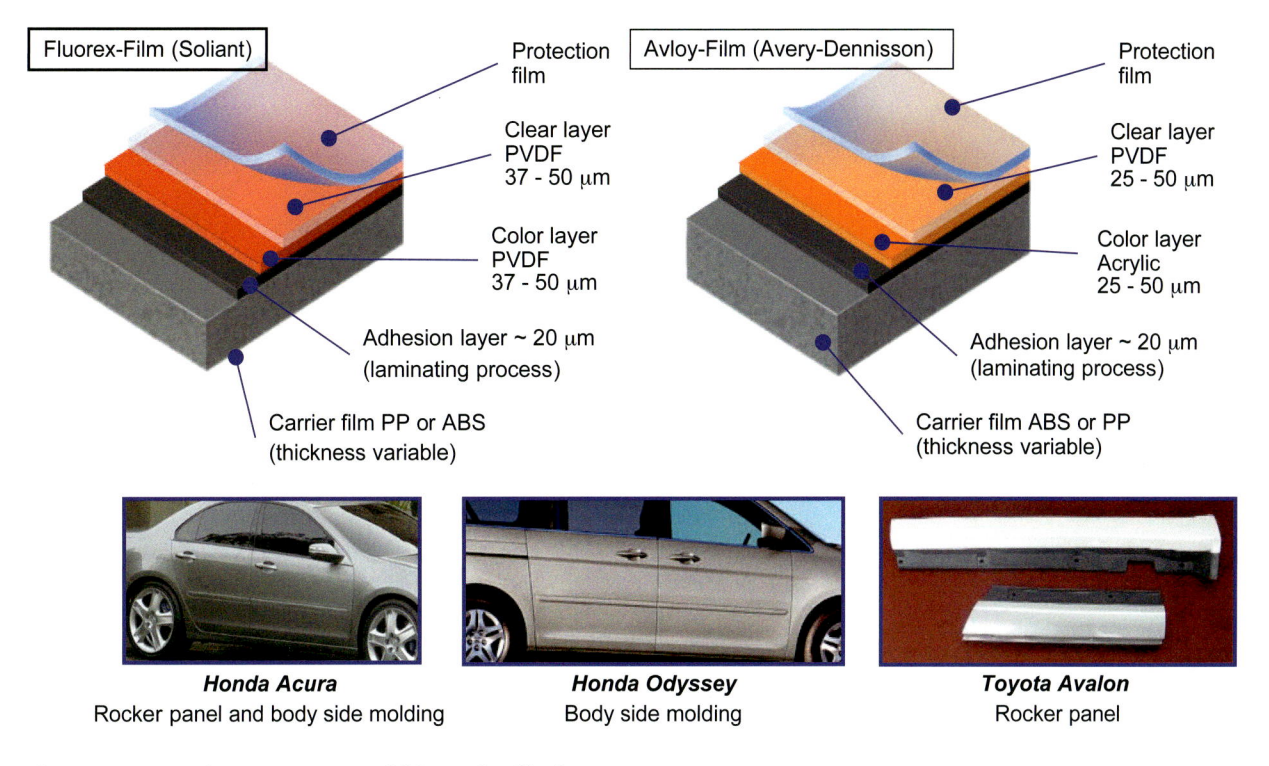

Figure 6-4 Laminate structure of "dry paint film"

Ionomeric material based on polyolefinic EEA is used by Mayco as a color and top coat in the bumper area of the Dodge Neon/Chrysler Neon.

Senotop films are equipped with PMMA color and top coats. As a clear coat, GEP also uses a thermoplastic material (PC copolymer) on a pigmented PC layer (Lexan SLX film). For foam-backing with polyurethane with a PC base film – as in the roof panel of the Smart – an adhesion layer of PC ABS must be coextruded at the same time (Figure 6-5).

This survey demonstrates that there is no film yet that is suitable, or can be used, for all body areas with their various requirements. In the view of Decoma, the vehicle body needs to be subdivided into three main sections: bumper and trim parts; vertical body components; horizontal body components (Figure 6-6).

Bumpers, rocker panels and body side molding have component-specific requirements with regard to impact strength, stone impact resistance and low weight coupled with adequate rigidity and temperature stability. Virtually all of those components are made of polypropylene materials. For this reason, a film for in-mold application must also, as a base material, consist of polypropylene (Figure 6-7).

Vertical body component such as fenders, door panels, tailgate panels or trim panels require impact strength, especially at low temperatures, rigidity, low thermal expansion in order to maintain control of gaps and tolerances, and finally, of course, low weight. Currently the base films used here are styrene copolymers (ABS/ASA blends). Polypropylene-based films are also undergoing trials. The back-molding material should be selected with respect to the film base (Figure 6-8).

The tailgate panel of the DaimlerChrysler A class (W168) is used at Decoma as a development part for vertical applications.

The thermoformed film blanks are inserted in the injection mold and backmolded. Processes being tried out include conventional injection molding, injection-compression molding and compression molding.

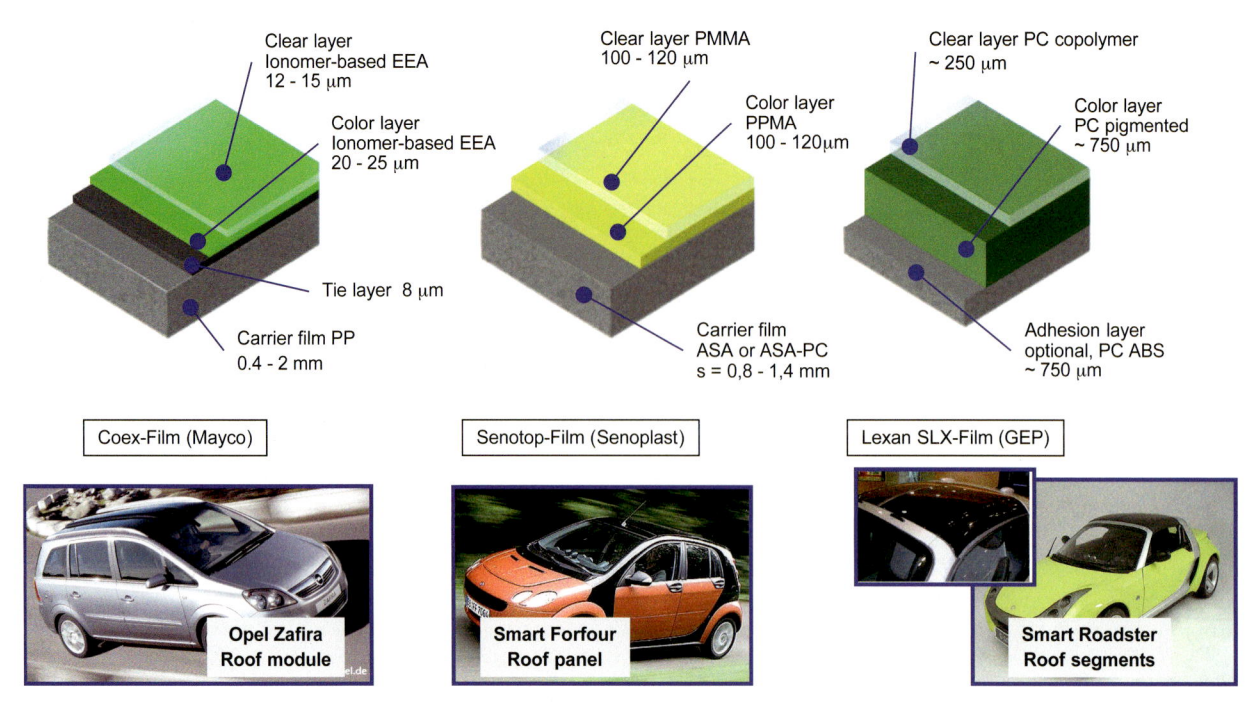

Figure 6-5 Laminate structure of coextruded films

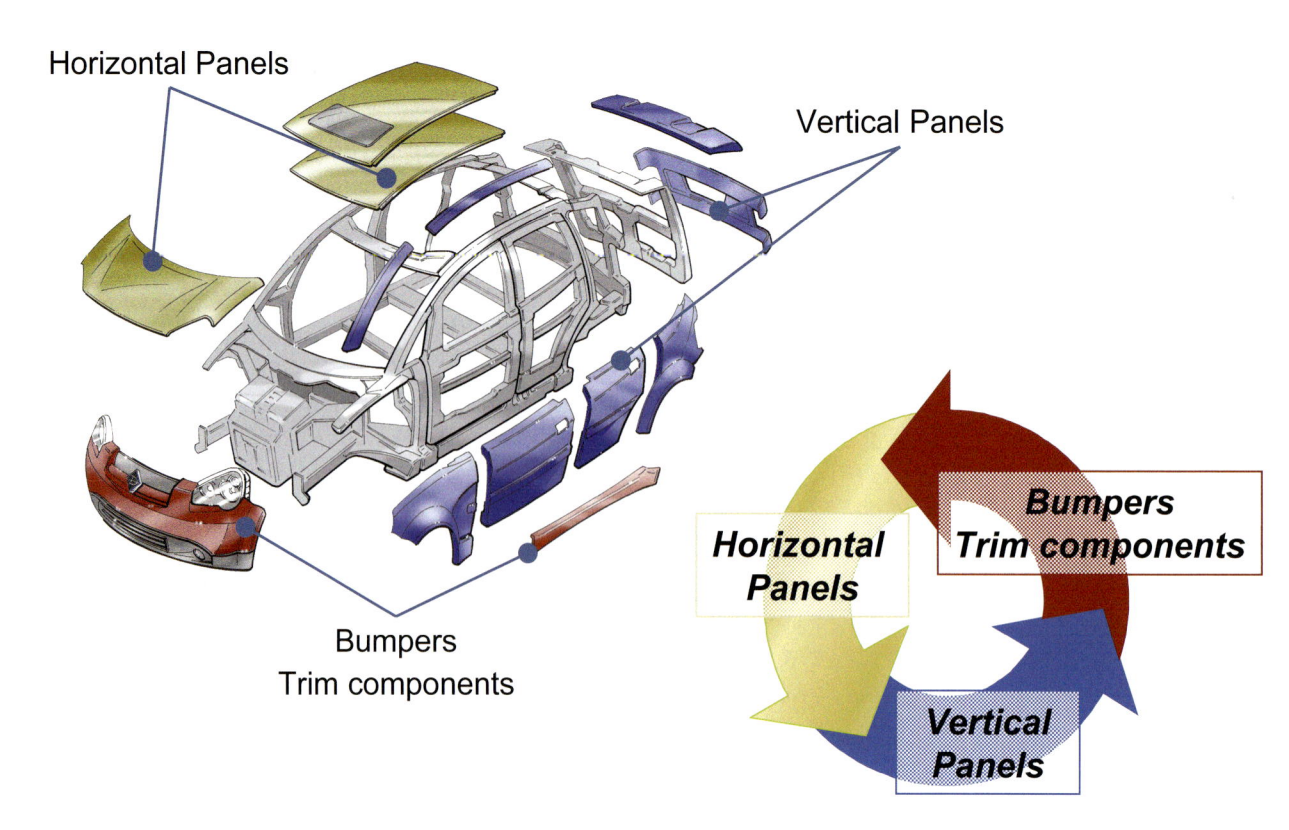

Figure 6-6 Main sections – car body panels

6.1 Structures and Body Panels

Technology:
- Class A thermoplastic paintfilm with PP carrier film
- Backmolded with TPO (PP-EPDM MF)
- Thickness carrier film 0,6 - 0,8 mm

Characteristics:
- Impact behavior
- Stone ship resistance
- Lightweight solution

Bumper System

Trim components
- Rocker panel
- Body side molding

Figure 6-7 Bumpers and trim parts

Technology:
- Class A thermoplastic paintfilm with carrier styrene copolymere or PP carrier film
- Backmolded with PC-blends, ABS or TPO
- Inj. Molding / Inj. compression molding
- Thickness carrier film 0.8 – 1.0 mm

Spoiler

Tailgate panel

Fender

Side door panels

Characteristics:
- Good low temperature impact behavior
- Stiffness
- Low CLTE
- Lightweight solution

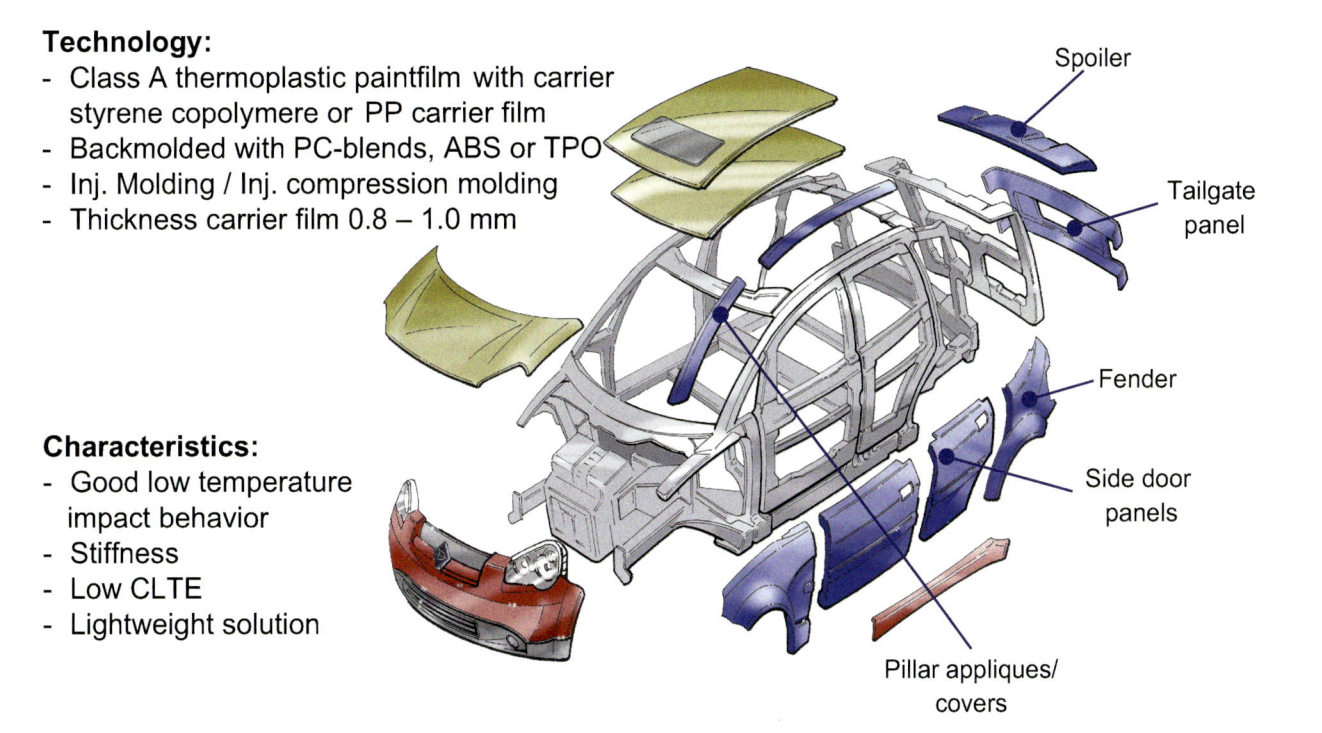

Pillar appliques/covers

Figure 6-8 Vertical panels

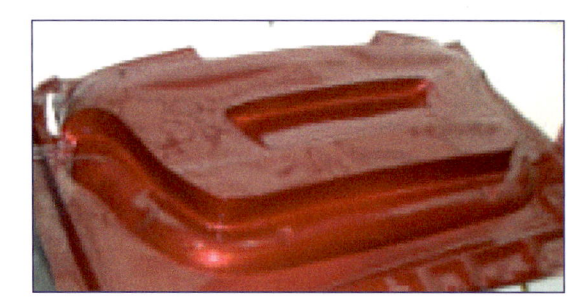

Part properties tailored to application

Material	E-Modulus	CLTE
PP MF15 offline-painted	2200 MPa	$45 \cdot 10^{-6}$
PP MF30 offline-painted	3200 MPa	$55 \cdot 10^{-6}$
Paint film + PP MF20	1800 MPa	$60 \cdot 10^{-6}$
Paint film + PP LGF40	4600 MPa	$\sim 32 \cdot 10^{-6}$
Paint film + PC/PBT LGF30	3700 MPa	$\sim 36 \cdot 10^{-6}$
Paint film + ABS GF 20	4200 MPa	$45 \cdot 10^{-6}$

Figure 6-9 Vertical panels – development example tailgate panels

In order to obtain high rigidity and low expansion values, long-fiber materials are also used as pellets or in in-line compounding (Figure 6-9).

These materials cannot be painted by spray-painting methods to achieve class A quality. The film insert molding technique is suitable for producing components to the required quality with high modulus values, excellent rigidity values, low expansion values and also, due to the film, the necessary impact strength.

Horizontal components – hoods, roof panels, trunk lids – are exposed to the highest temperatures. Rigidity, low weight and low thermal expansion are basic requirements, as is the wish for functional integration and, especially in the roof area, for simplicity in creating new versions (Figure 6-10).

Polycarbonate-based films are back-foamed with fiber- or mat-reinforced polyurethanes by the PUR-LFI or PUR-SRIM methods. These components are characterized by low weight due to their foam sandwich structure, high rigidity, stability even at high temperatures and a coefficient of linear expansion in the range of aluminum.

Functional integration in order to reduce system costs is on a path that promises success. Antenna modules can ideally be installed beneath horizontal plastic components and the various systems can be accommodated in a previously unthinkable location (Figure 6-11).

6.1.1.2 Paint Films – Structure, Manufacture, Properties

For about three years now, the DaimlerChrysler companies (starting at the Ulm research division), Wörwag and Decoma – and in part additional project partners – have been working on producing films with class A capability, which would satisfy the requirements and specifications. The result of this work is a painted film. With the paint film technique, components can be manufactured by different processes and with different materials to satisfy component specifications and deliver the surface quality required (Figure 6-12).

6.1 Structures and Body Panels

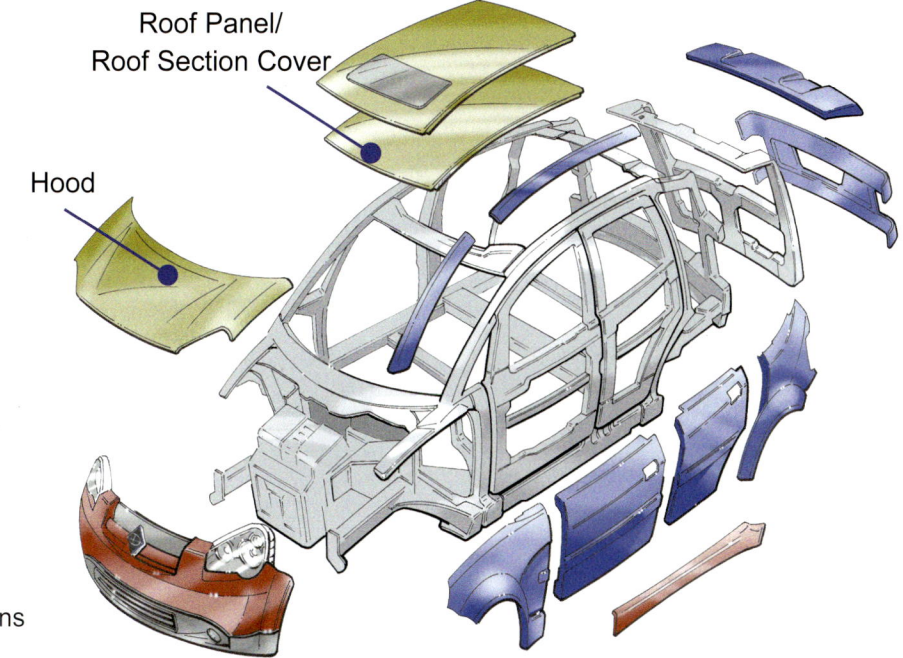

Roof Panel/
Roof Section Cover

Hood

Characteristics:
– Heat resistance
– Stiffness
– Low CLTE
– Integration of functions
– Lightweight solution

Technology:
– Class A thermoplastic paintfilm with PC carrier film
– Backfoamed with PUR/glass sandwich
– Technology: PUR-SRIM or PUR-LFI
– Thickness carrier film 1.0–1.3 mm

Figure 6-10 Horizontal panels

Joint development project:
Decoma / Hirschmann / DaimlerChrysler

Figure 6-11 Horizontal panels – antenna integration

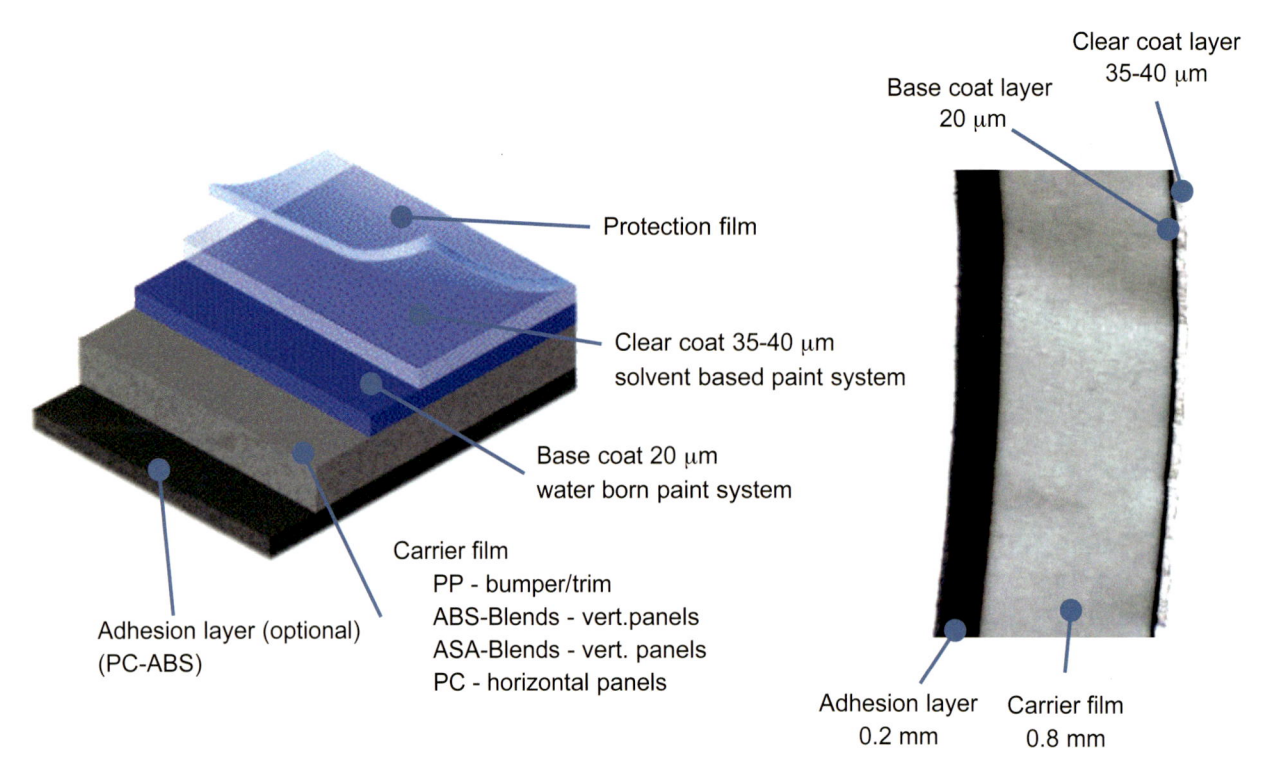

Figure 6-12 Laminate structure of paint film

Mono- or coextrusion films are used as carrier films. Polypropylene-based films are used for bumpers and rocker panels and also for body side moldings. Styrene-copolymer or polypropylene-based films are used for vertical components, while polycarbonate-based films are used for horizontal applications.

The base coat and clear coat are applied in a Linecoater unit. The base coat system is water-based – the paint is environment-friendly. The clear coat is a UV-curing system which is dried in the coating process but not chemically crosslinked.

The manufacturing process for the paint film starts with the extrusion of the carrier films. The basic requirements for a high-quality final product are purity of the raw materials and absolute cleanliness during extrusion (Figure 6-13).

During the coating process the paint is applied continuously. In the process chain, coating is followed by thermoforming to the desired component shape after which the clear coat undergoes ultraviolet curing in order to obtain the desired functional properties. The next steps in the process chain are back-injection molding and back-foaming of the films and, if necessary, trimming and assembly.

Unfortunately, inclusions in the backing film cannot be detected until after thermoforming when they appear as small dents. Dust and dirt particles must be avoided, as is the case with wet-coating (Figure 6-14).

A further quality criterion is the thermoformability of the films and paints. Cracks and crazing must not develop in either the base coat or the clear coat, even at higher degrees of elongation (Figure 6-15).

Elongation values of more than 35% have been secured, thus enabling production of small radii, edges and corners (Figure 6-16).

Color, gloss and texture are changed during processing from a two-dimensional film into a three-dimensional component.

6.1 Structures and Body Panels

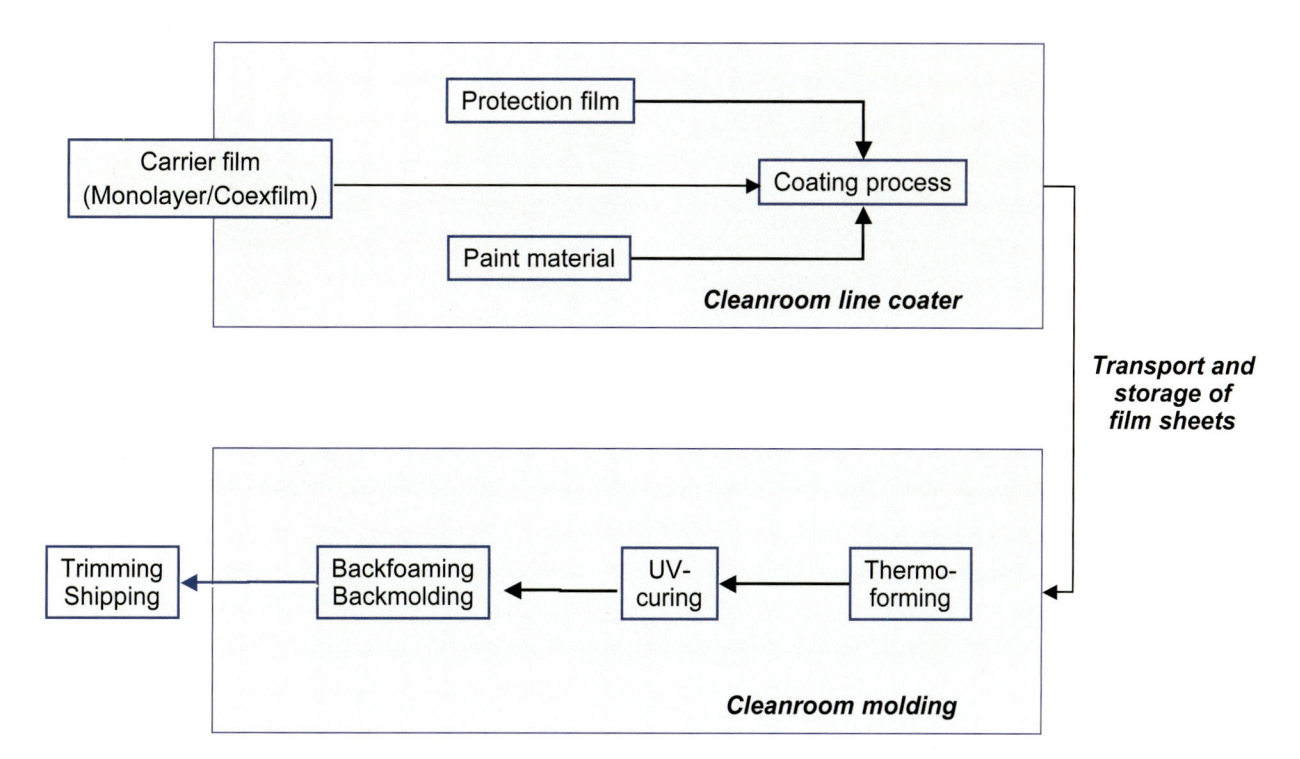

Figure 6-13 Process chain – plastic body parts with paint film

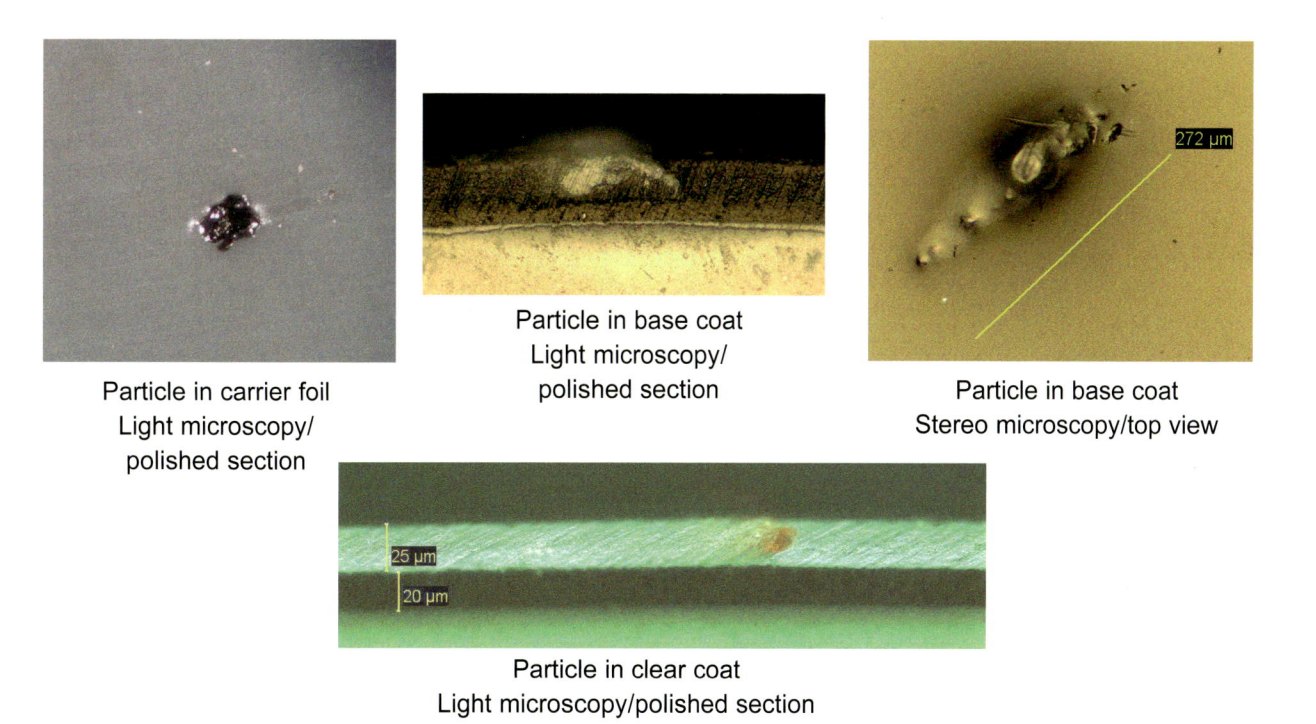

Particle in carrier foil
Light microscopy/
polished section

Particle in base coat
Light microscopy/
polished section

Particle in base coat
Stereo microscopy/top view

Particle in clear coat
Light microscopy/polished section

Figure 6-14 Surface defects – paint film development

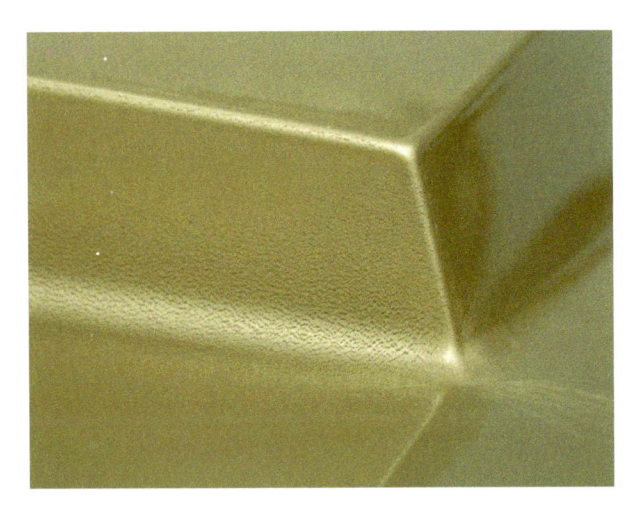

Crazing in base coat
Thermoforming - deep draw ratio ~ 35%

Microcracks in clear coat
Thermoforming - deep draw ratio ~ 30%

Figure 6-15 Surface defects in paint film development – thermoformability

Paint film
Thermoforming - deep draw ratio > 35%

Figure 6-16 Paint film technology – thermoformability

The color layers are inevitably thinned by thermoforming, depending on the amount of elongation (Figure 6-17). The temperature load is critical in this process for light colors, particularly white tones, if the correct part shape is to be achieved. A special paint development was required to prevent the occurrence of yellowing during this process step. The pressure and temperature applied on the painted film during backfilling also have an influence on the surface. The quality of the mold surface also has a decisive effect on the surface texture of the finished component.

It is possible to produce components with a short wave of 2–5 and a long wave of 2–4, measured after heat ageing (Figure 6-18).

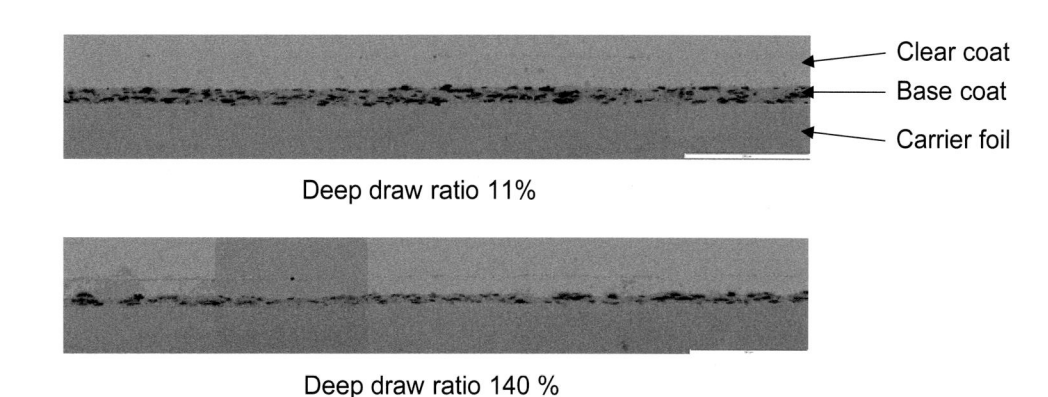

Figure 6-17
Base and clear coat shrinkage as a function of deep draw ratio

Deep draw ratio 11%
→ Clear coat
→ Base coat
→ Carrier foil

Deep draw ratio 140 %

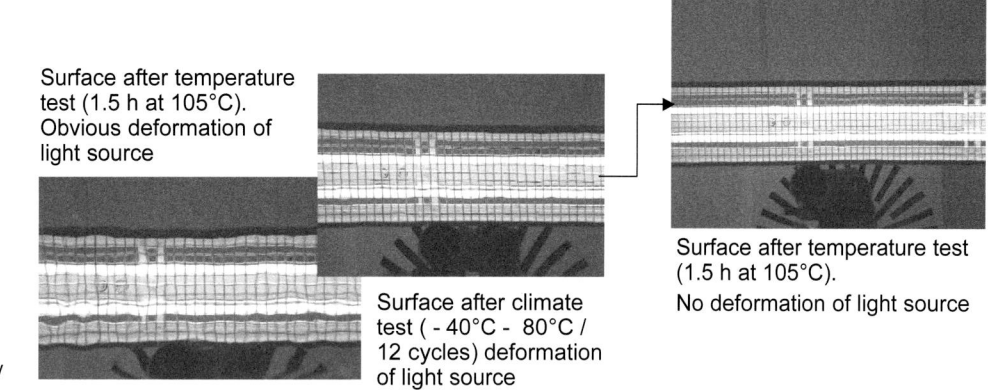

Figure 6-18
Development steps in surface quality – paint film technology

Surface after temperature test (1.5 h at 105°C). Obvious deformation of light source

Surface after climate test (-40°C - 80°C / 12 cycles) deformation of light source

Surface after temperature test (1.5 h at 105°C). No deformation of light source

Surfaces produced by spray painting have a short wave of about 15 and a long wave of about 5. Mirror-like surfaces can be produced with the paint film technology.

Layer adhesion testing, multiple stone impact testing, the steam cleaning test, as well as ageing and weathering stability tests were successfully carried out.

The specific chemical resistance of the component surface has been achieved, something which is not achieved by many sprayed-on clear lacquers (Figure 6-19).

Even the scratch resistance of the tested components attains the values of the latest generations of clear lacquer (Figure 6-20).

Color-matching with the original specimen was successful for the component using four colors: plain white and the metallic tones of black, beige and brilliant silver.

The reparability of paint film components was also part of the development work, as were off-line film paint systems, to allow custom colors to be painted by conventional methods.

6.1.1.3 Outlook

Currently, Decoma is setting up a prototype production line on the basis of the process chain we have described. The aim is to provide quality assurance of freedom from dust and foreign particles and lay the groundwork for the specifications and the economic feasibility of a serial production installation. Another objective is to supply a small production series from this prototype line in order to be able perform actual tests in practice (Figure 6-21).

The alternative to off-line painting – the paint film – is being systematically further developed by Decoma. The use of plastic components is coming closer, as can be seen from the exploded view in Figure 6-22.

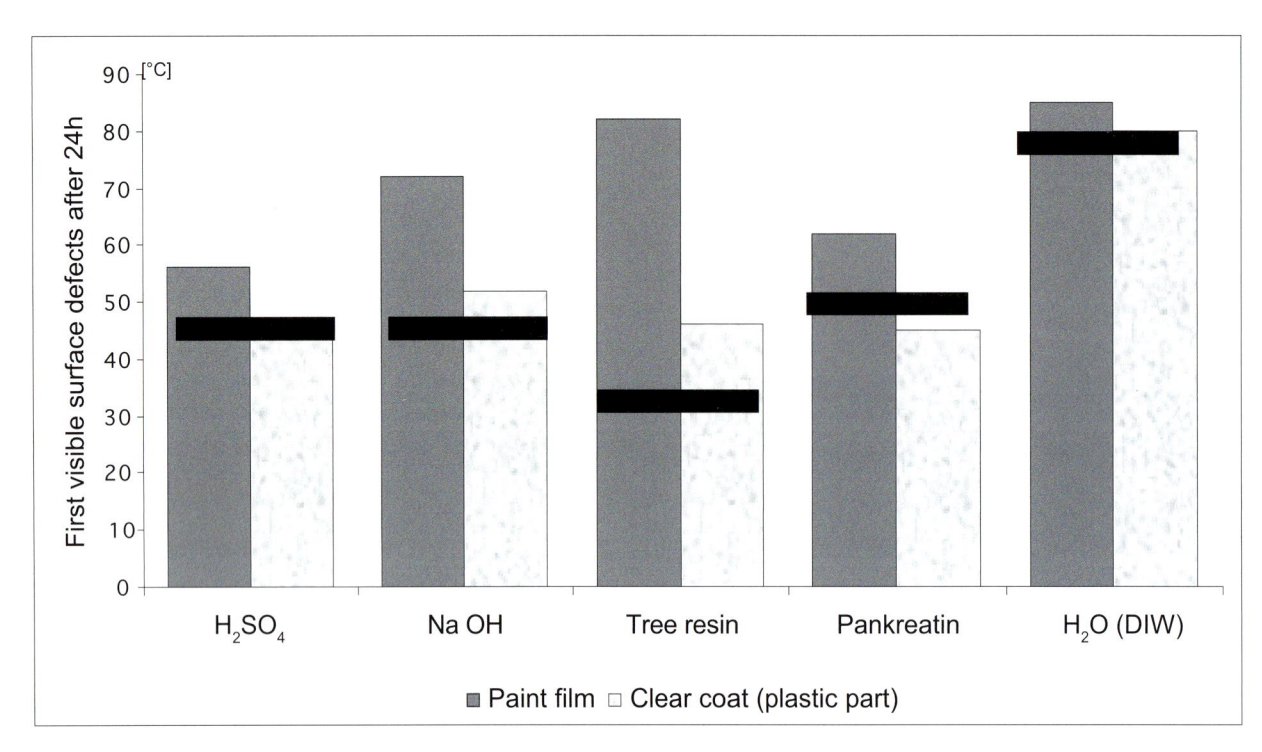

Figure 6-19 Chemical resistance

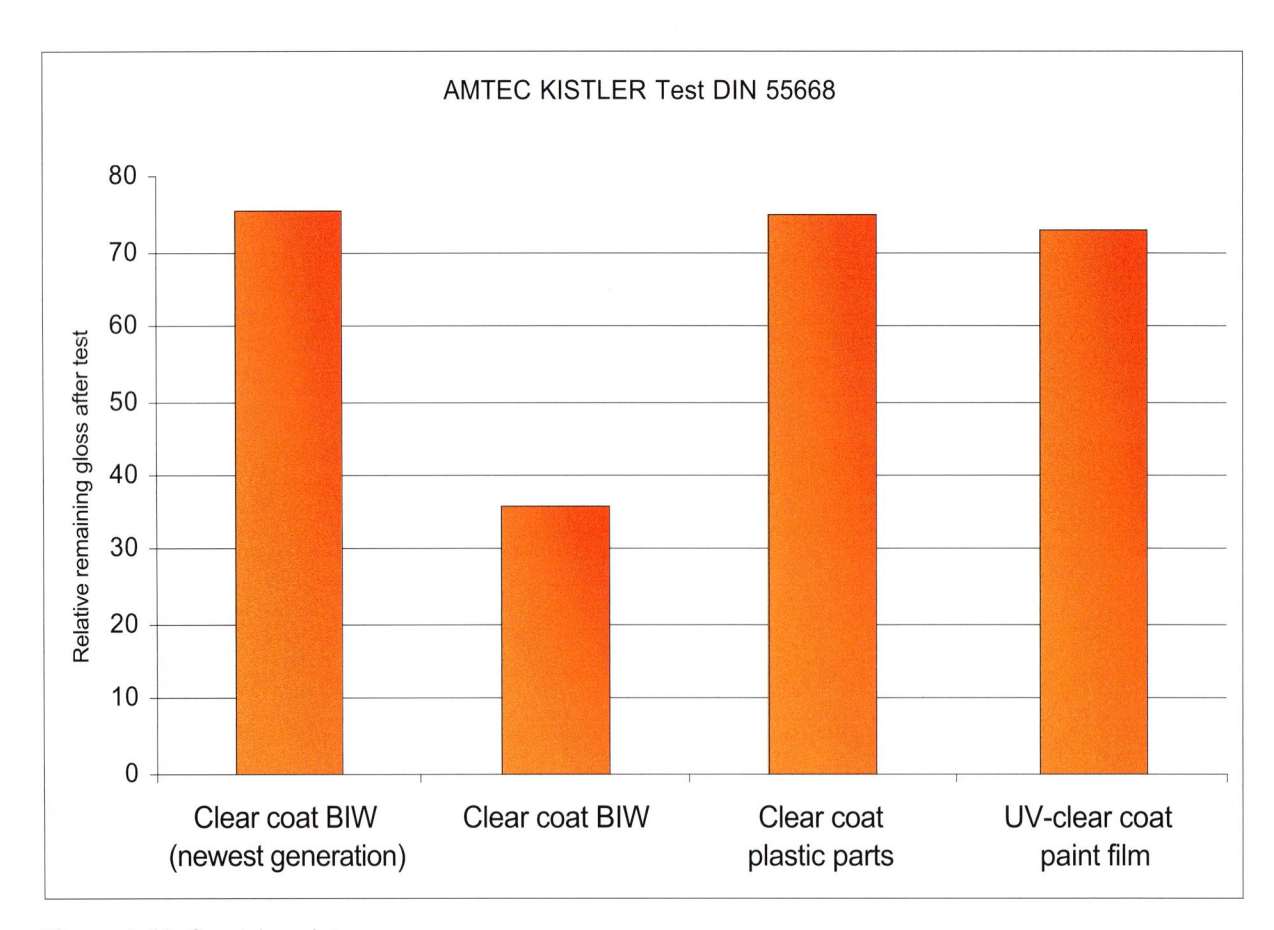

Figure 6-20 Scratch resistance

6.1 Structures and Body Panels

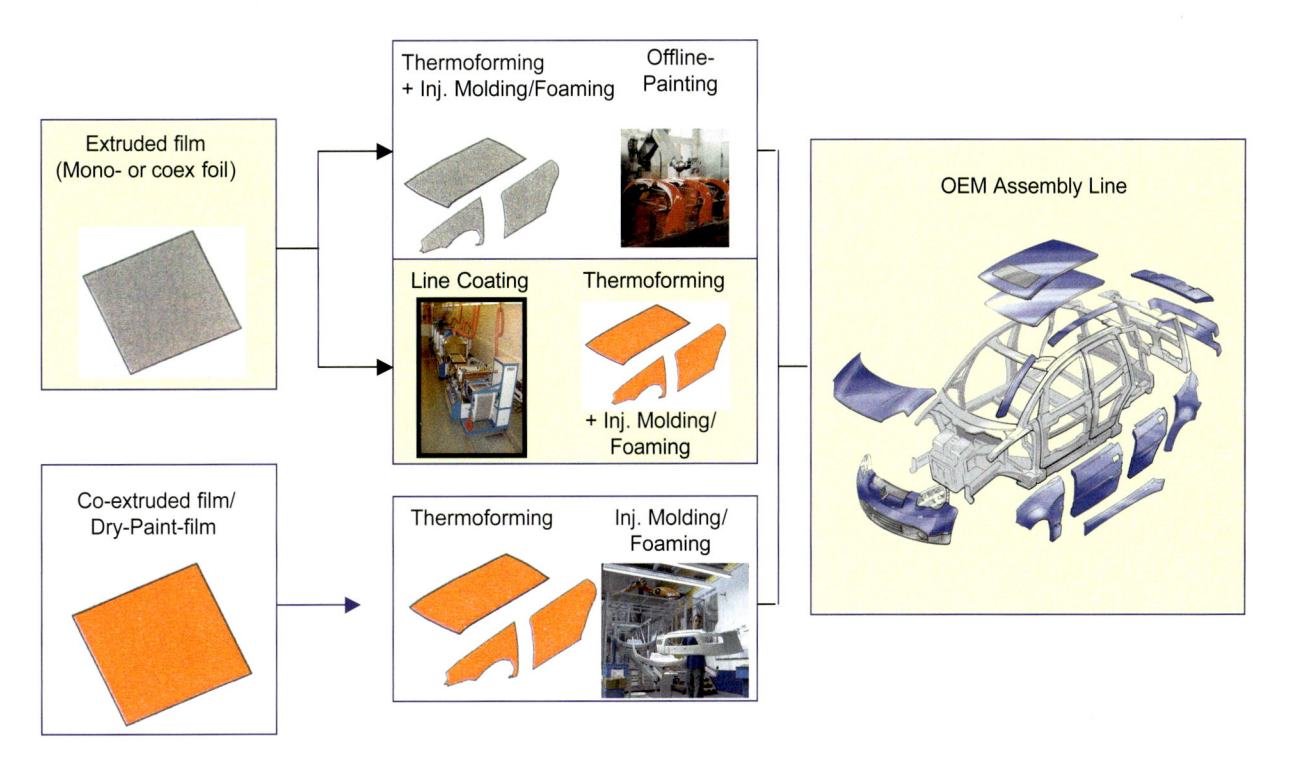

Figure 6-21 Production flow chart – in-mold film technology

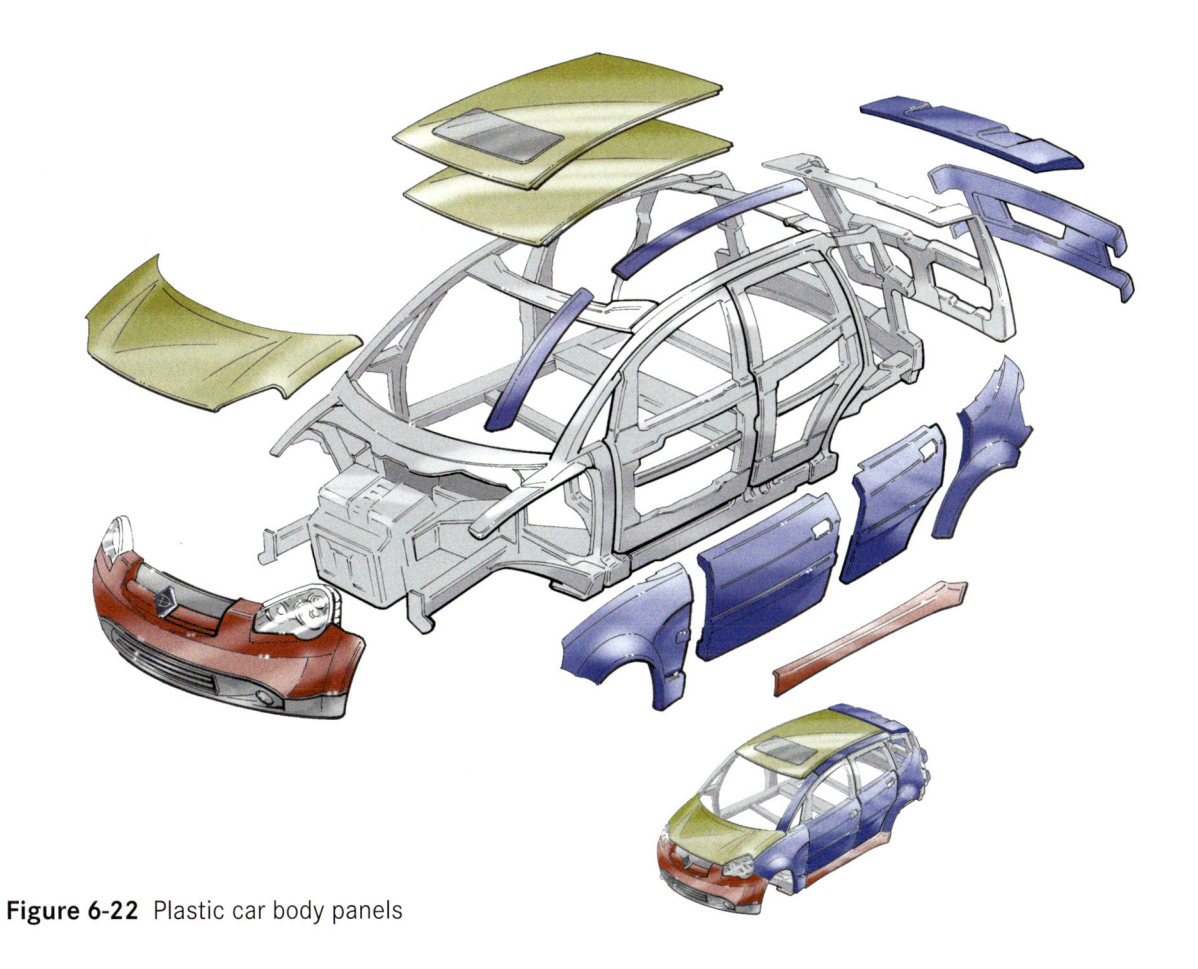

Figure 6-22 Plastic car body panels

6.1.2 Hybrid Tailgate: Getting the Best From Thermosetting and Thermoplastic Materials

Jean-Paul Moulin

The advantages and limitations of the two main families of plastics, thermosets and thermoplastics, are well known. Although combining the properties of different materials in "hybrid" solutions is a classical way, there are few examples of thermoset and thermoplastic combinations in the automotive industry. We will present here a first concept and validation prototype of a tailgate using these two plastics. A thermoplastic outer panel is molded using an innovative injection process to overcome traditional limitations on clamp tonnage and appearance on big parts.

After off-line painting, the tailgate is assembled by bonding the outer panel to an in-color molded thermoset modular part. This unique combination produces a low weight impact-resistant panel with class A finish on a stiff and geometrically stable structure. The full-scale validation of this concept will be presented as well as possible applications.

6.1.2.1 Introduction

Alternatives to steel for tailgate and trunk lid applications have been considered for more than 20 years. Weight saving, styling freedom and function integration were the main objectives. The first application of a full plastic tailgate appeared in 1982 on the Citroën BX in BMC (Bulk Molding Compound based on a polyester resin) material. In 1999, Mercedes introduced the technology for high-end cars on its CL coupe (C215 program) for the trunk lid.

Since then, most of the applications have been based on SMC (Sheet Molding Compound) technology (Renault Espace and Laguna, PSA 807 and Evasion, Volvo V70 and XC 90, Audi A4 convertible, BMW 6 series trunk lid). Thermoset technology offers a very good compromise between mechanical performance and cost for tailgate applications. However, SMC technology is limited by the difficulty to achieve class A finish in high-volume production (more than 500 per day).

That is why thermoplastic solutions have been considered in the last 10 years. Thermoplastic technology and material have been improved sufficiently to allow large outer parts and systems (bumpers, fenders, front-end module) to be designed and produced. Thus the Mercedes A class tailgate was manufactured in 1997 in thermoplastic (outer panels in thermoplastic alloy, inner structure in GMT). However, production of this tailgate has been abandoned on the new Mercedes A class (2004). It is difficult to overcome limitations of thermoplastic materials, such as geometric stability, thermal expansion and rigidity for ever more demanding automotive applications. It limits the use of full thermoplastic solutions on smaller applications, such as the back door of the new Renault Modus (injected polypropylene compound for the outer skin bolted on an injected polypropylene long-glass-fiber-reinforced inner panel).

An obvious solution, according to the material data sheet of thermosetting SMC and thermoplastic polypropylene, is to combine the properties of both materials (Table 6-1).

Table 6-1 Basic characteristics of thermosetting SMC and thermoplastic PP

		Thermoset SMC	Thermoplastic PP
Young's modulus	MPa	12 000	1 800
Max. stress	MPa	160	20
Thermal expansion	10^{-6} $°K^{-1}$	17	45
Low-impact resistance		■■	■■■■
Class A finish		■■■	■■■■
Functional integration		■■■	■■■■

6.1 Structures and Body Panels

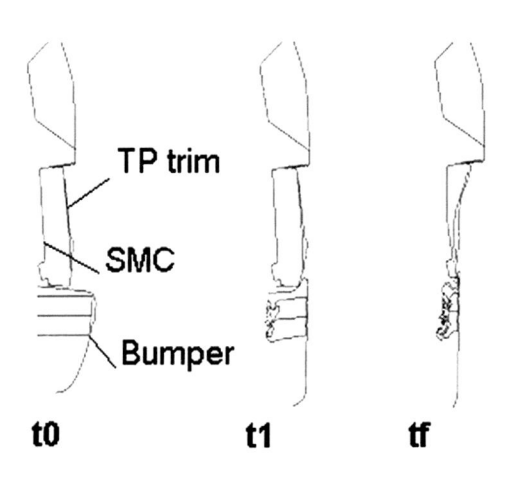

Figure 6-23 Impact simulation at 15 km/h on Renault Espace IV tailgate. The thermoplastic trim protects the SMC structure.

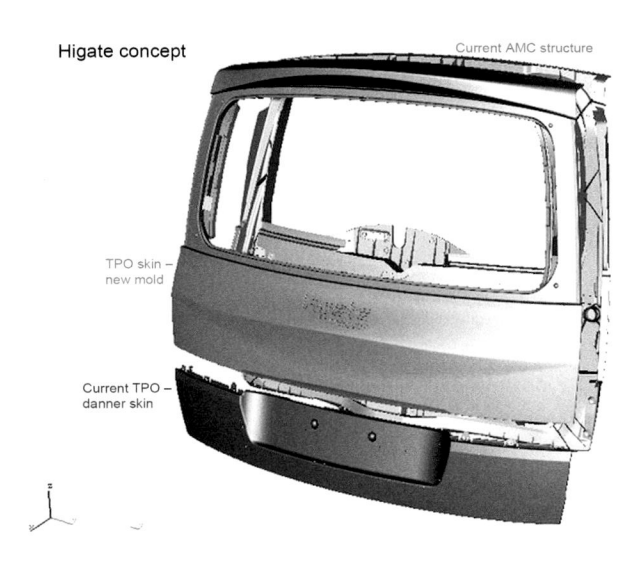

Figure 6-24 "Higate" tailgate concept

However, there are a few examples of applications in which thermosets and thermoplastics are combined. Thermoplastic covers and trims are assembled onto a SMC structure, such as on the tailgate of a Buick Rendezvous or a Renault Espace IV.

In the latter example, thermoplastic lower trim protects the SMC structure, preventing damage in the insurance test at 15 km/h (Figure 6-23).

Although thermoplastic material is used for its impact properties, this tailgate is basically an SMC part integrating TP trim and not a hybrid product engineered to make the best of both families of plastics.

6.1.2.2 Hybrid Tailgate

The objective of the Higate hybrid tailgate concept is to use the materials at their best for a tailgate application in order to optimize the cost/performance ratio. The Higate has a thermoset inner module with a bonded thermoplastic injection molded outer panel. It can be delivered as a fully assembled module to the OEM main assembly line. Equipment such as latches, hinges, wiper systems, rear lights, cables, and wiring can be attached to the inner module prior to outer panel bonding. Then the rear window is bonded onto the outer panel.

The inner module is made of SMC or AMC (SMC injection process [1]). The main requirements for this part are structural behavior (rigidity in torsion and flexion) and geometric stability over a broad range of temperatures. This inner module can be either painted with a self-graining appearance or molded in color in order to avoid the use of additional plastic inner trims.

The outer panel has to have class A finish. It is injection molded from a polypropylene compound specifically formulated to limit heat expansion. The material has a CLTE (coefficient of linear thermal expansion) of 45 µm/m/°C. It is painted using a classical paint system for polypropylene body parts. Finally, this outer panel is bonded to the inner thermoset structure by a newly developed bonding process that we will describe later.

The main advantages of this innovative hybrid tailgate in comparison with standard SMC tailgates are:

- Function integration possibilities (inner trims, license plate mount, etc.) thus reducing cost.
- Paint quality improvements (better distinctiveness of image – DOI, glossiness, etc.)
- Part cost reduction for an equivalent investment
- Low-impact resistance improves reparability
- Weight reduction
- Easier facelift with limited tooling investment.

Figure 6-25 Re-designed Higate tailgate

The objectives of the project were to fully validate this Higate hybrid concept. The Renault Espace IV tailgate was chosen as a validation basis. This tailgate is already being manufactured from thermoset materials. SMC inner module production parts were used to reduce the cost for validation. The outer panel was re-designed, an injection mold built and the associated bonding process developed to validate this new part.

6.1.2.3 Design

Styling

Some typical styling features were included in order to demonstrate the possibilities of thermoplastic injection molding in terms of styling freedom (sharp edges, deep drawn shape for the spoiler, small radii, lettering, etc.)

Dimensioning

Finite element calculations were run in order to simulate the rigidity of the tailgate especially in torsion, transverse rigidity and load under gas struts.

Gas Struts Load

The load exerted by each gas strut on the structure is 100 daN. The structure has to be dimensioned for a maximum deflection of 2 mm, which was obtained without any major modification to the SMC structure (see Figure 6-26).

This specification ensures that the deformation under gas struts load will stay in the elastic range and allow a perfectly flush fit to the car body by anticipating this 2 mm deflection in the mold geometry.

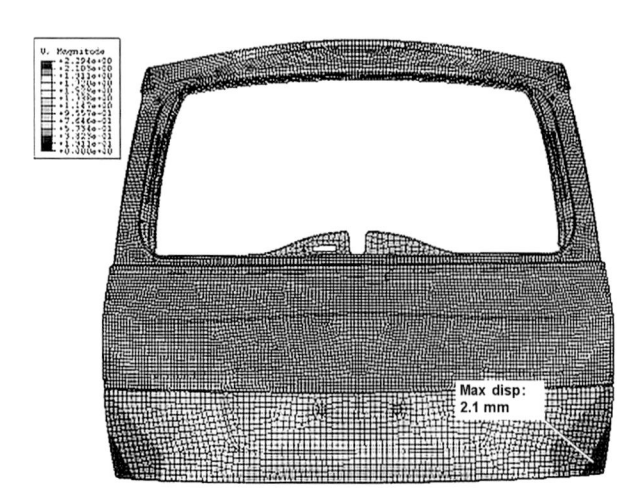

Figure 6-26 Finite element simulation for a 100 daN load on gas struts, displacement map

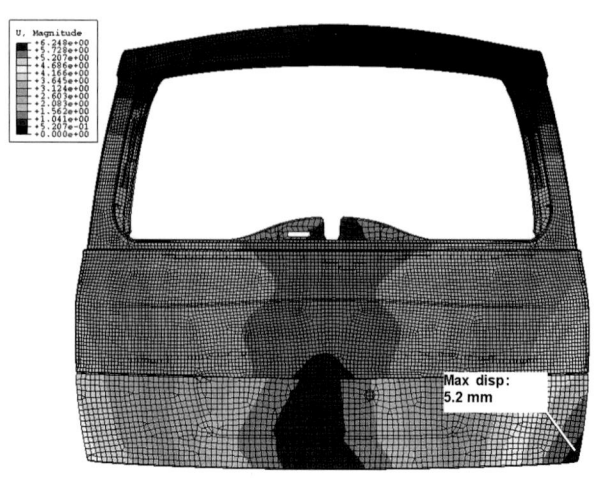

Figure 6-27 Finite element simulation for a 10 daN torsion load, displacement map

6.1 Structures and Body Panels

Torsion

When a 10 daN torsion load is applied along the main tailgate axis (Z axis), maximum displacement has to stay below 5 mm. This specification has to be fulfilled for a tailgate mounted without window. First results on the existing inner panel were 10 mm maximum displacement. The inner panel had to be ribbed in order to achieve the additional stiffness required. The weight increase on the inner panel is 15%. This extra weight is compensated by the weight saving on the skin and the function integration leading to an overall weight saving detailed at the end of this paper.

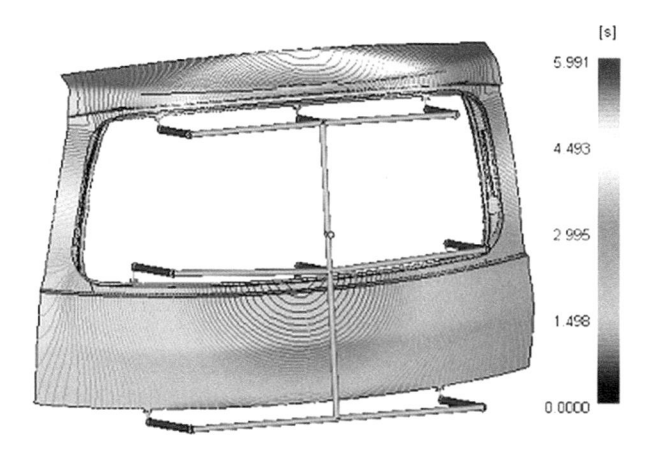

Figure 6-28 Injection simulation for the PP outer panel; flow patterns and filling time

Transverse Stiffness

A 10 daN load is applied on the latch area in y-direction. The calculated displacement was less than 2.5 mm which is within the specified limit of 3 mm.

6.1.2.4 Process

Thermoplastic Outer Panel

The outer panel is molded from a polypropylene 30% talc-filled compound. This material has good rigidity and a low coefficient of thermal expansion (CLTE < 50 µm/m/°C) while keeping excellent resistance to low impact. The mold has been designed with a valve gate injection system (8 drops) to fill the cavity and reduce flow and weld line visibility. Figure 6-28 shows the flow patterns simulated by finite element calculation.

The parting line for the outer panel mold has a special design capable of injection by three different molding processes: injection, injection-compression, injection-compression short stroke. This unique design allows the part to vary between 3.2 and 4.5 mm in thickness. This parting line also avoids the use of ejectors or sliding cores that may create visible defects.

Using sequential valve gate injection, parts were produced at 2200 t clamping force without visible weld or flow lines. Using the injection-compression process,

Figure 6-29 Outer panel injection mold; core side

parts have been molded under 1200 t clamping force and with reduced tiger-stripe visibility.

Outer panels are then painted using a classical painting process (flaming, primer, basecoat, clearcoat) on a controlled-geometry jig to eliminate warpage and distortion problems during paint application.

Thermoset Inner Module

The thermoset inner modules are made using the current series of molds produced for the current Renault

Espace in AMC (injectable SMC). A special resin was developed to produce the inner panel molded in color. This avoids the use of interior molded trim, which would increases both cost and weight. This resin has properties similar to the low-profile resin used in production (E modulus: 12,000 MPa, stress limit: 160 MPa) without porosity. Parts injected with this new material deliver the required mechanical properties together with a homogenous color finish (light gray). Sink-mark visibility is linked to the grained surface quality. Scratch and UV resistance are currently under investigation.

Bonding Process

The specifications for the bonding process were to equal existing production bonding performance in an industrial and cost effective process. To do so, the reticulation time had to stay below 3 minutes for production volumes from 300 to 1000 vehicle per day. The bonding joint has to be dismountable to allow for easy repair.

Glue Selection

Several types of glue had been tested on sheets to select the suitable ones. Lap shear tests were conducted systematically on different thermoset and thermoplastics samples. Out of this first phase, 4 polyurethane two-component (PU 2K) glues were selected for process and design optimization.

Process Parameters

The main issue when bonding thermoplastics and thermosets is to avoid any visible application marks on the outer thermoplastic panels. The glue has to be tough enough for bonding, but soft enough to limit stress in the outer panel.

The process parameters have to be carefully identified: surface preparation, heating technologies, time and temperature control, bonding joint profile. In order to monitor all these parameters, a specific apparatus was developed to allow tight control of the process.

The results of this second phase were the selection of a 3 PU 2K bonding system, the surface preparation for the polypropylene (degreasing, flaming, primer), the surface preparation for thermosets (degreasing, primer) and design rules for the bonding joint (thickness, angle, etc.)

Product Validation

Using the selected glues and the optimized parameters, several tailgates were bonded using a six-axis robot and a compression jig (Figure 6-30).

All the specifications for a final bonded product were met (stiffness, torsion, falling dart impact, sun test, dismountability, water tightness). Only a few marks on the outer panels were noticed after heat testing at 80 °C and cooling back to 20 °C. To solve this problem, the glue was reformulated (softer), the process and the joint geometry were optimized to avoid any marks after the heat resistance test.

6.1.2.5 Validation

Table 6-2 summarizes the tests and the results obtained so far for the validation of the hybrid tailgate concept. Most of the specifications are met, although some improvements need to be made during the tuning loop. We will detail the main results afterwards.

Figure 6-30 Bonding station prototype installation

6.1 Structures and Body Panels

Table 6-2 Validation summary on the Higate hybrid concept

		Target	Results
Geometry	Outer panel	Volume Schrinkage validation	Volume OK Shrinkage value to be tuned with injection process
	Inner panel	1 mm maxi deformation on sealing area	OK
	Tailgate flushness	Dispersions conformity	2 mm dispersion on prototype To be reduced with re-designed bonding jigs
	Tailgate gaps	Dispersions conformity	OK with tuned process for shrinkage ± 1 mm
Bonding	Ageing	No degradation Dimension variation < 0.5 %	Humidity resistance OK Humid ageing in progress
	Adhesion	100% Cohesive rupture	OK
Aspect	Outer panel	A class quality	OK similar to bumpers/claddings ...
	Inner panel	OK Design	Inner AMC standard + self graining paint: OK AMC moulded in colour inner: colour homogeneity to be improved
Module validations	Rigidity	Similar to current full SMC	OK with redesigned inner panel
	Perceived quality	Dent resistance: no deformation Low speed impact: no break	Dent resistance OK Low speed impact: falling weight 3 kg/50 cm without damages
	Thermal behaviour	Sun test (80 °C): no deformation after test, no functionality loss High heat test (110 °C): no deformation, no colour modification	**Sun test:** • No aspect or functionality degradation • No gaps modification back at room temperature • Max Y disp = 0.6 mm, Max X disp = 3 mm at 80 °C **High heat test** OK
	Acoustic transparency	Same as full SMC	Same level as serial tailgate: 76.2 dB vs. 77.3 dB for higate
	Slam/fatigue	Slam test at 23 °C	In progress

Sun Test

Figure 6-31 gives the maximum deformation measured on the tailgate during the sun test at 80 °C.

The choice of a low-expansion grade polypropylene together with the soft glue allows expansion on each side to be limited to 0.6 mm. Most of the expansion is translated into a deformation in the x-axis of 3.5 mm in the middle of the tailgate. This volume expansion remains limited and is not visible on the vehicle.

Acoustic Test

Tests have been conducted with the existing tailgate and a Higate prototype. A slight drop of 1 dBA has been

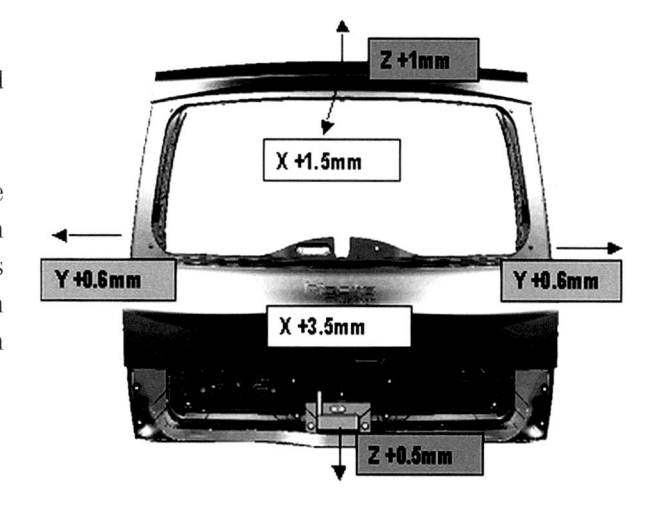

Figure 6-31 Maximum deformation of the tailgate during the sun test at 80 °C

Figure 6-32 Measured acoustic mapping on the Higate; max. 77 dBA in pillar area

measured with the new tailgate concept. Figure 6-32 shows that most of the noise leakage is situated in the tailgate pillar area.

This result is due to some holes in the pillar area required for prototype assembly but not for a production part. Without these holes, the noise level could be reduced to an average 75 dBA.

6.1.2.6 Higate Hybrid Assessment

Quality

The results displayed in the last paragraphs show that the hybrid tailgate meets most of the required specifications even if some improvements could still be made. Especially, the molding of the thermoplastic outer panel has demonstrated good feasibility without marks, flow lines, etc. The bonding to the injection molded SMC structure (AMC) has also proved to be industrially feasible and tough, while the outer panel finish is unblemished. Finally, the overall behavior of the complete tailgate is good and its excellent correlation with finite element predictive models will allow the reliable design of new hybrid tailgates.

Weight

Table 6-3 compares the weights of a traditional steel tailgate, a thermoset tailgate and a hybrid tailgate. From this table, it is clear that most of the weight saving is made by function integration into the injection molded outer panel: –40% compared to the traditional steel solution, –15% compared to the current thermoset solution. It also shows that a hybrid tailgate is particularly well-suited for highly equipped vehicles.

Cost

Compared to full thermosets solutions, cost savings can reach 10%, mainly due to easier paintability and function integration.

When compared to a steel tailgate, cost savings depend on the tailgate functionality level and the production volume. For a complex shaped tailgate with maximum equipment the savings can reach 20%.

6.1.2.7 Perspectives

Tailgate Applications

Future research work will focus especially on the process of the outer panel in order to limit clamping tonnage and improve appearance. New formulations of SMC resins will provide better mechanical properties while improving the molded-in color-grained finish. Automation of the bonding process will also reduce cycle time. These advances will allow efficient off-line production of the complete tailgate.

Table 6-3 Weight comparison between steel, thermoset and hybrid tailgates

	Tailgate in white	Equipments					Total
		Upper spoiler	Lower ext trim	Interior trims	Licence plate support	Others	
Steel	100	13	12	5	6	1	**137**
Full thermoset	85	int.	8	5	int.	int.	**98**
Hybrid "Higate"	80	int.	int.	int.	int.	int.	**80**

6.1 Structures and Body Panels

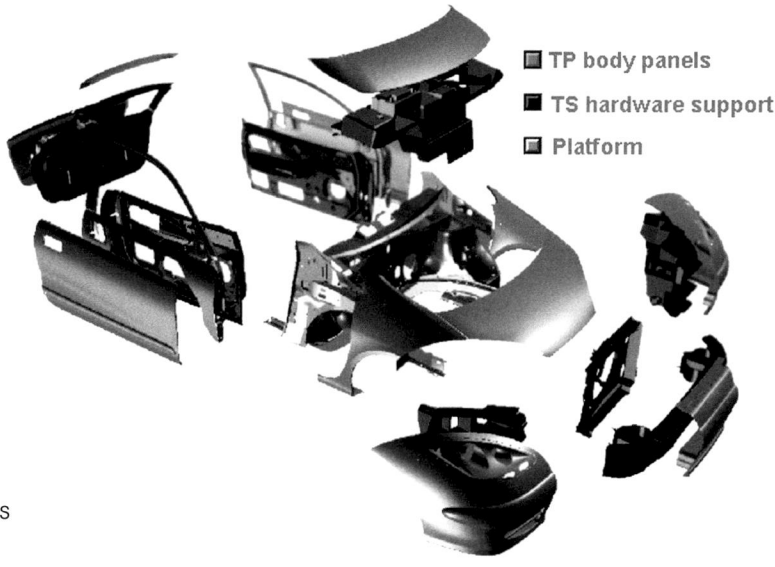

Figure 6-33 Some potential applications of hybrid TP/TS modules on vehicle

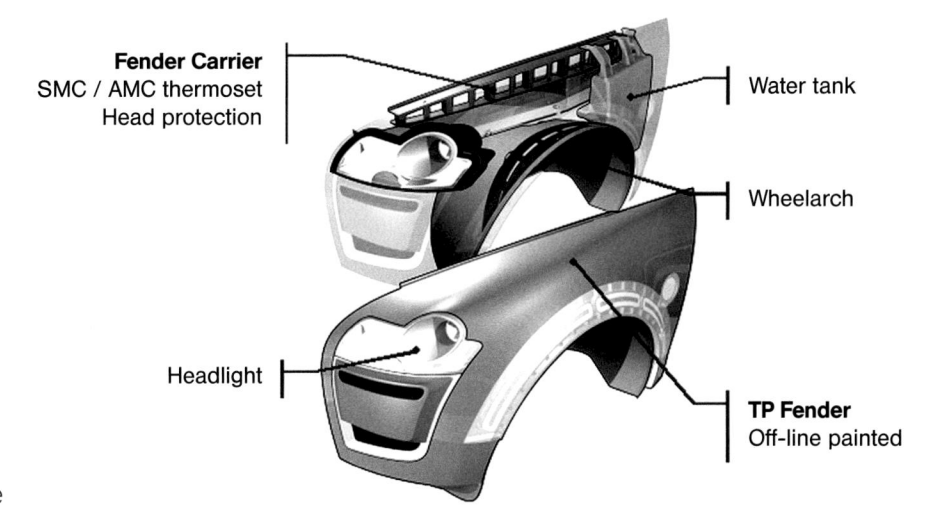

Figure 6-34 Fender module

At he same time, materials and glue are tested for on-line/in-line paint application in a temperature range from 140 °C to 200 °C.

Others Fields of Application

The successful implementation of hybrid thermoplastic/thermoset solutions for tailgates can be employed in other potential automotive applications.

Front-End Module

Some front-end carriers have already been manufactured in thermoset materials (SMC, BMC or AMC). Although it is difficult to anticipate crash behavior with composite beams, their use can lead to lightweight, high-performance solutions. The bonding of thermoplastic bumper cover offers significant advantages in terms of perceived quality (stiffness, geometry control, etc.) but leads to high repair costs.

Fender Module

An off-line painted thermoplastic fender can be mounted or bonded on a thermoset composites fender carrier, which also holds the headlamp. The fender carrier, injection molded in AMC or BMC technology, holds the fender and also absorbs energy efficiently under head-on impact so as to conform to the head-impact criteria stipulated by the pedestrian safety regulations.

Hood Systems

As stated above, thermoset materials conform to the pedestrian safety regulations for head impact with limited impact. A thermoset hood structure can be used in combination with steel or thermoplastic panels. The use of plastic materials can make the integration of additional functions (i.e., air ducts and inlet, engine access, etc.) in the hood easier in addition to energy absorption in case of head impact.

Cowl Grille

The cowl grille can be engineered with a thermoset structure that will sustain the engine temperature. This structure can be functionalized for air inlet, water separation and to support equipment such as windshield wipers or engine accessories. It will also absorb the energy of an adult head impact to ensure pedestrian safety.

Roof

The modularization of the roof is enabled by a roof frame from thermosets (SMC/AMC) that is bonded and/or bolted to the car upper body. Different roof panel options can then be selected such as painted steel or thermoplastic, glass or polycarbonate glazing. These panels can be fixed (bonded, riveted) for static versions or mounted on a sliding sub-frame for opening versions.

Doors

Based on the same approach used for the rear gate, the side doors can be made of a thermoset SMC/AMC structure that integrates door functions (window regulator, handles, armrest, loud speakers, interior trim). The thermoplastic outer panel can then be bonded. Most of the time, a metal frame will be required to conform to drop-test standards and to accomodate the window frame.

6.1.2.8 Conclusion

The design, process and validation of a first prototype thermoplastic/thermoset hybrid tailgate have proved the feasibility and attractiveness of such a concept. The range of properties of the two material families has been used in an appropriate way. Some pre-development work has already begun in which the initial design and process have been improved to achieve a better cost/performance ratio.

This research project has reached its objectives thanks to the use of integrated resources and means for innovation. An innovation team has been set up and is working on a project platform environment in cooperation with technology experts and with direct access to processing machines and testing facilities. All the design, simulation, tooling, process and validation tasks were accomplished in 9 months, achieving a faster time to market and hence saving money in comparison with "traditional" incremental innovation programs.

Reference

[1] European patent: EP0776754

6.1.3 Experience with On-Line Paintable Plastics on Vehicle Exteriors

THOMAS SCHUH

Under the headline "The plastic body is in sight" the Swiss magazine *Automobil-Revue* wrote in 1953: "This year, the manufacture of vehicle bodies from plastics has entered the stage of industrial production." Vehicle bodies made of plastics reinforced with glass fiber were announced and it was predicted that they would only weigh one-third of conventional metal bodies [1].

Looking at the current situation regarding the design and production of automotive bodies, it is clear that, even if this forecast was not entirely accurate, plastics are nevertheless firmly established in automotive engineering. However, plastic bodies remained limited to a few niche applications. Figure 6-35 shows the amount of plastic in various DaimlerChrysler products. It ranges from about 6% by weight in commercial vehicles to around 20% in the Smart, for which the concept of a plastic outer skin has been realized.

Of the 15–20% of plastic material typically found in sedans, 60% is inside the vehicle. The general public is not necessarily aware of the minor "revolution" that has occurred in the automotive exterior. In bumpers, rocker panels or radiator grilles plastics have replaced metals virtually without exception. The typical disadvantages of plastics, such as the higher cost of raw materials, higher coefficient of thermal expansion, low heat resistance, among others, have been outweight by their obvious advantages, such as low weight, high integration potential, high insensitivity to minor damage as well as extreme freedom of design.

The bumper is an impressive example of how, with the right design and suitable general conditions, an optimal product can be created. Since the mid-1970s, plastic bumpers have thus become established with all automobile manufacturers in a relatively short period of time.

Plastics are occasionally used today in rear lids, rear doors or hoods, but there is a very controversial discussion of the merits and demerits of the different material concepts. The number of fenders made of plastic is rapidly increasing, meaning that it is can be called a genuine trend.

Figure 6-36 shows some of the vehicles available on the market today featuring fenders made of plastic. It is noticeable that their number includes not only vehicles in the premium and niche ranges, but also vehicle series with very high production numbers.

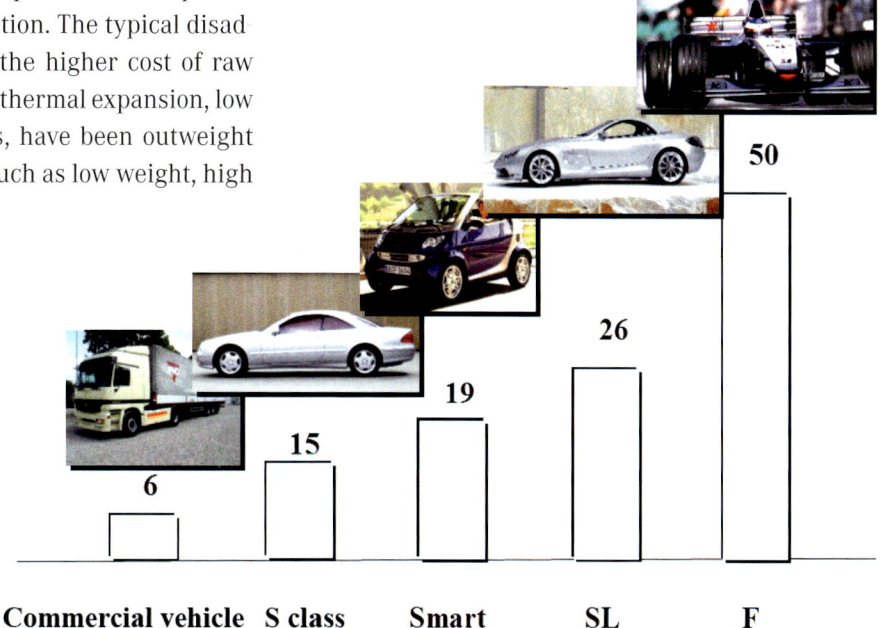

Figure 6-35 The amount of plastic in various DaimlerChrysler products (weight %)

Figure 6-36 Use of plastic fenders

Equally noticeable are the serious differences in size; for example, from a fender of the old A class to the wing of the S class coupe or the BMW 6 series. The vast majority of the applications are based on polyamide blends. However, there is also sporadic use of SMC[1], polyurethane or polypropylene.

6.1.3.1 The Road to On-Line Painting

Since today most of the plastic components in the vehicle exterior are painted in the body color, the painting concept takes on decisive importance. As a rule, bumpers, rocker panels and radiator grilles are painted off-line – in other words, independently of the body. The advantage of this is that special paint systems and different painting parameters (such as dryer temperatures, for example) can be employed to best suit the requirements of the plastic component. This is balanced by the disadvantages of the very high costs of separate painting, as well as the additional logistical expense required for supplying the component to the assembly line in the correct color and at the right time.

A further serious disadvantage is the effort needed to ensure a good color match between different painting installations and color systems.

One possible way of circumventing this problem is to paint the plastic components directly when they are already in place on the vehicle body. Due to the high temperatures in the electrophoretic coating dryer (temperatures as high as 200 °C), only thermosets – and SMC in particular – have been considered for this procedure so far. Examples of the implentation of this technology are the rear lids of the S class coupe, the CLK or the Maybach.

Due to their lower heat resistance, so far the vast majority of thermoplastic materials have been painted in-line – i.e., the components are fitted to the vehicle after electrophoretic dip coating and thus subjected to temperatures reaching "only" approx. 165 °C during the painting process itself. The other possible way of solving the problem – lowering the electrophoretic dip coating temperatures to a similar level – results in considerable reduction in quality of this important corrosion protection measure and has therefore been rejected by DaimlerChrysler.

---

[1] SMC: Sheet molding compound

Figure 6-37
Comparison of process chains

> **Previous process chain (wing of A class / S class coupe (worst case))**
>
> Injection molding - deflashing (- tempering) - priming (- sanding) - fitting after electrophoretic dip coating - painting
>
> **Ideal process chain (fuel filler cap / wing)**
>
> Injection molding (- deflashing) - fitting before electrophoretic dip coating - painting

Compared with on-line painting, in-line painting does, however, exhibit two considerable disadvantages:

- The "interruption" in the painting process after electrophoretic dip coating calls for additional areas in the painting lines where the components can be fitted. As a rule, such areas can only be included in existing installations at high additional cost if at all.

- In addition, fitting the components after electrophoretic dip coating disturbs the straight-ahead course of the paint lines due to the introduction of contamination this would involve.

6.1.3.2 New Materials and Processes

In the past, new material developments have provided ways of solving these problems. We refer to the developments of PA/PPE[2] blends and also to new developments based on polyamide materials. Their increased heat resistance allow them to pass through the electrophoretic dip coating drying process at temperatures of up to 200 °C. This means that even thermoplastic components can now be painted on-line.

The electrostatic form of painting common today calls for conductive components. So far, plastic components were coated with a conductive primer in a separate painting step. In the future, conductive types of plastic, such as are available today, will eliminate this additional process step as well.

Figure 6-37 presents a comparison of the previous process chain as exemplified by the wing of the Mercedes A class or S class coupe and the new process concept. Eliminating the tempering step (where required) and the separate coating with conductive primer and post-treatment result in considerable cost advantages despite the use of very expensive raw materials. This extremely lean process is, however, currently only feasible with water-based paints. Powder coatings, due to the water absorption of the polyamide blend, require additional process steps.

6.1.3.3 Experience with a Fuel Door

At first glance, a fuel door seems to be a simple, secondary component. However, due to its mostly exposed position, often in a central location on the side wall, a closer examination reveals requirements that are not easy to satisfy. These include:

- High surface quality and perfect color match with the surrounding side wall
- A high degree of dimensional stability in all directions
- Appealing tactile properties
- High tolerance to stress caused by misuse

For a long time, fuel doors were made of plastic in order to benefit from the associated cost and design advantages. Oval shapes with curved surfaces (concave or convex) are very difficult to realize in metal in compliance with the required tolerances and would thus be expensive. Here, plastics offer clear advantages. The requirements relating to dimensional stability are not only very demanding as far as the component

---

[2] PA/PPE: Polyamide/polyphenylene ether

Figure 6-38 Plastic fuel door of the Mercedes CLS

In the case of the recently introduced Mercedes CLS (see Figure 6-38), a plastic fuel door was painted in-line for the first time, following the process we have described.

The material used was a talc-filled Noryl GTX type from General Electric Plastics and the component was supplied by Sarnamotive. A filled grade was selected because of its superior mechanical properties and also its better expansion characteristics at elevated temperatures. In addition, the lower mold shrinkage results in improved dimensional stability. However, the filler content involves increased sensitivity to surface defects.

outline is concerned, but in addition the door must also follow the surface contour of the side wall. This is only possible with a very high reproducibility of all component dimensions.

In order to achieve perfect color-matching with the surrounding painted sheet metal, these plastic fuel doors were painted on-line.

To keep the entire process as lean as possible, a conductively formulated material grade was selected. Eliminating the step of coating the fuel door with conductive primer, which to certain extent can compensate minor surface inadequacies, increases the demands on the injection molding process further.

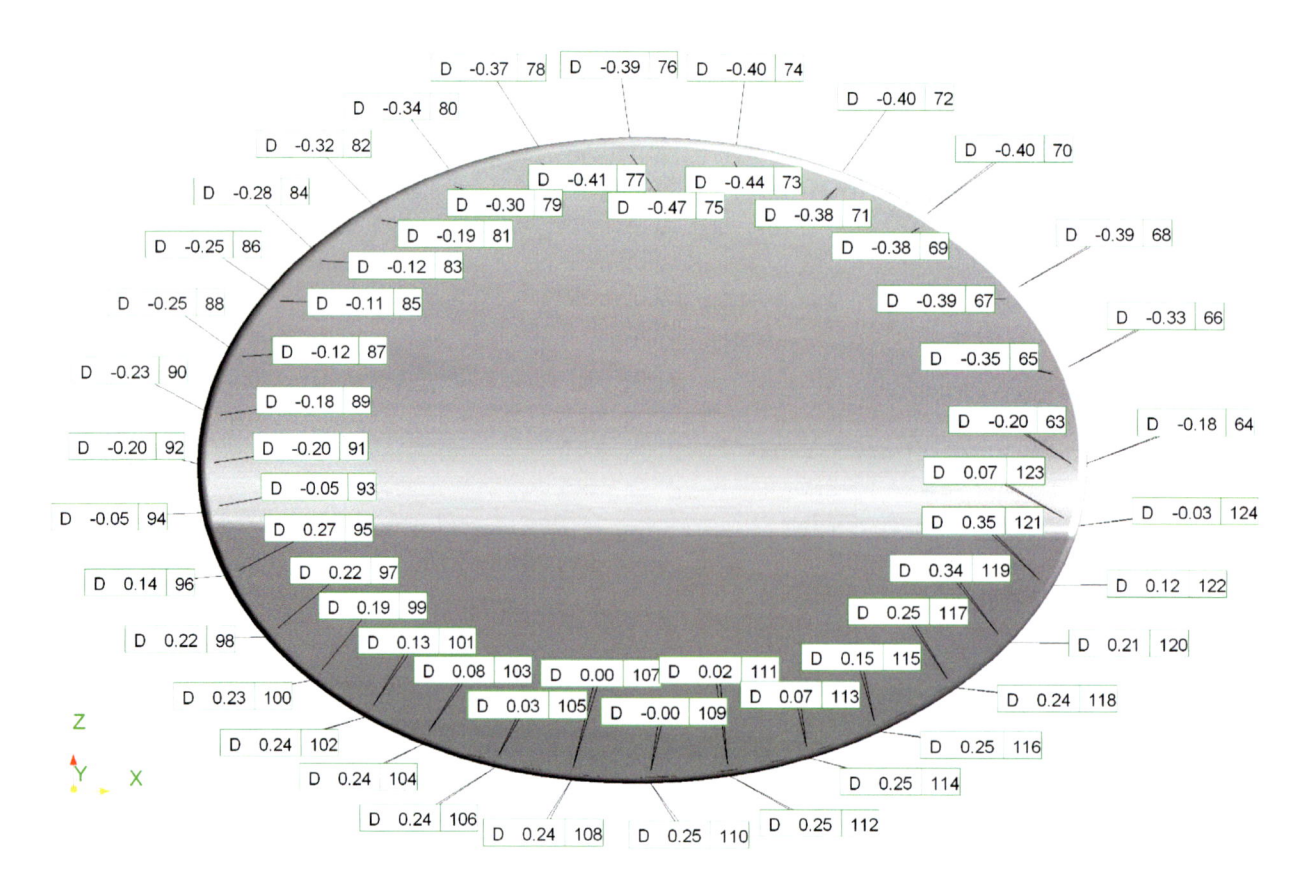

Figure 6-39 Measuring points on the circumference and the y-axis

The first preliminary trials with a pre-production mold from a different vehicle series had considerable problems in meeting the dimensional tolerances required. Besides the purely technical problems, even defining and adjusting a practical measurement concept turned out to be a challenge. Because of the specific surface topology of this fuel door, even the very smallest deviations in the x- or z- axis would have a considerable effect on the value obtained for y. Therefore, the securing and alignment of the component took on a great importance. Figure 6-38 shows the measuring points specified along the circumference and on the component surface.

In the beginning, considerable deviations from the target dimension occurred after the injection-molding step. In addition, a very marked scattering between the individual parts was evident. This problem was further exacerbated by warpage and additional aftershrinkage, caused by the drying conditions of the electrophoretic dip coating, i.e., 30 minutes at approx. 195 °C.

A gradual increase in mold temperature resulted in considerable improvements here, as can be seen in Figure 6-40.

Process parameters, such as melt temperature, injection speed, holding pressure profile etc. have a much higher influence here than with other material systems. Because the unpainted part must already have a perfect surface, mold design plays an important role. The hot runner, the gating system and the splits call for a high level of knowledge, expertise and care. Currently, it is still mandatory that every processing step is kept within a narrow process window in order to meet these challenging requirements. Figure 6-41 shows an example of the very good geometric reproducibility that was finally achieved.

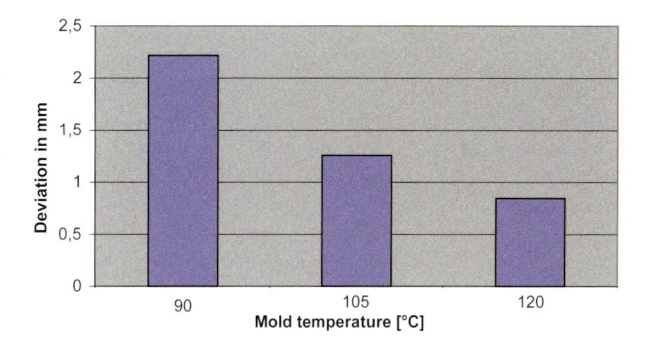

Figure 6-40 Dimensional change before and after painting as a function of mold temperature

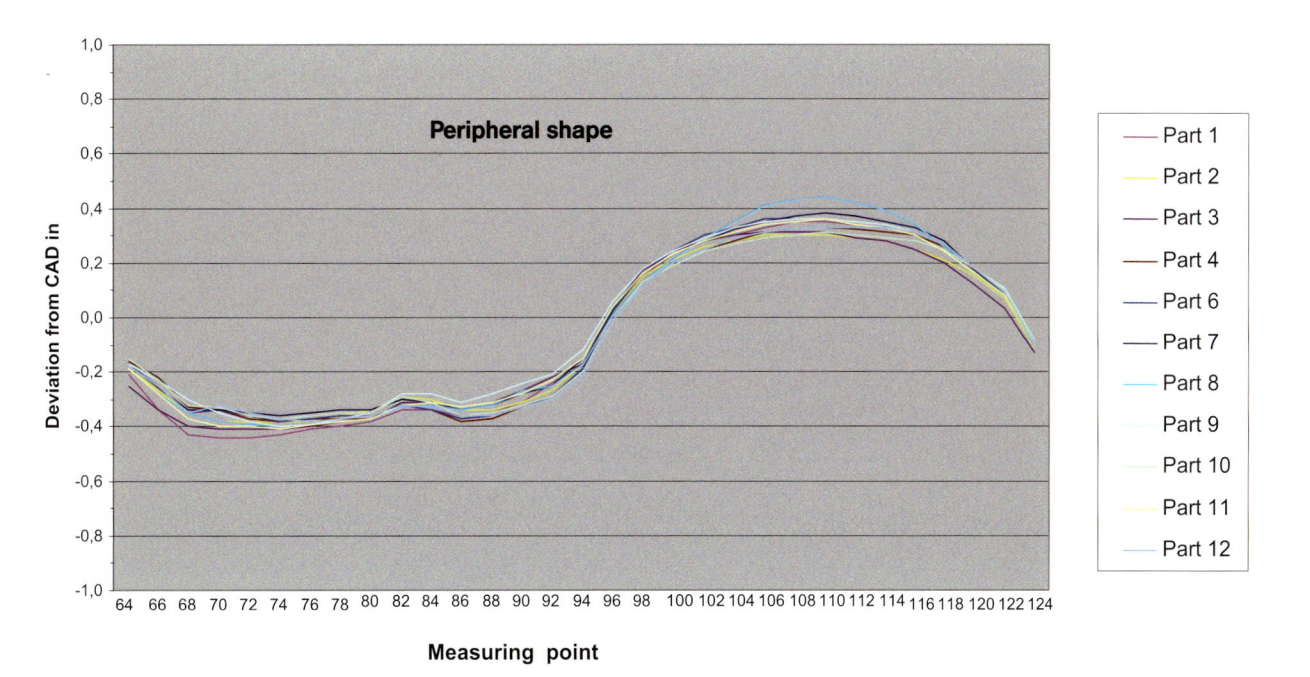

Figure 6-41 CLS fuel door, measurements after on-line painting

Figure 6-42 On-line painted thermoplastic fender of the Vito and Viano

6.1.3.4 The On-Line Painted Thermoplastic Fender

The potential for cost reduction, as exemplified by the fuel door, due to a leaner overall process is particularly attractive for larger components. It is obvious, therefore, that fenders should be produced by this technique as well. Examples of this are the Vitos and Vianos available on the roads today which have on-line painted fenders. Here, esthetic design and low weight were the decisive factors in the concept development.

Following injection molding and attachment of mounting elements, the plastic component is fitted onto the body and painted at the same time as the body. However, due to the size of the moldings and the gating system used, they need to be deflashed first.

The high susceptibility to split markings typically exhibited by the material meant that a lot of care had to be taken in the mold design. The use of a non-filled type of material and the roughly 10 °C lower maximum temperature loading in the Spanish production facility permitted a larger process window than for the fuel door. The experience gained from more than 100,000 vehicles produced to date is positive and also confirms this concept.

6.1.3.5 The Wish List of Automobile Manufacturers

The experience gained from these first successful examples of applications does, however, also indicate possible approaches for optimization measures:

- "More stable" processes that permit larger processing windows throughout all process steps (including the raw material)

- Improved simulation possibilities in order to shorten development times, reduce the scale of trials and provide faster optimization of process limits

- Materials with a considerably lower coefficient of thermal expansion (examples from the field of polypropylene for bumpers and rocker panels point the way here)

- Material types with a lower water absorption so as to reduce the associated disadvantages or additional effort required (particularly in the case of powder coating)

- An improved impact strength in order to increase resistance to damage further even at very low temperatures

- Materials less sensitive to split marks

- Measurement methods allowing holistic quantification of surface quality.

6.1.3.6 Outlook

Although this wish list is still long, the examples we have presented show that even thermoplastic components can be painted on-line without the need to make modifications to the paint lines common today. There is a great attraction to the idea of painting injection-molded components directly at the same time as the vehicle body and without additional work steps. This makes it possible to realize the potential of aluminum for lightweight design with a markedly increased freedom of esthetic design coupled with lower costs.

To what extent these developments will lead to fundamental changes – similar to those for the bumper – will depend in particular on whether further applications will provide more positive experiences and whether there is readiness to adjust the boundary constraints to suit these changes.

The plastic bodies of the Mercedes SLR McLaren as well as of the Smart are points of orientation for possible plastic concepts. Whether, when and how on-line paintable thermoplastic materials will start another "revolution" similar to the example of the bumper will be revealed in the not too distant future.

Reference

[1] Walter, G.: Kunststoffe und Elastomere in Kraftfahrzeugen, Verlag W. Kohlhammer 1985

6.1.4 Application of Plastic Bodywork Components in a Modern Bus

Lutz Ginsberg

6.1.4.1 Introduction

In the bodywork of modern busses, plastics – particularly glass-fiber-reinforced plastics – are state-of-the-art. Today the front and rear paneling and also the complete roof of the vehicles are made of fiber-reinforced composite plastics. As soon as the styling designer has laid a hand on the exterior, then in most cases the use of sheet steel pressings is not economically feasible. Typical annual bus production figures for a particular series range from just one unit to around five hundred units. In addition, there are different versions within a series that are tailored to the nature of the service the customer intends for those vehicles. A model cycle covers about eight years and in most cases a facelift occurs within this period.

The quality required of the outer body panels is constantly increasing. Higher quality in most cases means even higher tooling costs which often cannot be offset due to the low production numbers. The switch from a hand-built laminate or RTM to SMC in many cases is successful from the economic point of view only when components are used several times in the vehicle or even, going beyond the vehicle, in other series. This makes it increasingly difficult for the designer to achieve differentiation between the series.

Today the body shell is still predominantly designed as a self-supporting steel frame structure. At Neoplan towards the end of the 1980s the daring step was taken of making the entire bodywork structure out of fiber-reinforced composite plastics. It soon became clear, however, that as far as customer wishes were concerned, it was possible to respond to them much more flexibly with a steel body than with a full plastic body shell. It was therefore decided to use structural plastic components only in those locations where they could deploy their advantages fully and remain as unaffected as possible by body shell flexibility.

Taking the Starliner 1 series as an example, we shall describe some applications of fiber-reinforced composite plastic and the corresponding experience gained in more than three thousand vehicles to date. In addition we shall show how these applications are implemented in the Starliner 2 which is currently being developed up to readiness for full production.

6.1.4.2 The Integral Front-End of the Starliner 1

Back in 1996, the Conference on Plastics in Automotive Engineering included a report on an integral front-end which was then under development. This component has been exhaustively tested in a test vehicle of the

Figure 6-43 The new Neoplan Starliner 2

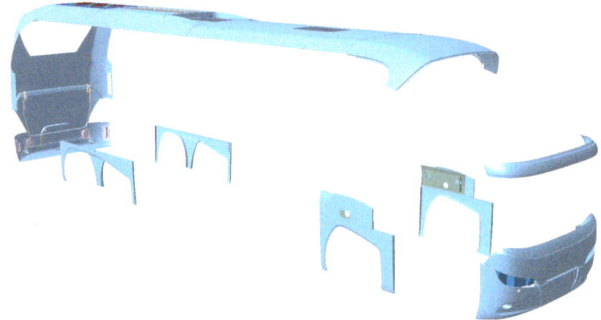

Figure 6-44 Applications for fiber composite plastics in the exterior paneling of the Neoplan Starliner 2

Cityliner series and served as the basis for the decision to equip the new Neoplan Starliner series (to be introduced in 1997) with an integral front-end as standard.

The new design of the Starliner series meant that the integral front-end, originally a single piece, had to be separated into an upper and lower part. The size of the component was no longer practicable for a single-part design.

The upper half of the integral nose forms the forward roof section and thus the transition between the three-dimensionally designed front part and the cylindrical cross-section of the vehicle. Both the A pillars and the nose cross-member between the upper and lower windscreen sections are integrated into this component. The A pillars have an extremely three-dimensional curvature. The usual commercially available square-section tubing could never have been bent into this shape using simple bending tools. To increase the rigidity of the A pillars, carbon fiber rovings were inserted into the corner areas of the profile cross-section between the top layers and the core material. This was necessary, since not only the upper windshield but also the two forward side windows were mounted in the upper part of the integral nose. The glazing makes a considerable contribution to the overall rigidity of the bus body shell and for this reason the load transfer points had to be given an appropriately sturdy design.

The complete component is flexibly bonded to the body shell at the roll bar and at the nose cross-member. In the vehicle interior, the longitudinal profiles of the inside ceiling and the complete nose dome are fitted to an intermediate frame made of aluminum. The windshield wiper system is mounted on laminated-in holders. The fixing points in the component are positioned exactly with respect to each other by means of a gluing device. It would not have been possible to actually build the exterior vehicle design in this way using any type of steel construction so far available.

The lower part of the integral nose forms the partition wall between the driver's station and the nose paneling elements. It is also flexibly bonded to the body shell. A very large number of functions has been integrated into this component. It constitutes the lower nose cross-member on which the lower windshield is mounted. In addition, we attached to it the windshield wiper unit, pre-assembled on a base frame. On the inside of the integral nose there is also the air duct between the air-conditioned front box, located beneath the driver's seat and the driver's station.

The steering column, like the air duct to the instrument panel, is mounted via defined interfaces. The component is a sandwich design consisting of top layers 3 mm thick, a matrix of unsaturated polyester resin, and a core of rigid PVC foam. It is manufactured by the liquid resin press molding process. A gluing

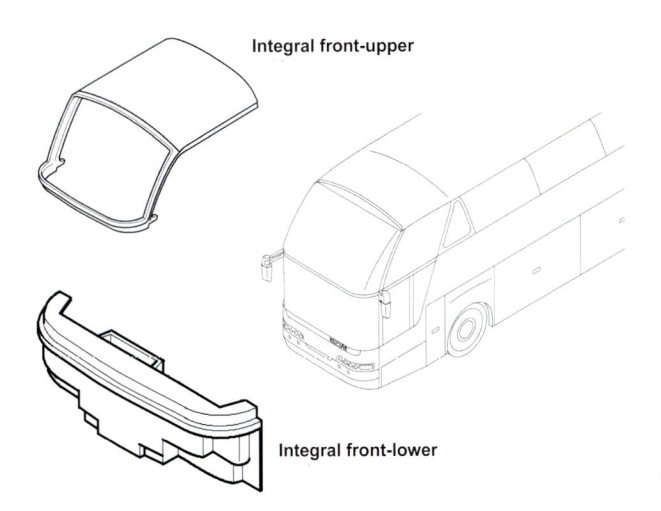

Figure 6-45 Integral nose in two sections

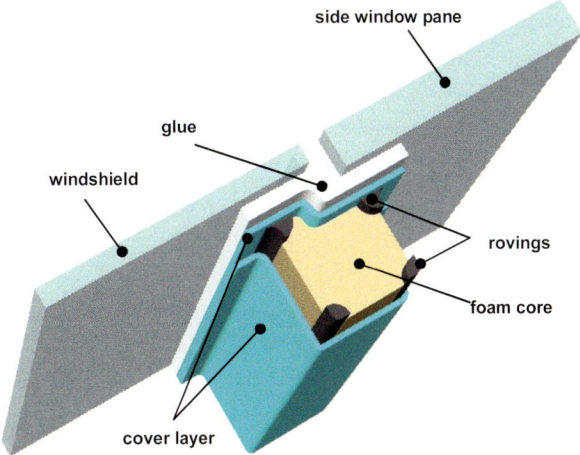

Figure 6-46 Section through the A pillar

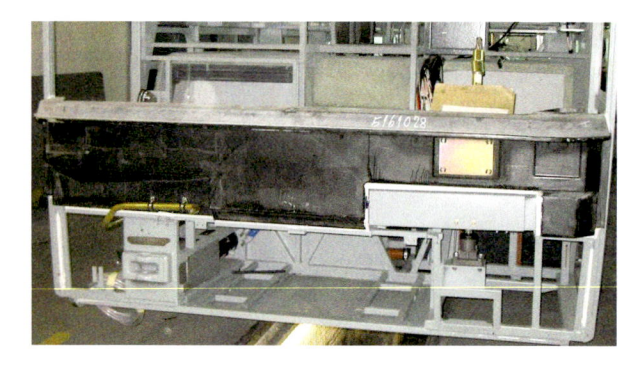

Figure 6-47 Lower part of the integral nose

device is used to position the lower part of the integral nose exactly and it is flexibly bonded to the top line of the nose section with a one-component polyurethane adhesive.

In contrast to the usual construction method whereby the fissured partition wall is made up of square tubing to which sheet metal panels were then manually fitted – a process requiring a great deal of skill – in this case it was possible to gain some major improvements. We were able to achieve full sealing against drafts. Production times per unit could be reduced significantly. The steel sheeting with all its tolerancing problems was completely eliminated.

6.1.4.3 Integral Rear End of the Starliner 1

As far as its surface area is concerned, the rear paneling is the largest three-dimensionally shaped fiber-composite-plastic component in the bus. Previously it was subdivided into several small components which resulted in higher tooling costs. More mountings had to be provided on the body shell and the individual components aligned exactly with each other and sealed. The body shell had to satisfy extremely strict requirements as regards dimensional conformity. The rear end panels used to be bonded onto the complex curvature of the body shell profiles and sheet steel paneling affixed to these on the inside by hand.

A decorative carpet was then glued onto the interior metal paneling. The supply lines for the air-conditioning unit and the in-ceiling electrical systems were routed between the outer skin and the inner metal

Figure 6-48 Integral rear-end of the Neoplan Starliner 1

paneling and the remaining cavities insulated. Fitting the metal paneling, particularly at the rear end corners, is a very complex and expensive process.

With the development of the integral rear we hoped to solve these problems. The new single-part component closes off the vehicle interior completely. There is absolutely no need for inner metal paneling in the rear end corners, since an FRP inner shell is used here instead. This is glued into position at a defined distance from the outer shell, thereby achieving a high level of component rigidity. The hollow space thus created is also used for routing various lines. The rear window frame is completely omitted, since the window itself can be mounted very well on the integral rear-end with its sandwich-structure reinforcement. Nor is any complex adjustment work needed, since the rear end has no more than one defined interface with the body shell on which it is mounted by means of a flexible bonded joint. We were also able to integrate the rear light supports which were previously mounted as separate units.

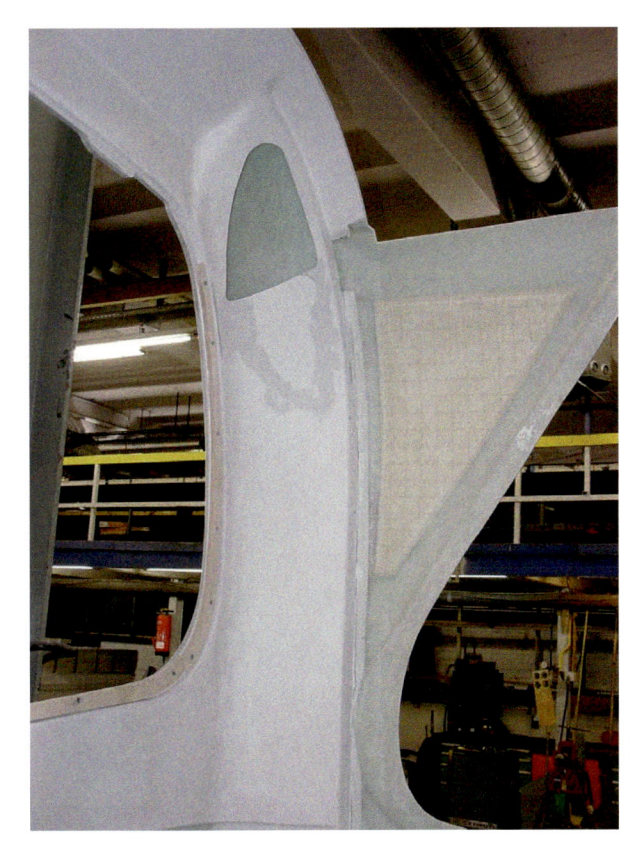

Figure 6-49 Inner shell of integral rear-end

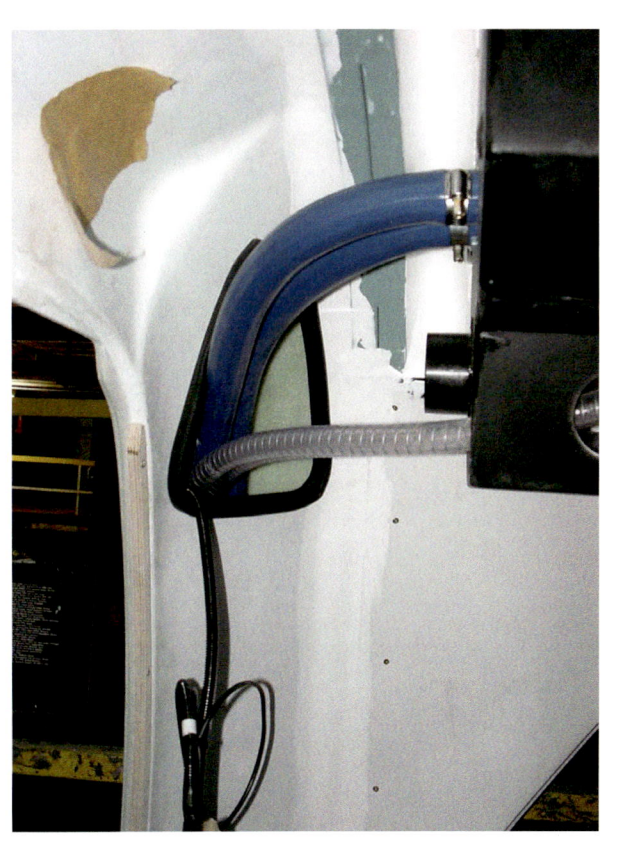

Figure 6-50 Cable glands in the integral rear-end

6.1.4.4 Experience with the Starliner 1

The integral components have also made a decisive contribution to making the design of the Starliner 1 a reality and they have also proved themselves in long-term service. To date, no complaints have been received with regard to the properties of these components, and the same is true of the bonded joints. During the course of development, more and more functions were integrated into the components – for example, air guidance was not possible without additional components (such as a separate air duct which was glued into the lower part of the nose section). This meant that a large number of assembly operations were simply transferred to the supplier. This made the costs for the component correspondingly higher. A cost comparison of the lower part of the integral nose between the fiber composite plastic design and the steel design was clearly in favor of the steel version. In the case of the upper part of the nose section, there could have been no better solution for the complex geometry of this component as regards comparative costs, even if the fiber composite plastic component is very expensive. Due its size and the integration of the inner shells with regard to dimensional stability, the integral rear-end is difficult to control fully. Its double-shelled design does however make it so rigid that this defect often cannot be corrected by adjustment work at the vehicle and the component lands in the rejects bin.

In the new development of the Starliner 2, attention was paid as early as the design phase that the manufacture of these components could be returned to conventional methods if need be. It was therefore an important criterion in the design of the upper windshield that the A pillars could also be bent to the required shape if they were made of steel. This meant that the forward roof section could once again take the form of a simple single-shell fiber-composite plastic component. The rear end, too, is no longer a self-supporting component. The quality of present-day welding equipment guarantees exact positioning of the paneling elements.

As regards body shell design, a large part of the rigidity and energy absorption required in the roll-over test was provided by the rear end. It would not have been economically possible to achieve this with a pure, self-supporting fiber composite plastic component with the installation space available.

6.1.4.5 The Nose Module of the Starliner 2

In the development of the Starliner 2, great attention was paid to optimizing the product sequence. At the time when the interior trim and fittings are installed, the body is pretty much already closed off. A lot of materials must therefore pass through the two doors into the interior of the vehicle. The instrument panel is installed at this time also. This means that it cannot be ruled out that door 1 is blocked by persons working on the vehicle and the entire flow of materials and other workers must use door 2.

The best solution for this appeared to be to keep the front end of the vehicle open for as long as possible and to leave fitting components such as the instrument panel, windscreen and front-end until the end of the production line. All of these components should be pre-assembled as a group and only then be fully connected to the body. For this to be possible, a completely new design was required for the front part of the bus.

Whereas installation space investigations were still carried out with a full-scale wooden model during development of the Starliner 1, the Starliner 2 has enjoyed the benefits of virtual packaging. Starting with the conventional nose section design following the conventional production sequence, all components impeding frontal access to the front end were removed in the CAD program.

These include the instrument panel with center console, refrigerator, instrument panel and air duct, the steering bracket with steering column and retainer, the partition wall, the windshield wiper system and the front cross-member.

The next step was to isolate all of the bodywork profiles on which all of these components are mounted. Here there were two different concepts for building the new integral front-end which was to perform the functions of interfacing, support, sealing and insulation.

Concept 1 envisaged using the steel profiles removed for the study as a base frame for the nose module. The partition wall and the wooden floor were to be glued to them, with the wooden floor serving as a support for the instrument panel, since it provided adequate rigidity. Since the module is bonded on and the ribbed structure should be changed as little as possible, the idea arose of dividing the rib tubes in the front-end floor area. The tubes in this area have a cross-section measuring $40 \times 40 \times 2$ mm. If they are divided into two profiles, the interface would already be defined. Using a $15 \times 40 \times 2$ mm profile for the module frame meant that a layer thickness of 5 mm was gained for the glueline. This step already created a relatively large bonding area in the form of a bearing face which, if the A pillars were included, could be further enlarged to a considerably extent as well. By giving the connection of the A pillar the form of a bracket on the module side, it was possible to restrict the two degrees of freedom still remaining to the module.

Concept 2 is a logical further development of what was already state-oft-the-art in the Starliner 1. On the basis of the steel structure, the load-bearing steel profiles and also the nose wall were merged into a self-supporting component made of fiber composite plastic. Here, in the area of the window bearing surface, it was necessary to compensate lower rigidity with respect to the steel nose cross-member by increasing the section modulus.

Figure 6-51 Bottleneck at the center door

6.1 Structures and Body Panels 239

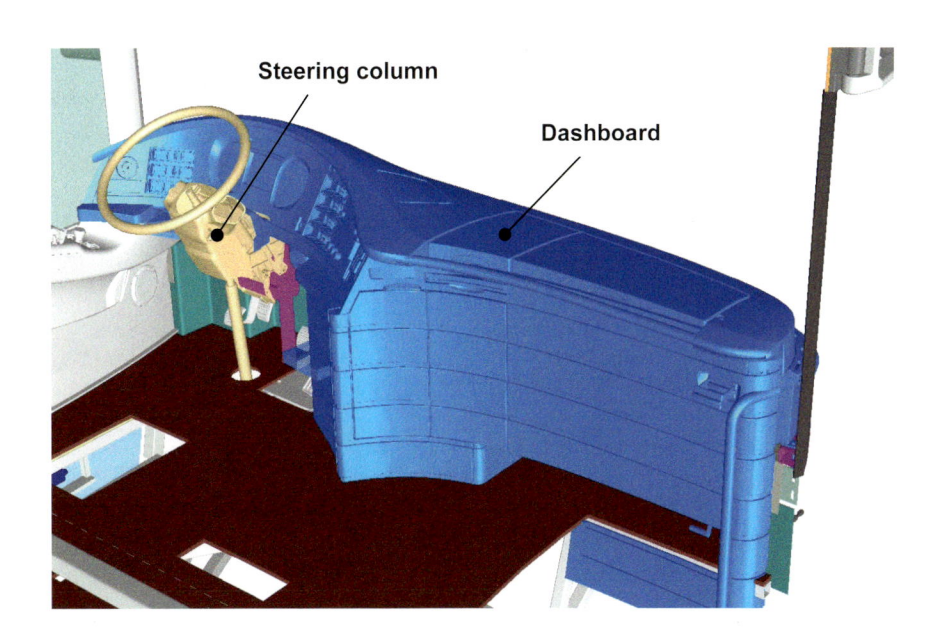

Figure 6-52 Front-end components, interior view

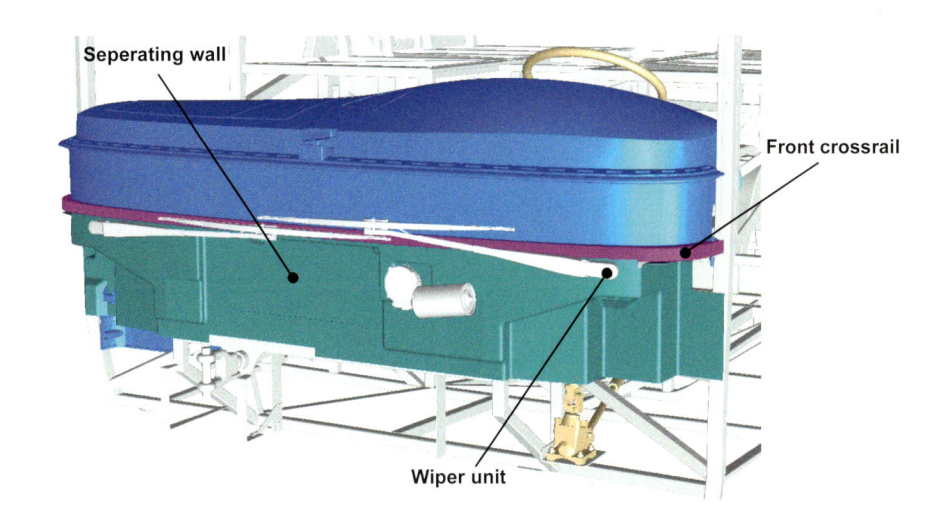

Figure 6-53 Front-end components, outside view

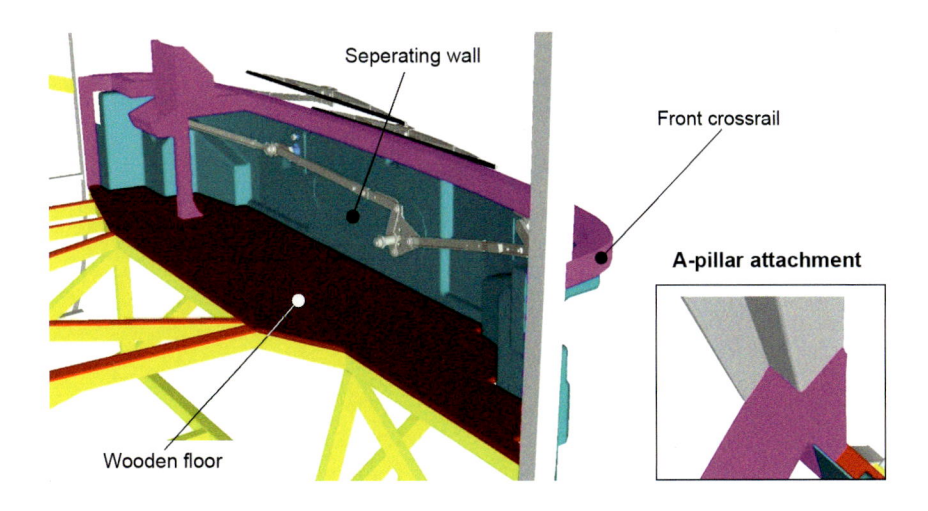

Figure 6-54 Integral nose – steel

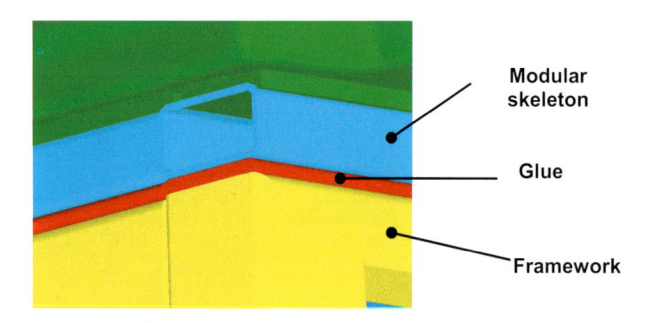

Figure 6-55 Interface with the body shell module

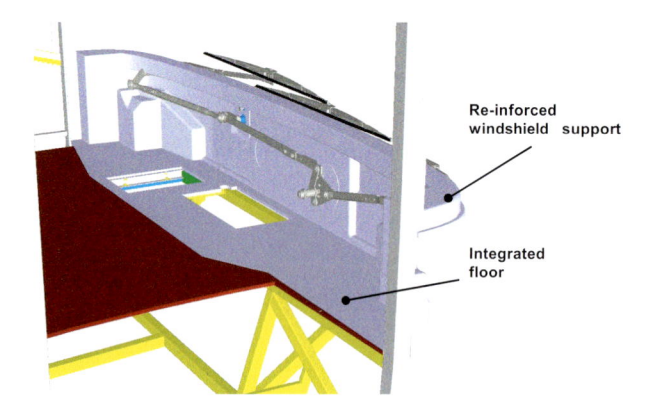

Figure 6-56 Integral nose – fiber composite plastic

The component is reinforced over its entire surface area by sandwiching a 10 mm thick layer of rigid PVC foam. In the vicinity of the window bearing area, its core thickness is 80 mm. It was possible to integrate the floor (which simultaneously serves as a bearing surface for the instrument panel) directly into the component. Here, great care was taken to ensure the design of the component was as simple as possible.

The connection between the module and the ribbed structure of the bus represents a very important part of production. Without it, a modular concept of this kind would not have been possible. There are problems with making the connection by means of screws, soldering or welding, nor is it even possible in some respects. Since a lot of experience and above all very good results had already been obtained with adhesive bonding, the decision was made to glue the nose module onto the body shell. The adhesive used must have a high resistance to shearing off and high elasticity as well, thereby allowing easy compensation of manufacturing tolerances. It must also cure quickly so as not to hold up the production sequence. A thickness of 5 mm was selected for the adhesive. This thickness gave rapid curing and maximum strength and no basic modifications needed to be made to the rib structure to prepare for this step. The tolerances of the components to be bonded together lay in the millimeter range and this was the reason why a flexible bond was not selected.

To be able to test the new production sequence while avoiding major financial risk, Concept 1 was favored initially since this concept only involved minor tooling costs. As soon as all optimization measures had been incorporated in the design of the nose module, there would be a changeover to a fiber-composite plastic component, provided the costs could be reduced to below those of the steel version. The first components were made as hand lay-up laminates and later there was a changeover to RTM or liquid resin press molding.

6.1.4.6 The Front-End Module of the Starliner 2

All of the front paneling parts of the Starliner 1 were made by hand lay-up, being subject to relatively large tolerances as are also the corresponding vehicle-side retainers, where the components are mounted. Due to this, actually fitting the components to the vehicle itself proved difficult in many cases and took a lot of time. At an early point in the development process for the Starliner 2, there was already some concern as to how the tolerance problems could be reduced.

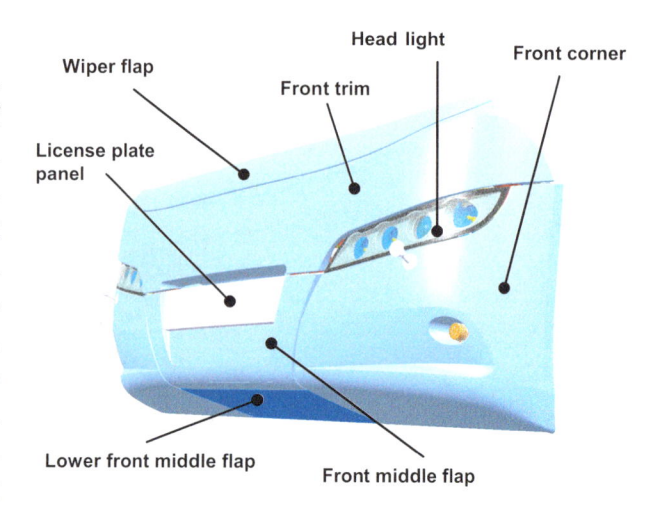

Figure 6-57 Front-end module

Although in the first Starliner series a spare wheel also had to be accommodated in the nose area, this was deliberately not done in the second generation for weight reasons. Today, in long-distance bus travel, flat tires are hardly ever changed by the driver himself. In most cases, the driver calls in the repair service. For customers who insist on a spare wheel, it is accommodated in the luggage storage compartment. This change meant that there was no need for a fold-up bumper. Only the headlights and the air-conditioned front box need to be accessible for changing bulbs or the air filters. The idea very quickly arose of preassembling the paneling elements on a separate frame and then mounting this as a complete unit on the body shell. The headlights can be accessed by jacking up the entire nose by just a few degrees.

In addition, special attention was paid to ensuring ease of access for repairs by splitting the front end into several components. The nose panel forms the central element of the front section. The mechanism for opening the windshield wiper panel is installed on the rear side of the nose panel.

This windshield wiper panel covers the windshield wiper system completely and opens automatically when the windshield wiper system is operated. This protects the wiper blades against the weather and even prevents the brushes of a vehicle washing facility from catching the wiper blades or arms and bending them.

The central nose flap can be folded upwards by 90° and detached. This provides optimum accessibility for maintenance work in the front box. In addition to the holder for the towing adapter, the radar sensor of the Adaptive Cruise Control (ACC) is found behind this flap. This results in special requirements for the material since it needs to be transparent as regards the radar beam.

The license plate panel is located on the central nose flap. The front fenders on the left and right accommodate the headlight arrays and blinkers. To obtain a better quality and also to improve the fitting precision of the nose panels we changed over from hand lay-up laminate to SMC.

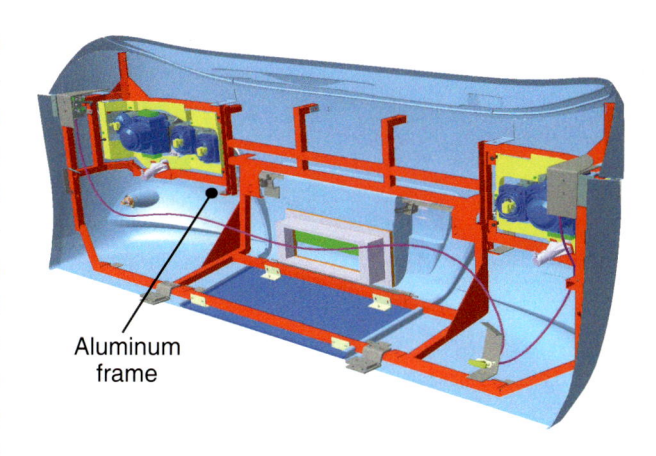

Figure 6-58 Inside the front-end module

Splitting the front end into several small components would have resulted in increased assembly and adjustment work compared with the first generation. In bus-building, too, more and more importance is attached to small gap dimensions. For this reason, the components are positioned exactly with respect to each other in an assembly device. Once aligned properly, they are fastened to an aluminum frame by screwed connections. The complete unit can now be screwed via two hinges to the body shell and a gauge is used to check dimensional accuracy. Compared with the previous assembly procedure, this represents a considerable time saving. Even the headlights which were previously connected directly to the nose panel are today mounted on the assembly frame.

Although Neoplan previously relied on typical separate spotlights in the form of self-contained type-approved headlights, for the second generation a custom headlight with a clear-glass lens was developed. In the past, the reflector served simultaneously as a headlight housing, which meant that, as far as the formal design was concerned, it was subject to the strict laws of optical engineering. Today, the designer has a great deal of creative freedom here, since reflection systems are increasingly yielding ground to projection systems. A headlight module was developed in collaboration with our suppliers at Hella. Standard projection systems could now be mounted on an individually styled panel by means of an adapter and the resulting unit closed off by a cover panel.

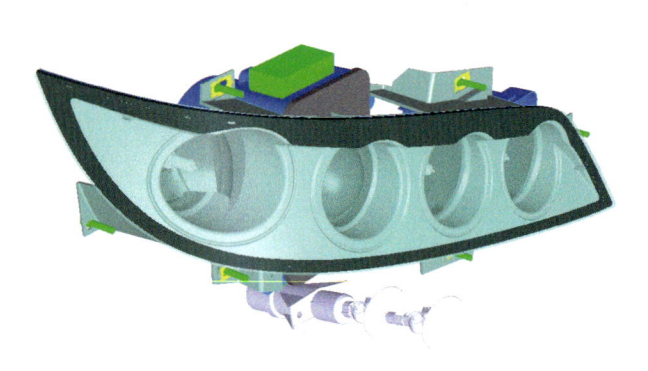

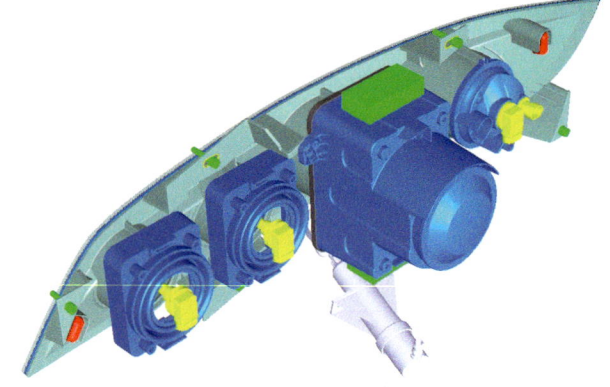

Figure 6-59 Headlight arrays

Although today, headlights typically have a cover disk made of polycarbonate, in this case the costs for the corresponding injection molding tool were much too high.

The planned annual production of 400 vehicles in which the headlights are to be installed means, assuming the service life of the series is about 8 years, a total of 3200 headlights on the left and right. Therefore, we went back to a glass cover disk which, although it had a higher unit price than a PC disk, nevertheless resulted in only minor tooling costs. The glass disk is flexibly bonded to the design panel.

With this concept it was possible to achieve markedly reduced tooling costs for a headlight, which contains not only a bi-xenon module but also a static curve light which is so far unique in bus-building.

6.1.4.7 Outlook

In the next few weeks volume production of the Starliner 2 will commence. The Starliner 1 will be phased out gradually and a number of applications for plastic will also disappear with it for which a great future was predicted eight years ago. It has not been possible to keep components of that kind within acceptable cost limits.

If modern touring busses are also to continue to be manufactured at a production facility in Germany, engineers will not only need to incorporate technical innovations in the vehicles, but they will increasingly need to concern themselves with seeing how they can manufacture the vehicles more effectively and to a higher quality. A step has been taken in this direction with the development of the nose and front-end module of the Starliner 2. Here, plastics and their processing continue to play an outstanding role and it is impossible to envisage bus-building without them.

6.1.5 Cab Body Panels and Parts: from Thermosets To Thermoplastics

Jacopo Corsi

6.1.5.1 Introduction

Iveco presents itself as a global company not only in the vehicular sector but as in fact involved with everything concerning the world of transport. For this reason the main themes Iveco focuses on are: safety, innovation, profitability and the environment.

The challenge is therefore to pursue excellence in developing products and processes tailored to customer needs, exploiting, of course, company know-how but also all of the expertise of parts suppliers, technology suppliers and raw materials manufacturers, without which playing a significant role in the present market scenario would be unthinkable.

With this premise, Iveco has pinpointed the innovative materials as one of the strategic items to be developed and applied in its products in order to improve performance in terms of cost and weight reduction, product quality, fuel economy and the integration of new functionalities.

A significant amount of plastic parts and panels (made by thermosetting or thermoplastic materials) is installed in heavy and medium truck cabs.

While intensive use in the truck cab interior is fairly similar to the car industry, as regards the external parts and panels there are some significant differences. These are mainly due to different vehicle purposes and dimensions but also to sales volumes.

Here, we will describe the trends that Iveco envisages for the next decade in terms of plastic materials, emphasizing the advantages and the implications related to appearance, design, the environment, structure and applications, in the move from thermosetting to thermoplastic materials. The planned trend is shown in qualitative terms in Figure 6-61.

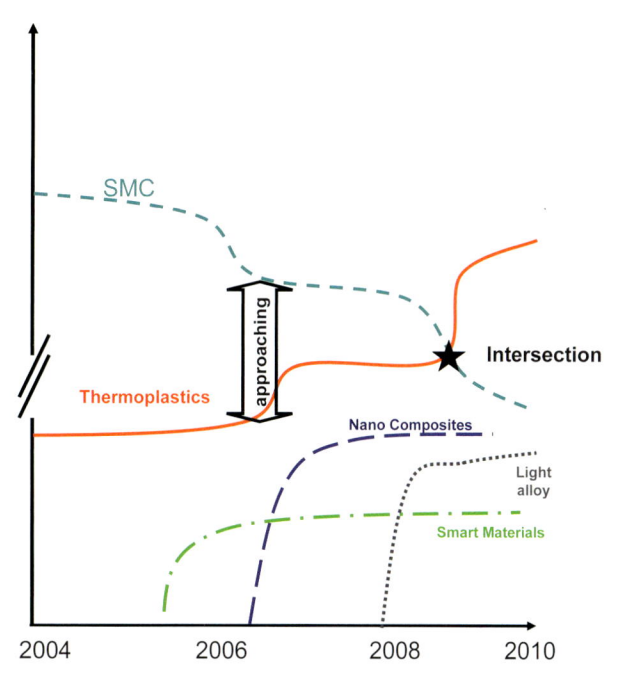

Figure 6-61 Planned trend for materials application

Figure 6-60 The Iveco product range

Thermoset components
- Roof shell and frame
- Frontal panel
- Bumper
- Steps
- Laeral extensions

Thermoplastic components
- Spoilers
- Sunvisor
- Side-deflectors
- Windshield side deflectors
- Bumper extension

Figure 6-62
Plastics on the exterior

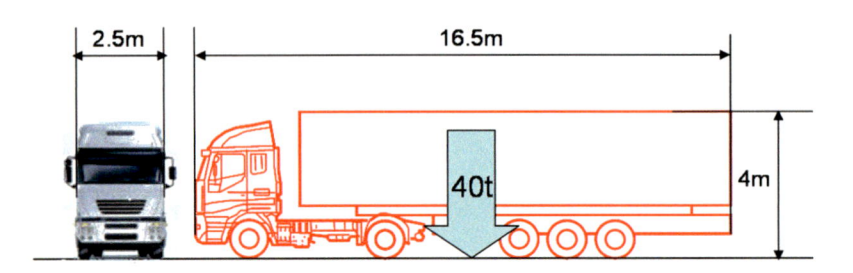

Figure 6-63 European regulations: weights and dimensions

Examples will be shown of components which have already "technologically migrated" from thermosets to thermoplastics. Also a comparison will be made between Iveco and other truck manufacturers in order to show different interpretations of plastic components on the basis of different materials and technologies.

6.1.5.2 Current Scenario for Truck Cabs

Weight

In the entire automotive sector, engineers typically consider achieving weight targets as one of the most critical factors in success. Indeed this parameter affects many other types of performance of the product. Fuel consumption, brake system effectiveness and payload are examples of parameters influenced by weight. Naturally, depending on the type of vehicle and its job, a weight reduction can bring about different advantages.

Moreover in recent years, all automotive producers have increased the amount of electronic components. The bottom line is that we can say that the global weight of a car or simply of a truck cab has increased due to the introduction of new functionalities. It is therefore clear that weight reduction and/or control is felt to be a major objective during product design and development.

Focusing more on trucks, we can evaluate to what extent weight reduction and/or control is significant or not.

In freight transportation today most trucks do not reach the maximum permitted weight since the payload is limited by the permitted volume and not by weight.

For this reason we can say that, even assuming a weight reduction, this improvement is rarely reflected in an increase in payload with the corresponding improvement in the economic efficiency of transportation. Obviously, weight reduction translates into lower fuel consumption, but as a rule of thumb, weight reduction of one ton in a typical European heavy truck leads to a fuel consumption reduction of about 0.5%.

Considering the Iveco Stralis AS cab, the amount of parts made of thermosetting materials (shared between SMC, BMC and RTM technologies) is about 100 kg.

The Stralis AS cab weighs about 1300 kg and therefore, even taking a 10% weight reduction into consideration, the corresponding benefit is only barely perceivable. To make this more clear, aerodynamic improvements in terms of CX deliver a more significant fuel consumption reduction ($\Delta CX = -5\%$ corresponds to Δconsumption $= -1\%$).

In other words, our strategy of moving from thermosetting to thermoplastic materials is not mainly driven by the weight reduction target, although this is a useful consequence.

This reasoning only focuses on the product and final customer, but if we move our attention to the fabrication process for a truck cab, perhaps we can then also see some other advantages arising from weight reduction.

As we have already mentioned, a typical truck cab has many large exterior parts which need to be included in the assembly process. Somehow either specific handling devices have to be installed to move these components or more workers are required to fit them. It is therefore clear that production cost efficiency can be obtained by reducing the weight of the parts to be mounted on the vehicle.

Cost

During the process for designing an automotive subsystem, there always comes the moment when decisions must be made about the materials and technologies to be used in its realization. In many cases, technical performance and quality to the customer are equivalent and the final decision is driven by costs.

	Stralis AS	Stralis AT	Eurocargo
Saving/Piece	31%	Directly Thermoplastic	50%
Investment payback time (yrs)	1.1		0.3

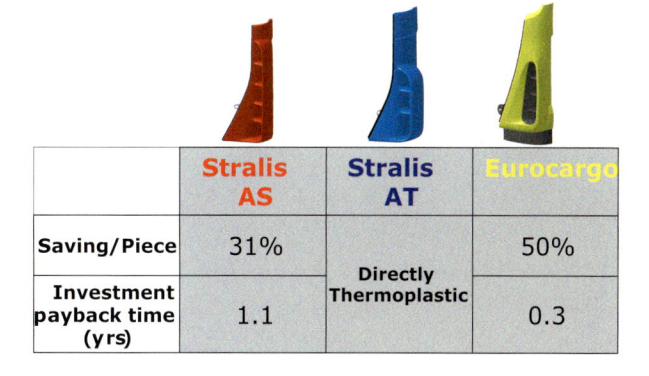

Figure 6-64 Medium and heavy truck cabs: side deflectors

In this scenario, Iveco has registered impressive results, in terms of cost reduction, by moving from thermosets to thermoplastics with the corresponding technologies. One typical application where we have obtained high efficiency is the re-engineering of the cab side-deflectors (see Figure 6-64).

For the Stralis AS vehicle, the original component was made of UP-GF by BMC technology and has been successively transformed into a new component with the same external shape but with the internal panel made of PP-EPDM and the external one of ABS.

The thermoplastic solution was directly applied also to the Stralis AT/AD, and now we are doing the same with the Eurocargo as well. The table below gives an idea of the savings achieved with this application.

Looking at these results, the potential of such a change in materials and technology is evident. Nevertheless, we can assume this as a "best case", since the shape, structure and dimensions were favorable. In other situ-

Table 6-4 Cost reductions resulting from re-engineering the side-deflectors

	STRALIS AS	STRALIS AT/AD	Eurocargo
Saving/piece (%)	31	Directly thermoplastics	50
Investment Payback Time (yrs)*	1.1		0.3

* including prototypes

ations it may be difficult to obtain the same structural behavior with one-to-one correspondence in terms of subsystem components without a consequent increase in overall investment and costs up to the point of exceeding the budget.

Moreover, in some other cases, dimensions become critical from the standpoint of feasibility. If we look at components such as the front panel or the bumpers, not the material but the technology (injection molding, for example) requires a compression force that is not readily available.

6.1.5.3 Plastics Product Performance

If we consider the exterior of truck cab, we can see several parts made of plastic material (see Figure 6-62). All of them must fulfill specific requirements that can be grouped into four main categories depending on the particular application. These are:

- Appearance   (visual properties)
- Styling   (functional properties)
- Environment   (life-cycle properties)
- Structure   (performance properties)

In the past, due to the large dimensions involved, requirements in terms of stiffness and strength almost always resulted in the use of thermosetting materials in conjunction with SMC/BMC technology. With some minor differences, all European truck manufacturers have taken this road. Nowadays, thermoplastic materials in conjunction with a proper design approach and fabrication technology are able to meet the structural gap when compared with thermosetting materials and can also bring other advantages and performance not available with thermosets.

Appearance

In the design of plastic products, especially those involved in visual quality evaluation, appearance is one of the performance factors that cannot be ignored. Naturally the exterior of truck cabs is not subjected to the same style evaluation as cars are. This is because customers for cabs are rarely driven by an emotional evaluation of the product, but are mainly interested in other properties of the product, such as productivity and suitability for specific tasks. Nevertheless, plastic components play a significant role in terms of the volume of a truck cab. For this reason, the esthetic properties of exterior panels, although not directly related to the major product performance factors required by customers, do nevertheless represent a quality indicator.

Class A surfaces, environmental resistance (UV radiation, acid rain, pigeon attack), scratch resistance, stone impact resistance, wash resistance, dirt repellence are all important aspects to be considered. Moreover, as in the car industry, customers are continuously refining their tastes to include esthetic properties and make comparisons between products from different brands, feeding a judgment driven by "perceived quality".

On the other hand, truck manufacturers always demand materials with improved qualities, not only for the final product but also for product development and manufacturing processes. In this context, painting must surely be a key process for costs and final product quality. Typically when dealing with painted plastic components, technicians have to face different problems in order to achieve the best results containing the cost of the product:

- Off-line painting vs. in-line or on-line painting
- Preliminary treatments (that is, primer application, flaming, plasma, corona, UV, fluorination, sanding, etc.)
- Porosity, blistering, dust or dirt inclusions, pops, runs, sags, fish-eyes and more
- Color-matching

The opportunity to paint the part directly during the cab manufacturing process represents a significant cost reduction and a guarantee for the final result.

Unfortunately for truck cabs, it is not easy to find exterior panels well suited to such a process – sometimes due to the large dimensions involved, or since production volumes rarely justify applications of proper materials for in-line painting. Also, the mechanical

6.1 Structures and Body Panels

properties of the latter, apart from thermal properties, are not qualified for the final task.

Painting is moreover also a matter of logistics; Iveco SMC components are produced in Brescia and then delivered unassembled to two other plants (Ulm and Madrid), depending on the specific part and specific vehicle. In this situation we have to manage the correct sequence in three different plants and also avoid large stocks building up. The resulting logistics are shown in Figure 6-65.

Although UP-GF as a thermosetting material is well suited for in-line painting, it mainly suffers from the long preparation time: mold flash removal, sanding and primer application. Moreover, in some cases fiber orientation determines porosity on the painted surface with consequent quality problems.

On the other hand, thermoplastics offer a wide range of materials for both in- and on-line painting and also off-line painting as well.

Table 6-5 shows different project solutions we can use.

The motivation that pushes us towards an intensive use of thermoplastic materials is the opportunity of finding the proper trade-off between market demands and the specific needs of the manufacturers.

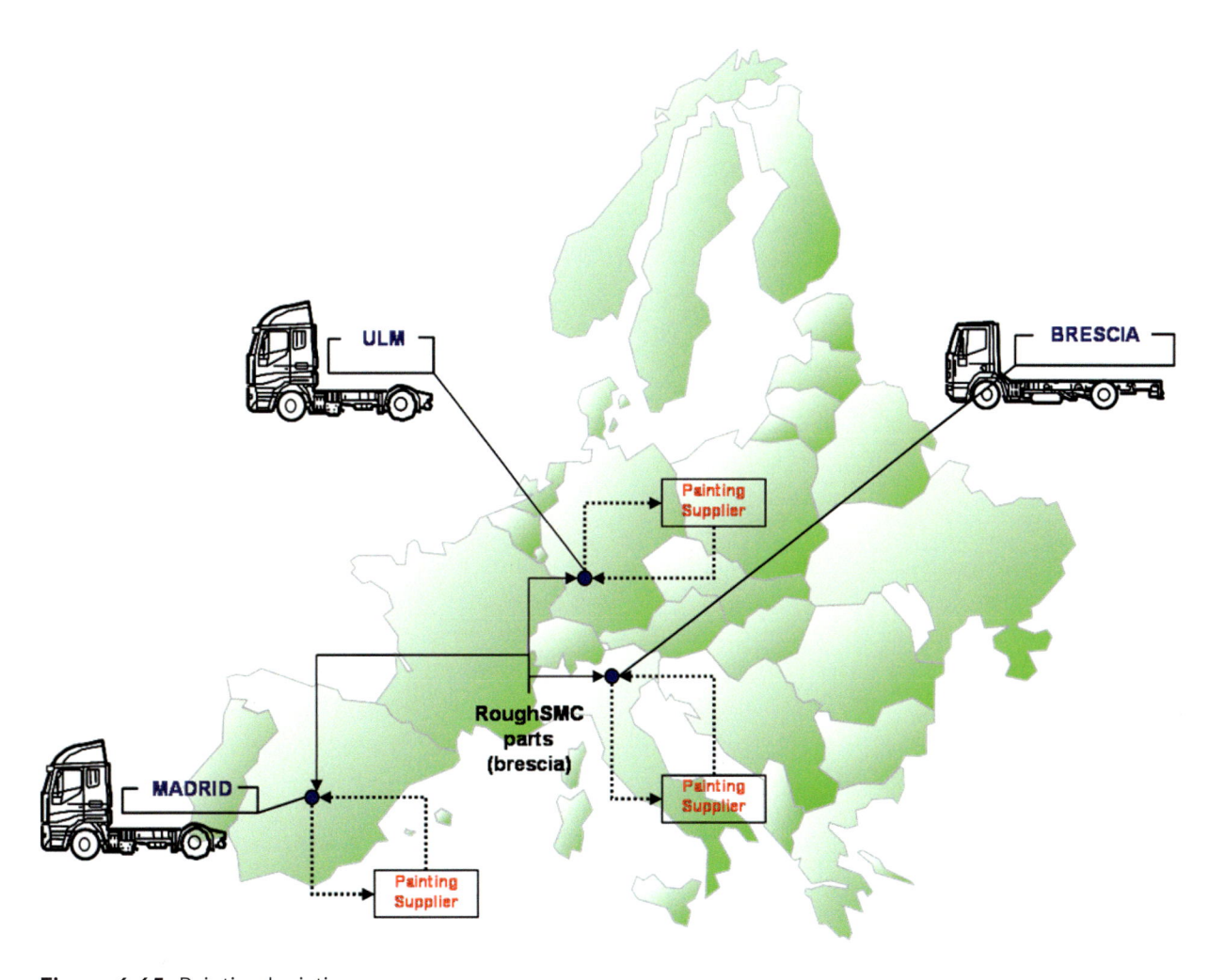

Figure 6-65 Painting logistics

Table 6-5 Materials vs. painting process

Plastics and Technologies	Materials	Surface Finishing
Thermoplastics	ABS ABS+PA ABS+PC ASA+PC PC+PBT PP+EPDM	Off-line painting
	PPE+PA	In/on-line painting
Charged Thermoplastics	ABS+PA ABS+PC ASA+PC PC+PBT PP+EPDM	Off-line painting
Reinforced Thermoplastics	ABS+PC-GF PBT+PC-GF PET+ASA-GF PC+PET	Off-line painting
	ABS+PA-GF PA6-GF	In/on-line painting
Molding in colors (MID) thermoplastics	ABS+PA ABS+PC ASA ASA+PC PC+PBT PP+EPDM	Possible clear coat application
Reinforced nano composites thermoplastics	ABS+PA PA6 PP	Off-line painting

Styling

In the design of cab parts and panels made of plastic material, one of the first problems to face is naturally the shape. In fact, when an exterior part is styled, the designers start to analyze feasibility in terms of production technologies, vehicle interfaces and integrated functions. In this context, decisions regarding production technology are crucial for the development of the component and are based on the substantial differences between thermosets and thermoplastics.

SMC and BMC technologies are often compared to steel in applications such as body panels which exhibit performance levels equivalent to steel or in some cases

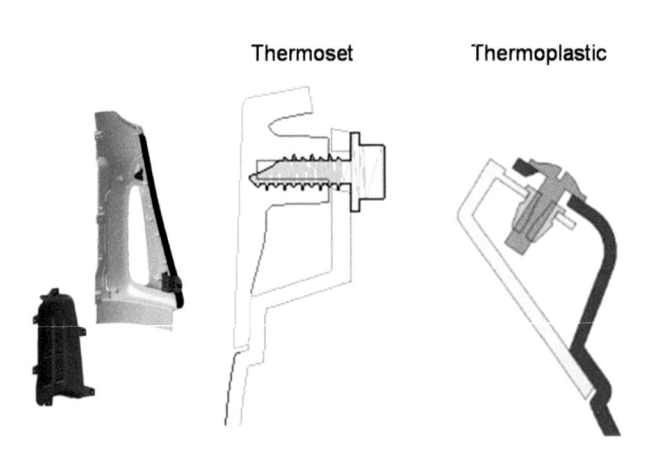

Figure 6-66 Solutions for mounting side-deflectors

even better. In recent years, due to injection molding technology, thermoplastics have also become an alternative to the former. The major advantages, as we have already shown, are in cost and weight reduction, but in many ways, shape complexity is a further benefit. On the other hand, different approaches to body interfaces and mountings are required and also the thermal effect must be taken in account in product design. This means more complex tooling or added parts required for mounting. Figure 6-66 shows the different solutions we had to adopt when changing over from thermoset to thermoplastic side-deflectors.

Thermosetting materials used for body panels have very low shrinkage, thereby enabling attachment directly in the material. Injected thermoplastics suffer from thickness changes exhibiting unaesthetic shrinkage; therefore a more complex solution is required.

The most effective advantage arising from the use of thermoplastics as against thermosets is the broad choice of blends that can be precisely tailored to the needs of the project. Moreover, in many cases, even with tooling already realized, it is possible to change the material to improve some specific performance features that are not satisfactory (thermal elongation, stiffness, dimensional stability, etc.).

Another significant aspect – here Iveco has experimented with Stralis A-pillar side deflectors – is the bi-injection of thermoplastic PP-GF and thermoplastic elastomer TPE (see Figure 6-67).

Figure 6-67 A-pillar deflector as example of bi-injection

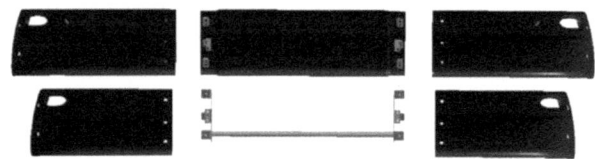

Figure 6-68 Three-piece sun visors made by injection molding

Injection molding + decomposition + modularity

One of the problems that can arise from injection molding technology involves the dimensions of the component. Indeed, although problems such as rigidity can be resolved by means of proper design, this high-pressure technology becomes critical when compression forces exceed 3500 tons with regard to the availability of required equipment.

In addition, viscosity can bring further problems. For example, if we look at truck cab sun-visors, it is not easy to produce one-piece components by injection molding.

In such circumstances, an intelligent solution would be to break down one part into an assembly of smaller parts or use alternative production technologies. Figure 6-68 shows a brand-new sun visor realized in three pieces of PMMA material for the Stralis AS/AT with induced modularity and structure standardization.

In the following we will show how thermoforming is a good alternative for large components such as aerodynamic spoilers.

Environment

Today more than ever, suppliers are requested to develop products and processes that are environmentally compatible. The entire life-cycle of a product must be managed from its manufacture and use through to its recycling or disposal.

The current EU regulative instrument 2000/53/EC which is concerned with the recycling of M1 and N1 class vehicles clearly defines the following concepts both qualitatively and quantitatively:

- Prevention
- Treatment
- Re-use
- Recycling
- Recovery
- Disposal

Although the legislation is limited to the standpoint of vehicle types and materials, it is clear that the issue is becoming more important. For this reason, life-cycle assessment will become a systematic approach in the development of products and processes so that materials innovation and the entire life-cycle cost evaluation must be considered strategic items in the near future. Moreover, energy consumption and related emissions must be taken into account in product life-cycle analysis.

In the light of these considerations, even in the field of plastic material applications, a comparison is needed between thermosets and thermoplastics (and steel should also be included).

When a thermoset composite is recycled, invariably there is a grinding process to break the composites into fragments. These fragments may then be used directly as a filler or reinforcement fiber in a new component. Alternatively, they may be thermally or chemically treated to produce energy or a base monomer.

Current methods of recycling thermoset composites include: the shredding or regrinding process which has concentrated on compound composites such as SMCs or BMCs. This is largely because these materials have wide applications in the automotive industries that have stringent recycling targets for vehicles at the end of their lives. The regrinding process is generally used for uncontaminated composites from trim and waste created during the manufacturing process. The shredded materials undergo different degrees of shredding and several more metal separation processes. These materials can then be further treated to separate out the fibers and resin or used directly as filler in new composite materials. The finer the material is ground, the better the finish that can be obtained. Compression and injection molding of new components using recycled filler often requires simple molds and high pressures.

Thermoplastics composites have some advantages for recycling especially since they can be reshaped upon heating. In addition, the approaches used for recycling thermoset composites also apply to thermoplastic composites. As for thermoset composites, the composite is generally ground to smaller fragments for a subsequent recycling process. Thermoplastic composites also allow reprocessing. The reground fragments of composites can be reground with new thermoplastic material to produce pellets for various molding procedures. These processes can include injection molding of parts or compression molding. The mechanical properties of these reprocessed materials should be evaluated and not assumed to be the same as the new material.

Structure

Migrating from thermoset to thermoplastic materials means that structural performance has to be taken into account. The main approaches are:

- Designing for stiffness
- Designing for strength

In the first case, we have to deal with Young's modulus of the material in order to limit large deformation of the component, while in the second attention is focused on loads that can become critical for the job the component is to perform.

Without going into the details of these design approaches, which typically have to be managed concurrently, it is interesting to underline some of the different behaviors and impacts between using thermosets rather than thermoplastics.

For example, in the case of panels, ribs often have to be incorporated to obtain improved rigidity in the part. In such situations, this is easy to obtain with SMC/BMC parts without any kind of collateral effects, but with thermoplastics this is not the case: sink marks on the sight side cannot be avoided and the resulting unaesthetic grille is unacceptable.

Furthermore, to obtain proper modulus values, it is often necessary to introduce glass-fiber reinforcement but, whatever the case, compatibility with painting requirements must be evaluated.

Table 6-6 shows possible materials for structural applications.

Table 6-6 Examples of structural thermoplastics

Plastics and Technologies	Materials	Surface Finishing
Short fiber reinforced thermoplastics	PA6-GF PA66-GF PBT-GF PET-GF PP-GF	embossing/painting/IMD
Long fiber reinforced thermoplastics	PA66-LGF PBT-LGF PET-LGF PP-LGF	embossing/painting/IMD
Hybrids (Thermoplastic + Steel)	PA6-GF PBT-GF PET-GF PP-GF	embossing/painting/IMD
	PA66-LGF PBT-LGF PET-LGF PP-LGF	embossing/painting/IMD

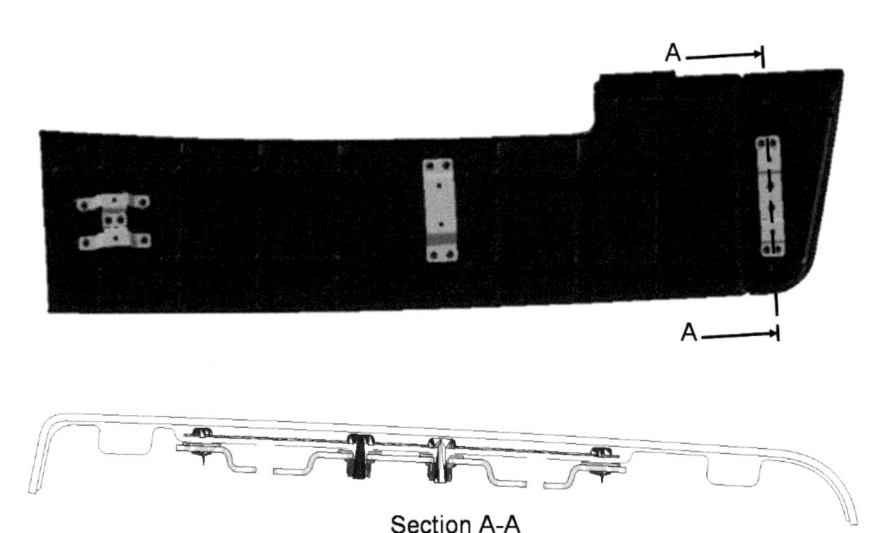

Figure 6-69 Stralis spoilers made of a PC+ABS double shell

Section A-A

If esthetic risk is present due to glass-fiber outcroppings, an acceptable solution could be to split the panel into two shells, the external one with an esthetic function while the internal one performs a structural function. This kind of approach has been used in the Stralis range with the aerodynamic spoilers. The material chosen for both external and internal panels is a PC+ABS blend.

Since these components have very large dimensions, thermoforming technology was used. Figure 6-69 shows the panel and a typical section. No weight increase has been registered in comparison with SMC components and the investment is also comparable.

Another critical aspect that may occur when using thermoplastics for body-panels is thermal loads. In this case, depending on the component and the material, different solutions can be adopted. If the material has low thermal expansion, sometimes there are no specific tricks available, in other cases proper interfaces must be applied that leave the degree of freedom for the thermal deformation. In the car industry examples exist of such systems that have been especially designed and patented for in-line painting of thermoplastics.

So far we have considered structural performance in the competition between thermoplastic and thermosetting materials where the use of thermoplastics requires the use of tailored solutions. Nevertheless there are also advantages that come "free of charge", putting thermoplastic materials one step ahead. Impact, shear stress and strain, failure and damage mode, are all properties typical of plastic behavior which are not only a purely structural advantage but also somehow translate directly into another major item, namely, safety.

In fact, if we examine accidents between vehicles and pedestrians, it is well known that the use of plastic can significantly reduce the severity of the occurring injuries. Additionally, legislation is working towards obtaining better performance from the design of vehicles and in the future the requirements will also become more restrictive.

Table 6-7 shows a ranking relating to the performance properties already discussed.

6.1.5.4 Benchmarking

Now we can present a comparison between the main truck manufacturers in the European market. It is possible to see that each one interprets the materials and the technologies to be applied for cab panels and parts in its own way. Naturally the reasons for these approaches to each sub-system may depend on the specific needs and internal strategy for the individual product.

Table 6-7 Comparative structural ranking of materials

	UP-GF	Short fiber reinforced thermoplastics	Long fiber reinforced thermoplastics	Hybrids (Thermoplastic + Steel)
Light Weight	5	7	7	8
Stiffness	8	5	6	7
Strength	6	8	6	9
Impact behaviour	5	7	6	8
Dimensional stability	8	6	7	7
Thermal expansion (CLTE)	8	6	6	7
Heat performance	8	7	7	7
Surface finish	5	8	7	8
Integration	4	9	8	8
Design freedom	5	9	8	8
Damage resistance	3	8	9	8
Recycling	4	7	7	6

1	2	3	4	5	6	7	8	9	10
Terrible			Marginal			Good			Excellent

Table 6-8 Comparison among truck manufacturers

	A-pillar	Air deflector	Front panel	Door extension	Side wall extension	Fender	Bumper	SPOILER AEROD
IVECO • STRALIS AS	PP-GF	PP-MD ABS+PC	UP-GF	UP-GF	UP-GF	UP-GF	UP-GF	ABS+PC
DAF • XF	STEEL	UP-GF ABS+ASA	UP-GF	ABS+ASA	UP-GF ABS+ASA	UP-GF ABS+ASA	STEEL ABS+ASA	UP-GF
MAN • TGA	UP-GF	UP-GF	UP-GF	UP-GF	UP-GF	UP-GF	UP-GF	UP-GF
MERCEDES • ACTROS	PPE	PPE+PA	UP-GF PA6-GF	UP-GF	UP-GF	– PC+PBT	UP-GF	FILM PUR-GF
RENAULT • PREMIUM	PP-MD	UP-GF	UP-GF	–	UP-GF	UP-GF	UP-GF	UP-GF
SCANIA • SERIE 4	STEEL	PP PC-PBT UP-GF	STEEL	PC+PBT	PC+PBT	PC+PBT	PC+PBT	UP-GF
VOLVE • FH	–	UP-GF STEEL	UP-GF PC+PBT	PC+PBT	PC+PBT	PC+PBT	UP-GF	UP-GF

6.1.5.5 Conclusions

We have presented an overview of the application of plastic materials in Iveco medium and heavy truck cabs. The focus has been on the advantages of using thermoplastic materials in comparison with thermosets. Discussion centered on four main categories: appearance, design, environment and structure. Many examples of parts and panels developed with materials previously used were considered. Iveco has planned the course it will pursue in terms of innovative materials to be applied throughout the product range. These items will be developed over the next few years in order to improve safety, productivity and environmental compatibility.

6.2 Front Modules, Crash Elements, Safety Concepts

6.2.1 Development of a Thermoplastic Lower Bumper Stiffener for Pedestrian Protection

STEFFEN FRIK

During recent years, the development of new vehicles has had to cope with ever new requirements. They arise from new regulations to ensure occupant and pedestrian safety and also from insurance classification tests (e.g., RCAR, GDV) and in-house requirements. Parallel to this, there is tremendous pressure to reduce both development time and the number of physical prototypes.

Pedestrian protection requirements have been mandatory for vehicle type approval since October 2005 and have been part of consumer assessment tests (EuroNCAP) for several years. In order to fulfill these new requirements, significant structural changes in the vehicle front-end are necessary.

An important additional component required for pedestrian lower leg protection is the so-called "lower bumper stiffener". The intensive application of numerical simulation techniques was the key enabler behind the development of this component within the available timeframe. As the main concept decisions were based on simulation results, there was an urgent need for highly accurate simulation results. In particular, accurate modeling of the material properties, including for high strain rates, and capturing the manufacturing effects, has played an important role.

In order to predict the material properties for highly dynamic impacts with the required level of detail, a special user-defined material model was developed by BASF.

During the development of the component, many different and partly conflicting boundary conditions had to be considered. Besides the primary pedestrian protection requirements, the concept also had to fulfill package and weight limitations as well as other full vehicle load cases. The component and the manufacturing process were designed so that the fiber orientation resulting from the mold injection process follows the main loads, in order to obtain the maximum benefit of the potential material strength.

To enable controlled fracture of the component during the insurance classification test, the material model must also be capable of capturing material rupture. It was thus possible to develop a component that would withstand the loading during pedestrian leg impact, but would break at the much higher loads during the insurance classification test. The aim is to keep the front rail and other expensive components undeformed.

6.2.1.1 Introduction

The pedestrian protection requirements led to new challenges for the design of vehicle front ends. They have a significant impact on vehicle styling, packaging as well as on design and, therefore, on the complete vehicle development.

Pedestrian protection tests are performed with a lower legform, an upper legform and head impactors. Lower legform and head impactors are free-flying, whereas the upper legform is guided until the impact with the vehicle front. In order to achieve the required performance targets, the vehicle structure must be designed so that it absorbs the kinetic energy of the impactors and provides sufficient clearance from all harder components.

As well as these pedestrian requirements, there are many other load cases which must be fulfilled, such as high-speed frontal impact, insurance classification, dent and polishing resistance as well as durability.

During recent years, the vehicle development process has changed dramatically. In contrast to the past, performance assessments during the concept and development phase are mostly based on simulation results. This means that all major concept decisions rely mainly on the accuracy of the simulation techniques applied. Physical prototypes are used in later development stages to confirm the design. Of course, this approach is only feasible if the simulation engineers are able to predict the structural performance for all relevant load cases with the necessary level of accuracy. Otherwise, high modification costs and timing delays are likely to occur.

Here, we describe a new integrative approach used in developing a lower bumper stiffener made of thermoplastic that improves the pedestrian protection performance of a new vehicle. It covers special numerical prerequisites, the detailed description of the relevant material characteristics, and describes implementation in the vehicle development process, starting with the concept phase and continuing through performance validation.

6.2.1.2 Pedestrian Protection: Regulatory Requirements, EuroNCAP

Since October 2005 pedestrian protection requirements have become part of the type approval regulations [1].

In the European Parliament Directive 2003/102/EG, the motivation for the pedestrian legal requirements is set out as follows:

(1) In order to reduce the number of road accident casualties in the Community, it is necessary to introduce measures so as to improve the protection of pedestrian and other vulnerable road users before and in the event of a collision with the front of a motor vehicle.

(4) Pedestrian protection objectives can be achieved by a combination of active and passive safety measures; the recommendations by the European Enhanced Vehicle-Safety Committee (EEVC) of June 1999 are the subject of a wide consensus in this area; those recommendations propose performance requirements for the frontal structures of certain categories of motor vehicles to reduce their aggressiveness; this Directive presents tests and limits values based on the EEVC recommendations.

In addition to these, much more stringent requirements are expected to become effective starting September 2010.

For some years, EuroNCAP has performed consumer tests assessing the pedestrian protection performance of new vehicles [2]. In comparison to the regulatory tests, more severe requirements must be fulfilled in order to achieve the maximum EuroNCAP point score (see Figure 6-70).

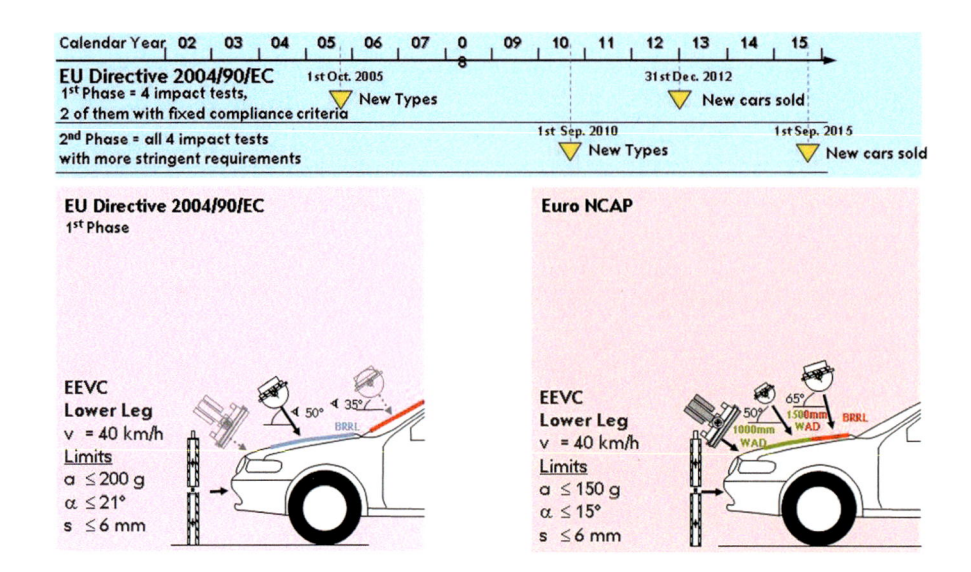

Figure 6-70 Regulatory and EuroNCAP tests

6.2 Front Modules, Crash Elements, Safety Concepts

In order to fulfill the requirements stated in European regulations 2003/102/EG and the even more stringent ones as applied in consumer tests, additional measures need to be taken with the vehicle front structure. Usually, additional parts such as the lower bumper stiffener, which will be described in detail, are required to achieve the necessary performance.

As the lower bumper stiffener does not affect the other pedestrian protection requirements, this paper will only focus on the lower leg impact test.

Legform Impactor

The legform impactor consists of several elements. Two metal tubes, connected by ligaments that represent the bending properties of a human knee, form the core of the impactor. In order to capture the shear properties, a spring-damper element is also implemented in the knee area. The stiffness of the ligaments must be well-defined, since it determines the performance of the impactor. Both metal tubes are surrounded by a 25 mm thick layer of Confor foam that is finally covered with neoprene.

During impact, the acceleration in the upper part of the legform, the bending angle and the shear displacement in the ligaments are recorded. The lower legform is shown in Figure 6-71.

As specified in the regulations, the impactor is propelled at a velocity of 40 kph against the vehicle front. Shortly before the impact, the legform is allowed to travel without guidance.

For the given vehicle, the legform knee impacts the front absorber system for the insurance classification tests. During the leg impact, acceleration, bending angle and shear displacement are recorded. None of the maximum values is allowed to exceed the limits described in 2003/102/EG.

6.2.1.3 Legform Impact Test – Overall Concept

The performance of the vehicle front-end during leg impact is mainly determined by two components. In order to enable deformation in the longitudinal direction, EPP foam parts are located in front of the bumper beam. Additionally, a lower bumper stiffener is

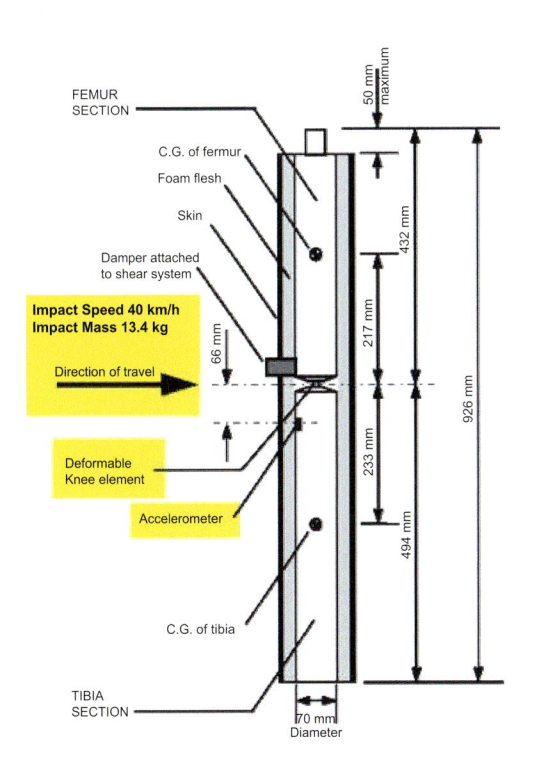

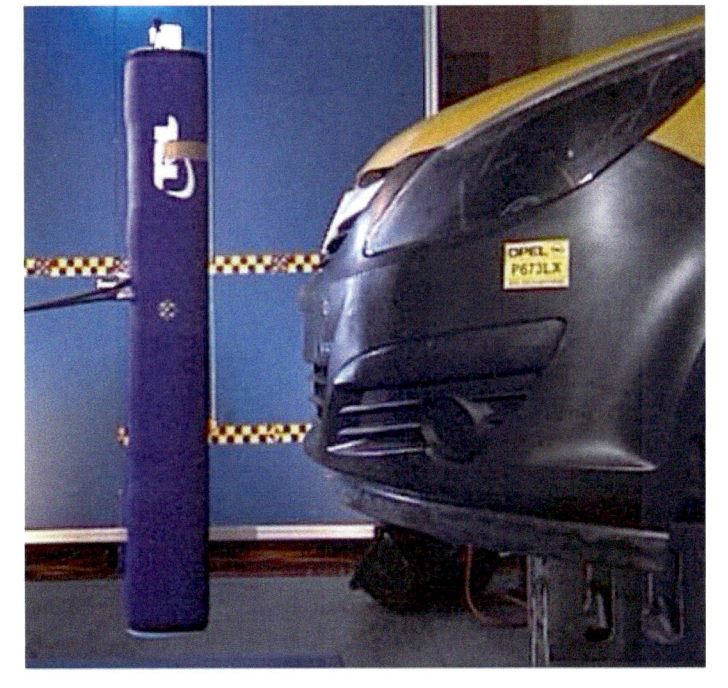

Figure 6-71 Lower legform impactor structure

introduced in the lower area of the bumper fascia. This latter part is intended to support the legform impactor, in order to reduce the bending angle of the knee. Simultaneously, the acceleration of the lower leg must be limited. This means that the mechanical properties of the lower bumper stiffener must be precisely balanced against the deformation properties of the foam absorber (see Figure 6-72).

In order to provide the required functionality for other vehicle load cases, the pedestrian protection concept must be as efficient as possible to cope with the limited package space. The entire available deformation space must therefore be utilized as intelligently as possible. This is a considerable challenge, especially for vehicles with short front overhangs.

w/o Lower bumper stiffener

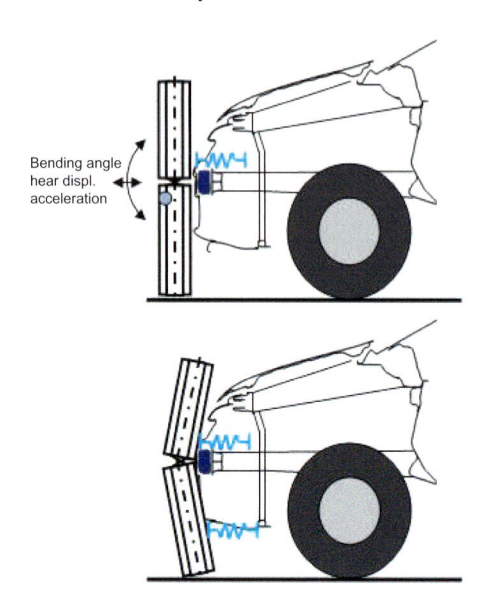

High load in knee area of the impactor

with Lower bumper stiffener

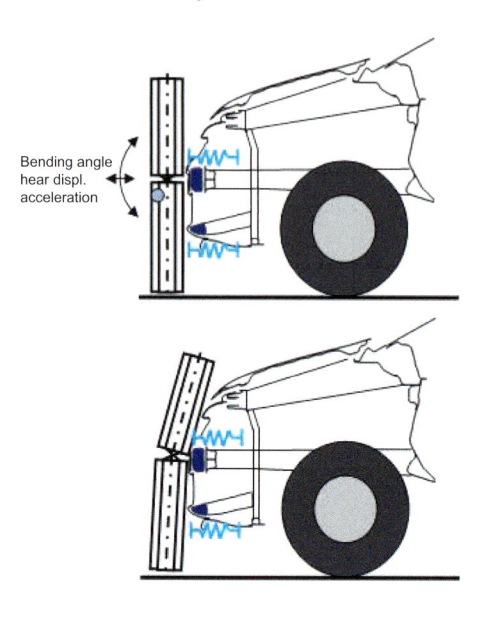

Lower bumper stiffnener supports rotation over the front of the vehicle

Figure 6-72 Leg impact kinematics

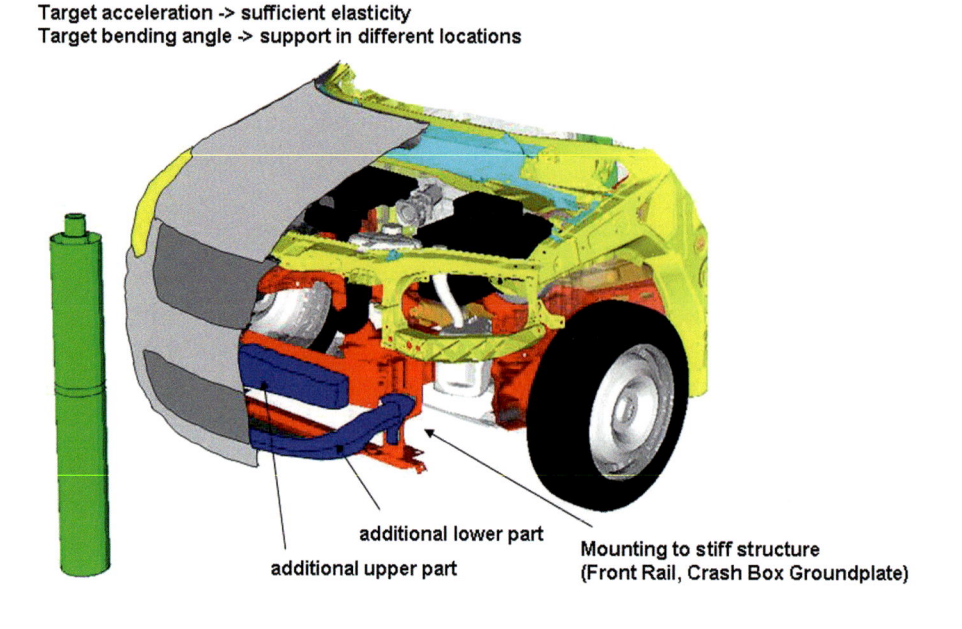

Target acceleration -> sufficient elasticity
Target bending angle -> support in different locations

additional lower part
additional upper part
Mounting to stiff structure
(Front Rail, Crash Box Groundplate)

Figure 6-73
Vehicle front-end

A further aspect is the package situation in the vehicle front-end. In addition to the free space between the individual components, such as cooling unit, charge-air cooler, AC condenser, coolant and refrigerant pipes, wiring harness and so on, there must also be sufficient space to allow these components to be installed or removed.

The lower bumper stiffener has been designed as a injection-molded part made of thermoplastic in order to exploit maximum design flexibility.

6.2.1.4 Requirements on the Lower Bumper Stiffener

The lower bumper stiffener has to fulfill a large number of requirements:

- For pedestrian protection purposes, the part needs to be stiff enough to support the lower leg during the impact such that bending angle, acceleration and shear displacement remain below the specified limits.

- In contrast to this, it must deform in a desired manner during the low speed insurance classification tests. Otherwise it could damage additional parts, leading to higher repair costs and thus a worse insurance classification.

- As the lower bumper stiffener is an additional part in the vehicle front-end, it must fit into the given vehicle package. This means that there must be sufficient clearance to all existing parts under all driving conditions and that the required space for all installation and removal activities must be maintained.

- In order to achieve low product costs, the component must be cheap and should not require additional parts.

- As an example, the component should be designed so that no foam blocks are required to fulfill the performance targets.

6.2.1.5 Material Modeling

The challenge in crash analysis for thermoplastic parts is the numerical description of the material properties. This behavior is often highly dependent on strain rate. At higher strain rates, the material can reach much higher yield stresses compared with quasi-static loading. Furthermore, with many plastics the yield stress under compression is much higher than under tension. In addition, with large strains, inelastic regions remain that do not fully relax on stress removal. Plastics therefore exhibit very complex, nonlinear viscoplastic behavior.

Further difficulties in the numerical treatment of the material for such components are caused by its short-glass-fiber filling. Due to the orientation of the fibers during the transient filling process, the mechanical properties of the material are anisotropic (direction-dependent). This means not only anisotropic stiffness, but also anisotropic values for yield stress and elongation at or near a breaking point.

The numerical material description of fiber-reinforced plastics under crash loads therefore has to account for the following effects:

- Anisotropy due to glass fibers
- Non-linearity in stress-strain
- Strain-rate dependence
- Anisotropic failure
- Coupling with process simulation
- Tension-compression asymmetry.

These requirements for material modeling clearly exceed those which were applicable some years ago. Figure 6-74 summarizes some of these characteristics of polymers.

The stress-strain curves shown in tension vary as functions of strain rate as well as of orientation (longitudinal and transversal). It is worth mentioning that the well-known strain-rate dependency can have an effect of about the same size on the stress values as the orientation effect! It is therefore essential for the simulation of parts made of short-fiber-reinforced

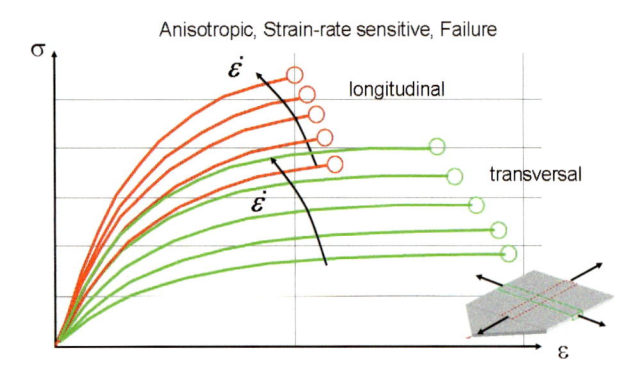

Figure 6-74 Material behavior at different strain rates

materials to incorporate both effects. The belief in a generally applicable global factor, which scales the longitudinal stress-strain curves down in order to include anisotropy, turns out to be wrong. Such a scale factor can only be found to describe a specific response of the system (e.g., one displacement) in a particular load case. If the same scale factor is applied to the same part, but to another load case, it can be noticeably wrong. The reason for this effect is that, in the first load case, a transversally oriented area in the part can have the highest effect on the behavior, while in the second load case, a longitudinally oriented area can be responsible for the part's performance. Figure 6-75 shows the variation of the failure strain with both strain rate and orientation:

The failure behavior of a specific material tends to be strongly dependent on micro-mechanical data of the fiber/matrix compound under consideration.

A numerical material model capable of accounting for those effects has been developed and integrated into FE analysis programs by BASF. The lower bumper stiffener described in this report is a plastics application where such a detailed modeling of the material behavior is a crucial factor in a proper description of the part's behavior in a crash.

The material law developed is based on a viscoplastic formulation for the thermoplastic phase and an elastic model for the reinforcing fibers, which has been combined with a micro-mechanical model for describing the material composite. It contains an anisotropic failure model based on energy principles. The input parameters for the model consequently consist of data for the thermoplastic matrix, as well as data for the fibers. In the case of the glass fibers, additional data such as fiber content, geometry and orientation distribution density are necessary for a full characterization of the material.

The orientation distribution density of the fibers in the component is non-homogeneous and heavily dependent on the filling process. It has to be calculated from the results of the injection-molding simulation. For the project described in this paper, this was done using "integrative simulation" techniques, described in detail in earlier sections.

Figures 6-76, 6-77, and 6-78 display in graphic form the main ideas behind the "integrative simulation" approach:

Orientation tensors

$$\mathbf{a} = \int_\omega \mathbf{p} \otimes \mathbf{p}\, \psi(\mathbf{p})\, d\omega$$

$$\mathbf{a}^4 = \int_\omega \mathbf{p} \otimes \mathbf{p} \otimes \mathbf{p} \otimes \mathbf{p}\, \psi(\mathbf{p})\, d\omega$$

(Tucker 1987)

Taylor expansion of ODF

$$\psi(\mathbf{p}) = \frac{1}{4\pi} + \frac{15}{8\pi} \cdot dev(\mathbf{a}) : dev(\mathbf{p} \otimes \mathbf{p}) + \frac{315}{32\pi} dev(\mathbf{a}^4) :: dev(\mathbf{p} \otimes \mathbf{p} \otimes \mathbf{p} \otimes \mathbf{p}) + \ldots$$

Figure 6-75 Orientation distribution function

Mean field theory (Mori and Tanaka, Tandon and Weng)

$$\sigma_0 = \mathbf{E}_0 : \varepsilon_0$$
$$\sigma_1 = \mathbf{E}_1 : \varepsilon_1$$

Homogenization

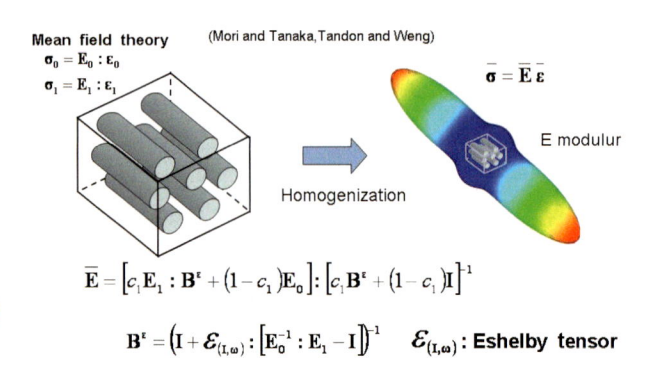

$$\overline{\sigma} = \overline{\mathbf{E}}\, \overline{\varepsilon}$$

E modulur

$$\overline{\mathbf{E}} = \left[c_1 \mathbf{E}_1 : \mathbf{B}^\varepsilon + (1 - c_1)\mathbf{E}_0 \right] : \left[c_1 \mathbf{B}^\varepsilon + (1 - c_1)\mathbf{I} \right]^{-1}$$

$$\mathbf{B}^\varepsilon = \left(\mathbf{I} + \mathcal{E}_{(I,\omega)} : \left[\mathbf{E}_0^{-1} : \mathbf{E}_1 - \mathbf{I} \right] \right)^{-1} \quad \mathcal{E}_{(I,\omega)} : \textbf{Eshelby tensor}$$

Figure 6-76 Homogenization of fibers and polymers

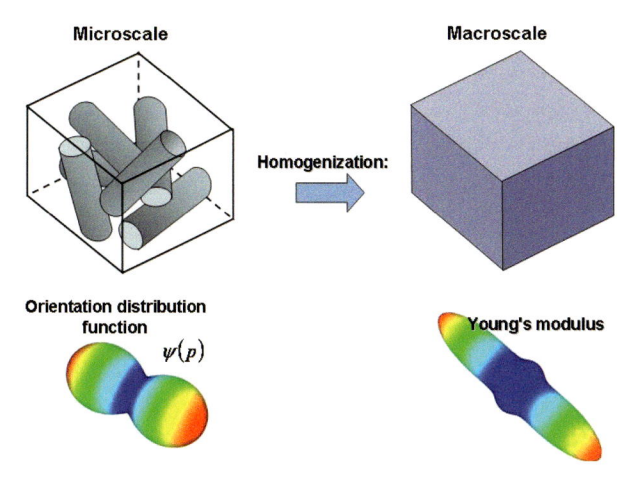

Figure 6-77 Material modeling of composite materials

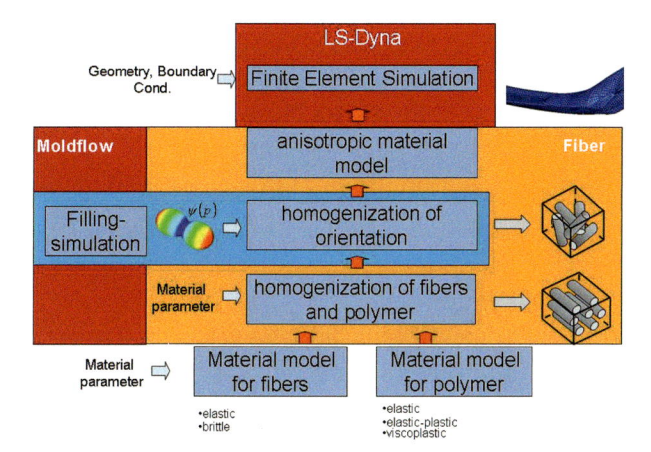

Figure 6-78 Integrative simulation: implementation

The statistical orientation distribution function (ODF) describes the number of fibers found in any given direction at any material point of the plastic part. In the case of FE analysis, it has to be evaluated at every integration point of every finite element in the plastic part. Based on the ODF, the fiber and polymer material models can be homogenized into a single material model for the compound.

6.2.1.6 Material Data Acquisition

Obviously, every material model can only be as good as the underlying data measurements. To determine the anisotropic nature of the material, an optical high-speed measuring system, including control and evaluation software, has been developed at BASF. The principal data flow of the measuring system is shown in Figure 6-79.

The system consists of a high-speed hydraulic tensile test machine, in which the specimen is mounted. The specimens are machined from injection-molded plates (150 × 150 mm) with different inclinations to the flow direction.

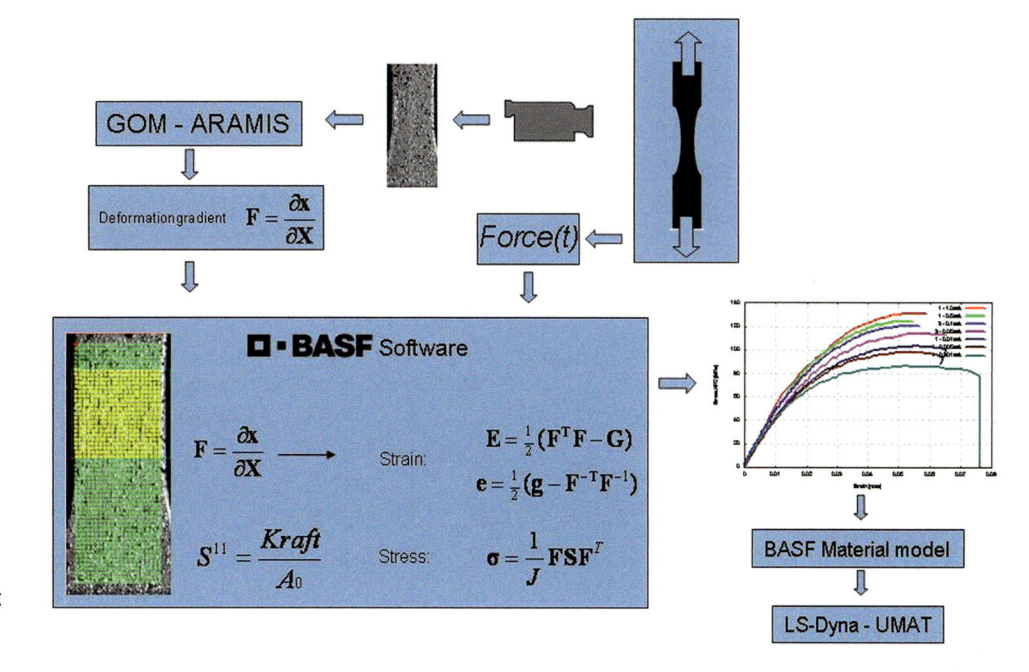

Figure 6-79
Measurement of material properties: data flow

Before the specimen is mounted on the machine, an irregular colored pattern is sprayed onto the specimen's surface. While the test is in progress, the relevant area of the specimen is digitally filmed by a high-speed camera at up to 50,000 frames per second, depending on the frame resolution. The image file containing the filmed frames is post-processed with the ARAMIS program from GOM. Using gray-scale correlation methods, ARAMIS is able to calculate the 2-dimensional strain field on the specimen with reasonable resolution. In the system developed at BASF, the deformation gradient $\mathbf{F}$ is output by ARAMIS. Thus, in conjunction with the machine force measured, this makes possible the calculation of true stress-strain relationships. Another advantage of the system is that it is possible to use only segments of the specimen's strain field to calculate the curves. This feature extends the use of the system, especially when localization occurs. Furthermore, based on the deformation gradient information, the transverse strain is computed and used for the calibration of the material parameters (Poisson's ratio). This procedure is illustrated in Figure 6-80.

As with every high-speed measurement technique, the highest useable tension test speed is limited by inertia effects. Starting from a specific traction speed, the measured curves start to show oscillations, which are caused partly by the material behavior, but mostly by inertia effects (see Figure 6-79). It is worth mentioning that it is not only the machine inertias (clamping device, hydraulic acceleration, control valves) which contribute to those oscillation curves, but also to a great extent the specimen itself. If one performs an FE analysis of an ideal tensile test with a prescribed constant pulling speed at one end of the specimen, it can be seen that, at higher pulling speeds (> 1 m/s), wave propagation along the specimen becomes important. Due to the sound velocity limit of the material, one end of the specimen can already show noticeable deformations (and strains), while the other end has not detected any force at all.

6.2.1.7 Using Integrative Simulation for the Thermoplastic Lower Bumper Stiffener (LBS)

The FE model for the LBS filling analysis can be derived from the crash FE model by simply splitting all quad-elements into the triangular elements which are used for process simulation. The use of the crash mesh for deriving the process analysis mesh is possible but not necessary. This is because the definition of the mapping process from one mesh to the other in the Integrative Simulation approach is totally independent of the underlying FE meshes.

Figure 6-81 shows the data flow in the project under consideration.

The process simulation for the LBS helped in obtaining preliminary machine settings and enabled the characterization and improvement of important values such as pressure loss, fill time and resulting warpage. The part is gated via a single gate on the symmetry line of the part. This single gating location is necessary to ensure that fiber orientation is adapted to the mechanical loads in the part, although the resulting pressure loss means higher locking pressure and a larger machine. A different gating scenario, one which would create weld lines in the part, would directly predetermine breaking points due to the resulting fiber orientation in the weld line.

After the fiber orientation results from the filling process have been mapped to the FE crash mesh using BASF's FIBER software, crash calculation can

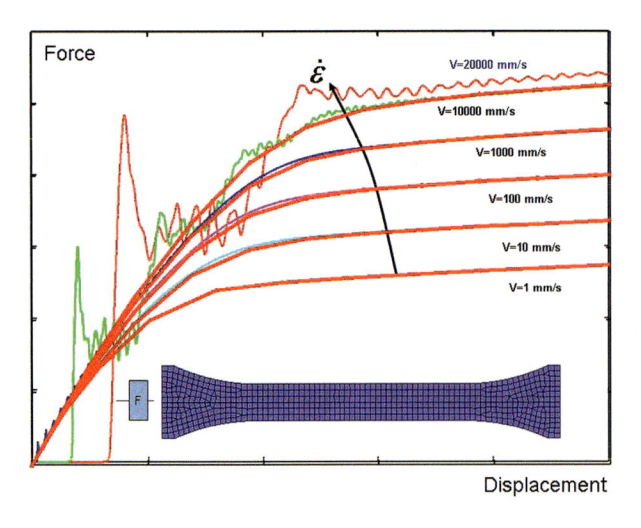

Figure 6-80 Dynamic tension test and simulation

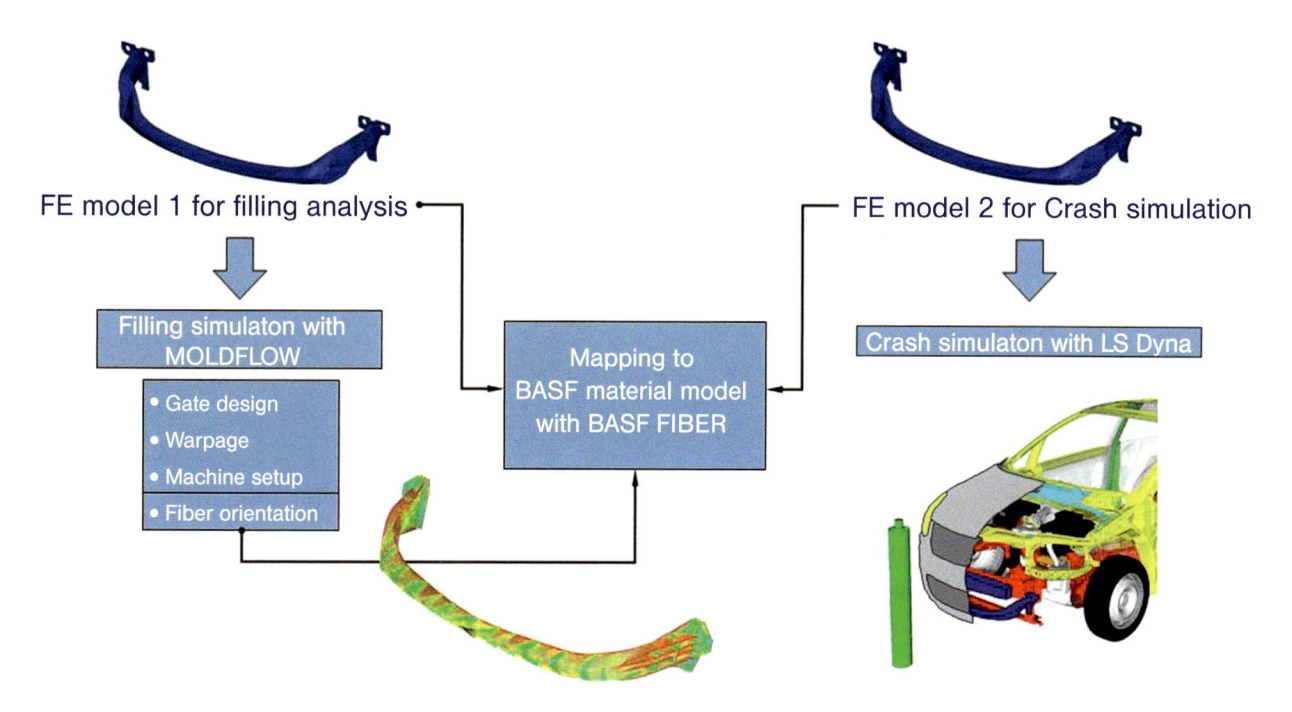

Figure 6-81 Integrative crash simulation: data flow

start with LS-Dyna. As mentioned above, the material model developed is integrated into LS-Dyna by means of a so-called "user material". The pedestrian crash model used for the lower leg impact simulation is a specially validated submodel of the full car crash model. At the beginning of the project, the lower leg crash calculations with the plastic part were done solely at BASF. After the project had reached a certain stage of maturity, the user material software libraries were transferred to Opel, so that certain variations could be investigated directly at Opel. After various geometry changes, the integrative simulation link was maintained by creating INCLUDE files for the part mesh. After filling simulation and result mapping, those files contained the results of the mapping process. They could then be seamlessly incorporated into the crash analyses at Opel.

6.2.1.8 Design

The lower bumper stiffener consists of a C-shaped profile with horizontal and vertical ribs. The opening of the C profile is oriented backwards. Both side areas are also stiffened by a ribbed structure. The LBS is mounted to the left and right front rail and the bumper fascia.

At each side, three positions are weakened in order to initiate a controlled LBS failure during the insurance classification tests, so that the LBS does not impact other components and hence cause additional repair costs. However the LBS system can easily withstand the much lower impact loads for pedestrian leg impacts.

6.2.1.9 Molding Principle

For the mold, one central gate was chosen. This enabled the fibers to be oriented in the loading direction and hence the optimal utilization of the higher strength and stiffness of the fibers.

The side areas were demolded by two slides on each side. The slide positioning enabled the creation of a homogeneous transition from transverse orientation in the front LBS areas to longitudinal orientation in the side regions.

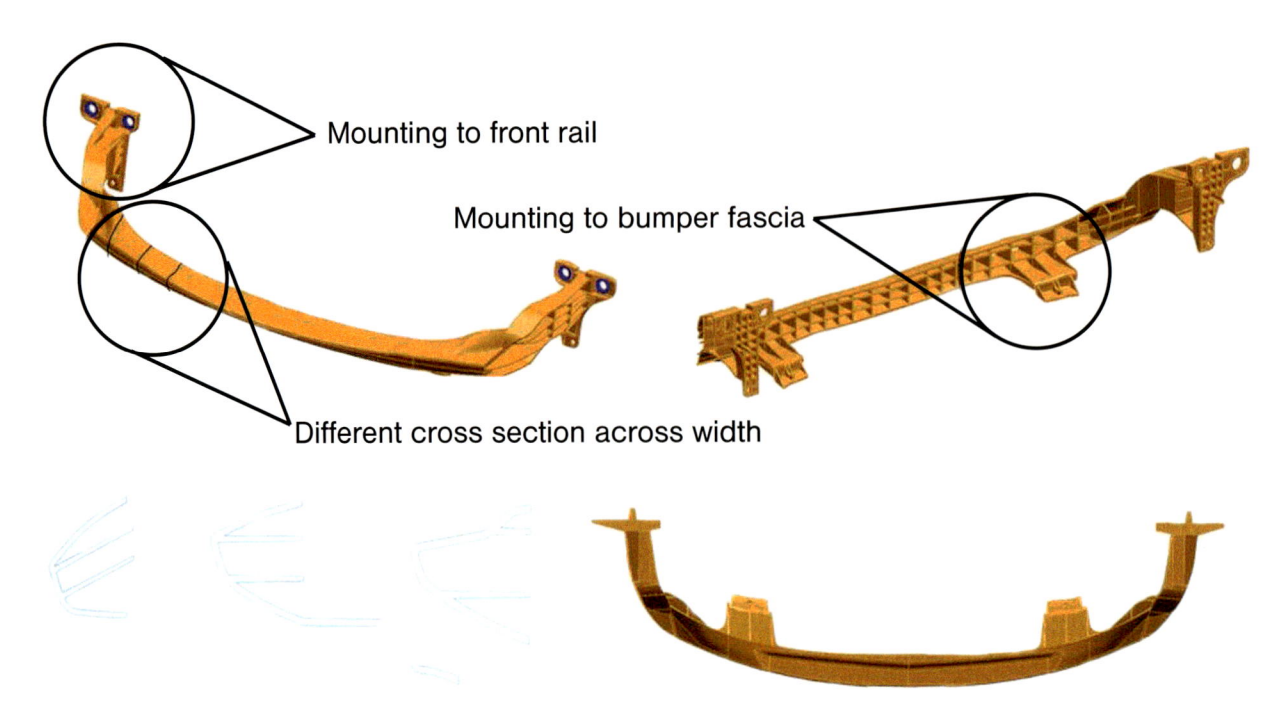

Figure 6-82 Lower bumper stiffener design

Mounting to front rail

Mounting to bumper fascia

Different cross section across width

Homogeneuous force flow required:
- homogeneous transitions
- multipart tool concept

Different demoulding directions

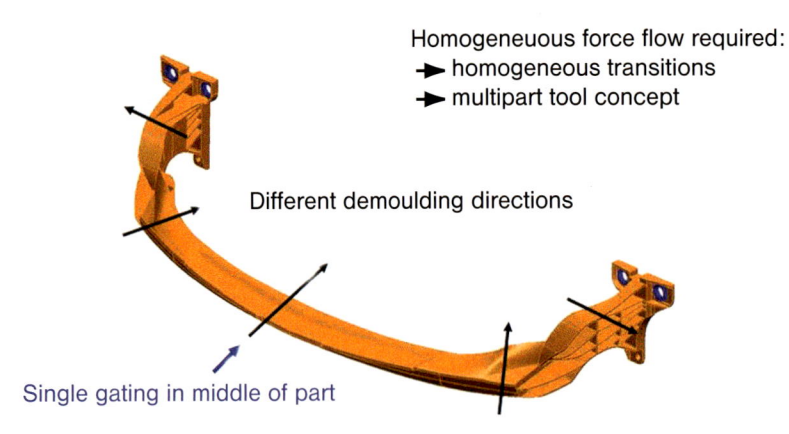

Single gating in middle of part

Figure 6-83 Molding principle

6.2.1.10 Test Results

Leg Impact Tests

All pedestrian-related targets were achieved for all potential impact locations. Since the material properties were described with such detail and accuracy, the simulation results correlated very well with the physical validation tests (see Figure 6-84).

The shape of the time history curves and the maximum value of the leg acceleration is predicted very well. The bending angle is slightly over-predicted, whereas the shear displacement is well within the range of the various test results.

Insurance Classification Tests

During full vehicle impact for the insurance classification assessment, the loads are much higher than during the pedestrian impact. Thus, as intended, the lower bumper stiffener cannot withstand these loads and breaks in a controlled manner, hence avoiding damaging any expensive parts in its vicinity.

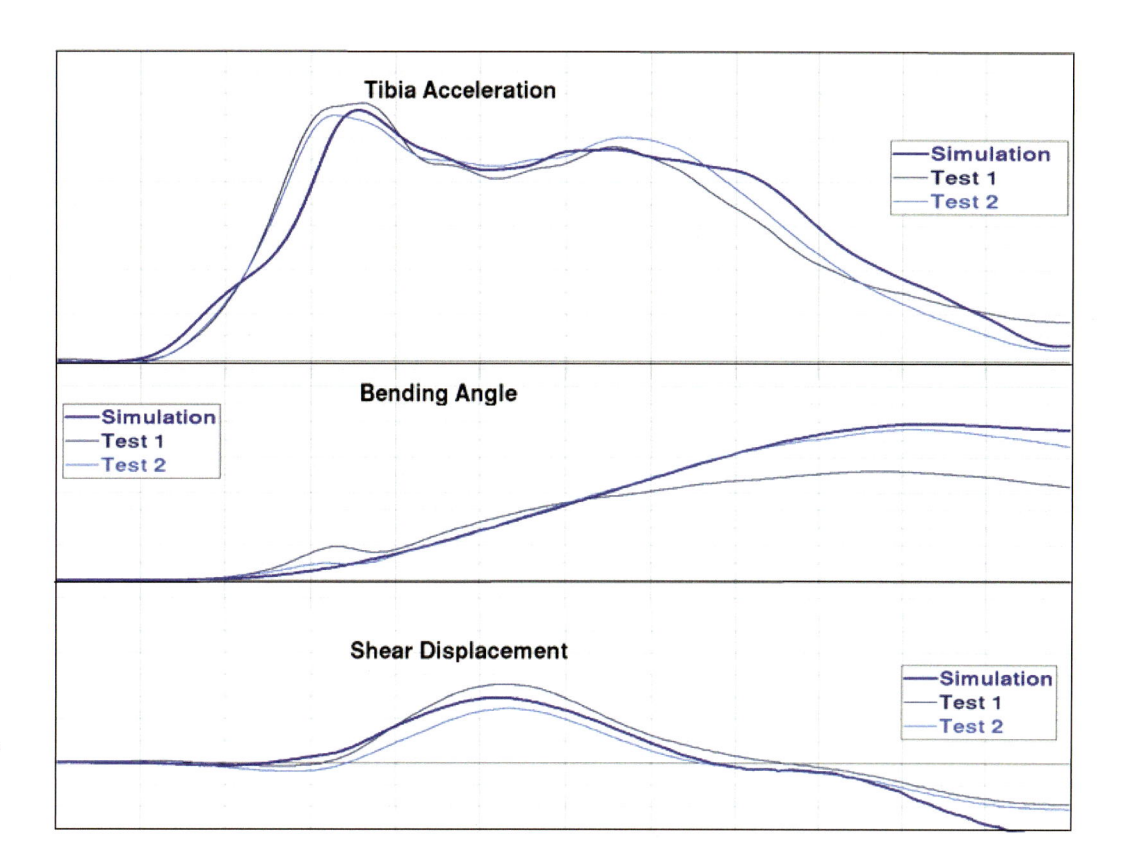

Figure 6-84
Legform test results

6.2.1.11 Conclusions and Outlook

The joint project between Opel and BASF described here has resulted in a much better representation of the material properties of plastics for crash simulations. It is thus now possible to predict the behavior of plastic components to the level of accuracy required for today's vehicle development processes.

It was found that the key enabler for this capability is the successful application of the integrative simulation approach. By performing mold flow analyses, the impact of the local fiber orientation on the material properties can be determined in advance. The material model developed by BASF is an essential step in taking advantage of this information and making it available for the subsequent simulations with explicit programs such as LS-DYNA. With this approach, correlation with physical tests could be improved significantly.

Applying the integrative simulation approach, the lower bumper stiffener was developed successfully in a very short time frame. All of the test results using physical prototypes validated the performance that had been predicted by simulations.

References

[1] EU Regulation 2003/102/EG.
[2] EuroNCAP Test Protocol.
[3] EEVC, WG 17, Final Report.
[4] Glaser, S.: SFR Parts, FEM Computation of Dynamic Behavior. *Kunststoffe* Vol. 91 (2001) p 7.
[5] Glaser, S.: Integrative Simulation. Vom Polymer über den Prozess zum Bauteilverhalten. VDI Conference, Baden Baden 2004.
[6] Glaser, S., Wüst, A.: Integrative crash simulation of composite structures: the importance of process-induced material data, 4th LS-Dyna Forum 2005, Bamberg
[7] Junginger, M.: Charakterisierung und Modellierung unverstärkter thermoplastischer Kunststoffe zur numerischen Simulation von Crashvorgängen, Dissertation, Fraunhofer-Institut für Kurzzeitdynamik, EMI, Freiburg, 2002.

6.2.2 Crash Simulation with Plastic Components

Steffen Frik

The current development of motor vehicles is characterized by an enormous increase in requirements. On one hand, there are additional legal requirements, especially in the pedestrian and occupant protection area. In addition, consumer tests (e.g., EuroNCAP), insurance classification tests (e.g., RCAR, GDV) as well as in-house requirements must be fulfilled. Furthermore, there is a strong need to reduce the development time and the number of physical prototypes, raising an additional challenge for the application of simulation tools during the vehicle development process.

In order to fulfill these requirements, crash simulation models contain an increasing number of plastic components that must be described accurately. Since these components – in contrast to other structural parts – are manufactured and even developed by suppliers, the integration of the allocated development activities into the OEM's vehicle development process causes additional challenges.

This paper describes how this is performed during Opel's vehicle development projects. It will also cover the development of additional material models and the determination of parameters for polymer materials that are required for crash simulation purposes. Recent progress and future challenges will be summarized.

6.2.2.1 Introduction

During recent years, the automobile industry has continually reduced both development time and the number of prototypes, in order to reduce time-to-market and development costs. Thus the efficient application of simulation tools is of vital importance in achieving these challenging targets. When crash simulation applications were first employed in the automobile industry, they were mostly run to prove the feasibility of predicting structural performance during crash events. As the amount of time needed for these simulations was quite high, their impact on vehicle development was limited. Today, simulations are run on a day-to-day basis and no physical prototypes are built without first checking the anticipated crash, static, and dynamic performance by means of simulation.

All simulation activities must be integrated into the development process, in order to achieve an efficient collaboration and to leverage synergies between virtual and physical tests. Simultaneously the effort that is required to create and manage the simulation models must be reduced dramatically in order to deliver the required results in time.

The following section describes the integration of virtual development activities in the overall development process. The integration of simulation activities performed by suppliers will be of special interest.

6.2.2.2 Integration in the Vehicle Development Process

Figure 6-85 shows Opel's current vehicle development process which consists of different development phases. At the beginning, the majority of development activities is performed virtually by means of simulation tools. As soon as physical prototypes are available, more and more development work is done with hardware tests.

Nowadays all simulation activities are an integral part of the overall development plan. All hardware development phases are prepared by virtual development phases (Figure 6-85). This means that prior to the physical building of prototypes, the required performance must be proven by means of simulation. This performance status is reported at so-called Virtual Vehicle Assessment Gates.

All simulation and design groups work together intensively during the entire development. Thus simulation

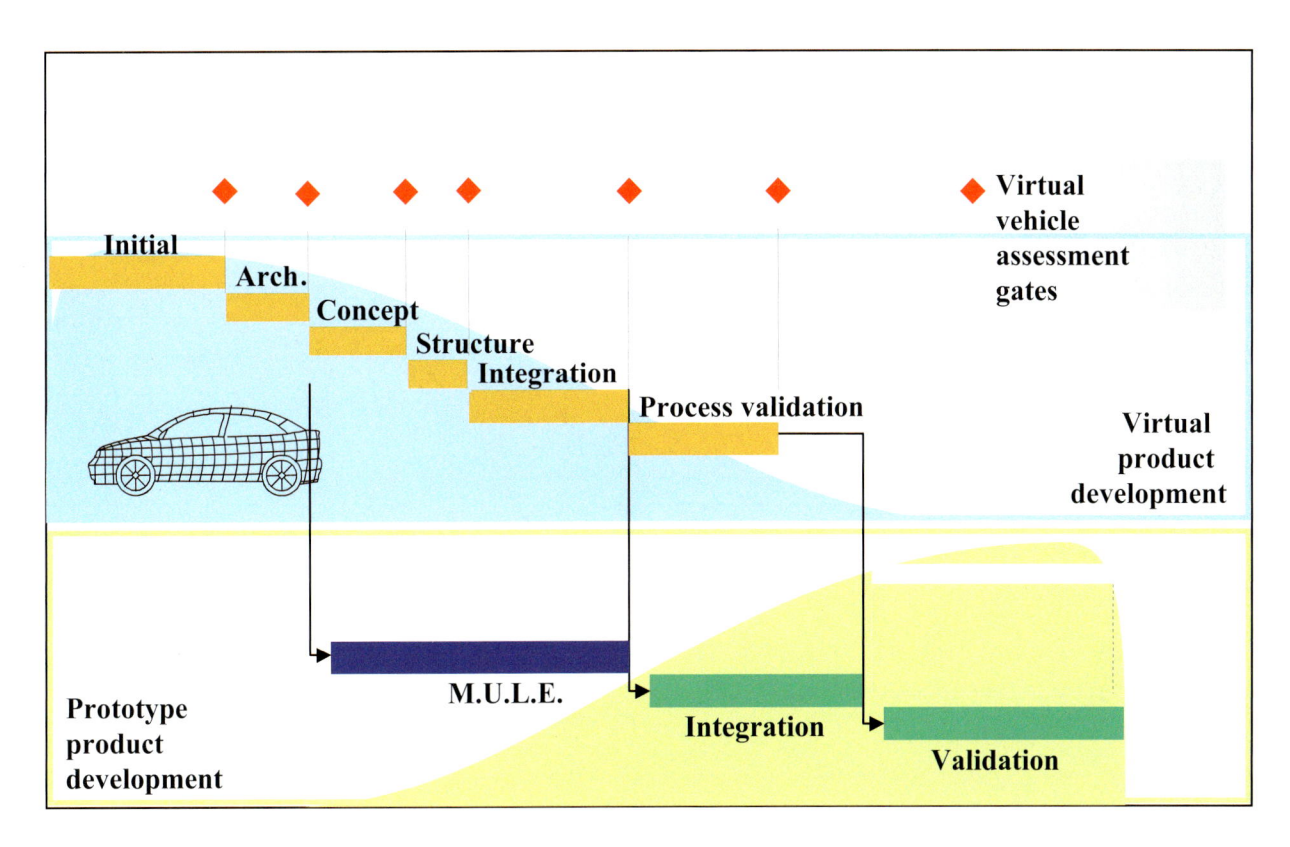

6.2 Front Modules, Crash Elements, Safety Concepts

Figure 6-85 Vehicle development process

results can be applied directly to improve the design. On the other hand, design as well as manufacturing issues can be introduced into new proposals created by the simulation groups and assessed by means of simulation tools. The different design proposals must be synchronized in order to achieve a common understanding as to the specifications of the virtual prototypes to be assessed during the Assessment Gate.

Prior to the respective Assessment Gates, CAD synchronization points are implemented. They are intended to unify the individual designs and concepts that are worked on during vehicle development (Figure 6-86). These CAD data are used to create a common CAE model that is used as a baseline for all simulation load cases. Of course, application and load case specific modifications must be applied to this model. Concurrent simulations are performed with the already existing models to drive the development. These results are used along with the new results to assess the achieved performance level at the Assessment Gates.

During the vehicle development, all simulation models are continuously validated with all available test results, in order to improve their quality and the confidence level of the simulation results.

As well as the CAD synchronization points, which unify the existing geometry information, CAE exchange points have been established. They are intended to enable the required exchange of CAE models as well as corresponding information. Since an increasing share of the overall development is done by suppliers, this became necessary to ensure that the required simulation models and simulation results are available to all development partners.

This information exchange must be bi-directional, because less and less development hardware is being installed at the OEMs. Thus they can provide much less reference hardware for use by the suppliers. As an example, during certain development phases, vehicle bodies are no longer available for the suppliers to run

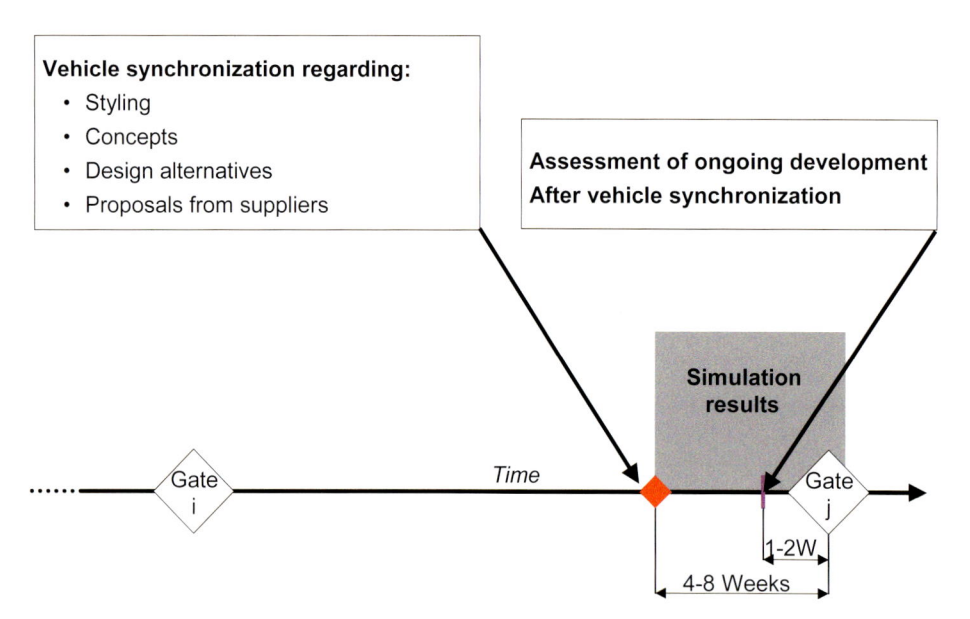

Figure 6-86 Synchronization process

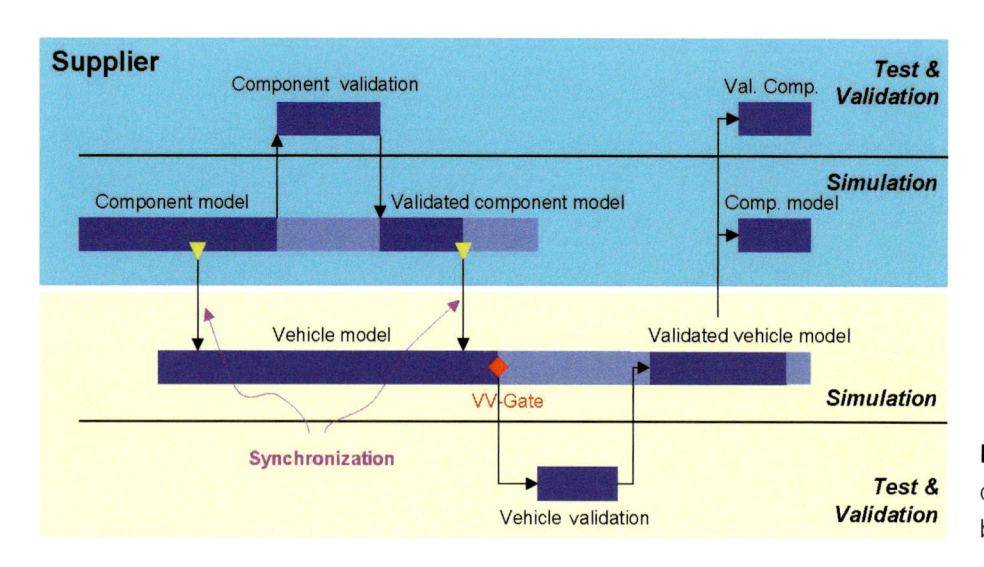

Figure 6-87 Integration of simulation activities by OEM's and suppliers

their component tests. This means that the suppliers no longer have access to physical development hardware and are forced to perform their development activities by means of simulation during the respective phases.

During Opel's vehicle development process, this information flow is written down in the ASOR (Analytical Statement of Requirements) that is part of the official order documents. It describes the required information and the points in time when this information is needed. According to the supplier integration level this information content may range from simple property data for the raw material up to complete, detailed CAE models of complete components, e.g., the instrument panel.

The ASOR describes required CAE model content as well as modeling guidelines defining element types, element sizes, numbering schemes for components and elements. The CAE models must be created using the software defined in the CAE software portfolio, in order to allow an easy integration of all component models into the complete vehicle model. At the CAE exchange points the CAE models are exchanged, along with the respective simulation results proving the required component performance.

6.2.2.3 Crash Simulation of Polymeric Materials

In the past, crash simulation activities focused on predicting the structural performance of the vehicle frame. Now, the overall performance of the complete vehicle including all dummies is of interest. Therefore, the dummies and all relevant interior components, such as the IP and trim parts must be included in the simulation models.

Since these parts include a significant amount of plastic components, accurate modeling of these parts is crucial to achieving the required level of accuracy. Polymer materials are simulated for the interior and some other components. Table 6-9 shows some examples of these applications.

In contrast to metal materials, the simulation of polymer materials for crash applications is relatively new. Due to the limitation of computer and modeling resources, crash simulation activities were basically focused on the vehicle's structural performance. Therefore, comprehensive databanks containing complete material data, including strain-rate-effects and forming process impacts such as work or bake hardening, have been established for metals.

It is a very difficult task to create a similar database of polymers due to the extremely varied behavior of such materials. Table 6-10 shows the stress-strain-characteristics for some groups of polymers. As the behavior may range from brittle fracture (type A) to entropy elastic deformation (type E), it is obvious that completely different material models are required.

Since some polymers exhibit volumetric change under loading, this effect must be measured, bringing additional difficulties for the execution of the tests.

In addition, there is still no general viscoelasticity and viscoplasticity theory so that the theoretical basis for the development of material models must be improved as well [2].

Some polymers show that the ambient temperature significantly impacts material performance even within the temperature range for which vehicles are designed. Therefore, additional test series are required to determine the material properties for all relevant temperatures.

Additionally, material properties of polyamides depend tremendously on the water content of the material. The E-modulus may vary by a factor of 3 to 4 if the water content varies by a few mass percent.

This shows that there is still a strong need for intensified material testing and further development and user-friendly implementation of material models in the simulation software codes.

Table 6-9 Automotive application of polymeric materials

Thermoplastics	Trim parts
PU and EPF Foams	Seats, bumpers, padding, dummy components
Elastomers	Mounts, dummy components
Adhesives	Windshield adhesives

Table 6-10 Stress-strain characteristics of polymers [1]

Type	σ-ε Diagram	Polymer materials
A		Polystyrene, Polymethylmethacrylate, Styrene-acrylonitrile
B		Impact polystyrene
C		Polyamide, polypropylene, Polybutylenterephtalat
D		Polyethylene-LD
E		Elastomers

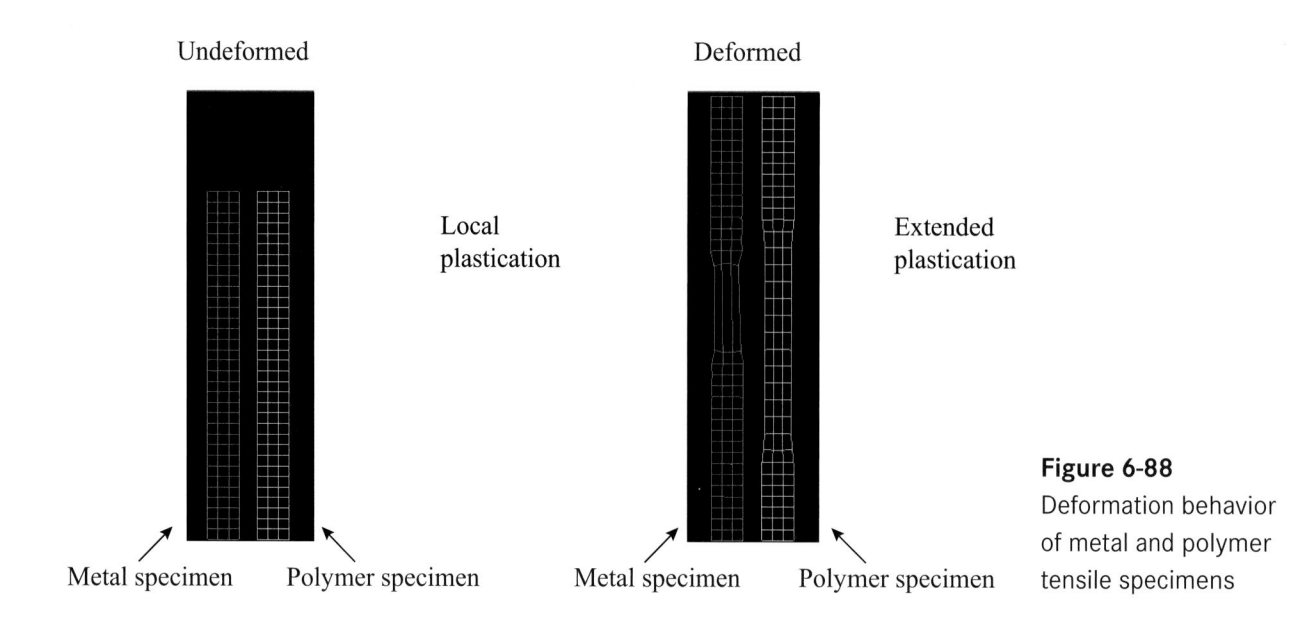

Figure 6-88 Deformation behavior of metal and polymer tensile specimens

6.2.2.4 Material Properties of Polymers

The first attempts to model polymers for crash simulation purposes utilized the existing material models for metals, but applying measured polymer material data [3].

As metals and polymers show significantly different material behavior, this approach is only valid for small deformations, if at all.

As shown in Figure 6-88, a loaded metal specimen exhibits local initial plastication followed by local necking. If the load is increased further, the specimen breaks in that area. For polymer specimens, the initial plastication area is stabilized and then extends over a much larger area if the load is increased further.

Elastomers

Elastomers are hyperelastic materials and allow large, reversible deformations. They are employed for engine and subframe mounts and for some dummy parts. The choice of a suitable material model depends strongly on the individual load case and the type of material loading. As this choice may have an extreme impact on the accuracy of the simulation results, it must be done with special care.

When choosing a suitable material model, several factors including accuracy, the numerical stability of the material model and the amount of required input data are important issues that must be taken into consideration. As complete material data tests are rarely available during daily development work, this may be of importance as well.

LS-DYNA offers 4 different material models that can be applied to describe the characteristics of elastomers. In [5], these models were assessed regarding their capability for simulating complex loading situations, including superposition of uniaxial, biaxial, and planar tension. The impact of the number of available material tests (uniaxial tension results versus a complete set of tests for all loading conditions) was also analyzed.

All material models show sufficient accuracy for uniaxial loading. If these models are applied to biaxial or planar tension, some of them show significant deviations with respect to the test results. For cases where material tests for all potential loadings (uniaxial, biaxial, and planar tension) are available, all material models still lead to reasonable results for small strains. If only tests for uniaxial tension are available, as most often happens during vehicle development, and larger strains are present, only the application of the model using the reduced polynomial form is applicable. All other models show unacceptable deviations (Figure 6-89).

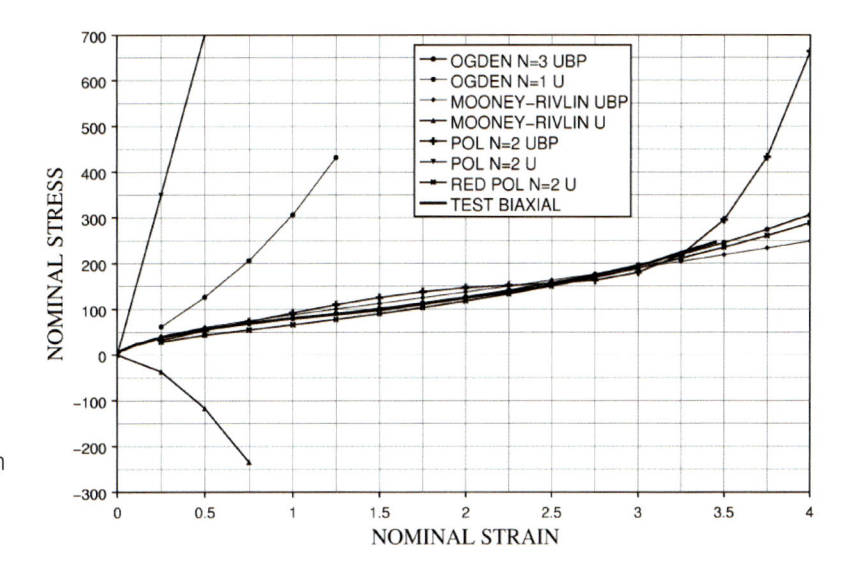

Figure 6-89 Calculated stress-strain characteristics of elastomers for planar tension

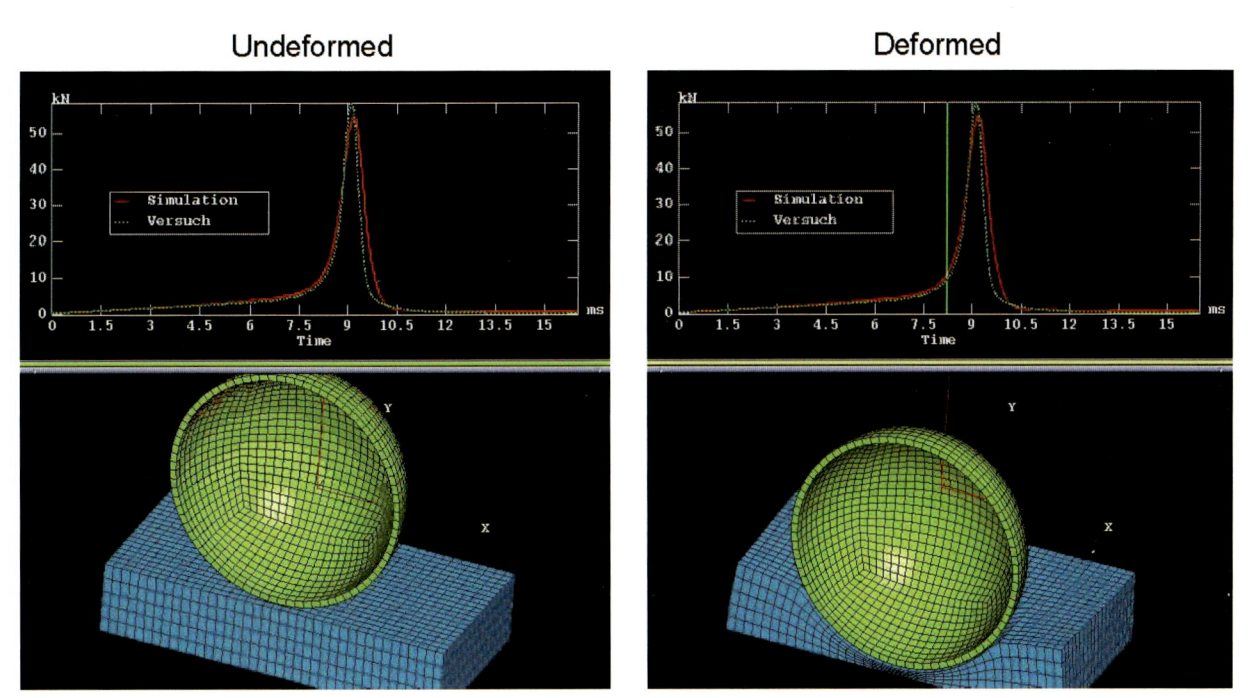

Figure 6-90 Deformation of a foam brick impacted by a sphere

Foam Materials

During recent years, a research project of the Forschungsvereinigung Automobiltechnik (= Automobile Engineering Research Group) was dedicated to the modeling of foam materials. This project was focused on the analysis of seat and bumper foams. Comprehensive tests were performed to measure the material behavior for different types of loading at numerous strain rates. This material data was then used to develop material models that allow numerically stable simulation of foam materials even in the case of large local deformations [2].

Figure 6-90 shows an example of the tests that were performed during this project: A sphere falls onto a foam brick and deforms it. Despite large local deformations with a compression rate of almost 98%, the simulation does not develop any numerical problems. The correlation with test data is very good.

6.2.2.5 Application During Vehicle Development

Obviously, not all required simulation data are available from the beginning of a vehicle development. Therefore all simulation areas are restricted to a very limited set of data that can be used for the initial simulation runs. From our experience, the following approach should be used to overcome this situation:

1. The material properties of the raw material used to manufacture the respective component must be determined by means of quasi-static or dynamic tensile or bending tests. Normally, this data is provided by the raw material manufacturer.

2. As soon as the first component prototypes are available, the component performance is validated by means of component tests, in order to validate the baseline models and to detect geometry and manufacturing effects. The validation tests are designed such that the material loading is similar to real load cases performed with the complete vehicles. Depending on the supplier integration level, these tests are performed by the OEM or the supplier.

3. Validation of all component models in the complete vehicle model by the OEM, applying the real load cases.

Based on recent experience, the introduction of manufacturing effects is of crucial importance to achieve more accurate simulation results. Some of the suppliers already perform flow analyses for thermoplastic materials, in order to determine the local orientation of the fibers introduced to improve the material strength. Additionally, the impact of different coatings on component performance may be significant. In order to obtain better validation data and to sort out uncertainties, the specimens should be treated with the coating that will be used for mass production.

Results that are achieved during phases 2 and 3 are compared with the initial results for the raw material data. Thus this model can be updated and improved. If there is performance deviation, the modified data will be stored for use in future projects in order to retain the knowledge acquired.

6.2.2.6 Conclusions and Outlook

Since accuracy requirements are continuously increasing, there will be a strong need for further development of all simulation methods. This also includes the improved modeling of individual components, as well as an enhanced modeling of material behavior. Despite all efforts during the last few years, the material databases for polymer materials are still much less comprehensive than the respective databases for metals. In order to fill this void, joints efforts of OEM's, material manufacturers, and the plastics industry will be required. In addition to obtaining accurate material properties, it is essential to improve the material models currently available in the simulation codes.

References

[1] Erhard, G.: Designing with Plastics. Carl Hanser Verlag, Munich, 2006.
[2] Du Bois, P.: Recent Challenges in Crashworthiness Simulation. LS-DYNA Update Forum, Stuttgart, 2003.
[3] Du Bois, P.: Crashworthiness Engineering with LS-DYNA, 2001.
[4] LS-DYNA User's Manual. Version 970, 2003
[5] Gosolits, B., Visinescu, R. and Maucher, R.: Numerische Simulation von Elastomerlagern mit LS-DYNA. LS-DYNA Anwenderforum 2002, Bad Mergentheim.

6.2.3 SLR Crash Element: from Concept to Volume Production

MICHAEL BECHTOLD, BRUNO MÖLTGEN

6.2.3.1 Introduction

During the mid-1950s, three letters became a legend: SLR. A Mercedes-Benz SLR racing car set new standards for superior sedans. With their new Mercedes-Benz SLR McLaren, Mercedes-Benz and their formula 1 partner McLaren have once again demonstrated their competence in the development, design and production of high-performance sports cars.

Visually, this car is breath-taking and boasts with innovative technology in both power and handling characteristics.

In the field of body and safety technology, the new Mercedes-Benz SLR McLaren justifies its reputation as an innovation bearer among modern GT's. Hightech materials from aviation have now arrived in mass automobile production.

For many years, fiber-reinforced composite plastics have performed sterling service in Formula 1 and in the aerospace industry. The objectives staked out for body strength and lightweight design have meant that the specifications of the Mercedes-Benz SLR McLaren call for very extensive use of carbon-fiber-reinforced plastics.

Components made of carbon fiber – at the same level of strength and rigidity – weigh up to 55% less than comparable steel parts and are up to 25% lighter than aluminum components. This makes CRP the material of choice in the manufacture of high-performance automobiles. In addition, this ultramodern lightweight material has very good weight-specific energy absorption; its values for CRP are four to five times higher than those for metals.

The partnership established between DaimlerChrysler and McLaren that began in Formula 1 was a very important factor in the decision to collaborate. In vehicle design, it meant that on the one hand it is possible to draw on the experience gained with CRP in Formula 1 and the McLaren F1 Road Car and on the other hand to make use of the automotive knowledge and expertise of a high-end car maker. But DaimlerChrysler's own experience with CRP from the fields of production and materials technology, not to mention research and development, was also of considerable importance in development of the vehicle.

Figure 6-91 Mercedes-Benz SLR McLaren

Figure 6-92 Carbon-fiber-reinforced body shell structure of the Mercedes-Benz SLR McLaren

Although the Mercedes-Benz SLR McLaren is a series application in the automotive niche market, not solely the technically feasible, but rather the economically sensible was the central aspect in selecting its manufacturing technology. That is what distinguishes the application discussed here from many previous cases.

The Mercedes-Benz SLR McLaren introduced in September 2003 included a large number of innovative applications for plastics. For example, the vehicle structure consisted predominantly of carbon-fiber-reinforced plastic (CRP).

Development work focused not only on component function, but also on aspects of production engineering, firstly to meet cost targets within the project and secondly, however, also to open up potential for future applications with higher volumes.

The aim of also manufacturing the energy-absorbing components in the car front end in fiber composite technology represented a particular challenge. Although these materials are known to offer high weight-specific energy absorption, this was the first time that collision-related structures made of CRP were to be built into a standard production vehicle with worldwide type approval.

For this application, with its high relevance to safety, development up to readiness for volume production demanded intensive collaboration between all Mercedes-Benz experts. Not only its function, but also reproducibility and process reliability during production had to be demonstrated beyond any doubt.

6.2.3.2 Idea and Concept

Virtually every weekend at motor racing events and Formula 1 in particular, carbon structures demonstrate their high capabilities as regards passive safety.

In real accidents, the front-end structure is affected in more than half of the cases. For this reason, frontal collisions carry special importance in the accident scenarios taken into consideration during vehicle development work at Mercedes-Benz.

The proven energy-absorption properties of metal structures are actually exceeded by CRP. What is advantageous for CRP is its very even pace of destruction and thus of energy absorption. Metal structures on the other hand tend to crumple; only little energy is absorbed during this folding process.

The high weight-specific energy absorption of CRP means that a significant weight advantage can be obtained in comparison with a metal component. CRP collision structures can also be specifically adapted to the deceleration curve required.

The actual challenge lies in implementing these findings originating from research work and from Formula

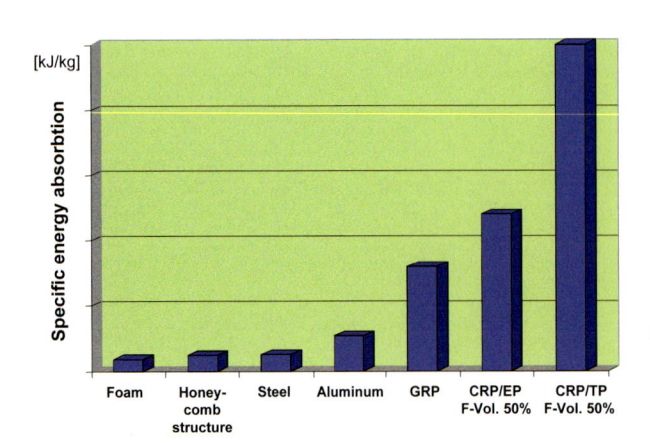

Figure 6-93 Weight-specific energy absorption potential

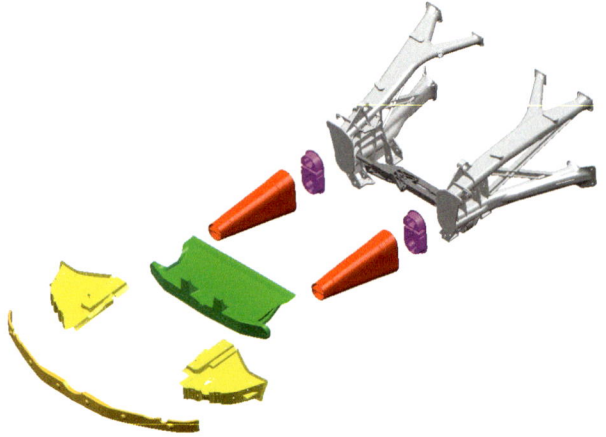

Figure 6-94 Front-end structure elements of the Mercedes-Benz SLR McLaren

1 in low volume production. The SLR project with its prominent promise of "high performance" offers an ideal platform.

At DaimlerChrysler and McLaren, a special front-mid-engine concept was developed for the SLR front end/central engine concept. It consists of an engine support of cast aluminum while the front crash structure (FCS) is made of carbon and mounted in front of it.

The heart of the front crash structure is the crash element itself, an integrated longitudinal member also known as the FCS side member or FCS tube.

6.2.3.3 Component Development

In component design, adaptation to the required deceleration curve occupied a central position in development work. The goal was not to achieve the maximum energy absorption possible, but rather such deceleration levels as could be tolerated by passengers. Taking into consideration the most varied accident scenarios in frontal collision (100% or 40% overlap as well as 30° oblique-angle impact), dimensioning of the FCS side member is always to be seen as involving interplay with the restraint system.

The FCS side member must be dimensioned such that in a frontal collision with a 40% overlap it is "hard" enough to absorb most of the collision energy, but when there is 100% overlap (both side members under load) it will be "soft" enough to prevent deceleration values exceeding those tolerable for passengers.

Before the collision behavior of a CRP component could even be simulated, the available software had to be developed further. To cover destruction of the material, dynamic collision computation must employ a very complicated calculation algorithm and also a complex model of the material. The basis for virtually ideal energy absorption characteristics of the FCS side member was developed with the aid of this simulation software, a large number of additional component tests and intensive parameter studies. In addition to component design relating to frontal collision, the following points were among those also taken into consideration in the component design of the FCS side member:

- Installation space and packaging
- Retaining front bumper, hood hinge, head lights, coolers etc.
- Towing loads
- Corrosion protection
- Fatigue strength.

The symmetrical cross-section shown in Figure 6-96 with oval shell and central I-beam was obtained from simulation work and function trials. The conical shape of the component with the small end in front is determined both by functional and installation space requirements.

Figure 6-95 FCS side member after collision

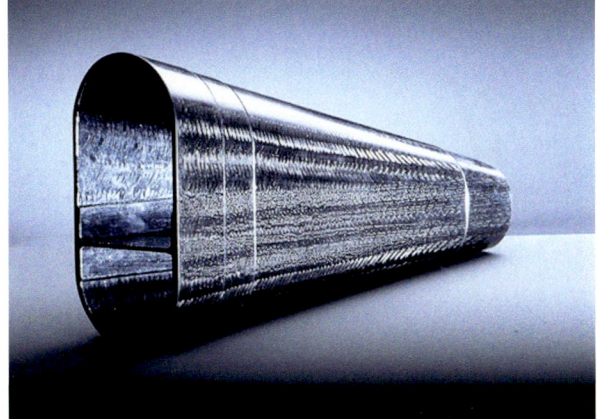

Figure 6-96 FCS side member

6.2.3.4 Prototype Production (Development Phase)

As one would expect, the early development phase made use of classic techniques such as sequential layer structuring using fiber woven or non-woven pieces, manual impregnation with resin followed by curing in the autoclave. This procedure, typical of prototype construction, allows a high level of flexibility to be exploited during component development.

However, at no time was the need for implementation in actual production left out of consideration. To avoid premature selection of procedures, right from the start the developers aimed at the simplest possible geometry for the collision side members and an even and symmetrical layered structure. To safeguard process capability, the individual process steps in prototype building were carried out using the techniques envisaged for volume production. In the definition of the quantity production concepts, not only geometrical requirements but also the following considerations were of decisive importance:

- High process reliability
- High reproducibility
- The highest possible degree of automation, and
- Short production cycle time.

The production strategy which appeared the most promising as regards meeting these requirements was subdivided into:

- *Preform production* – i.e., making a dry fiber structure, followed by

- *Resin injection* – i.e., impregnation with resin in a separate process step.

6.2.3.5 Preform Production

For the production of the FCS side member, two different preforming concepts were employed in volume production. They can be distinguished based on the fiber material as sequential and direct preform production.

The sequential method uses semi-finished fiber products in the form of woven or non-woven fabric. Here, the layered structure of the component is built up step by step and layer by layer. With the direct method, the preform is made directly from the fiber material. The production technology required for this derives from the textile industry.

An automated sequential procedure is used to produce the side member rib. Here, several layers of a multi-

Figure 6-97 Laminate structure at prototype stage

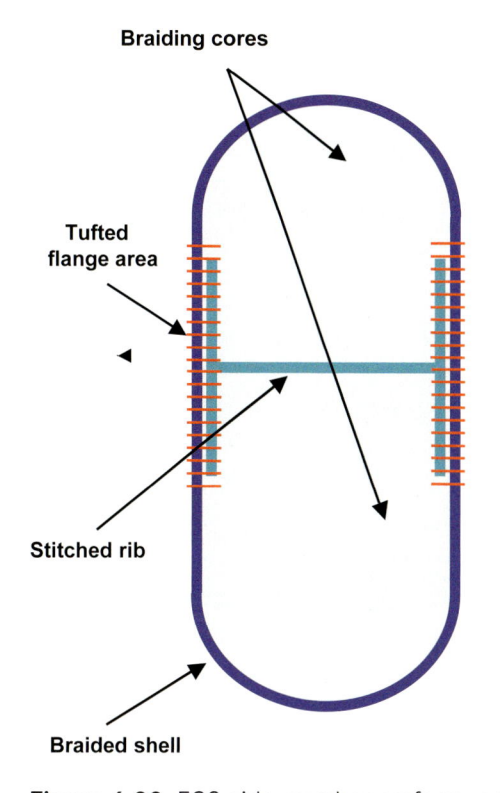

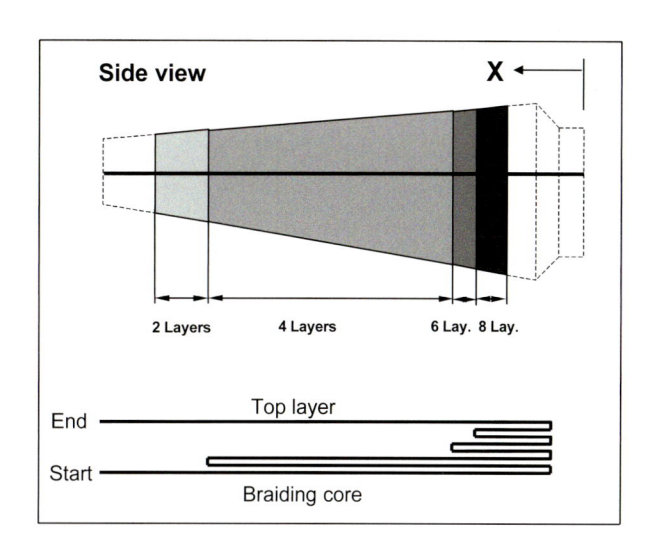

Figure 6-98 FCS side member preform: cross-section and laminate structure

axial non-woven made of carbon fibers are stacked as required by the layering structure previously defined. The stitching in the middle part, which is required to prevent delamination, is done by the manufacturers of the non-woven fabric using a multi-stitch machine. The fabric is supplied on rolls and rib blanks are cut out with the aid of templates. Employment of a CNC cutter means that this step can be automated without any problems for higher production numbers.

In order to obtain the geometry required for the shell and the specified fiber orientation, the direct preforming method using a circular braiding unit was selected.

Since conventional circular braiding equipment does not provide for layered structures that change longitudinally, the specified increases in wall thickness could not be achieved until the braiding machine had been specially modified for that purpose.

The required sequence of layers was achieved by a forward and backward movement of the core through the braiding ring. As a prerequisite for accurate and repeatable positioning of the wall thickness transitions, additional clamping devices had to be developed and connected to the braiding machine in both its mechanism and control system.

Figure 6-99 Circular braiding machine

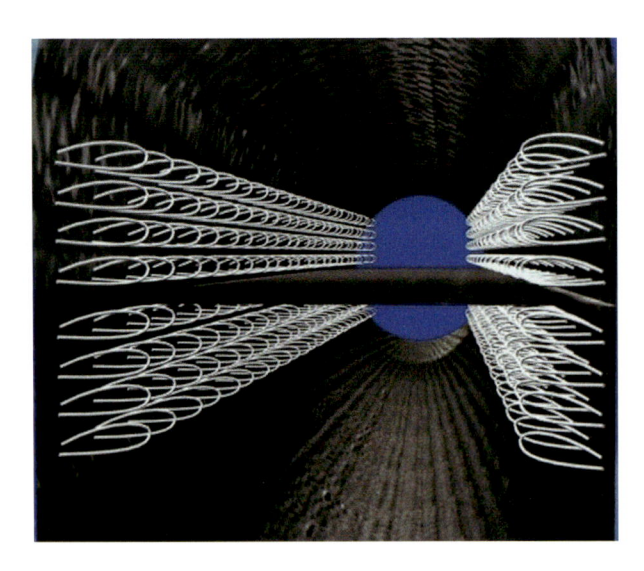

Figure 6-100 Path of tufting threads in FCS pre-form

A two-part core made of polystyrene particle foam was used for braiding while simultaneously holding the rib blank in the correct position. During the following tufting process, it stabilized the preform and even provided protection against deformation and damage during transport to the next process step.

If controlled and reproducible collision behavior is to be secured, it is of outstanding importance to have an effective connection between the rib sides and the woven outer shell. By itself, the resin applied during impregnation to provide this connection is inadequate, since delamination may neutralize the strengthening effect of the rib. Stitching the braided areas as is done when the web material is made cannot be employed in this production stage, since the braiding core prevents access to the preform interior. Instead, tufting of glass fiber yarn was selected as the technique of choice.

The characteristic path of the tufting thread is shown in Figure 6-100.

In order to achieve high precision in reproducibility of parameters relevant to the results, such as stitch width and seam spacing, this sub-process used a newly developed robot-guided and fully automated tufting head. A thin layer of glass fiber attached by means of a binder protects the tufted area from accidentally pulling out the relatively loose loops.

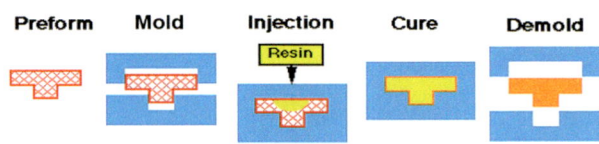

Figure 6-101 Impregnation technique: Resin Transfer Molding (RTM) principle

6.2.3.6 RTM Resin Injection Method

The preform is impregnated with epoxy resin by the resin transfer molding technique. During the prototype phase, a mold consisting of aluminum cores and a CRP outer shell was used.

The production molds, in contrast, are made entirely of steel. In order to allow smooth handling of the preform and component and also to enable an automatic cycle, a special, hydraulic closing device was built for opening and closing the mold and also for performing relative movements of the two mold halves.

In addition, an ejector system, also hydraulically powered, ensured that the component was removed from the mold in a defined manner. Machine specifications were drawn up on the basis of an FMEA analysis taking into account the monitoring of all parameters important for component quality. In addition to monitoring and documenting temperatures, times and pressures, a flow measurement device was installed to ensure the correct mixing ratio.

6.3.2.7 Trimming the Component

If the desired fiber content is to be guaranteed in both the front and rear ends of the FCS side members, these side members initially have to be made with overlengths. Trimming to their final length is done on an automatic trimming machine employing two diamond-toothed, rotating cutter disks.

6.3.2.8 Connection Method

In order to allow for easy access for servicing and repair, the front-end collision structure must be connected to the vehicle sections behind it so that it can be

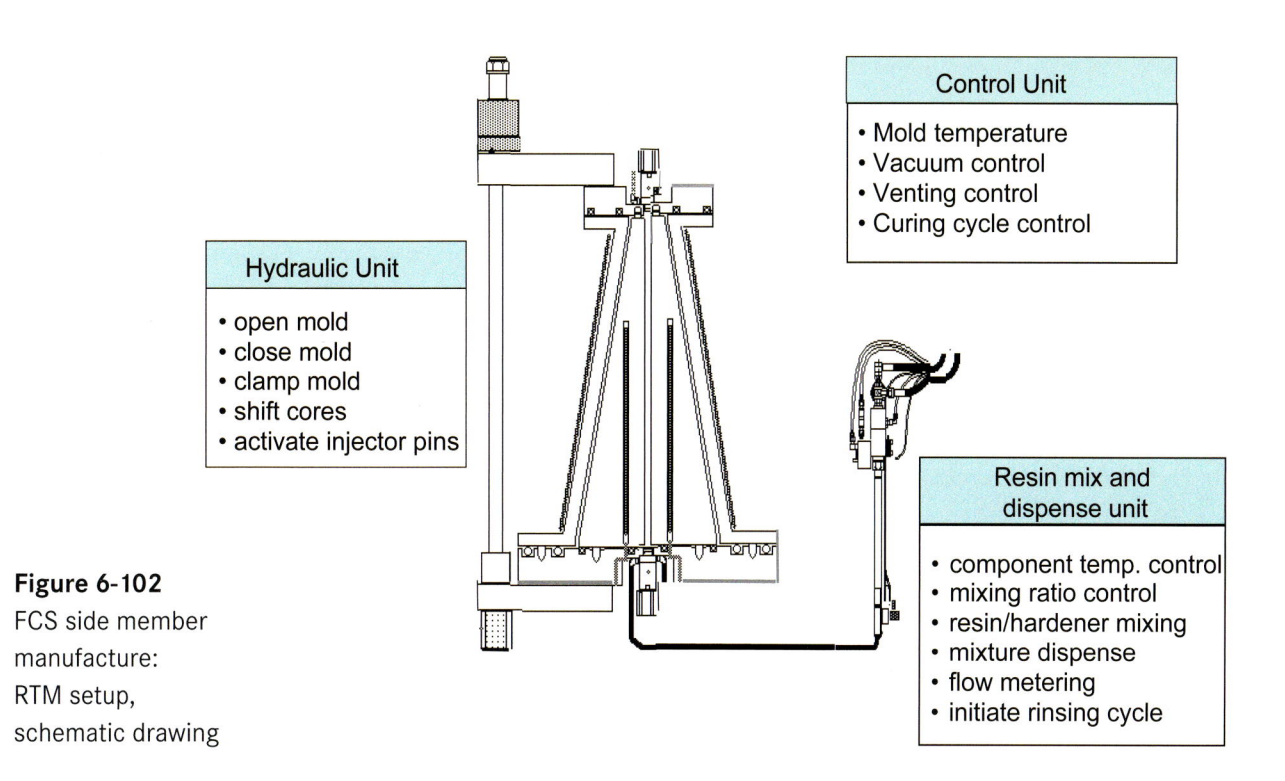

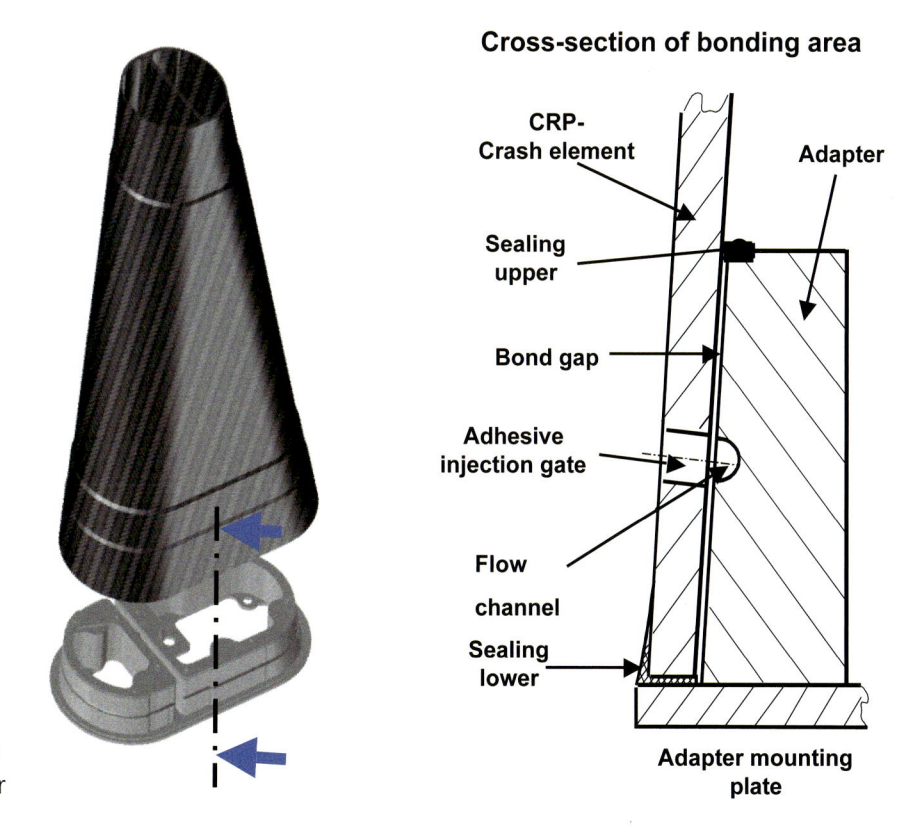

6.2 Front Modules, Crash Elements, Safety Concepts

Hydraulic Unit
- open mold
- close mold
- clamp mold
- shift cores
- activate injector pins

Control Unit
- Mold temperature
- Vacuum control
- Venting control
- Curing cycle control

Resin mix and dispense unit
- component temp. control
- mixing ratio control
- resin/hardener mixing
- mixture dispense
- flow metering
- initiate rinsing cycle

Figure 6-102
FCS side member manufacture: RTM setup, schematic drawing

Cross-section of bonding area
- CRP-Crash element
- Adapter
- Sealing upper
- Bond gap
- Adhesive injection gate
- Flow channel
- Sealing lower
- Adapter mounting plate

Figure 6-103 Joints between FCS side member and adapter

removed. Since strict operational strength and collision requirements apply here, a screw connection was selected. To route the forces into the CRP crash elements, adapters made of aluminum were developed, which were then structurally bonded to the elements. Since this is a safety-related connection, the highest level of process reliability was a focus of the development.

Applying adhesive to one or both mating surfaces and then joining them did not seem to be an appropriate method of securing the required adhesive gap width and adhesive distribution. With the process technique actually selected, the components are positioned on a holding device and the adhesive is then injected into the adhesive gap. A channel in the aluminum adapter contributes to rapid and even distribution of the adhesive. A rubber seal with precisely defined hardness prevents unwanted flash of adhesive and simultaneously allows for the required venting.

Small notches in the seal result in a controlled, minimal discharge of adhesive and thus permit simple and dependable visual monitoring of adhesive distribution.

6.3.2.9 Corrosion Protection

Fiber-reinforced plastics are usually not susceptible to corrosion. However, due to the electrical conductivity of carbon fibers and their comparatively high position in the electrochemical series for metals, there is still the risk of galvanic corrosion if metal surfaces are in contact with carbon fibers and if an electrolytic solution (such as water with dissolved salts) is able to reach the interface. Tried and tested preventive measures include the use of an insulating layer between the CRP and the metal, or the use of stainless steel.

Both methods are used in the SLR collision structures. Stainless steel inserts are used as mountings for these structures. A (knitted) glass-fiber layer is attached at the rear part of the interior faces of the preforms before they are put into the RTM mold and impregnated with epoxy resin at the same time as the carbon shell. This creates an effective insulation between the collision structures and the aluminum adapters bonded-in later.

6.3.2.10 Quality Assurance and Process Control

At the start of the project, investigations were carried out using various non-destructive component inspection methods, such as ultrasound, computer tomography and thermography. However, apart from the high expense involved, the useful information obtained from the results was not very convincing for this particular application. The quality concept was therefore defined by the intention to do away with component inspections as much as possible and instead to monitor and document the process parameters of decisive importance for function and quality. For automated processes, the documentation was integrated into the machine control system; for steps carried out manually, execution was logged on control sheets by the responsible personnel.

Including the required receiving controls, a total of 105 parameters are monitored from the making of the rib material to the bonding of the FCS side members and the adapters.

6.2.3.11 Summary and Outlook

Collision structures made of carbon fiber provide a very high weight-specific energy absorption. In the case of Mercedes-Benz SLR McLaren, this was the first time that collision-related structures made of CRP were incorporated in a standard production vehicle with worldwide type approval.

Collision structures made of fiber composite plastic, particularly reinforced carbon fiber, have a high potential for applications, not only in high-performance sports cars, since they contribute to meeting the increasing requirements relating to collision performance but without the disadvantage of a high level of additional weight.

Innovations such as the combination of automated textile preform technologies, further technical development of the RTM resin injection technique and of adhesive injection have the technical and commercial potential to transform CRP applications from their niche market to the next higher production volume league of car manufacturing's small series production.

Acknowledgements

This development was awarded the Grand Innovation Award 2004 "Body Exterior" by the Internationale Gesellschaft Für Kunststofftechnik e.V., the German section of the Society Of Plastic Engineers, Inc. (SPE).'

To conclude this report, mention should be made of the numerous suppliers involved in the development of materials, production processes, installations and component manufacture. Their contribution to this challenging innovation was obviously essential to accomplish the start of production on schedule.

6.2.4 Thermoplastic Crashboxes

Holm Riepenhausen

6.2.4.1 Influence of the New EURO NCAP Guideline on Bumper Development Work

Within the context of new vehicle development work, suppliers of bumper fascias see themselves confronted by the requirements of EU directive 2003/102/EC. This defines the limit values for the protection of pedestrians and other road users before and during a collision with motor vehicles. With effect of 1 October 2005, no vehicle certifications or type approvals will be issued in the European Union for vehicles of 2.5 tons or below which do not satisfy the technical provisions relating to pedestrian protection set out in Sections 3.1 or 3.2 of Annex 1 of this directive. From 1 September 2010, vehicles will have to comply with the limit values in Section 3.2 to obtain type approval within the EU.

This section lays down the limit values for knee bending angle, knee shearing displacement and maximum acceleration.

Once the vehicle contours have been largely defined by the vehicle type and styling, the control of the movements of the lower impactor (or also the leg impactor) required for securing compliance with the limit values can be worked out via the elasticities of the impact zones, for example. Defined braking of the impactor usually means that more installation space is required, and as a rule this is not available. Our task, therefore, is to make the most efficient use possible of the space between the bumper fascia and the flange plate of the longitudinal member, in other words, to increase the efficiency of crash management.

6.2.4.2 How Crashboxes Work

Metal crashbox systems of very different kinds have established themselves on the market, using approaches such as tapering tubes, telescopic tubes, folding boxes, foams or shear boxes. Some solutions even use plastic in the form of honeycomb structures or composite fiber boxes, for example. Each and every solution is a compromise between function, weight and costs. This is where the new "thermoplastic crashbox" starts.

6.2.4.3 Working Principle of the REHAU CMS Crashbox

The REHAU CMS crashbox dissipates collision energy by utilizing a cutting process. The shear force of metals is expressed in simplified form by:

$$F_C = b \cdot h \cdot k_c$$

whereby k_c is the specific cutting force (a function of the rise per tooth, material-dependent).

With thermoplastics, the influence of temperature also needs to be quantified. All other relevant parameters, such as cutting speed, cutting material, cutting geometry, cutting wear, etc., are available for metals as empirically acquired characteristics and correction factors. But for plastics, data of this type are simply not available in the literature. For this reason, all relevant influencing factors have to be acquired empirically.

Table 6-11 Definition of the statutory requirement for the impact of the lower legform against a bumper as specified in EU Directive 2003/102/EC Annex 1

Specification	Year	Impact speed	Knee bending angle	Knee shearing displacement	Acceleration
Section 3.1	2005	40 km/h	21°	6 mm	200 g
Section 3.2	2010	40 km/h	16°	6 mm	150 g

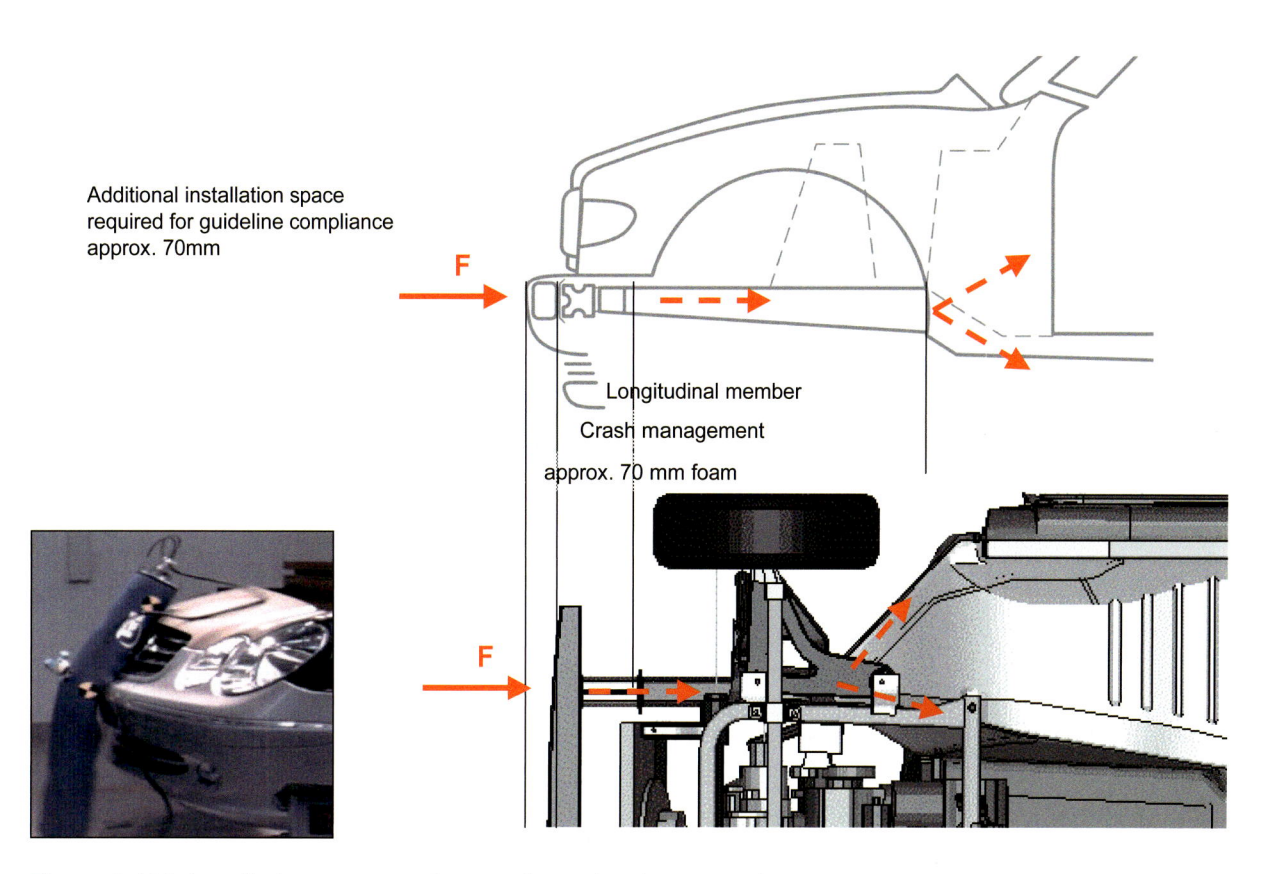

6.2 Front Modules, Crash Elements, Safety Concepts

Additional installation space required for guideline compliance approx. 70mm

F

Longitudinal member
Crash management
approx. 70 mm foam

F

Figure 6-104 Installation space requirement for pedestrian protection

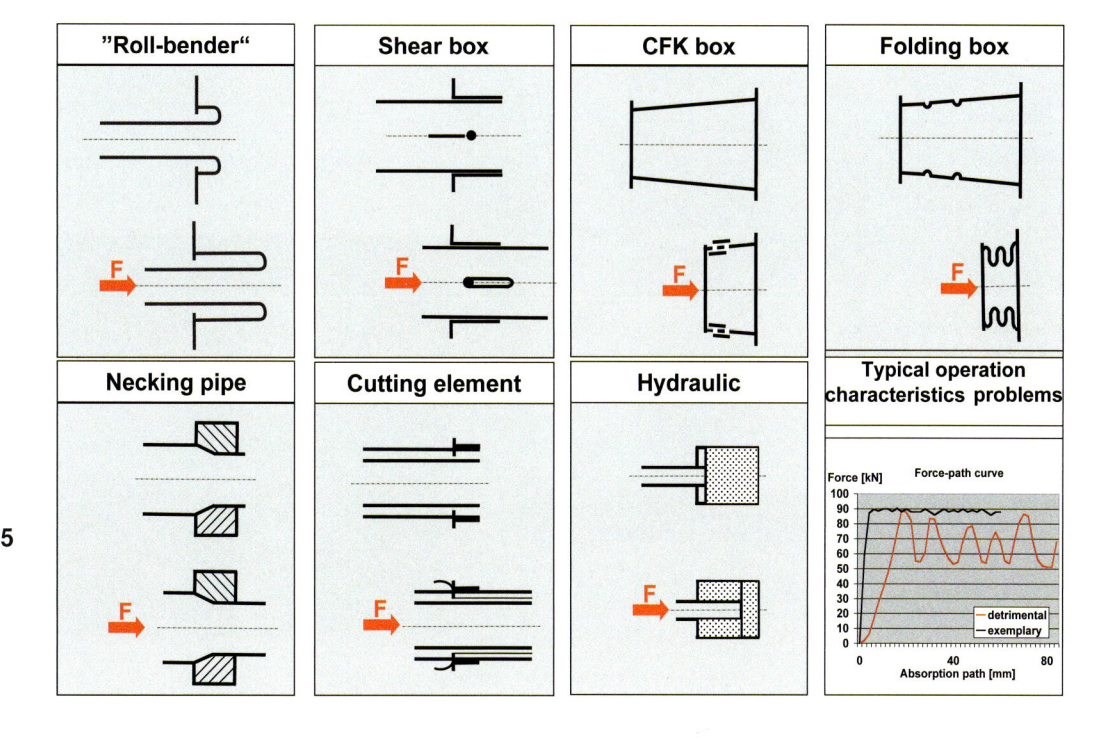

"Roll-bender" | Shear box | CFK box | Folding box

Necking pipe | Cutting element | Hydraulic | Typical operation characteristics problems

Force-path curve
— detrimental
— exemplary

Figure 6-105 Overview of energy absorption principles

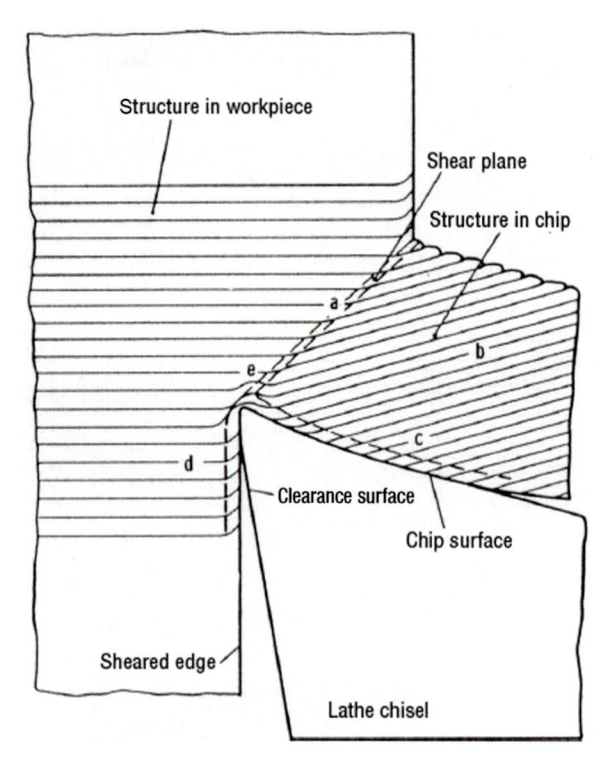

Figure 6-106 Working principle of crashbox

6.2.4.4 Test Setup

For the test setup, a cylindrical cross-section was selected for the thermoplastic absorber together with a coaxially mounted cutting plate with a variable number of cutting elements whose shape and cross-section can also be altered.

6.2.4.5 Determination of Relevant Influencing Factors

Influence of the Cut Section

In a first step, it was possible to demonstrate that the linear relationship between cutting force and the cut section known for metals also apply here. This linear dependence could be proven in both static and dynamic tests. This assumption simplified dimensioning at an early stage as well as subsequent fine adjustment of the cutting forces and thus of the energy level via the pipe cross-section and the cutting disk.

Figure 6-107 Test setup

Figure 6-108 Cutting plate

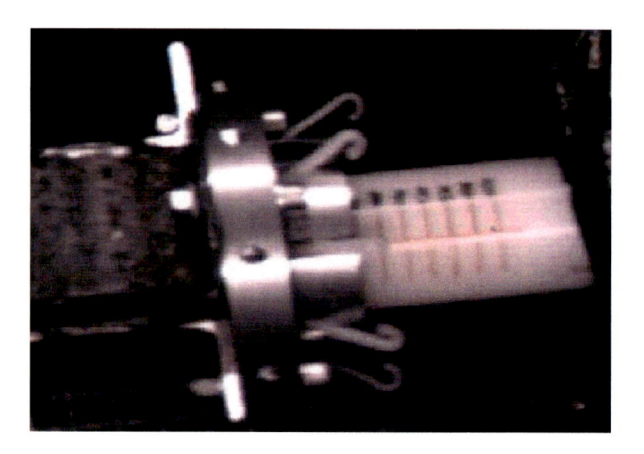

Figure 6-109 Dynamic test

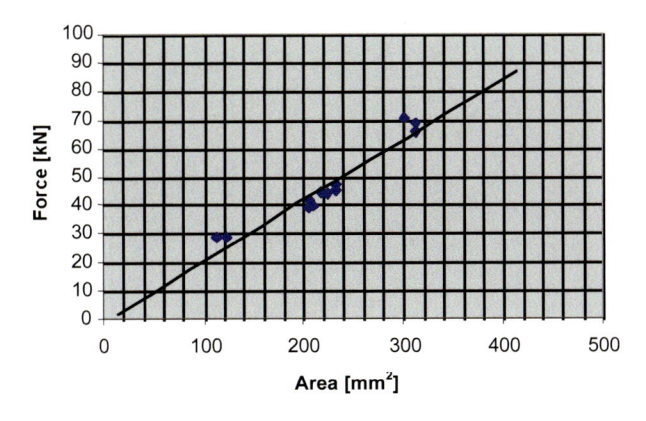

6.2 Front Modules, Crash Elements, Safety Concepts

Figure 6-110 Force-to-area ratio (dynamic), force [kN], area [mm²]

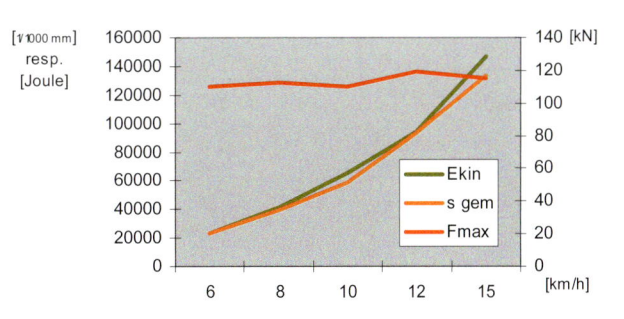

Figure 6-111 Theoretical energy over measured cutting path and at maximum force measured

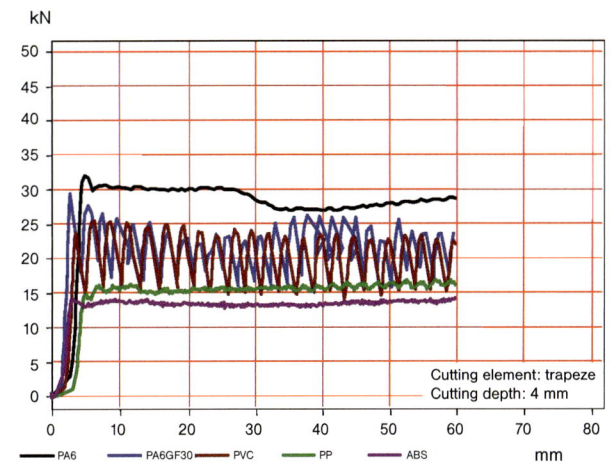

Figure 6-112 Comparative curves of various plastics

Influence of Cutting Speed

In the real application case, the cutting speed between the crash tube and the cutting plate cannot be controlled: cutting speeds as a function of the impact velocity of the vehicle range from 1.6 and 4.2 m/s for a low-velocity collision and from 15 and 18 m/s in the high-speed range. What would be desirable would be a more or less constant force level over the entire speed range.

In fact, the test series indicates that the dependence on speed is negligible. Only in the case of static loads does a drop in the force level occur and this depends on the material. Here, a first requirement for obtaining a satisfactory efficiency is met, since the force level does not change significantly during the course of the energy consumption process.

Influence of the Material

Analogous to machining metal, the cutting force may be expected to be dependent on the material applied. Indeed, the type of material exerted a significant influence that qualitatively affected not only the force level, but also the course of the curve.

Embrittled areas have a recognizable effect on the force level curve, depending on the material. In general, fiber-filled materials appear unsatisfactory due to the amplitude fluctuations that occur in the force level.

However, it was quite possible to identify materials with homogeneous force curves in the dynamic test (up to 40 km/h). The steep increase in force combined with a homogeneous force curve (a constant force level) is a guarantee for high crashbox efficiency.

Due to the satisfactory nature of the design properties and the availability of relevant material data (ageing, for example) the tests were subsequently continued using polyamide. The use of other materials is, however, perfectly conceivable.

Influence of Temperature

The very word "thermoplastic" indicates that its material properties are dependent on temperature. However, since it has not yet been possible to identify any necessary relationship between a specific material property and shear loads, here too it appears that such a dependence would have to be determined empirically.

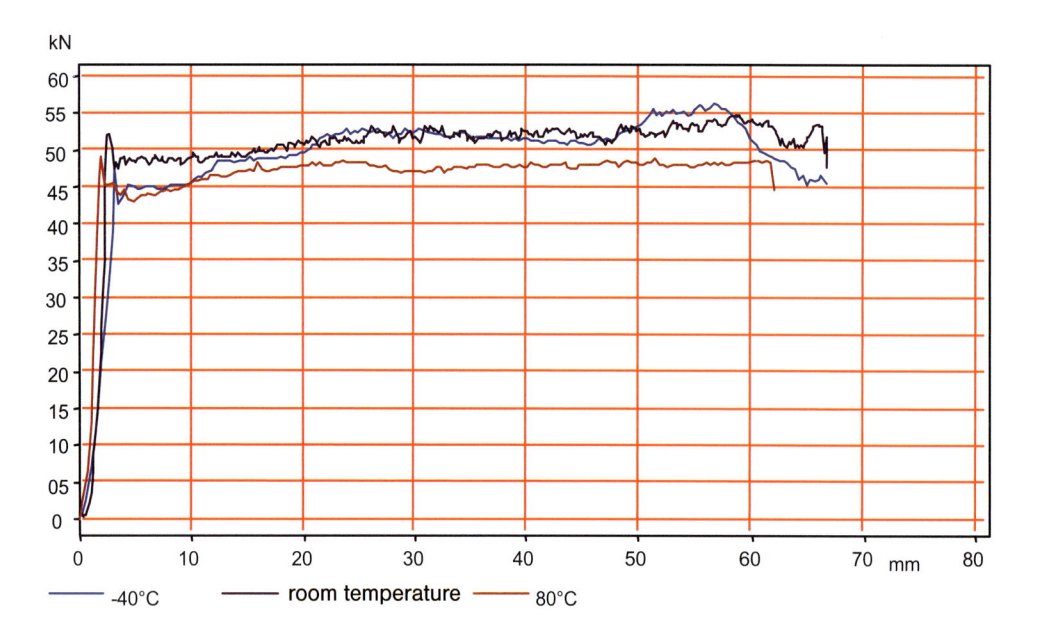

Figure 6-113 Comparison of curves at various temperatures (PA6)

It was astonishing that, in the relevant temperature range of –40 °C to +90 °C, we were able to detect only relatively minor temperature dependence during the dynamic test.

Summary of Influencing Factors

In analogy to machining metals, the shear force ($F_c = b \cdot h \cdot k_c$) of plastics is heavily influenced by chip size and other material-specific properties. The desired energy level can be tailored very well by means of chip size and number of chips.

The force curve and thus the energy level are more or less constant over the relevant speed range.

The dependence on temperature is relatively low, depending on the material, and can be controlled further by making the appropriate changes to the formulation of the plastic. In all tests carried out with individual absorbers, the crashbox was found to have an efficiency of 85–95%.

6.2.4.6 From the Crashbox to the Energy Management System

No conclusions can be drawn regarding the functionality of an entire system solely on the basis of the properties of a crashbox. Not only collision-related requirements but also the individual requirements of the various OEMs are of decisive importance for the further design and development of the overall system

For this reason, tests were continued on complete systems, not only with aluminum bumper beams (high-strength extruded profiles), but also with steel top-hat sections (high-strength cold-formed steels), first using a dolly and subsequently with complete vehicles.

Collision Requirements

- Pendulum impact, externally flush 4 km/h, no damage
- Pendulum impact, corner 30° at 2.5 km/h, no damage
- Pendulum impact, central, 4 km/h, no damage
- Wall impact, 4 and 8 km/h, damage according to design specifications
- AZT test 15 + 1 km/h onto 10° barrier, without damage to vehicle structure
- Pole test, frontal, 15 km/h, with preservation of bond
- IHS test, 30° barrier, 8 km/h
- ODB crash, 30 km/h
- High-speed collision, with preservation of bond

6.2 Front Modules, Crash Elements, Safety Concepts

Cutting element combined with aluminum bumper beam (F 65 kN)

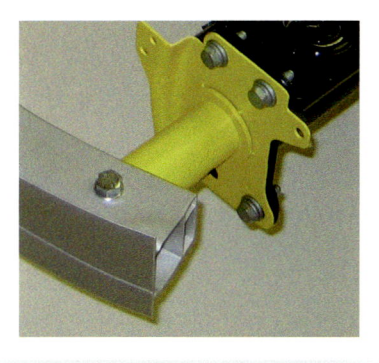

Cutting element combined with steel top-hat sections (F 110 kN)

Figure 6-114 Two design versions

Supplementary Requirements from Customer's Product Specifications

- Alternating climate test
- Hydropulser, alternating climate test
- Towing test
- Lashing over towing lug
- Contribution of structural rigidity as function of drive concept

Individual Targets

- Costs
- Weight

6.2.4.7 Example of Total System Configuration

In order to absorb the occurring shear forces, an absorber tube with an inner metal reinforcement sleeve is welded directly to the bumper beam. The metal cutting plate, which can also serve as a flange for a threaded connection with the 1a beam, is molded onto the absorber tube in the encapsulation process.

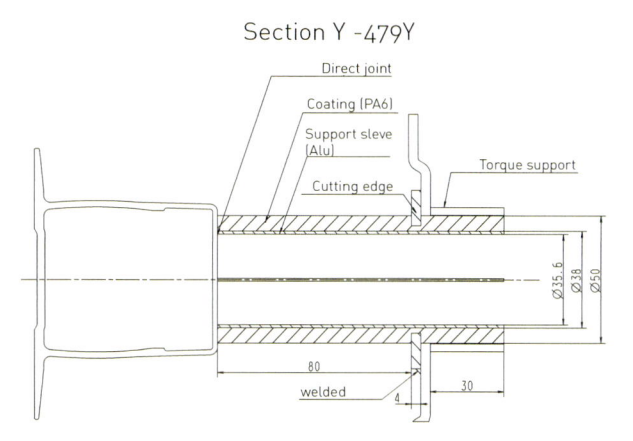

Figure 6-115 Energy absorber and beam to which it is welded

The functionality of the illustrated configuration of a component (hybrid) with inner support sleeve, plastic encapsulation as sacrificial material, and connection to the flange plate has been validated using a testing tool.

| Core end | Inside-thread tow lug | Encapsulation |

| Steel inner sleeve | Aluminum inner sleeve |

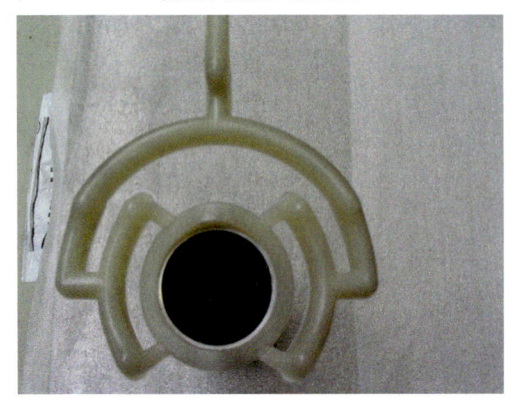

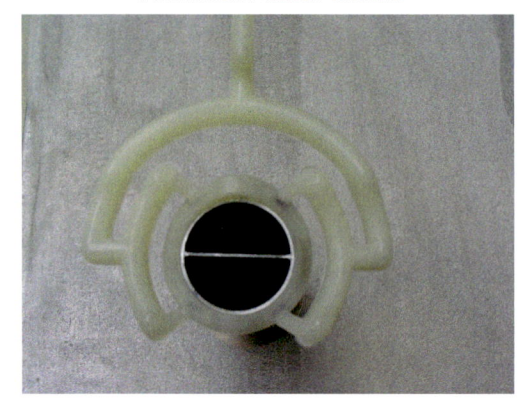

Figure 6-116 Absorber tube details

The advantages of such a design:

- Tensile and compressive forces can be transferred equally via the encapsulated cutting plate

- The plastic acts as electrical insulation between the different metals, thereby enabling free selection of flange and beam materials

- The plastic portion of the component offers economically attractive potential for integrating tow lugs, mounts for headlight wash-fluid nozzles, cable fasteners, etc.

- The plastic portion also has outstanding surface slip properties, resulting in comparable force levels for straight and diagonal force transfer

- The system does not require any residual block length worth mentioning at full compression

- The weight of a crashbox with flange plate and inner sleeves for a force level of 100 kN is only 1000 g

- If the top-hat profile is open to the front, additional packaging advantages result

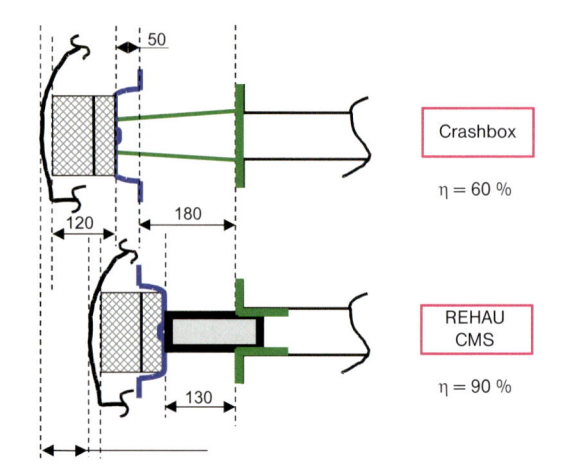

Crashbox $\eta = 60\ \%$

REHAU CMS $\eta = 90\ \%$

Figure 6-117 Packaging advantage from combined thermoplastic absorber and inverted top-hat profile

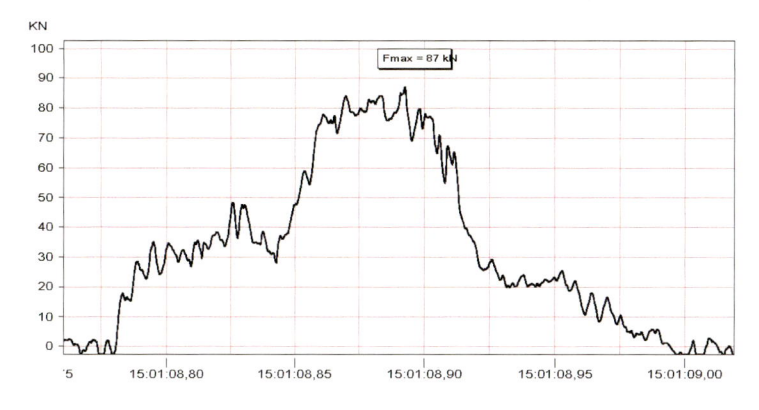

Figure 6-118 Preliminary ODB crash test

6.2.4.8 System Testing

All of the component properties found in the component tests are also present in the system tests. However, the achievable energy level is also determined to a decisive extent by the properties of the bumper beam.

Sensitivity to Shear Load

Under load application up to 20° off-center, the crashbox was functioning at the same energy level as with a straight application of force, even during component testing. Depending on the overall rigidity of the composite, even greater angles are possible in the system. Uncontrolled crumpling of the crashbox is not to be expected until angles are even greater.

This results in a favorable insurance classification even with type approval on the basis of the IHS test (10° barrier). One side effect is that at two-stage airbag deployment, the level of expected damage can be detected more easily and precisely by the acceleration sensors.

Composite Strength

With the corresponding configuration, structural strength is retained even in the rigid-pole test. Although the top-hat profile has collapsed and the high-strength steel beam has broken, the composite is preserved over the more ductile internal steel tubes.

In this loading case the plastic portion of the component is irrelevant.

Structural Rigidity

Due to the limited material strength of the plastic, but also due to the relatively small surface area where the internal metal tube is connected to the top-hat section, transmission of torsional and bending forces is limited. This means that, in direct comparison with closed and welded cross-members and folding sheet-metal boxes, we will find reduced strengths in the front end of the vehicle and at lower natural frequencies. This requirement is, however, not of primary importance in all vehicle concepts.

Figure 6-119 Dolly with vehicle weight

6.2.4.9 Concluding Assessment

The suitability of a thermoplastic crashbox depends to a significant extent on the customer's requirements profile. Let us now review the advantages and drawbacks of the system.

Advantages:

- Outstanding efficiency
- No remaining block length worth mentioning
- Which means outstanding packaging
- Constant force level without peaks even with oblique incidence of force
- Electrical isolation of the metal structures
- High integration potential (tow lugs, and so on)
- Absorber tube means shorter folding of the 1a beam, thereby increasing its efficiency

Drawbacks:

- Low contribution to the structural rigidity of vehicle
- Open 1a beam required to allow mounting of the absorber tube
- Cross-section of the 1a beam depends on the size of the absorber tube

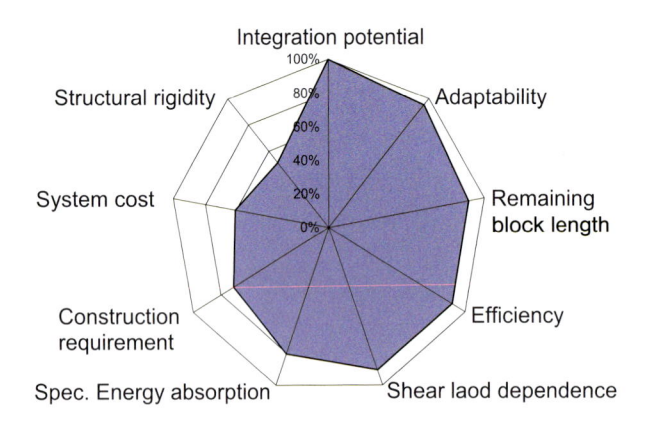

Features:

+ highly efficient (85%-95%)
+ no significant remaining block length
+ minimal installation space
+ very high specific energy absorption (relativ to weight)
+ easy adaption to force level
+ no significant characteristic curve change under partial shearing
+ thus better modulation of up-front sensors
+ integration of tow lugs possible
+ potential barrier due to encapsulation

− low contribution to structural rigidity of vehicle
− open 1a beam

Figure 6-120 Property profile cutting element

6.3 Roof Modules, Hardtops

6.3.1 Roof Module: As Exemplified by the New Opel Zafira

ALAIN LEROY

6.3.1.1 Introduction

Although only a few years ago the car buyer was still content with a simple sliding roof when it came to optional roof systems, the trend has been developing increasingly in the direction of a feeling of greater spaciousness in the vehicle interior. This trend is probably also due to the greater availability of air-conditioning units on the automotive market. Today's customer is also calling for greater variability even in the roof systems, particularly in the case of family and so-called lifestyle vehicles. In many cases, stowage facilities was lost in the instrument panel area, particularly as the result of ever more sophisticated cockpit features and accessories, not to mention the space required by the front-seat passenger and knee airbag systems. The previously unused space in the roof area lends itself to substitute for this loss. Some familiar examples of developments in this direction can be seen in the interior roof console of the Opel Signum and even more clearly in the Citroën Berlingo Spacelight and the Nissan Quest.

For this reason, Opel decided to follow this trend in its new Zafira. The goal in the development of the optional roof system for the new model was to be able to offer as much spaciousness as possible – in other words, large glazed areas in the roof – but also at the same time generous stowage capacity in the roof area.

In order to increase this stowage space, it was decided to equip the roof with an additional external roof box (like the Zafira Snowtrekker concept car presented at the Detroit Auto Show in 2000).

Figure 6-122 Citroën Berlingo Spacelight ceiling

Figure 6-121 Citroën Berlingo Spacelight top

Figure 6-123 Nissan Quest ceiling

Figure 6-124 Top view of Opel Zafira with roof module

Figure 6-125 Interior view of Opel Zafira ceiling

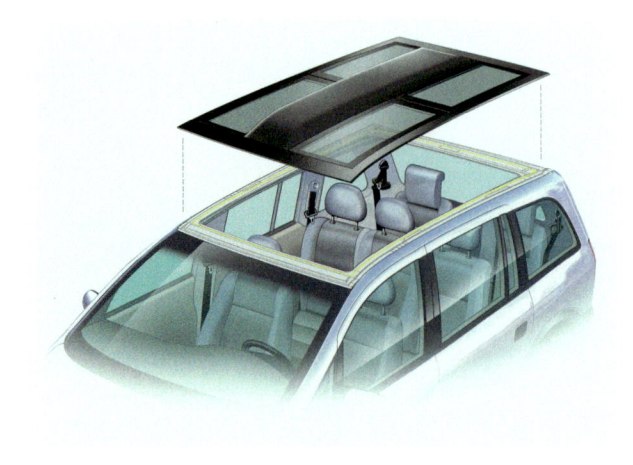

Figure 6-126 Roof module of Opel Zafira

This gives the vehicle with the roof module an independent, striking appearance, even seen from outside, underlining the functionality of special equipment by the car's appearance.

A system with this degree of complexity will necessarily mean more individual parts and thus increased assembly requirements. Taking the limited space available in the production plant into consideration, increased assembly requirements cannot be accommodated unless the complete roof system is delivered to the line pre-assembled – in other words, as a modular roof. Not only the complexity but also the design called for solutions which cannot be implemented using conventional designs, such as a steel roof. Therefore the concept of a plastic main support for the optional roof system was developed. Development work for this roof was carried out in collaboration with our partner Webasto until the concept could be implemented to mass production.

6.3.1.2 Overall Concept and Configuration

Basically, there are three alternatives for a roof module: (1) a solution with integrated roof interior lining, (2) a purely functional module, such as a sliding roof system without interior lining, or (3) a hybrid concept, including parts of the roof interior fittings. In principle, they can all be designed for assembly from above – a so-called "top-load" system – or from below – a "bottom load" system. The top-load roof module offers the following advantages: assembly is easier and the external styling can be extended right up to the roof strip or roof rack – in other words "rail-to-rail". However, with a "top-load" system, integrating the complete roof interior lining is problematic, since typically this lining is wider and longer than the available roof opening. This means that a relatively complicated folding system is required for the liner. Although a possible solution, this still represents an additional design and development risk. For this reason, a top-load hybrid concept was chosen for the new Zafira.

A frame made of continuous strands by the roving process was selected (LFI = long-fiber injection) for the

6.3 Roof Modules, Hardtops

main support for the roof module. By itself, this frame weighs about 8.3 kg. The visible part of the external surface is covered with a decorative plastic film. This concept means that the special design feature of the higher external box as well as the complex internal configuration can be achieved with a minimum of components. The glass panes, which are bonded directly to it, also make an important contribution to satisfying the structural requirements for the roof module. The extreme width-to-length ratio of the glazed areas did, however, demanded a division of the glass area in the region of the middle transverse member to satisfy the practicalities of manufacturing. Terminating profiles were provided at the front and rear both to protect the glass edges and for visual reasons. The front profile extends partially around the sides and is used for closing the gap between the roof module and the roof rack for aero-acoustic reasons. Rails made of extruded aluminum and also a transverse member made of sheet steel serve as guidance or support for the sunshade system. They also contribute additional reinforcement to the unit. A total of five stowage boxes, also made of plastic, are grouped together into two sub-assemblies which are screwed to the LFI frame.

The complete roof module is bonded directly to the body with a polyurethane adhesive using a method similar to that commonly used with automotive glazing. The only step taken to provide for mounting the roof module in the bodywork was to replace the steel roof used in the basic model of the vehicle with a welded-in adapter frame. This adapter frame serves structural demands and closes the roof frame profile at front and rear which would otherwise be open, and also ensures adequate flange length for the direct bonding of the roof module. Another advantage of this concept is that further roof module versions can be implemented without the need for body modifications; this in turn means that they can be built at any time without the need for layout changes in the production plant. Changes at short notice are thus possible, without interrupting production, or when the plant closes for a period or during model year changes. Since addition or replacement is carried out in a way similar to replacing a car window, new versions for the aftermarket or for special editions are entirely conceivable and are feasible without the need for conversion work on the body structure.

The advantages of a plastic frame for the roof module are obvious. Mounting elements for functional parts can be integrated into the frame without any problems. In addition, it would be possible to integrate complicated shapes of varying cross-section within one part. The wall thickness of the LFT, which was selected to suit local conditions, means that a dimensionally stable structure can be obtained without additional reinforcement. Behind the decorative film in the area of the centrally located longitudinal box, the wall thickness of the foam backing is about 7 mm.

As protection against the sun, tinted glass with a light transmission value of 10% is fitted, as are sun blinds with a knitted, visually opaque material. All four sun blinds beneath the corresponding windows are connected to each other mechanically and can be operated electrically via a switch in the front part of the roof frame. Outside the transparent area, the glass has been screenprinted black. The external color of the base frame in the visible area has been matched to the black glass fritting and is therefore also very glossy. This restriction to just one color thus rules out any color-matching problem, which could otherwise occur, and thus allows for one part number for the complete unit. This means there is no need for delivery to the production line in assembly sequence (SILS). This provides a

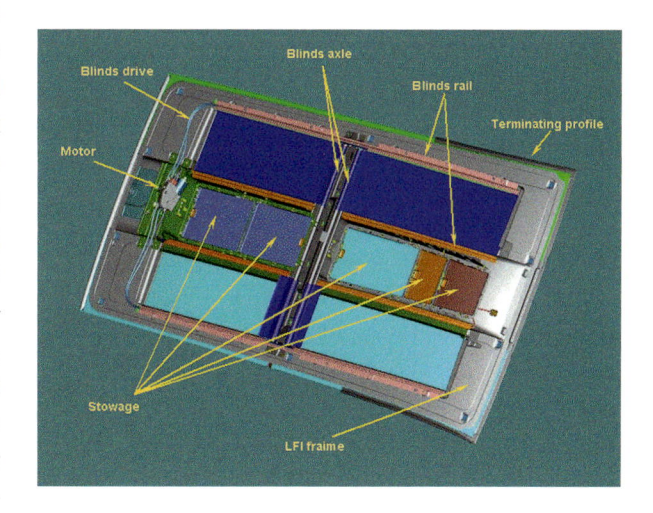

Figure 6-127 Roof module elements of Opel Zafira

significant logistical advantage. For the same reason, the various roof antennae are fitted in the production plant and not pre-fitted at the roof module manufacturing facility. On the other hand, the high-gloss black visual effect required makes very strong demands on surface quality, particularly regarding resistance to scratching and scuffing. This played a decisive role in the selection of the film uses for the outside box.

There are five stowage boxes of different sizes. For safety reasons, they are all provided with a twin-latching lock and an opening damping device.

It goes without saying that a specific headliner with the corresponding openings in the glazed area is required for this roof version. The edge foldovers of the headliner are used as guide elements for the sun blinds. The junction between the stowage boxes and the headliner is formed by two frame surrounds which serve to even out minor installation inequalities.

6.3.1.3 Collision Characteristics and Structural Rigidity

The vehicle and roof module meet all requirements regarding both the frontal collision and the rigid pole collision. This can be mostly attributed to the fact that the bodyshell structure alone already provided good results and the roof itself only needed to make a relatively small contribution to collision properties. Furthermore, the Zafira with its height typical for an MPV profits from the fact that the roof is less severely loaded in the pole collision than sedans with a lower overall height.

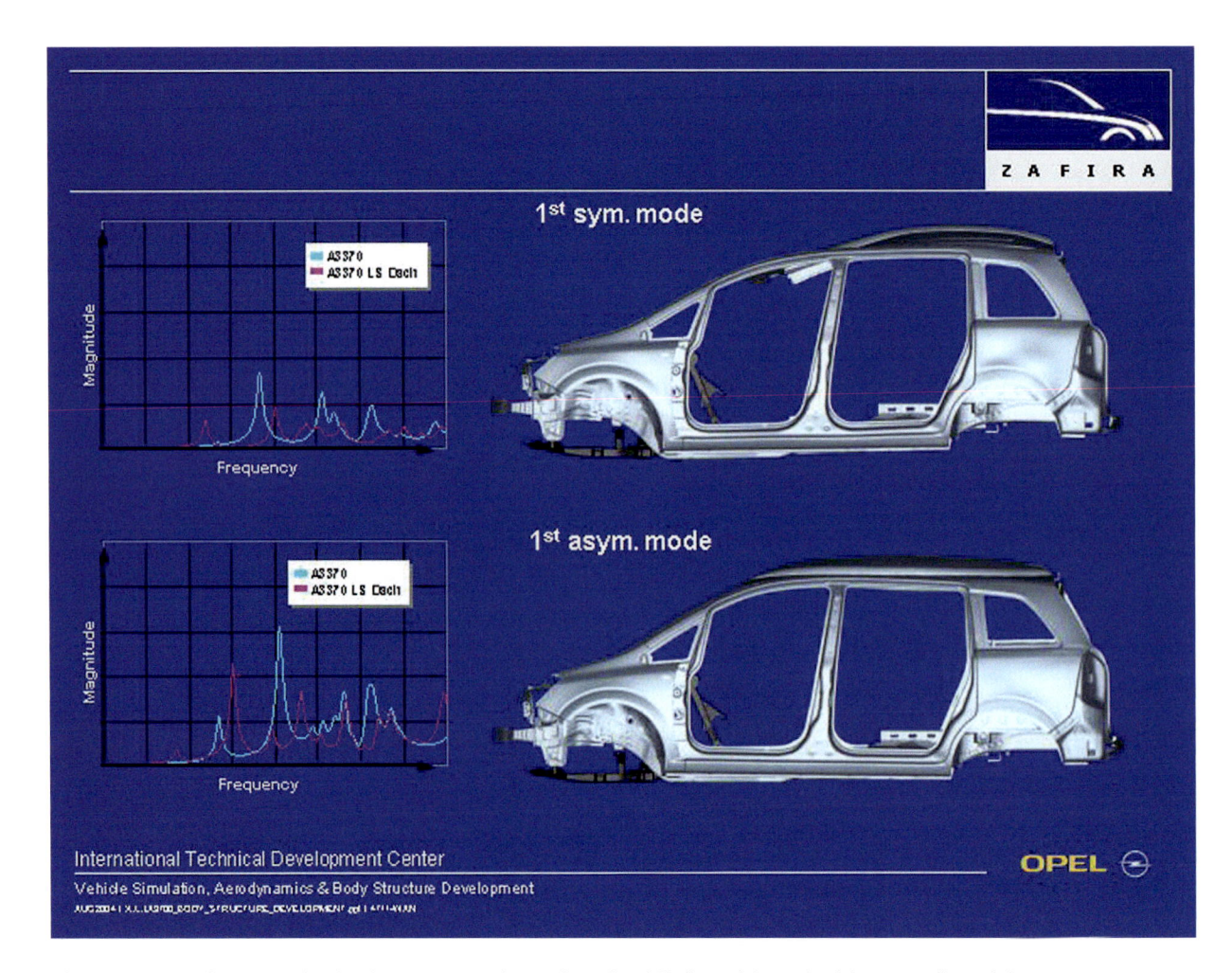

Figure 6-128 Comparative body structure dynamics: Opel Zafira with and without roof module

6.3 Roof Modules, Hardtops

Table 6-12 Structure comparison of full roof and roof module

Version	Full roof	Roof module
Dynamic simulation		
1. Torsion	33.3 Hz	32.8 Hz
1. Bending	38.3 Hz	32.4 Hz
1. Roof panel	< 50 Hz	21.0 Hz
Static simulation		
Structural rigidity (kNm/deg)	13.7	12.7

Both loaded boxes and their locks came out well in the collision simulation and also in the real collision test. The stowage boxes are designed for an ultimate load in the front two small boxes of 1 kg and 2 kg in the three larger rear boxes.

In the case of dynamic simulation, it emerged that the natural frequencies in both bending and torsion, although somewhat less than with a full roof, were still within requirements. The static structural rigidity of the vehicle with roof module is at 12.7 kNm/deg actually greater than that of the predecessor model with conventional sheet metal roof (12.4 kNm/deg).

6.3.1.4 Support Frame

The support frame is the core component of the roof module and with a weight of approx. 9.9 kg and external dimensions of 2100 mm × 1100 mm it is also the largest individual component. It serves primarily as an assembly support to which screwed, bonded and even foamed components are also fitted. It has a continuous adherent surface to which the polyurethane adhesive bead is applied and the roof module is then bonded to the body during vehicle final assembly.

What Kind of Material Structure is Suitable for this Support Frame?

When making the selection of a suitable material, a comparison was made among all technologies that were considered possible candidates. Various technical requirements were of decisive importance in choosing a suitable structure:

- Low weight
- No marks in the plastic skin film
- High integration capability
- High mechanical properties of the material
- High heat resistance
- Local strength adaptation
- Good dynamic vibration properties
- Single-part implementation of the frame (module requirement – Webasto)

Criteria such as product costs and product availability were also taken into account in selecting the material.

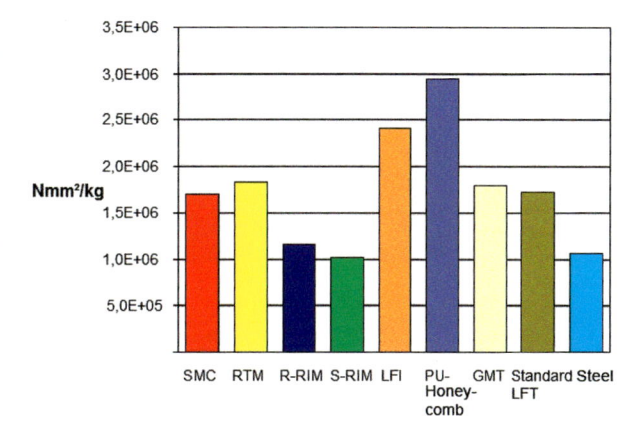

Figure 6-129 Lightweight design coefficients

proportion glass	22 %
max. temp.	120 °C
flexural stiffness	2,7 *E +06 N*mm²
elastic modulus of bending, logitudinal	2000 - 3000 N/mm²
elastic modulus of bending, transversal	2000 - 3000 N/mm²
density	800 g/dm³
impact resistance DIN EN ISO 179	31 KJ/m²

Figure 6-130 PU LFI structure and material properties

Figure 6-131 Comparative coefficients of thermal expansion

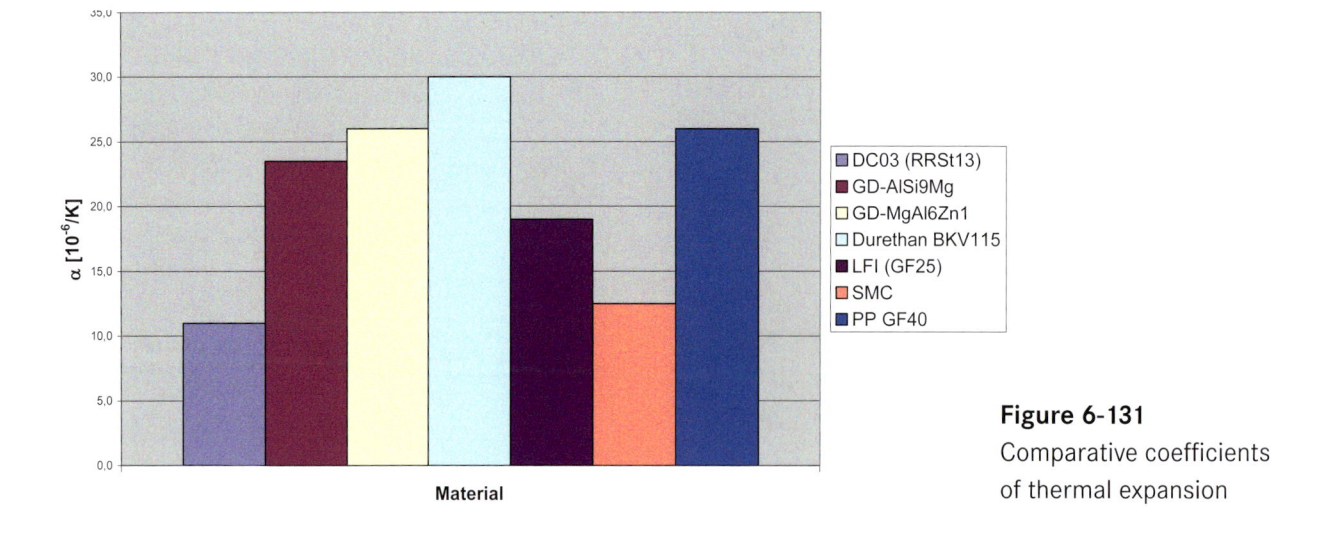

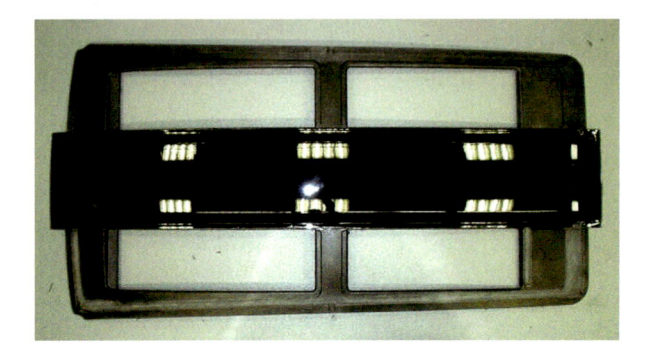

Figure 6-132 Exterior view of PU GF22 frame with GEP Lexan SLX plastic film

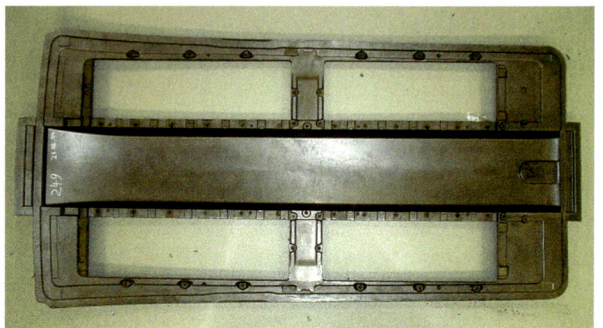

Figure 6-133 Interior view of PU GF22 frame

Support Frame in PU GF22/LFI

Following intensive examination of the possibilities mentioned, our choice fell on polyurethane reinforced with glass rovings, with a glass volume content of 22%. This material structure can be produced by means of an LFI (long-fiber injection) process.

From the engineering point of view, this PU-GF 22 system offers all important properties required for use as an assembly support. Good heat resistance (110 °C) and low thermal expansion of the structure (see Figure 6-131) are the key to reliability in use with a component of this size.

The PU GF material has proved itself an outstanding alternative to thermoplastics or thermosets (such as SMC solutions, for example), especially with regards to the integration of mounting elements and also the realization of abrupt geometrical changes. As an individual part, the frame still has little inherent rigidity.

Due to the subsequent installation of the windows, roof boxes and aluminum guide rails of the blinds, good rigidity properties are achieved nonetheless.

This LFI process is particularly suitable for small and medium-sized production runs.

6.3.1.5 Outer Skin

The outer skin of the roof module extends longitudinally over the entire roof of the vehicle and covers about one third of the entire vehicle roof area (2100 mm × 470 mm). The areas to the left and right are covered by glass panels.

6.3 Roof Modules, Hardtops

Since the outer paneling of a vehicle is noticed first, the main emphasis in our development was on securing an appearance similar to glass with mechanical properties similar to a vehicle paint system. These are characterized by a gloss > 100 gloss units and a scratch resistance according to the Amtec Kistler method (DIN 55 668) with a residual gloss > 40 gloss units.

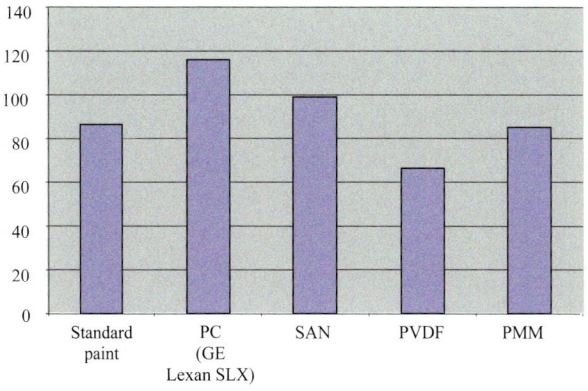

What material structure is suitable for this skin application?

In making our selection of materials, we compared all materials that we considered possible candidates.

6.3.1.6 Skin with Lexan SLX Plastic Film from GE Plastics

Compared with other products, the Lexan SLX polycarbonate plastic film from GE Plastics came closest to meeting requirements and was therefore the material of choice. The laminate structure of the coextruded Lexan SLX plastic film is shown in Figure 6-135.

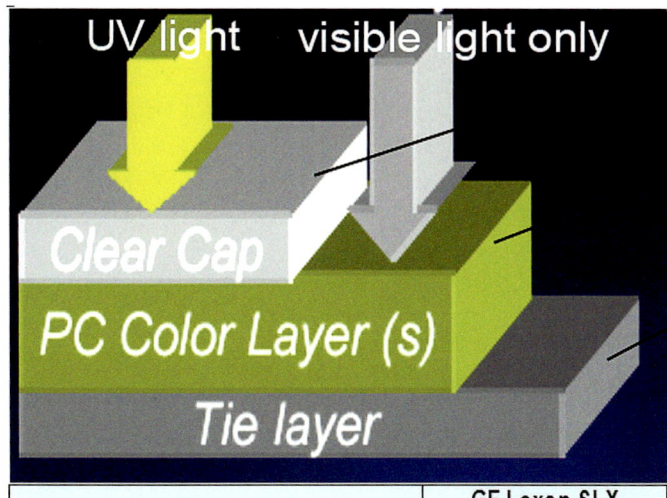

Figure 6-134 Gloss measurement at 20° in accordance with DIN EN ISO 2813

Material properties:	GE Lexan SLX XL 6856
Surface	mirrorfinish (Class A)
Build-up (from clear layer to tie-layer)	PC/claer cap PC ABS/PC
Thermal expansion	71 * 10^{-6} 1/K
Elongation at break	45,00%
Tensile modulus	2000-2100 MPa
Specific gravity	1,2 g/cm³
Thickness	1,3mm
Supplier of protective film	Bischoff & Klein

Figure 6-135 Film structure of GE Plastics Lexan SLX/XL6856 and material properties

- Lexan SLX has surface properties very similar to those of glass and offers the required resistance to any number of environmental influences. The film can withstand the chemicals usually encountered in the real world, its ultraviolet stability as per DIN 75 220-D-OUT-T and its temperature stability (110 °C) and also its scratch-resistance meet the requirements.

- Good heat resistance of the plastic film means that it will be reliable in use.

- Should repair work be necessary, the plastic film can be repainted. Microscratches can be corrected by polishing, thanks to the UV-protection layer. On the other hand, repair of partial areas of the surface is not possible.

- In addition, the plastic film has good processing properties that are of great importance to the process steps of "thermoforming, trimming and back-foaming".

From the technical point of view, it was possible to satisfy all requirements with the exception of "color-matching and color variety".

6.3.1.7 Description of the Process

The first step in the production process is the preparation of the external film for thermoforming. To ensure a clean surface, the film is first cleaned. It is then thermoformed under clean room conditions and cut to the required dimensions. The external film is then inserted into the back-foaming mold to make the

Figure 6-136 Chemical resistance of various surfaces

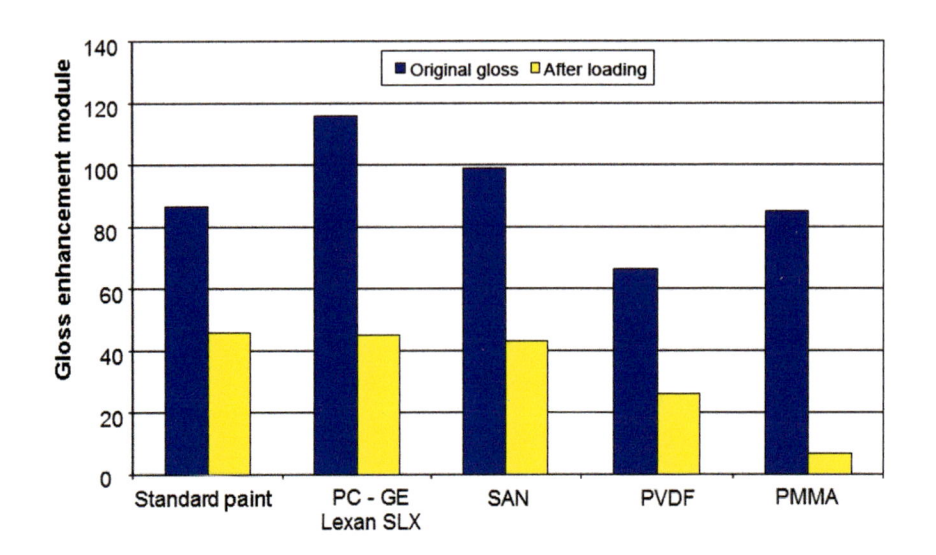

Figure 6-137 Scratch resistance as per Amtec-Kistler car wash test (DIN 55 668)

6.3 Roof Modules, Hardtops

PU-GF support frame and back-foamed. After demolding, the finished frames are deflashed. Next, the glass panels are glued onto the frame with an accelerated polyurethane adhesive. Selection of a fast glue means that the dwell time in the clamping device during the curing process can be limited. The individual components, e.g. the blinds and stowage boxes, can then be fixed by mechanical means on the rear of the module. Once the complete unit is finished, a comprehensive quality inspection is performed. This includes checks of all functions, such as the blinds and the stowage boxes. For dispatching, the roof module is then put into frames specially designed for this purpose and which also allow automatic removal at the assembly line in Bochum.

Figure 6-138 Slight scratches removable by polishing

6.3.1.8 Process Requirements for Perfect Glass Optics

Before a perfect surface quality can be achieved in the mass production process, an absolutely clean environment (clean room) and accurately repeatable processes must be ensured during all process steps. This also applies to the back-foaming process where attention must be paid to the highest degree of cleanness in the foam molds. A precisely repeatable foaming process, homogeneity in glass distribution and the absence of voids in the components all exert decisive influence on the long-term quality of the product, since it is not possible to detect deviations at the shipping stage. During the entire production process the visible outer skin of the roof module is protected by film.

Figure 6-139 Painting the surface after major damage

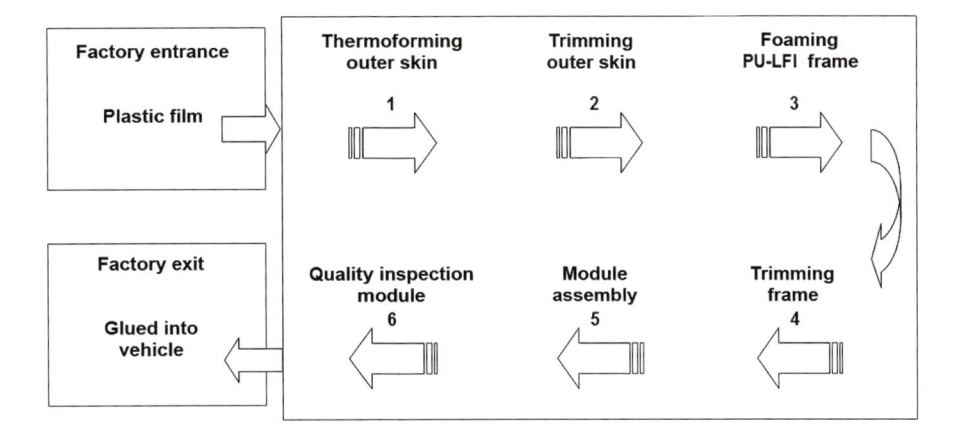

Figure 6-140 Manufacturing sequence

6.3.1.9 Stowage Boxes

Stowage boxes are fitted beneath the PU GF frame in the area of the vehicle front end, each consisting of two modules. The structure consists of a box carrier and movable inserts which are equipped with a damper and a latching element. The stowage boxes extend over the full length of the vehicle roof.

Particular attention was given to optical aspects in the selection of materials for the movable components and those visible to the passengers. In addition, the stowage boxes should have a load-bearing capacity corresponding to their volume. For this reason, the decision was made to use PA ABS/GF8, which also provided the strength required for a single-walled design while providing an increase in spatial capacity. By adding 8% glass, we achieved an adequate limitation of the thermal expansion and sufficient heat resistance.

High load-bearing capacity was of particular importance when selecting the material for the box supports. We choose PA-ABS/GF20. It was necessary to add 20% glass, not only to keep the thermal expansion within limits in adaptation to adjacent components, but also to ensure the required heat resistance.

The lock housing is made of a PA ABS GF20 material. The latch itself is made of POM, as this provides good sliding properties.

6.3.1.10 Outlook

It is highly likely that vehicle purchasers will expect more and more customer benefits. This will compel the manufacturers to offer even more functions and vehicle feature versions.

The standardization of vehicle interfaces, which was also carried out in this project, shows the way to a greater variety of modules and an easy way to manage component variability. By shifting to the supplier, a considerable reduction in logistical complexity at the automotive production facilities is achieved while at the same time providing the intended reduction in manufacturing costs.

In addition, beginning in 2007/2008, plastic paneling will greatly gain in importance due to the use of paint films on the vehicle exterior. Paint films with color-matching capability will be available at that time not just in niche applications but also for mass production. Improvement of scratch resistance is a further target.

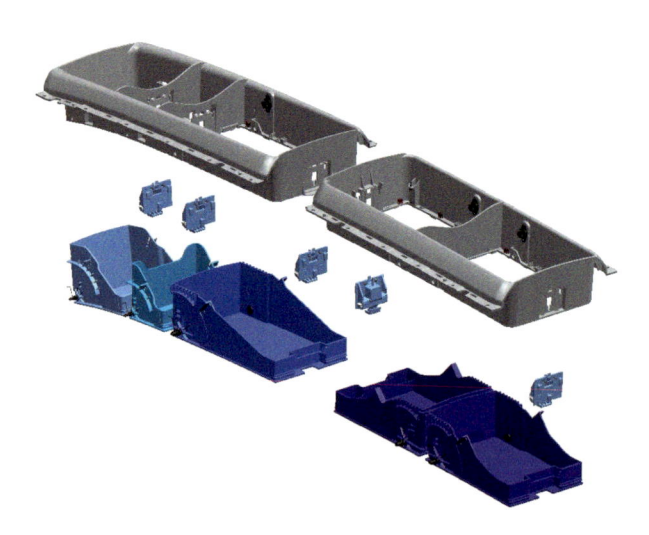

Figure 6-141 Stowage boxes consisting of two modules with PA GF20 for the structural parts and PA GF8 for the visible moving flaps

6.3.2 The Z4 Hardtop Made of SMC – a Self-Supporting Automotive Part Made of Thermosetting Material in the Context of CARB Legislation

Wolfgang Witek, Rudi Kühfusz

6.3.2.1 Introduction

As a self-supporting automotive part made of thermosetting material, the hardtop of the Z4 has to meet a wide variety of requirements (Figure 6-142).

The best way to satisfy these requirements is a solution using plastic. The primary challenge here, however, is that the stricter requirements for reducing hydrocarbon (HC) emissions are satisfied [1, 2]. This was not possible with conventional SMC material systems.

Figure 6-142 Requirements for the Z4 hardtop – extract from the product design specifications

6.3.2.2 Emission-Optimized Development of the Z4 Hardtop Made of SMC

SMC a Material for Low-Emission Components?

For visible components made of SMC on the exterior of passenger cars, special requirements apply to the surface. Depending on the particular system, conventional class A SMC's for flat components have a high potential for emissions with total hydrocarbon values of as much as several hundred mg/m². This must be reduced even when the fact that emissions usually decay exponentially is taken into account. On the other hand, the production of SMC components does have a large number of degrees of freedom which offer opportunities for reducing emissions.

Development Strategies and Results

Optimizing SMC emissions can on the one hand be approached chemically; on the other, the material can also be treated physically (Figure 6-143).

SMC Formulations

The properties of SMC can be strongly influenced via the formulation; however, each optimization measure for certain properties goes hand in hand with changes to other properties. This also applies for surface quality and emissions. Class A SMC for flat components in the visible area has emission values several times higher than those of SMC for structural components.

Cross-Linking Conditions

With reactive plastics, such as thermosetting materials, emissions depend in practice on a considerably greater number of process parameters than is the case with non-reactive plastics. In the case of the class A SMC used for flat components, lowering the mold temperature from 160 °C to 150 °C results in an

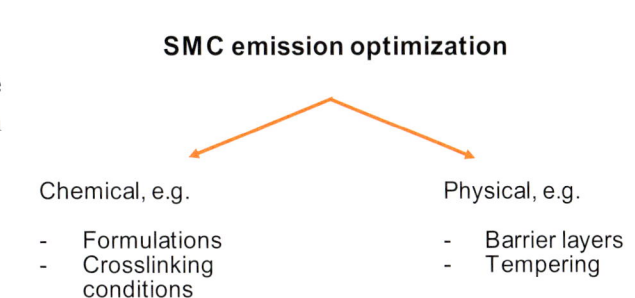

Figure 6-143 Possibilities for reducing the HC emissions of SMC components

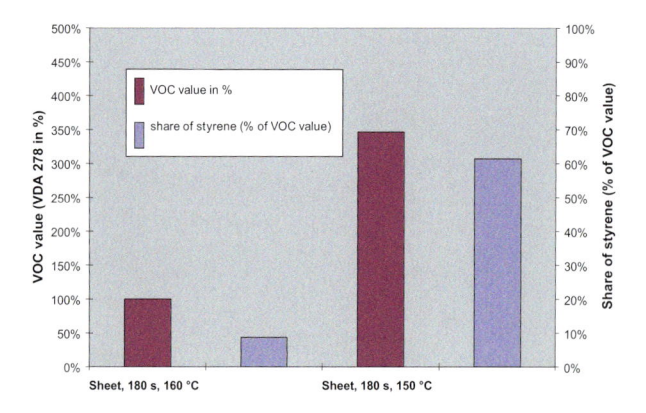

Figure 6-144 Influence of mold temperature on the emission behavior of a class A SMC for flat components

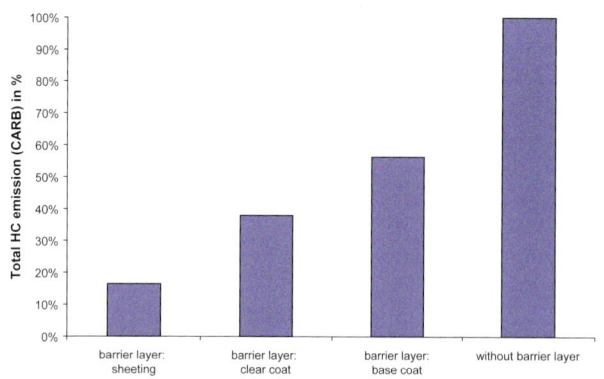

Figure 6-146 Total hydrocarbons per CARB of test components with different barrier layers for estimating the emission behavior of Z4 hardtops made of SMC

emission value that is more than three times higher (Figure 6-144). An analysis by individual substances of thermodesorption shows that styrene is primarily responsible for this increase.

Press time and homogeneity of mold temperature influence the emission values, too, but to a lower degree than temperature.

Barrier Layers

The possibility also exists of reducing emissions by means of barrier sheeting or suitable paint systems. Figure 6-145 shows a test component (bonded sheets) which can be used for investigating the effect of different barrier layers.

- Outer shell made of class A SMC for flat components, bonded to inner shell made of SMC for structural components.
- Geometrical relationships as in the original structure (=hardtop)
- Outer shell with four-layer paint system.
- Different barrier layers on the inner shell.

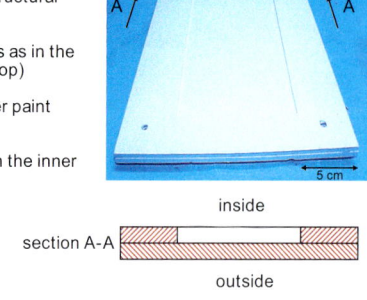

Figure 6-145 Test component for investigating barrier layers

In the case of the test components, use of SHED's (Sealed Housings for Evaporative Emissions Determination) has certain advantages, since the entire test component together with its interactions can be measured. It can be seen that the various barrier layers have effects of different intensities on the emission value (Figure 6-146).

6.3.2.3 Design and Assembly of the Z4 Hardtop

The Z4 hardtop consists of

- an SMC inner shell,
- an SMC outer shell which is adhesively bonded to the inner shell,
- a headliner, and
- other add-on parts, such as the spoiler, rear window, seals (Figure 6-147).

For the inner shell as a non-visible part, an SMC for structural components was chosen which causes minimal emissions. The outer, visible, shell consists of an emissions-optimized class A SMC for flat components. The emissions are further reduced by the use of a barrier-layer paint system. In addition, further emission-optimized materials are also used. For example, the adhesives used in the final assembly of a composite structural component can make a considerable contribution to overall emissions. The objective of implementing the new emissions optimum found

6.3 Roof Modules, Hardtops

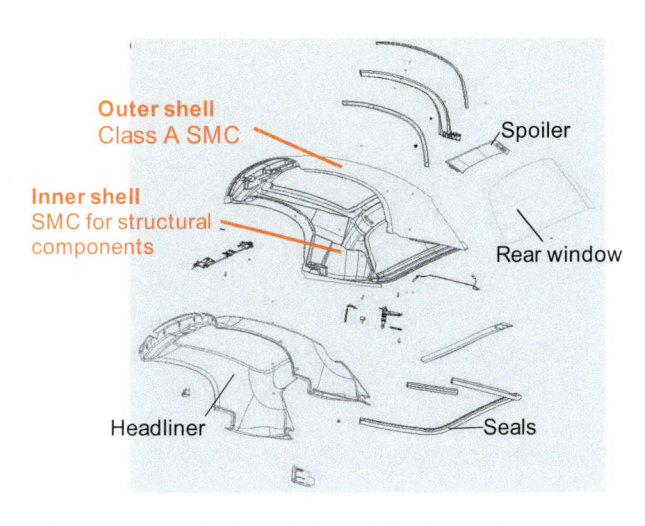

Figure 6-147 Exploded view of the Z4 hardtop made of SMC

for the SMC in volume production while also meeting other, sometimes opposing requirements, means that the processes must be run within defined limits. The following will detail the organizational and technical measures required to realize this.

6.3.2.4 SMC Production

The resin pastes required for making the SMC are produced discontinuously in batch sizes of 2–3 tons. The large quantities of raw materials involved (resins, thermoplastic additives, monostyrene, fillers) are stored in a closed circuit of tanks and silos. The individual components of the specified formulation are kept on the computer of the central weighing unit.

Workers involved with the emission-optimized SMC receive special documented training. It is a matter of making them conscious at all times that every step in their work must be safe. An additional source of help here are detailed work instructions explaining every individual step graphically. For emission-optimized formulations, only thoroughly and completely cleaned containers may be used. The so-called "cutting" of residual quantities with similar LP formulations which adhere to the walls of the containers is strictly forbidden for low-emission formulations. The resin paste is produced in programmed dissolvers, thus ensuring that every batch passes through the same stirring process. The temperature of the paste is limited to 30 °C. Before the batches reach the SMC unit, the following properties are checked: viscosity, gel time, curing time and also peak temperature. At the same time, the measured values are entered into the QM module, all values are checked to ensure that they are within specification.

The resin paste is dispensed continuously into heated doctor-blade units in order to ensure the temperature remains constant. Analogous to the required procedure already mentioned, the doctor-blade units, pipes and pumps must be thoroughly cleaned beforehand. A sample is taken from each batch to measure the amount of thickening over the service life of the SMC.

The length of the glass-fiber rovings is 25 mm. Passing them through a downstream impregnation zone ensures that the glass fibers at the end of the SMC unit are impregnated with resin paste (Figure 6-148). The weight per unit area is kept within tolerances of ±5% with the glass content varying no more than ±3%. In addition, a test strip of SMC is taken from each batch and these strips are then stored under the same conditions as the material produced.

The semi-finished product is rolled up into reels of approx. 400 kg, packed styrene-tight and then enters

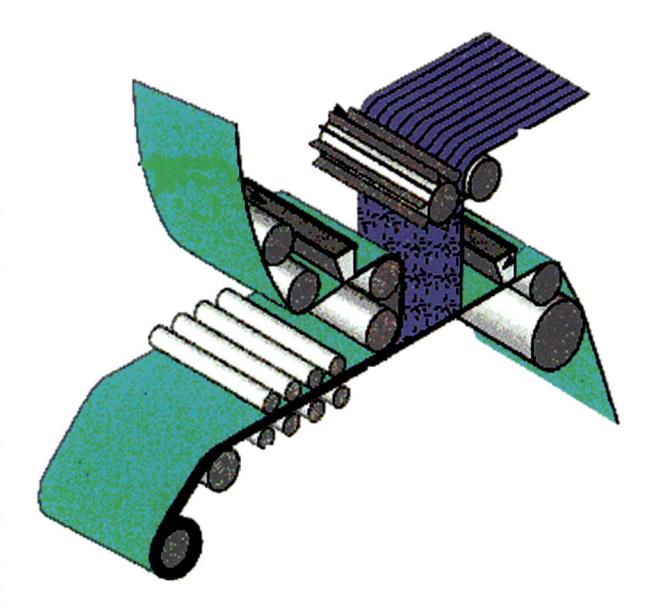

Figure 6-148 Simplified sketch of an SMC unit

the maturing station (whose temperature is regulated to 30 °C) for controlled thickening before finally arriving at the compression station. Following the specified maturation period before the SMC is press-molded, the plasticity of the sample strips is checked. The matured sample is also used for making test panels and for testing mechanical properties.

The Compression Molds

In order to satisfy the high requirements on this large class A component, the molds themselves had to meet special requirements. It was necessary to chrome-plate their surface on account of the class A LP SMC being used. Special care must be given to ensure an even vertical flash face clearance of no more than 0.05 mm to allow ease of deflashing.

Compression Molding of the Outer Part

The high requirements as regards the surface quality of the outer part mean that a synchronized high-speed press with vacuum support in the compression process must be used. The steps of the molding cycle such as fast closing or split control are executed by the press process computer on the basis of control inputs.

Workers involved in production have received special training and care is taken that the same teams are always employed. By visual displays of process operations at the workplace, compliance with the tight process window can be ensured. This affects not only the specified molding temperature and the curing time but also the preparation of the SMC blank. The cutting unit, which stands directly adjacent to the press, supplies the blanks cut to the lengths specified in the program. The blanks package is then placed in the compression mold. In order to deliver the minimum emission values possible, production must run at mold temperatures of 160 °C and with a curing time of no less than 180 seconds. If there were no special emission limits, the corresponding mold temperatures would be approx. 145 °C, while the curing time for the wall thicknesses in the present component would be below 110 seconds (Table 6-13).

Table 6-13 Production parameters of SMC components as a function of emission requirements

Production parameters of SMC components (for same wall thickness)		
	Without specific emissions requirement	Minimal emission limits
Mold temperatures	145 °C	160 °C
Curing times	110 seconds	180 seconds

Demolding is done manually using a vacuum suction device. The component is deflashed in a device that protects the component against scratching or other damage. Every pressing is labeled in the area of the rear window and then placed on a specially designed carrier for transportation within the plant.

Compression Molding of the Inner Part

The same process steps listed above apply to the compression molding of the inner part. Due to the frame-like configuration of the inner part, cutting it to shape is more complicated than is the case with the outer part. Various blank layers are prepared using templates and then laid in the compression mold by two workers.

Following demolding, the compression mold is carefully cleaned by removing molding remnants from the area of the split. The spacer pieces must be inspected for contamination and cleaned if necessary.

A special fixture is used for deflashing and then sanding the deflashed areas. The part is temporarily labeled in the area of the rear window that will be cut out later. The parts are then placed on a specially designed carrier for transportation within the plant and are taken to the milling station.

Secondary Finishing

Since the number of hardtops does not require the uninterrupted use of the compression molds, the inner-part and outer-part batches are processed alternately. The parts are stored in closed rooms in order to prevent

6.3 Roof Modules, Hardtops

Figure 6-149 Milling adapter for inner part and glued completed part

Figure 6-150 Inserting the outer and inner shells into the gluing unit

any contamination. Milling work is performed using a robot (Figure 6-149). More than fifty holes are drilled in different positions and in different diameters. During one processing run, the robot switches several times between drilling and milling tools. The cycle time is dictated by the subsequent bonding procedure for the two parts. An uncut inner piece is removed from the transporter and inserted into the milling machine adapter. At the same time, a ready-glued part is inserted into the second device delivered by the robot to the gluing unit. The exact position of the two parts is checked automatically and, once the door of the processing box has closed, the program starts.

The robot cuts the hole for the rear window out of the glued component and also drills any holes that may still be required. Once this process is completed, the inner part is removed from the cutting unit, placed on a work stand where the flashing is cleanly removed from the holes. The next step is to prepare the surfaces for gluing by cleaning with isopropanol.

Gluing

Gluing takes place in a closed unit. The same robot applies the adhesive and handles the component (Figure 6-150). First, two subcomponents with blind rivet nuts are riveted into the inner part on the assembly unit. For the subsequent assembly of the hardtop, a steel bracket for mounting is required on the left and right of the component. These are glued into the inner part. The robot applies the glue at the end of a glue application cycle. The brackets, with the adhesive applied to them, are removed from the gluing unit delivery station and then, on the assembly device, held down and riveted in place. To accelerate the curing of the polyurethane adhesive, a hot-air blower is used. The inner shell is then inserted into the holder of the gluing unit. The gluing surfaces of the outer part are cleaned in the same way as the inner part. Once the inner shell, outer shell and the mounting brackets have been set down in their proper places, their precise positions are checked automatically and the unit is given the go-ahead. The robot then applies the adhesive to the outer shell and the steel brackets. Once the glue has been applied, the robot puts down the gluing unit and changes to the transportation device. The inner shell and outer shell are picked up in turn by means of vacuum suckers and placed in the heated joining mold.

Once the adhesive has cured and the gluing unit has opened, the robot uses its vacuum device to remove the finished part, laying the bonded hardtop in the delivery station.

6.3.2.5 Preparations for Painting

As identification for the finished part, the SAP number is printed on it and also glued into the specified position in the inner part. For tracking purposes, the part numbers of the inner shell and outer shell are added to the specified list. Traceability is assured at every process step thanks to the internal records held on the SAP system. This applies not only to the manufacturing and further processing of the moldings, but also to the material used. Where applicable, information may be obtained as to the raw materials used, for example, to find out what catalyst batch was used in making the SMC for the component.

After unlatching the output holder, the glued hardtop is removed and placed on the cutting unit's input station. Once cutting of the part is finished, it is placed on the processing cart. Since painting is not carried out at the component producer, the processed hardtops are packed in closed containers that are also used for transporting them after painting. The parts are mounted on individual transportation frames, thereby ensuring that they cannot touch each other even if transportation conditions are not optimal. Development and optimization work carried out in conjunction with the system supplier Webasto resulted in the closed containers now used for volume production.

6.3.2.6 Painting

Painting takes place off-line at Wayand in Idar-Oberstein, a painting company experienced in working with car components. In order to ensure the same top-coat state as the painted metal bodywork, a multilayer structure consisting of a primer and base and a clear coat is necessary. The temperature of the paint drying unit is 95 °C. The primer is matched to the SMC substrate. The top coat is applied in the ten body colors of standard vehicle production and the interior side is given a clear coat. The emission-reducing effect of the clear coat demonstrated with the testing parts has been implemented in volume production.

6.3.2.7 Summary

The emission behavior of complex automotive parts is influenced today by not just one factor alone. The example of the Z4 hardtop made of SMC shows that the application of a package of measures can bring conventional SMC to the point where competitive, self-supporting automobile parts can be made that meet the strictest requirements. In addition to emission-optimized formulations with processing parameters adapted to and traceable for each part, the measures also include checks on raw materials and semi-finished products.

Further measures include the use of barrier layers to seal off volatile residues from the polymerized SMC as well as emission-optimized add-on components and assembly adhesives. In addition, the workers took training courses that informed them about hydrocarbon emissions and also showed the required work operations in visual form.

Low-emission components are today a requirement in automobile manufacturing. The correlation between material, processing parameters and emissions must be known. From this knowledge, suitable requirements can be derived for materials selection and production processes. If the processes are run within these defined limits, reproducible emission values will be set to a low level.

Figure 6-151 demonstrates this with the aid of the total hydrocarbon values of the emissions-optimized Z4 hardtop made of SMC. These lie within about 15% of a comparable hardtop that has not been emission-optimized.

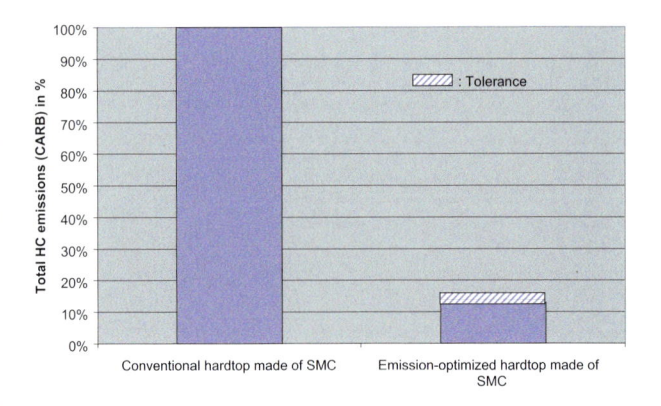

Figure 6-151 Total hydrocarbons per CARB for a conventional hardtop made of SMC and an emission-optimized Z4 hardtop

References

[1] State of California, Air Resources Board, z.B. Californian Evaporative, Emission Standards and Test Procedures for 2001 and Subsequent Model Motor Vehicles www.arb.ca.gov/regart/levii/to_oal/evaptp01.pdf

[2] Frank, U., Schwager, H.: Entwicklung HC-emissionsoptimierter Kfz-Bauteile. In Kunststoffe im Automobilbau. Düsseldorf, VDI-Verlag, 2002

6.3.3 Innovative Noise-Optimized Folding Top for the New BMW 6 Series Convertible

GÜNTHER ROTH

6.3.3.1 Introduction

The design and development partners for this project were Parat (polyurethane insulation and manufacture of the fabric covering), Huntsman (supplier of the polyurethane system), Edscha (system supplier). One of the major development goals was to develop a folding top that would deliver the best acoustic values in the passenger compartment of its class in order to meet increasing customer requirements with regard to passenger compartment acoustics. Another goal was to be able to manufacture the convertible top at a reasonable cost. A technique was developed that allowed the convertible-top material to be manufactured two-dimensionally. This meant a considerable reduction in tooling costs.

6.3.3.2 Requirement Profile

Acoustics

Acoustic measurements taken from the standard production vehicle, the 3 series convertible, both in the wind tunnel and on the track were used as basic values. These measured values were to be improved by about 3 DB.

Thermal insulation

The thermal insulation should be able to considerably reduce the time required to heat up the interior at outside temperatures below zero.

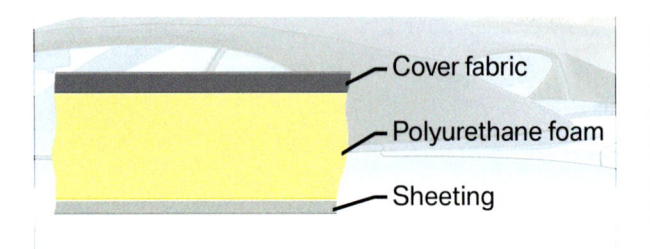

Figure 6-152 Structure of polyurethane insulation

Mechanical Aspects and Function

All mechanical functions, such as opening, laying back, and closing, have to be performed several thousand times and under all climatic conditions.

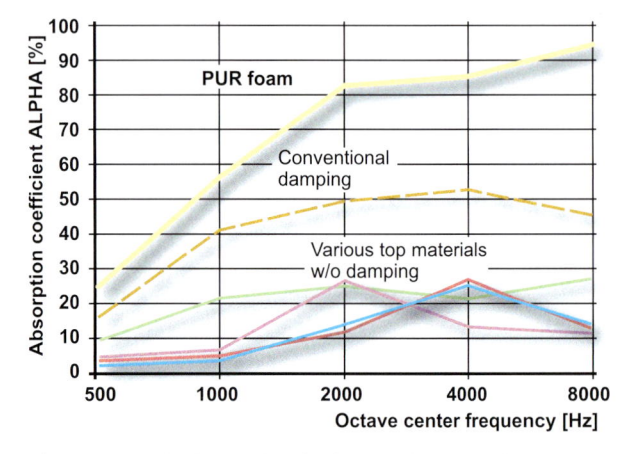

Figure 6-153 Acoustics (noise optimization)

Figure 6-154 Thermal insulation in the 3 series convertible top

Figure 6-155 Convertible top of the BMW 6 series convertible

6.3.3.3 Material: Polyurethane System

Selection of Materials

At the beginning of the development process, the acoustic properties of various polyurethane systems in the form of specimen pieces were investigated in the reverberation chamber. Since the available viscoelastic polyurethane systems in the component composite met neither the acoustic nor the mechanical requirements, the following optimization measures were taken.

Properties

The following acoustic requirements were to be met:

- High level of insulation (air-borne sound)
- High level of damping (solid-borne sound)
- Precise distribution of density

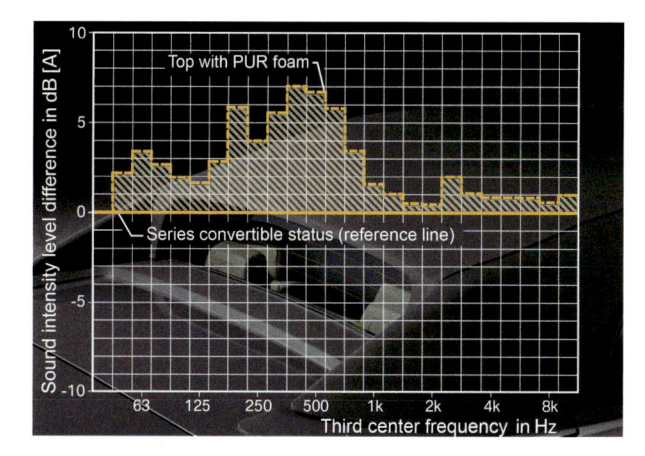

Figure 6-156 Comparative sound intensity

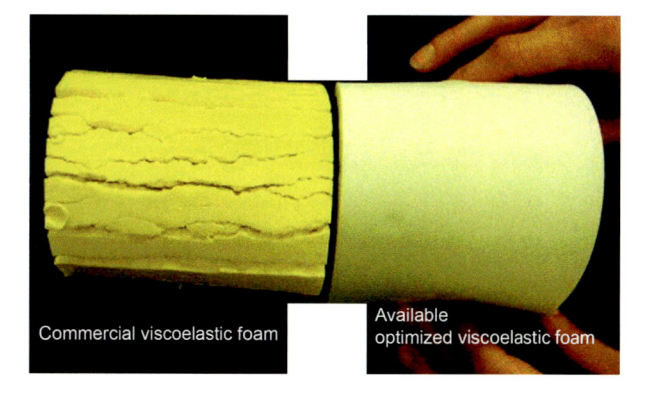

Figure 6-157 Optimized polyurethane foam

Table 6-14 Modification to the polyurethane foam system

Acoustic	Mechanical values/processing
Increase in density	Improvement in flow behavior (no formation of cavities)
Improvement in storage modulus	Optimization in density distribution
Improvement in loss factor	Increase in tearing strength Increase in tensile strength Improvement in recovery characteristics

The mechanical property requirements include:

- Good adhesion to the cover fabric
- High resilience (return to original shape)
- High abrasion resistance
- High wear resistance
- High tearing strength
- Low residual deformation
- No color changes in the fabric
- High dimensional stability

The climatic requirements include meeting all mechanical and acoustic requirements in accordance to:

- Climatic exposure test
- Solar irradiation simulation
- Hot-country trials
- Cold-country trials
- Endurance testing 5000 load cycles
- Providing low fogging values
- Compliance with required emission values according to the BMW standard
- Compliance with requirements relating to flammability characteristics

Processing

The polyurethane foam system is processed on a Krauss-Maffei high-pressure machine using the reaction injection-molding (RIM) method. The mixing head is guided by a multiaxial robot. The mold is closed and opened in parallel by a mold carrier.

Figure 6-158 Foam molding

Figure 6-159 Curing

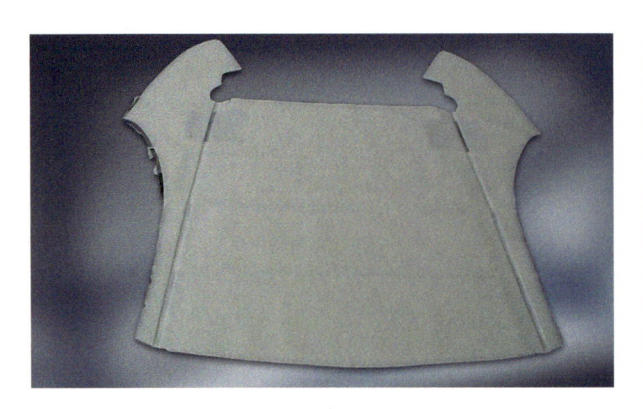

Figure 6-160 Foam-molded 2D component

6.3.3.4 Manufacture of the Convertible-Top Cover

Manufacture of the Fabric Cover (Packaging)

The raw material was prepared on the basis of the cutting data and other inputs determined during development work. In the same way, the cover fabric was cut to shape, fabric flaws sorted out, and the material passed on to the stamping unit for further processing.

The blanks, whose shape is created by parallel stamping, are joined by an overlap seam after which sealing tape is applied. The next step is the application of polyurethane insulation by means of a newly developed technique. The polyurethane is applied by a robot on the basis of the process parameters worked out during the development phase. At the end of a curing period, the covers go to final assembly. Here, all fasteners are sewn on, and the edging is sewn. After this, a quality check is carried out before – as the last process step – the rear window is foam-molded into the covering. Following final processing and appropriately documented quality inspection, the convertible top covers are packed and made ready for dispatch.

Mold Design

During the development phase, numerous tests were carried out to determine the best mold design. Since the top covering becomes a three-dimensional component once the overlap seam has been sewn in, it is obvious that a 3D mold is also necessary.

Figure 6-161 Cover of the series 3 convertible (left); cover of the series 6 convertible (right)

6.3 Roof Modules, Hardtops

Figure 6-162 3D mold for convertible top of Series 3 convertible, lower part on the left, upper part on the right

Satisfactory results were obtained right after the first trials. However, after several parts had been made, the following problems were noted:

- Foam incursions
- Flow paths
- Heights to be overcome
- Venting

As a result, the challenge was to design a 2D mold into which a 3D part had to be positioned. This was achieved very quickly and the advantages became immediately apparent. The main advantages of the 2D mold lie primarily in its low construction costs. At the same time, costs for mold modifications are far below those incurred with a 3D mold. The low weight of the mold is a major advantage, for example, when setting-up the machine. Even the ease with which the pre-assembled cover can be positioned in the mold is a great advantage, since this ensures economically efficient production.

**Production Technology:
Mold – Robot – Foam-Molding Unit**

During the course of mold development, it was also necessary to find and set up suitable production technology. With the aid of the first prototype molds, various trials were carried out on a semi-technical scale at the machine suppliers in order to determine and set up the production technology.

Figure 6-163 2D mold, top part

Figure 6-164 2D mold, lower part with cover blank

At this stage, it was necessary to decide how injection into the mold was to be carried out. In other words, should injection occur into the open mold with the aid of a robot-guided mixing head, or should injection be into the closed mold with the mixing head immovably fixed on the mold?

On the basis of these findings, the foam-molding unit from Parat and Krauss-Maffei, including the mixing head guidance system using an ABB robot, was designed taking, into consideration the most modern manufacturing technologies and processes. The decision to inject the foam into the open mold resulted from the product-specific molding concept.

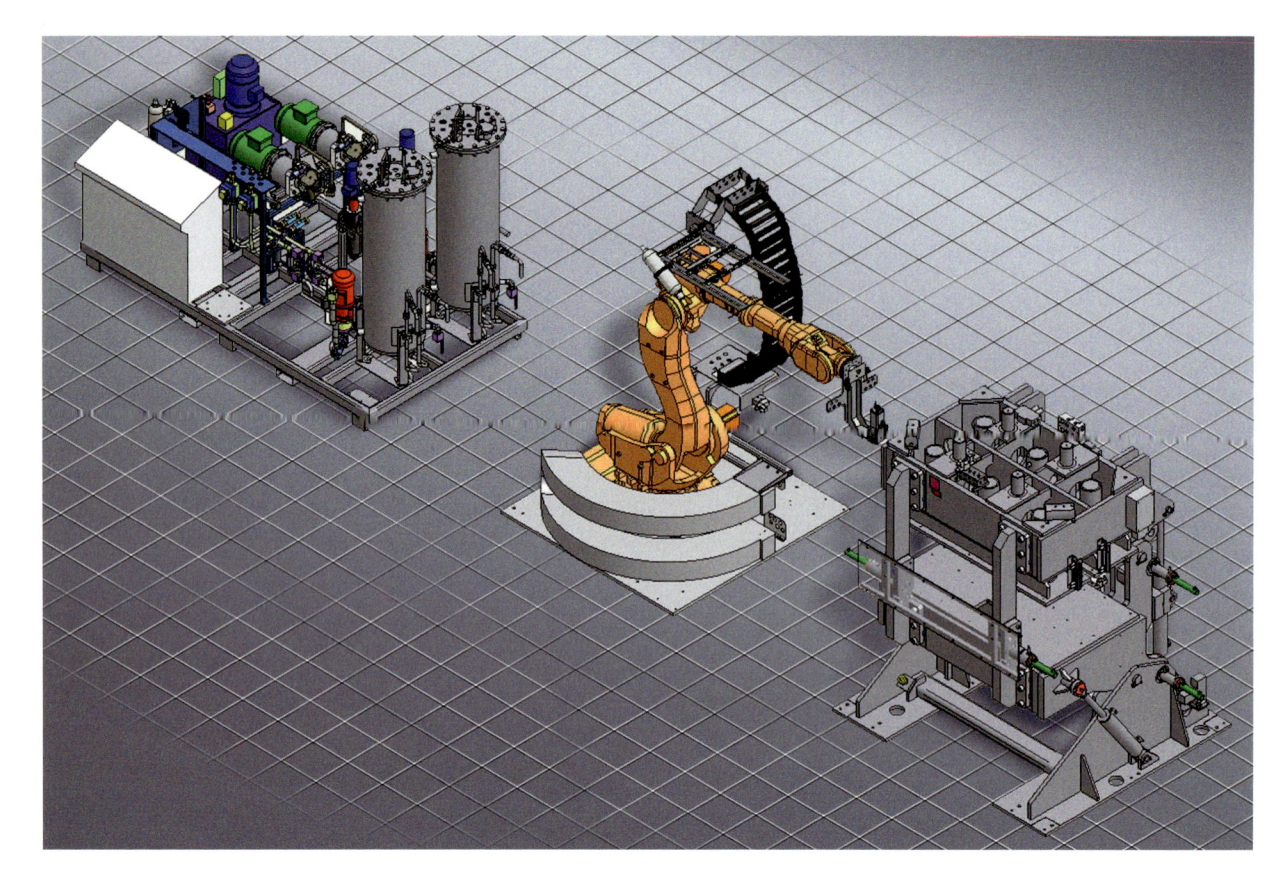

Figure 6-165 Production hardware

By evaluating various mold designs and carrying out several foam-molding trials in collaboration with BBG, this molding concept was successfully carried over into volume production.

Figure 6-166 Foam-molding unit

Production Sequence

- The cover is positioned on the lower mold half and fixed in place.

- The sheeting is positioned on the upper mold half and fixed in place.

- The robot injects foam into the pre-assembled convertible-top cover.

- Once the mold closes, the foam begins to cure. Curing time is around 180 seconds.

- With foam-molding completed, the convertible-top cover can be removed from the mold carrier and placed on a frame upon which it is taken to final assembly.

6.3 Roof Modules, Hardtops

Figure 6-167 Positioning the cover on the lower mold half

Figure 6-170 Curing period (180 seconds)

Figure 6-168 Sheeting positioned on the upper mold half

Figure 6-171 Removal of the cover from the mold carrier

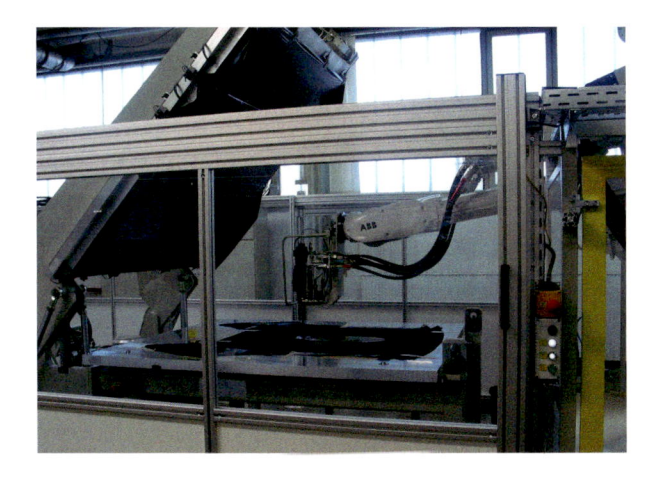

Figure 6-169 Starting foam injection

Figure 6-172 Packaging the foam-molded cover (Edscha)

Assembly of the Fabricated and Foam-Molded Cover

Once the convertible-top frame has been assembled at the system supplier, the liner and the outer covering are fitted. The fasteners sewn or welded onto the covering are screwed to the frame.

The next step is a quality check before the completed assembly is shipped out to the OEM.

6.3.4 Roof Module for Commercial Vehicles Made with SMI Technology

GÜNTHER MEDERLE

6.3.4.1 Brief Description

Changing market requirements during the life cycle of a product make modifications to components, systems, modules or entire cabs necessary. A comparison is required of the various manufacturing technologies with regard to function, weight and costs which also takes the remaining life of the product into account. In addition, the selection of tooling and manufacturing technology will also depend on implementation time, product quality and cost-efficiency.

6.3.4.2 General Constraints

Market Situation and Market Development

The following general conditions currently apply within the heavy truck segment.

- The market for long-haul operations is growing
- The cab interior volume in the long-haul segment is increasing
- Harder competition, increased price pressure
- Fleet business growing.

Figure 6-173 provides a comparison of the MAN product range (heavy series) with the competition: Despite the widely diversified product range of the TG program in the market segment of 18–50 tons maximum gross weight, when compared with the competition, the classic long-haul haulage fleet sector is not optimally covered by the top model, the XXL cab. MAN currently covers this sector predominantly with the top model, the XXL cab. As a result, with the XXL cab the customer is offered a version with an optimum of function and comfort that has its corresponding effect on the cost side.

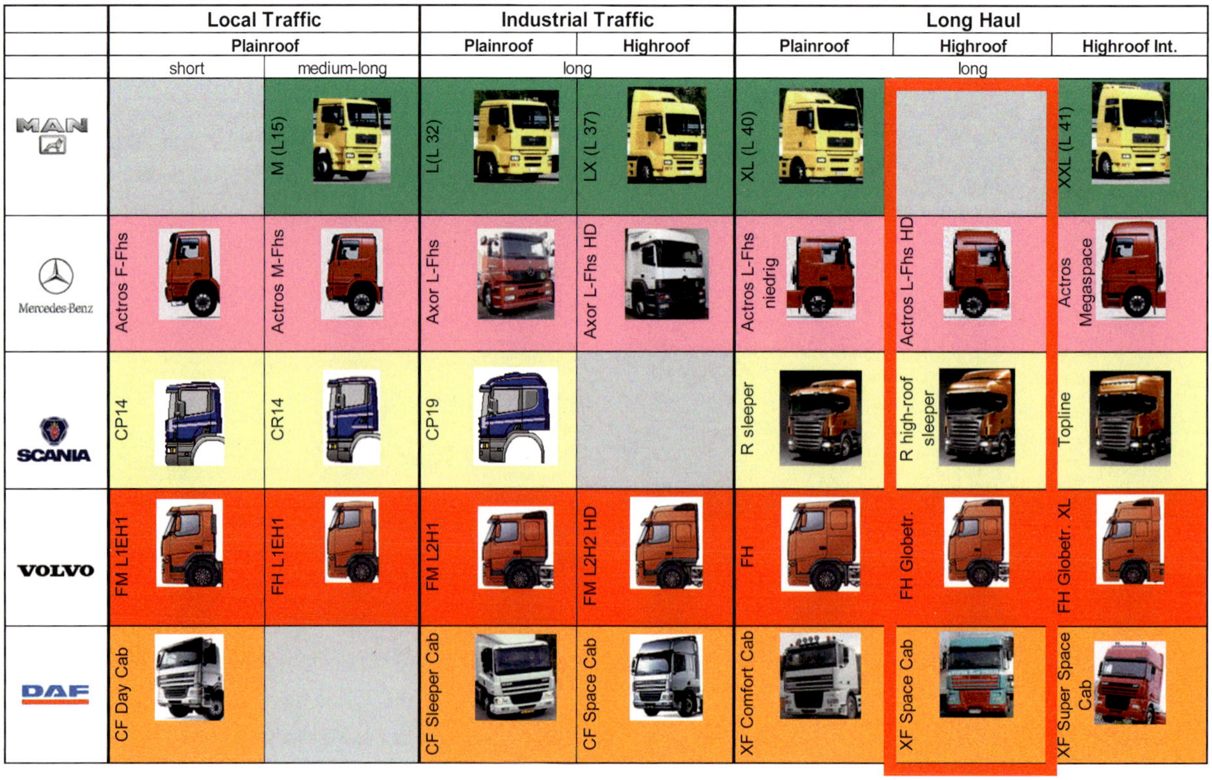

Figure 6-173 MAN product range and the competition – heavy series (source: MAN Nutzfahrzeuge AG)

6.3 Roof Modules, Hardtops

Comparing the XXL cab with the new XLX cab shows that the most important differentiating features are the roof height (difference: 225 mm) and the side window above the door in the XXL cab.

Economic Situation

In the fleet segment, price competition is fierce on account of the numbers of vehicles purchased by individual customers and the associated quantity discounting. In this business sector, the relationship between the sales revenue and the manufacturing cost of a vehicle is generally critical. This makes it necessary to define vehicles with special features for this service area and to select the best manufacturing processes for specific assemblies from the points of view of demand and cost-efficiency.

For this reason, MAN decided in 2004 to develop a cab version optimized for the fleet segment in long-haul operations. A project under a tight schedule was started using target costing.

Project Overview

The project is divided into two phases:

- Phase 1: Technology and concept development
- Phase 2: Series development.

6.3.4.3 Benchmarks in Manufacturing Processes

Preparation for Decision Making

On the basis of fundamental strategic considerations, market development and the existing range of models were compared. In addition, future development in the individual segments was estimated and an overall feasibility study carried out for various scenarios. At the same time, a wide variety of European and Asian manufacturing technologies were examined. On the basis of a market research study, work was begun on the package design and on preparing draft designs, while at the same time performing a technology analysis.

Figure 6-174 Cab comparison of the XXL and the XLX (source: MAN Nutzfahrzeuge AG)

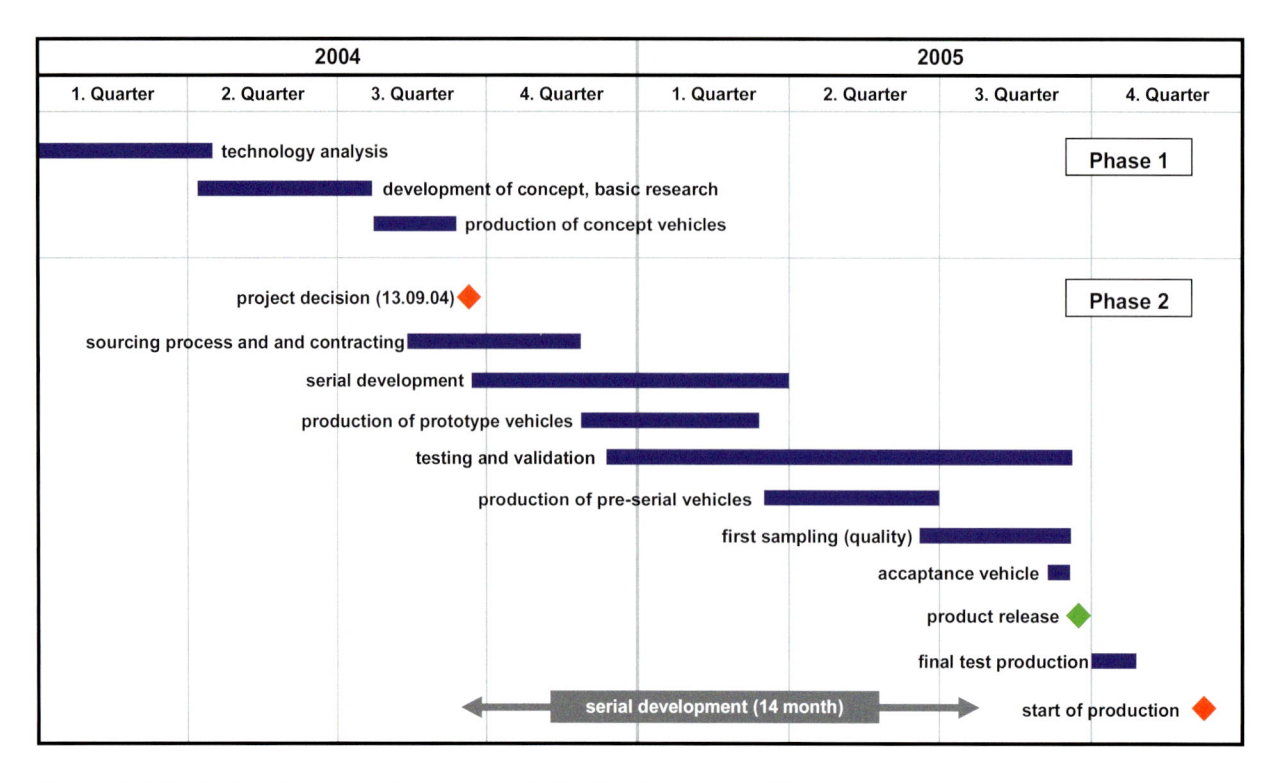

Figure 6-175 Project time schedule (source: MAN Nutzfahrzeuge AG)

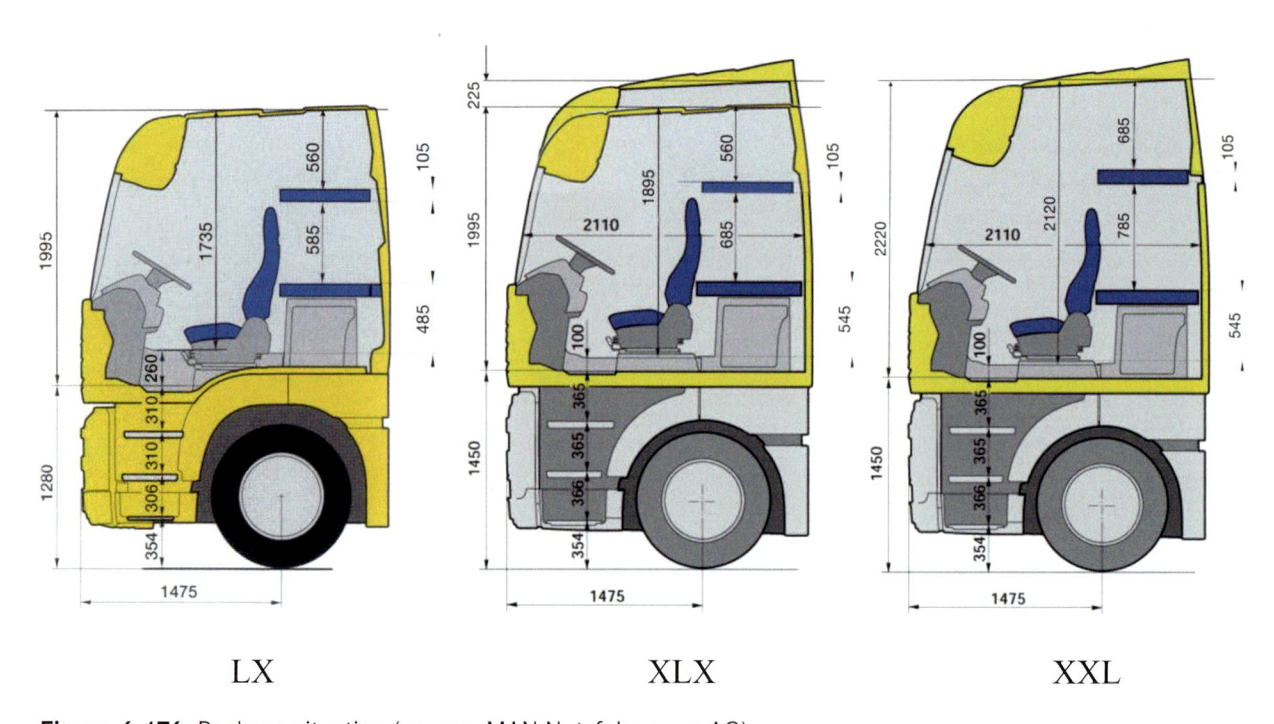

Figure 6-176 Package situation (source: MAN Nutzfahrzeuge AG)

6.3 Roof Modules, Hardtops

| XL | XLX | XXL |

Figure 6-177 Design development (source: MAN Nutzfahrzeuge AG)

In addition to comfortable head room and standing height of about two meters, the rest area behind the seats was also given a comfortable design. It accommodates two wide and most importantly comfortable beds with the LGA certificate "Ergonomic sleeping comfort", thus ensuring maximum regeneration after resting.

As a decision-making aid, a prototype vehicle was also built (center of Figure 6-178) and compared directly with existing models in an internal presentation.

In parallel with the decision-making and concept process, aerodynamic simulations and reference tests were carried out in conjunction with other aspects.

Figure 6-178 Prototype XLX cab in comparison with the LX and XXL cabs (source: MAN Nutzfahrzeuge AG)

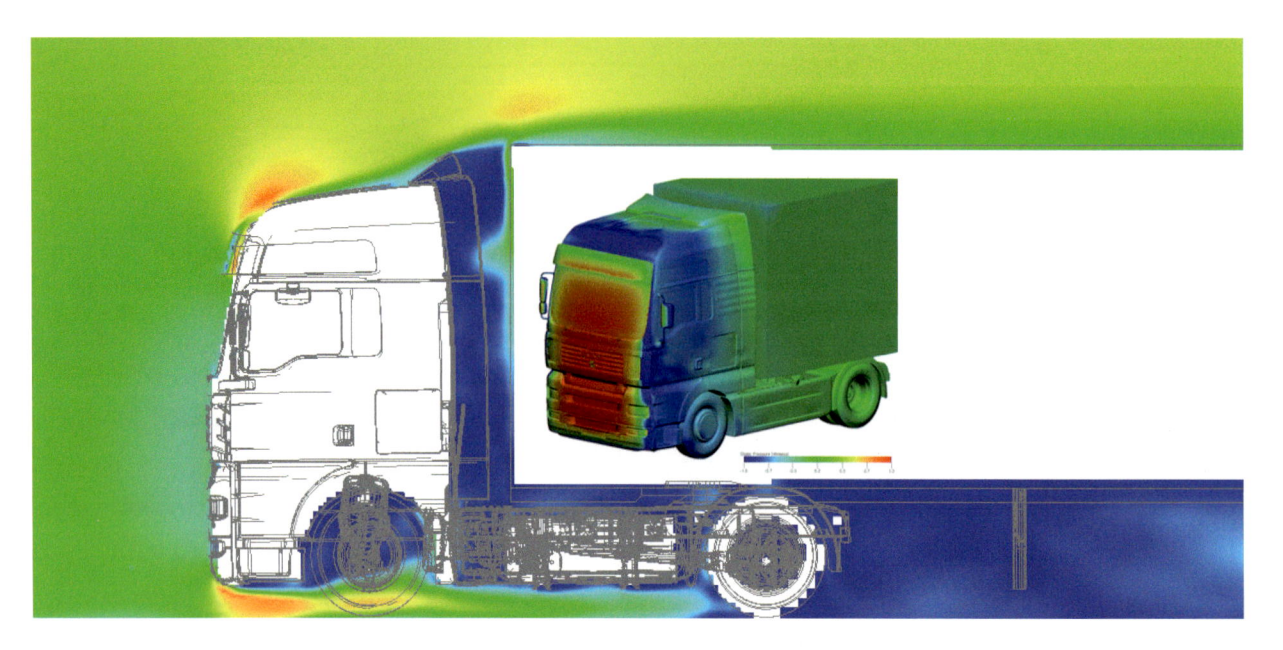

Figure 6-179 CFD simulation (source: MAN Nutzfahrzeuge AG)

6.3.4.4 Comparison of Manufacturing Processes

A comparison was the different manufacturing technologies compiled, considering cost, weight, function, target date, quality, logistics, assembly and technology risk.

The following versions were examined in more detail:

- Sheet metal (steel) in monocoque design (own production, Europe, Asia)
- RTM or SMI technology (Europe)
- Aluminum space-frame (Europe)
- Hybrid design: steel/aluminum – plastic (Europe).

Under the given general constraints, such as product life, production concept, painting concept and implementation lead time, the RTM/SMI version (Sandwich Monocoque Injection technology) was assessed as the best overall solution.

When the tooling and manufacturing costs were considered as a function of the estimated yearly production and given an amortization period of 6 years, the following cost situations were the result.

For numbers of up to approx. 10,000 cabs per year, the RTM/SMI version is clearly the least expensive alternative. At higher production quantities, a classic sheet steel monocoque construction has an advantage from the cost point of view. One decisive advantage of RTM/SMI technology is however the short development and tooling time which, at 14 months, had a decisive influence on decision making.

Product and Technology Decision

Following a six-month concept phase, in September 2004 selections were made for product and technology:

- Cab variant: XLX (wide, medium high)
- Method used to make the high-roof: SMI technology
- Target date: December 2005.

The product design specifications for a total of eight individual new or modified parts were prepared in parallel with the decision process and went into the inquiry stage following the concept decision. For the purpose of the project decision, items critical with respect to target date and technology risk (high roof and luggage rack) were defined. These items were awarded on the basis of strategic inputs while complying with

6.3 Roof Modules, Hardtops

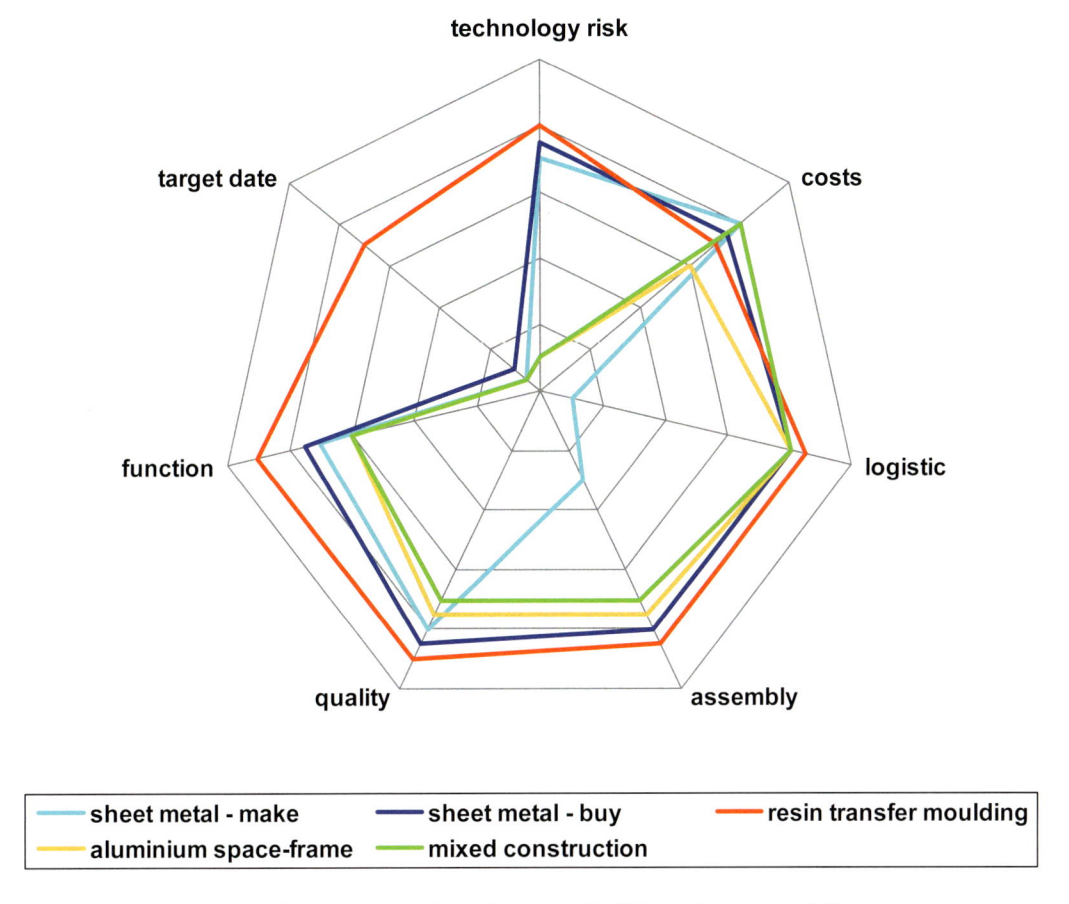

Figure 6-180 Evaluation of production technology (source: MAN Nutzfahrzeuge AG)

XLX High-roof (technology comparison)

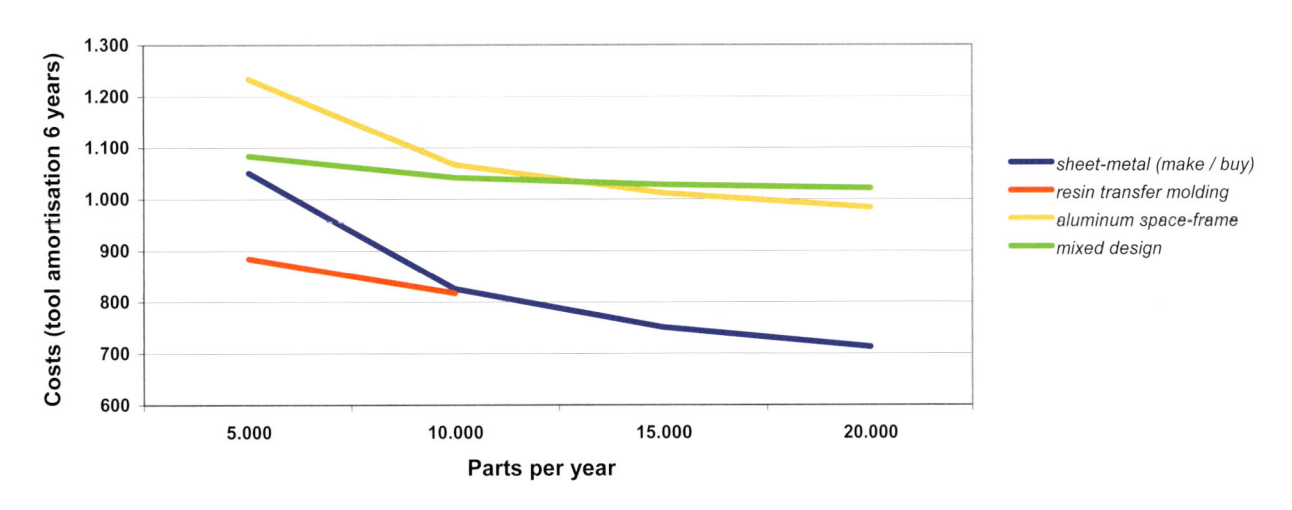

Figure 6-181 Cost comparison (source: MAN Nutzfahrzeuge AG)

Highroof (new)

Exterior:
roof spoiler (LX modified)
side plate (new)
sun shield (XL modified)

Interior:
luggage rack in LFI-Technology (new)
curtain (modified)
curtain bar (modifiziert)
rear panel carpeting (new)

Figure 6-182 New or modified assemblies in the XLX (source: MAN Nutzfahrzeuge AG)

cost-efficiency and competitiveness aspects. As specified in the SE Manual, the project was run by an interdepartmental team with regular project meetings. In parallel with this, specific development meetings were held with a total of 7 system suppliers.

6.3.4.5 Product and Tooling Concept

Product Concept

In addition to the project advantages under the given general constraints, SMI technology also offered some fundamental concept advantages such as

- Low weight
- High rigidity and strength
- Good thermal and acoustic insulation
- High part precision
- Possibility of integration
  (headliner, mounting points, etc.)
- Short preparation time for tooling
  (and also for additional molds).

The structure of the roof material (from outside to inside) consists of:

- Face sheet, outside, 1.5 mm
- Sandwich core, 17–45 mm
- Face sheet, inside, 1.5 mm
- Fabric lamination, inside

Mold Concept

The SMI technology is a further development of the RTM process. Considering size and weight, possible molding technologies are NVD molds (nickel vapor deposition) or galvanic nickel molds. To meet the high quality requirements, MAN highroofs are made solely with NVD molds made by NTT/Weber. The tooling concept, i.e., mold building process consists of the following steps:

- Preparation of CAD surface data (outside/inside)
- Preparation of male molds (mandrels) for inside and outside surfaces

6.3 Roof Modules, Hardtops

High strength

Best insulation

Low weight

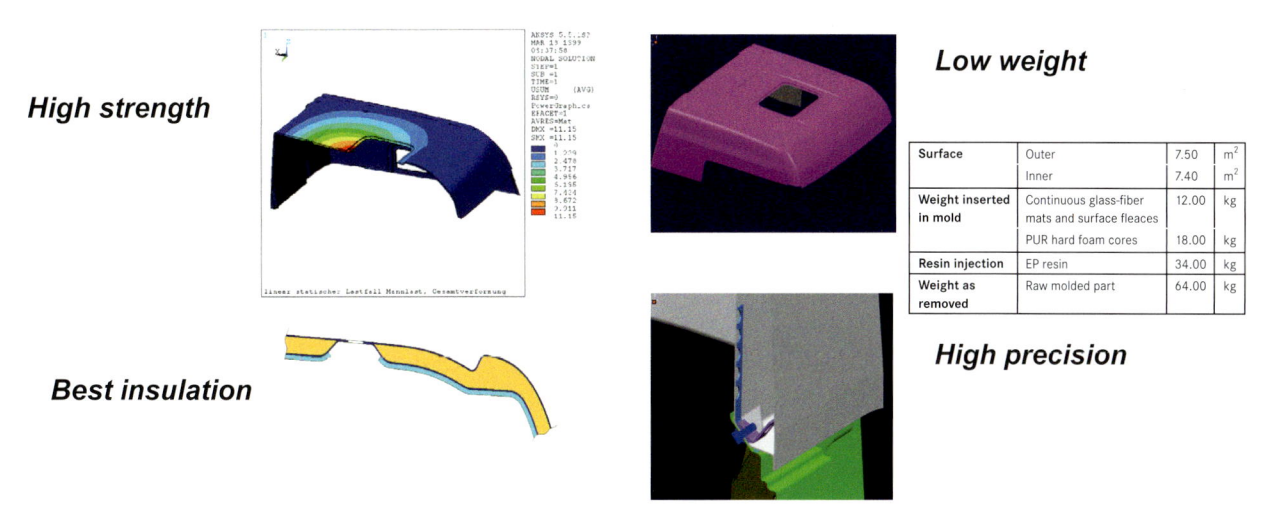

Surface	Outer	7.50	m²
	Inner	7.40	m²
Weight inserted in mold	Continuous glass-fiber mats and surface fleaces	12.00	kg
	PUR hard foam cores	18.00	kg
Resin injection	EP resin	34.00	kg
Weight as removed	Raw molded part	64.00	kg

High precision

Figure 6-183 Highroof in SMI technology (source: Fritzmeier-Composite)

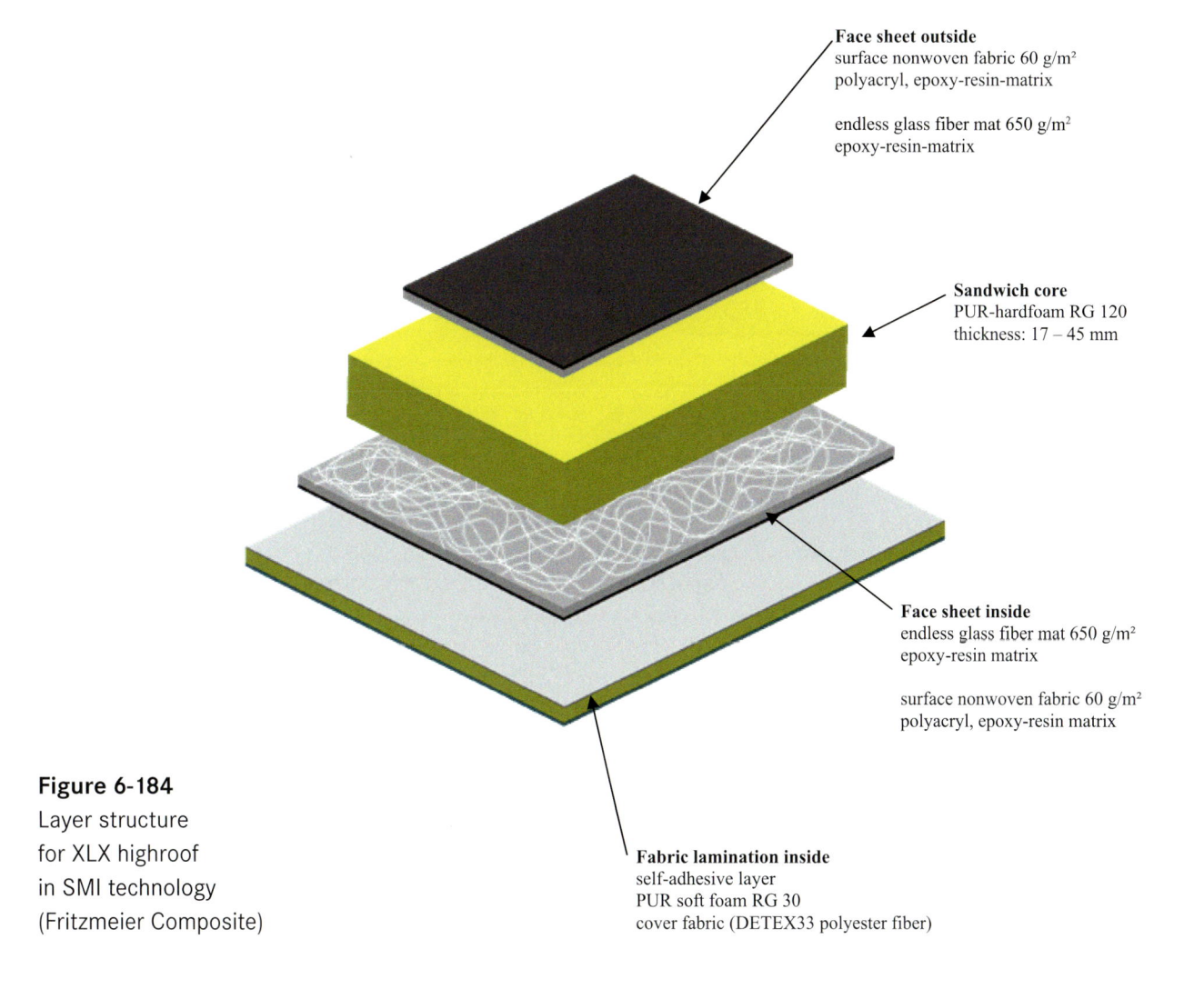

Face sheet outside
surface nonwoven fabric 60 g/m²
polyacryl, epoxy-resin-matrix

endless glass fiber mat 650 g/m²
epoxy-resin-matrix

Sandwich core
PUR-hardfoam RG 120
thickness: 17 – 45 mm

Face sheet inside
endless glass fiber mat 650 g/m²
epoxy-resin matrix

surface nonwoven fabric 60 g/m²
polyacryl, epoxy-resin matrix

Fabric lamination inside
self-adhesive layer
PUR soft foam RG 30
cover fabric (DETEX33 polyester fiber)

Figure 6-184
Layer structure
for XLX highroof
in SMI technology
(Fritzmeier Composite)

- Preparation of nickel shells (negative) by the nickel deposition process
- Preparation of the substructure for the nickel shells
- If necessary, preparation of further nickel shells with the aid of existing male molds.

For the male mold, a cast aluminum structure was made and then cut to shape on the basis of CAD data. Depending on the size of the component to be made, the complete model (mandrel) is made either in one piece or in two pieces, as is the case with the MAN highroof. Figure 6-185 shows the right-hand half of the master model of the high roof for the XLX cab.

The male mold serves as a master mold for the subsequent nickel deposition process in which a nickel shell is created layer by layer by a chemical process. Depending on requirements, several nickel shells can be made from the master mold within a few weeks. The shells made by the nickel vapor deposition process are absolutely stress-free, extremely strong and about 18 mm thick. Figure 6-186 illustrates the NVD process in simplified form [1].

Figure 6-185 Male mold "mandrel" (source: NTT-Weber)

Figure 6-187 shows the mold structure (nickel shell, intermediate layer and substructure).

Aluminum mandrel oil heated 175°C gas tight construction

Ni(CO)4 gas 100% controlled chemical condensation

Ni(CO)4 gas 60%

Nickel shell, thickness 18 mm growing 0.25 mm / hour removal in hot condition

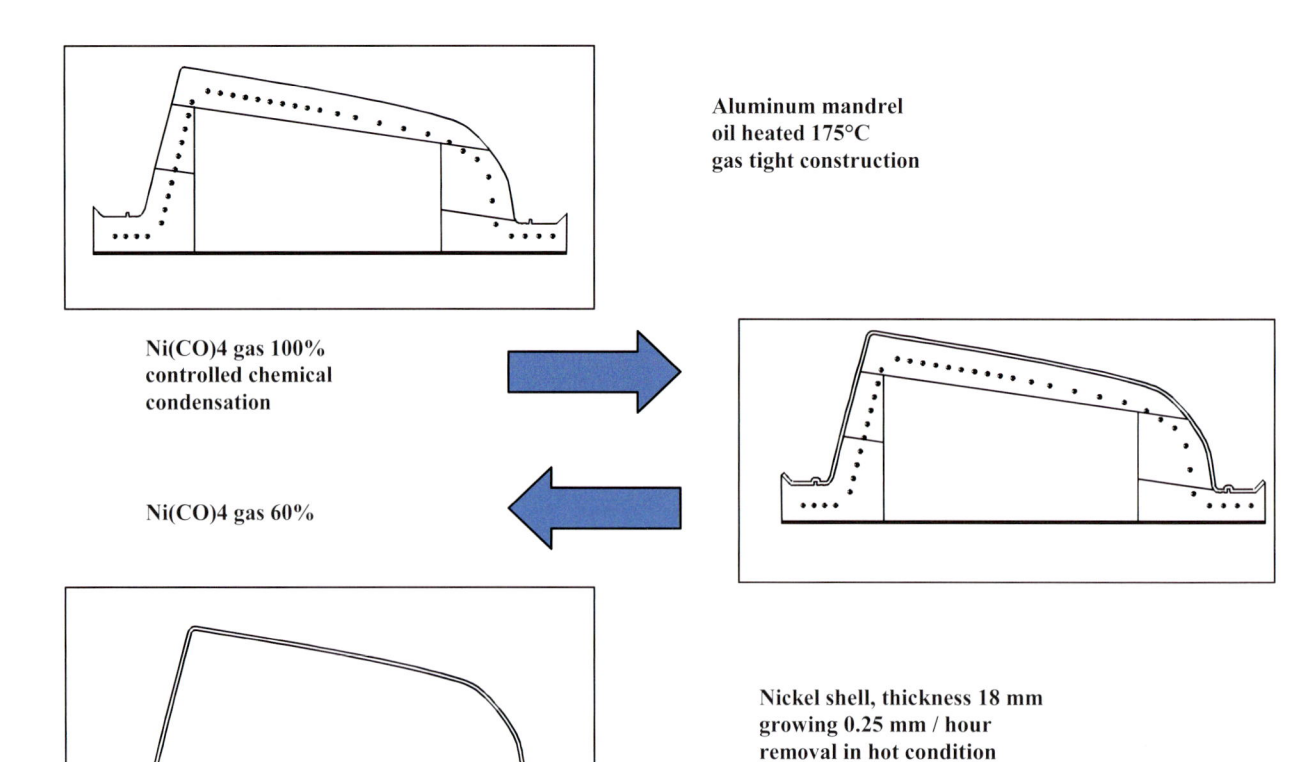

Figure 6-186 NVD process (source: Fritzmeier Composite)

6.3 Roof Modules, Hardtops

After completion of the nickel deposition process, the left-hand and right-hand mold shells are joined together to form a single piece (see Figure 6-188).

Next, the cooling system is installed and the substructure of the tool constructed. The nickel shell and substructure (steel-rib design) are joined together with a special casting compound.

Once the mold has been assembled, the cooling circuits are connected and calibrated so as to obtain a homogeneous temperature distribution (max. $\Delta t = \pm 2\ °C$). The mold surfaces are then finished, followed by acceptance on the part of the purchaser.

The mold is shipped out and put into operation at the customer's plant. After manufacture of the first parts

Figure 6-187 Nickel shell with mold substructure (source: Fritzmeier Composite)

- Nickel shell thickness 18mm
- Steel/copper/epoxy thickness 30-50mm
- Rib, thickness 25mm with heating duct
- Stainless steel tube
- Base plate thickness 45mm

Figure 6-188 Nickel shell "female" mold (source: NTT-Weber)

Figure 6-190 Nickel shell and substructure assembled (source: NTT-Weber)

Figure 6-189 Nickel-shell substructure (source: NTT-Weber)

Figure 6-191 NVD mold (source: Fritzmeier Composite)

and fine adjustment of the process parameters, the first validated and accepted parts are produced.

The Production Process

The process as a whole is subdivided into the following process steps:

- Fitting the metal reinforcements into the foaming mold
- Manufacture of the foam cores
- Preparation of the foam cores and attachment of the glass-fiber mats (UNIFILO, 600 g/m^2)
- Insertion of the foam cores and glass fiber mats into the NVD mold
- Filling the mold with epoxy resin, approx. 33 kg
- Reaction of the epoxy resin, approx. 25 minutes
- Removal of the rough molding from the mold
- Finishing of the rough molding on a CNC milling center
- Painting the finished part in the body color
- Fitting defined items of equipment (accessories) and shipping on a just-in-sequence basis to the customer.

Fitting Metal Reinforcements to the Foaming Mold

In the first step in foam core creation, metal reinforcements for built-in or surface-mounted components (air horn, spoiler, sun visor, etc.) are inserted into the foaming mold.

Figure 6-192 shows the XLX highroof with the corresponding reinforcement elements.

Production of the Foam Cores

The foam cores are made with aluminum foaming molds using a polyurethane foam system (Büfadur/Büfapur, RG 120–140).

The finished foam moldings are fitted out with the additional inserts, such as reflection grating for aerials and prefabricated glass-fiber mats and then forwarded to the RTM process.

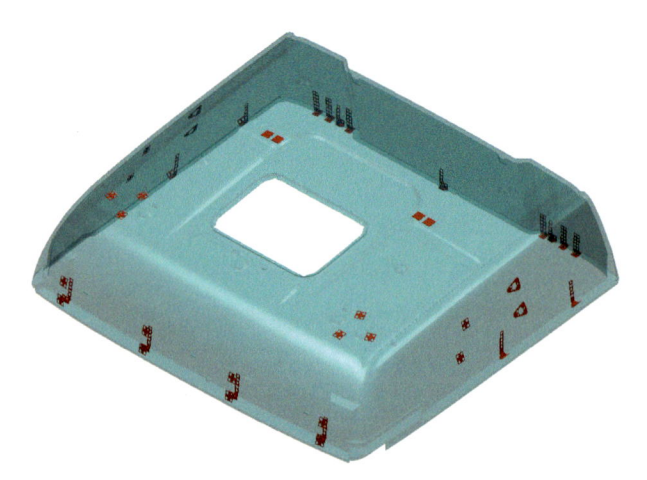

Figure 6-192 XLX highroof with metal inserts (source: MAN Nutzfahrzeuge AG)

Figure 6-193 Foam-core mold (source: Fritzmeier Composite)

Production of the Rough Molding

Production of the rough molding is subdivided into the following steps:

- Insertion of the prepared polyurethane foam cores into the NVD mold
- Closing the mold, application of the vacuum
- Adding the epoxy resin (approx. 3 min) and reaction (approx. 25 min)
- Opening the mold and removal of the rough molding.

These processes are illustrated in Figures 6-194 to 6-199.

6.3 Roof Modules, Hardtops

Figure 6-194 Inserting the foam cores with glass fiber mats (source: Fritzmeier Composite)

Figure 6-197 Venting the mold (source: Fritzmeier Composite)

Figure 6-195 Prepared mold prior to RTM process (source: Fritzmeier Composite)

Figure 6-198 Opening the mold after the RTM process (source: Fritzmeier Composite)

Figure 6-196 Addition of resin and reaction (source: Fritzmeier Composite)

Figure 6-199 Removing the rough molding from the mold (source: Fritzmeier Composite)

Finishing the Rough Molding

The off-mold rough molding is finished in a milling unit. Here, holes and contours are produced where necessary and the different equipment versions are created.

Painting the Finished Component and Pre-Assembly of the Roof Module

The highroof component is painted in the appropriate body color together with the other fitted components. Once painting is finished, defined equipment items can be fitted at the highroof unit suppliers.

Figure 6-202 shows the fitting of the roof internal trim (foam-laminated fabric) that is glued on by means of a positioning and holding device. This concept means that it is not necessary to fit a self-supporting roof liner.

Figure 6-203 shows the glued-in interior trim which follows once the roof has had its defined mounting parts or conversion parts fitted.

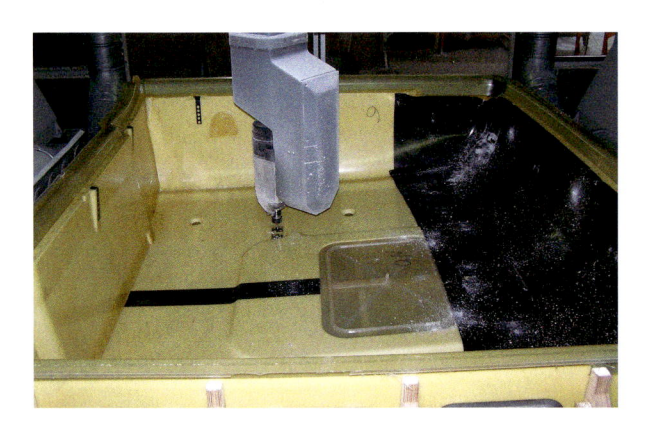

Figure 6-200 CNC machining of the rough molding – drilling function holes (source: Fritzmeier Composite)

Figure 6-201 CNC machining of the rough molding – milling outer contour (source: Fritzmeier Composite)

Figure 6-202 Gluing the interior roof trim into position (source: Fritzmeier Composite)

Figure 6-203 Glued-in roof trim (source: Fritzmeier Composite)

Just-In-Sequence Delivery and Installation of the Roof Module

The pre-assembled roof modules are shipped to MAN in Munich and Steyr (Austria), completed and attached to the lower part of the cab by a fully-automatic bonding process.

6.3.4.6 Summary

The use of SMI technology for the manufacture of roof modules offers decisive advantages in particular cases of application. However, it would not make sense to use this manufacturing process in all cases, since an overall consideration of the general conditions and constraints is required before a final decision can be taken.

Figure 6-204 Delivery to the assembly plant (source: MAN Nutzfahrzeuge AG)

Figure 6-205 Completing the interior equipment (source: MAN Nutzfahrzeuge AG)

Figure 6-206 Applying adhesive bead to the roof rim (source: MAN Nutzfahrzeuge AG)

Figure 6-207 Mounting the roof module on the chassis (source: MAN Nutzfahrzeuge AG)

Figure 6-208 Presentation of the XLX cab in Amsterdam 2005 (source: MAN Nutzfahrzeuge AG)

In hindsight, the use of SMI technology for the high-roof of the XLX cab was the correct decision, since the investment, implementation time and remaining product life all in all met the changed market requirements best.

The vehicle was first presented in October 2005 at the "International Road Transport Show" in Amsterdam and full production started on time in December 2005 in accordance with SOP planning following a total development time of 14 months.

Finally, we would like to thank our system suppliers, especially Fritzmeier Composite, for assisting us in the preparation of this chapter by generously supplying illustrations and detailed information.

Reference

[1] Häberle, H., Kneifel, E.: Neue innovative Werkstoffkonzepte für Fahrerhauskomponenten am Beispiel der MAN-Hochdächer [New innovative material designs for cab components as exemplified in MAN highroofs]. VDI Reports 1504, VDI-Verlag Düsseldorf 1999

6.4 Automotive Glazing

6.4.1 Organic Glazing in the Automobile

FALK ULLMANN

6.4.1.1 Introduction

At the beginning of the twentieth century there were only a few "horseless carriages" and they still traveled relatively slowly. Gradually, the number of these motorized vehicles increased – as did their speed – but drivers found the headwind thus created did not improve their driving. To make their job easier, a transparent sheet of glass was required: in this way the windshield came into being. Although it made drivers more comfortable, it did not provide adequate protection against stone impact which could result in glass splinters that could seriously injure the passengers.

At this time, the French chemist Edouard Bénédictus discovered that a glass flask containing a solution of nitrocellulose did not shatter upon breaking, but rather exhibited a star-like pattern of cracks. In 1919, Henry Ford realized the advantages of this and had "laminated windows" manufactured in France using nitrocellulose. This is an early version of today's laminated safety glass. Between 1919 and 1929, Henry Ford (Figure 6-209) provided his entire range of vehicles with these windows.

Today, laminated safety glass consists of sheets bonded by a polyvinyl butyral film.

In 1930, S.A. des Manufactures des Glases et Produits Chimiques de Saint-Gobain took out the patent for single-pane tempered safety glass, manufactured by pre-stressing glass to make single-pane tempered safety glass. The French term "verre trempé" means "quenched or cooled down suddenly" – an important part of pre-stressing. This type of glass provides better resistance to flexural failure and to temperature changes. When it does break, it breaks into the tiniest particles of glass. Until today, laminated safety glass has been used in the automobile primarily for windshields and single-pane tempered safety glass primarily for side and rear windows. The first vehicles in Europe with plastic glazing were the Messerschmitt KR 175 bubblecar between 1953 and 1955 (Figure 6-210) and its successor, the KR 200. The laterally folding roof was made of PMMA. However, at that time, uncoated PMMA exhibited inadequate scratch-resistance and impact strength with the result that this development was not taken further.

Figure 6-209 A 1925 Ford (Model T)

Figure 6-210 A 1953 Messerschmitt KR 153 bubblecar

The introduction of polycarbonate for transparent automotive applications was aided by the development of improved scratchproof paints. Although polycarbonate provides a high degree of impact strength, allows a fair amount of freedom in design, and also has considerably lower density than glass, its resistance to scratching, abrasion, weather and chemicals does not meet requirements.

As early as the 1980s, GE Bayer Silicones was marketing scratchproof paints specially designed for use with polycarbonates. Further developments came closer and closer to producing mature systems.

Once plastic headlight lenses were introduced in the USA, in 1993 Hella supplied the first European lenses made of polycarbonate for the headlights of the standard Opel Omega.

The motivating force behind the use of plastic with a scratchproof coating was not only weight savings but in particular the high freedom of design this material offered. Today, virtually every automobile light lens in Europe is made of polycarbonate. Designers are now making full use of the design possibilities provided by this polymeric material in order to create a distinctive front end.

In addition, polycarbonate with a scratchproof coating has established itself in niches such as:

- Safety glazing of police vehicles
- Armored cars and
- Racing cars.

Once the Smart Car with its rear side panels made of polycarbonate with a scratchproof coating was

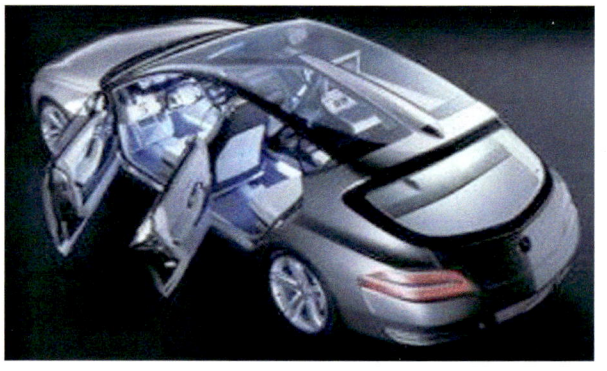

Figure 6-211 Mercedes-Benz E class (top); F500 Mind research vehicle (bottom)

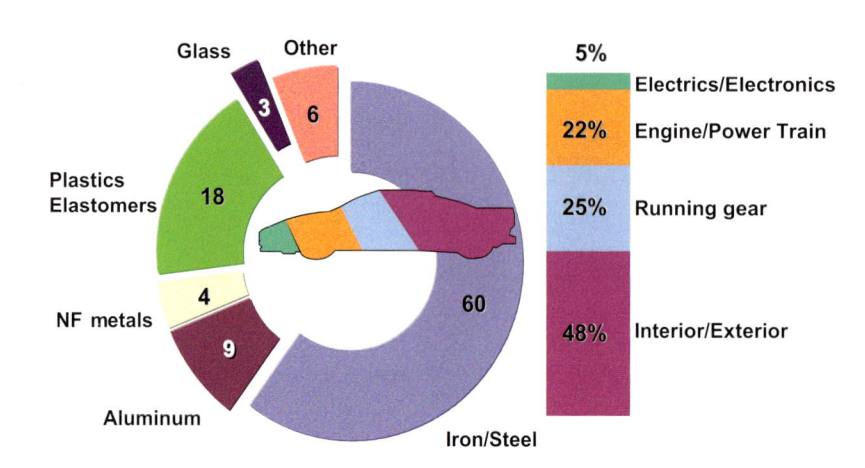

Figure 6-212 Proportions of materials in a mid-range vehicle (source: Mercer)

launched onto the market in 1998, the interest of automobile manufacturers and suppliers in Europe was finally aroused. In fact, it was in 1998 when Bayer AG and GE Plastics founded the joint venture Exatec, and in 2001 when Saint Gobain-Sekurit together with Schefenacker set up their joint venture Freeglass. The activities of these two joint ventures are aimed exclusively at developing and manufacturing automotive glazing made of polycarbonate with a scratchproof coating.

In the USA, the automotive supplier Lexamar is a producer of hardcoated exterior trim parts and other items. Another application is the transparent roof of the Corvette.

6.4.1.2 Automobile Glazing

For a number of years, a clear trend towards larger areas in automobile glazing can be obserserved (Figure 6-211).

A vehicle currently comes with up to 6 m² of glass installed. The windshield alone takes up an area of approx. 1.2 m². It contributes approx. 3% to the total weight of the vehicle and in the case of a mid-range vehicle may amount to more than 40 kg (Figure 6-212). The basic glazing here (without the panoramic roof) occupies an added-value share of more than € 200.

Connected with this development is an increasing influence on:

- Comfort or vehicle climate
- Noise and
- Weight.

Current areas of development include, for example, functional aspects such as antireflectiveness, self-cleaning, acoustic glazing, switchable (or "smart") windows, and thermal insulation.

Possible approaches to weight reduction include lightweight glass made of a glass/film/thin-glass combination and types of so-called composite glass made of glass/polymer/glass composites (Figure 6-213).

Some of these systems are not yet fully technologically mature and currently not competitive from an economic point of view. In comparison with these new developments, polycarbonate with a scratchproof coating offers a considerably higher potential for lightweight design and a much greater freedom for the designer.

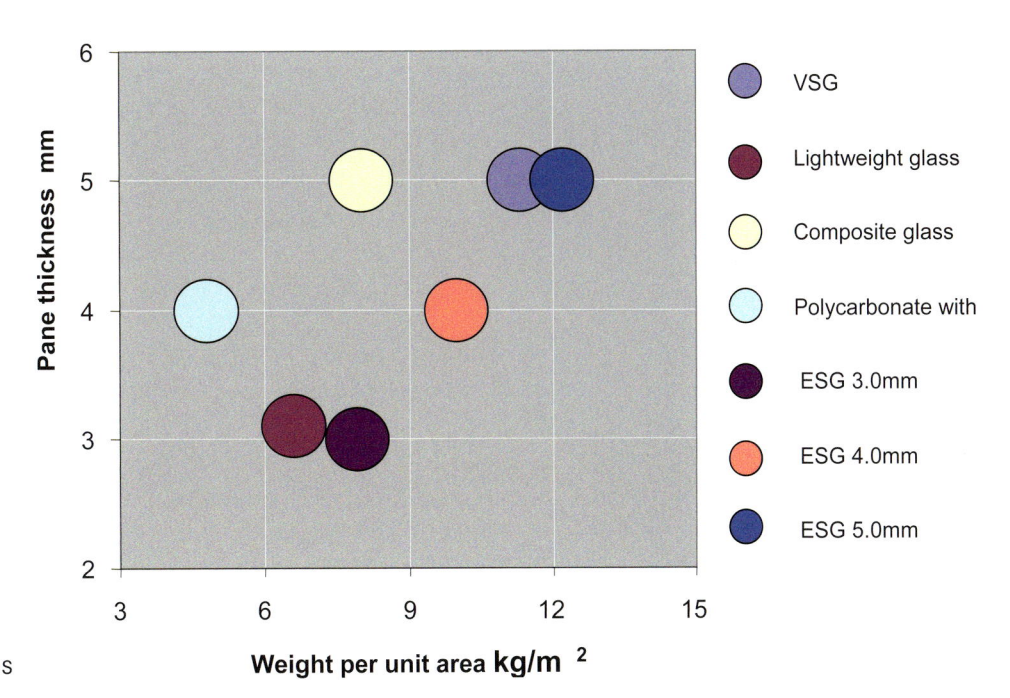

Figure 6-213
Wall thickness and weight per unit area for various pane types

6.4.1.3 Laminar Systems

Polycarbonate is very sensitive to scratching and abrasion. When exposed to ultraviolet radiation in sunlight, yellowing can be observed after only a few days. In addition, its chemical resistance against certain media is unsatisfactory. However, thanks to modern coating systems, these disadvantages can be reduced to a certain extent, depending on the actual requirements of the application.

UV-curing acrylic paints have already been used for for a considerable period for headlight lenses or covers. In order to meet stricter requirements, polysiloxane paints are used. Here, the systems PHC587 (one-pack paint) and AS4000 (paint plus primer) from GE Bayer Silicones have become industry standards (Figure 6-214). Current developments are aiming at improving weather resistance (the AS4700 system) and at increasing abrasion and scratch resistance by using nano particles.

While various paint systems have been available for years, vacuum deposition methods to coat large-scale free-form surfaces are still in their development stage. The coating systems created by plasma CVD processes certainly offer great potential for delivering the properties required. In comparison with paint systems, they offer considerably improved scratch and abrasion resistance. Depending on the process technology, even functions such as infrared coatings and optical filters could be implemented. However, in commercial applications, the ultraviolet resistance of the current systems is still inadequate. Developments in this area are a major focus in industry.

6.4.1.4 Advances in Manufacturing

The accelerated development that has taken place over the last six years can be attributed not only to the scratchproof paints currently available (which have been tailored to polycarbonate) but also to advances in:

- Plastics processing, enabling the manufacture of large-area components with a high degree of functional integration, and

- Greater process reliability in the application of the scratchproof paints.

Here, the technology developers profit from their experience in coating headlight lenses and from the years of expertise accumulated in two-component injection molding. Close collaboration between suppliers and the OEMs as well as machine manufacturers and toolmakers has also ultimately helped in making this technical breakthrough possible.

The technology for component fabrication basically includes two-component injection molding and the application of scratchproof paint in highly automated painting lines.

The swivel platen method uses two injection units facing each other, aligned with the machines' longitudinal axes. Between the fixed platen and the moving platen there is a sliding table on which a vertical swivel platen is mounted that carries the mold halves. By rotating about its vertical axis, it allows the mold cavities to be filled alternately with material from the two injection units (Figure 6-215).

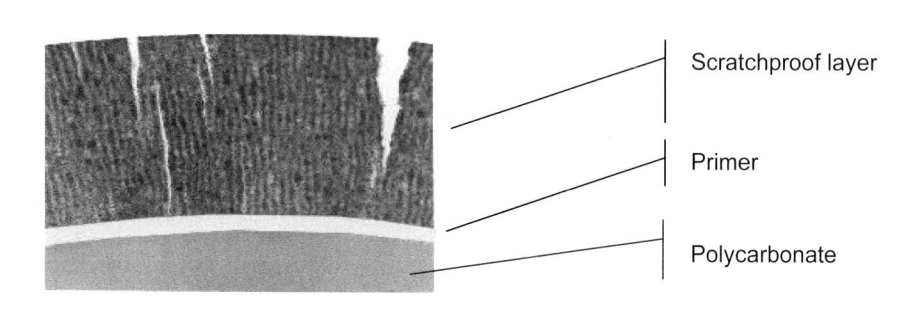

Figure 6-214
Photomicrograph of the AS4000 layer system (source: GE Bayer Silicones)

6.4 Automotive Glazing

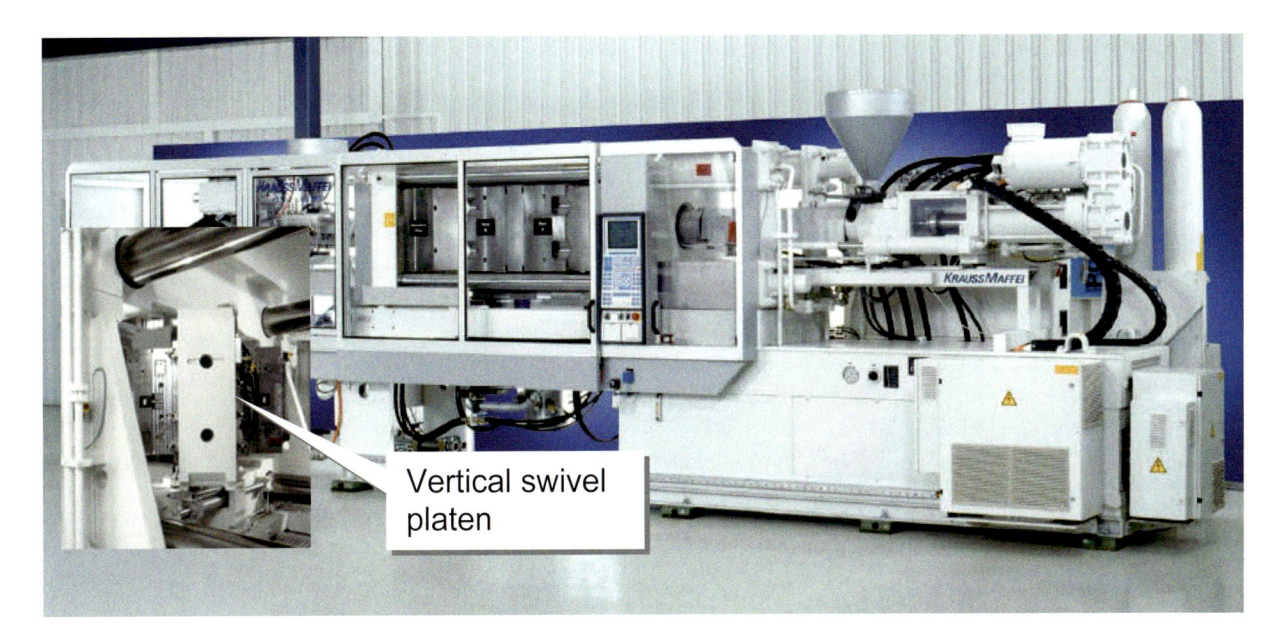

Figure 6-215 Two-component injection-molding unit – swivel-platen type (source: Krauss-Maffei)

(Vertical swivel platen)

In the two-component method, the first step is usually to make the transparent pane. Following rotation of the mold by 180°, the second step is in-mold laminating in order to integrate additional functional elements. Here, the two materials – both the transparent part and the part determining the function – must be adjusted to each other.

Particularly in the case of large-area components, the highest requirements are placed on precision in shape and dimensions, while low stresses and orientations are also demanded. With the various injection-compression molding versions available, such as clamping compression, breath and expansion compression and also the "IMPmore" method (in-mold pressing) developed by Summerer in conjunction with Battenfeld, it was possible to fulfill these requirements for the most part.

Not only the shaping, but also the process for providing the component with a scratchproof and weather-protection coating represent a technological challenge.

The polysiloxane-based paints available are thermal curing systems that, after being applied, must be cured at a temperature of 130 °C for a period of at least 30 min. In addition, the fact that glass the transition temperature for polycarbonate of approx. 148 °C requires exact temperature control.

In the circulation method, the paint is prepared and applied predominantly by flow coating (Figure 6-216). The line units are completely sealed off, and a particularly high degree of air cleanness has to be maintained within both the flow-coating cabin and the drying unit. This is particularly important, since the crystal clear components are very sensitive to dust.

Figure 6-216 Flow-coating a headlight lens (source: GE Bayer Silicones)

To secure the required quality, both component production and painting should be located in close proximity, preferably within the same operational environment. Every step required for handling and transporting the component not only involves the risk of damage, but also creates additional costs. It is also important to make sure that process know-how is shared throughout the entire production chain, ensuring a uniform quality concept.

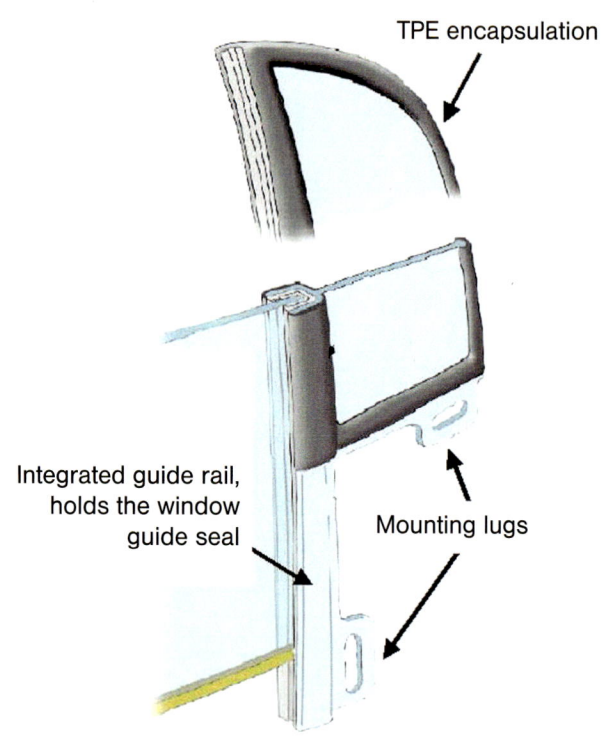

Figure 6-217 Design for a fixed side window (source: Freeglass)

6.4.1.5 Mass Production Applications

A decisive prerequisite for acceptance on the part of the automaker is not only technological but also economic competitiveness with mineral glass. During concept design work, close coordination between the automobile manufacturer's departments and the supplier is mandatory in order to take advantage of the potential offered by plastics.

The key word here is functional integration. Only when a large number of functions is integrated in one component, such as:

- Window guides
- Mounting elements
- Quarter-window surrounds and
- Gluing flanges

(Figure 6-217) the manufacture of polymer glass systems will become economically viable.

It is therefore a matter of fundamental importance that component suppliers provide not only the technological know-how but also techniques of product conception and design. Polycarbonate with a scratchproof coating was used for the first time in relatively large production quantities in 1998 in the fixed side window pane of the Smart Car and in its successor, the Smart City (Figure 6-218).

Figure 6-218 Fixed side windows in the Smart City (left) and the Smart Roadster (right)

Figure 6-219 Mercedes-Benz C class sports coupe with polycarbonate trunk-lid panel

The front triangular window of the Fiat Multipla was also manufactured from plastic. The next application was the Mercedes-Benz C class sports coupe in 2001 (Figure 6-219) with a transparent trunk-lid panel and the Smart Roadster in 2003 (Figure 6-218), which also has a fixed rear side window.

The black in-mold laminated area serves not only as accommodation for additional functions but also for incorporating mounting elements and creates a sense of depth that contributes to the impression of high quality.

6.4.1.6 Potential Applications and Outlook

The future realization of the full potential of glazing systems made of coated polycarbonate will only be possible if the advantages offered by plastic processing are consistently exploited. Prerequisites for successful implementation are ensuring process stability of these technologically very demanding production systems and the collaborative development of competitive component concepts together with the automotive manufacturer.

The advantages of plastic glazing are:

- Comparatively high degree of freedom in design
- Significantly lower weight
- Extremely high impact strength and damage tolerance, and
- Possibility to integrate functions not possible with a conventional glazing system.

The disadvantages are:

- Scratch, abrasion and weathering resistance, which are (still) lower than those of glass
- Lower rigidity, and
- High thermal expansion.

The material price, which is considerably higher than that of glass, can be compensated by suitable functional integration. Indeed, cost advantages are achievable when polycarbonate competes with laminated safety glass.

When plastic is used in roof systems, both the lowering of the vehicle's center of gravity and weight reduction play an important role. Coloring the polymer substrate is easier and less expensive compared to glass.

However, in the foreseeable future polycarbonate glazing will not find application in windshields or in moving side windows. For these applications, the requirements on the systems are far too demanding to be met by polymeric windows at the present state of the art.

To advance the development of rear windows, the scratchproofing systems must be improved to cope with the mechanical effects of the wiper unit. The integration of heating elements also represents a challenge to research and development that must not be underestimated. A decisive impulse could come from plasma CVD technology, once it is mature enough for volume production applications.

Beyond this, a number of development trends can already be detected:

- Increasing functional integration (even including lights and LED's; see Figure 6-220)
- The use of transparent polymer systems in roof modules (Figure 6-220)

- Integration in panoramic roofs (Figure 6-221), and
- Component designs for non-transparent applications (pillar trim, wind deflectors, etc.).

Intelligent solutions are required to meet future challenges, which can only be mastered in collaboration of all parties involved.

Figure 6-220 Integration concept for LED's and mirror base (left); roof module concept (right) (source: Freeglass)

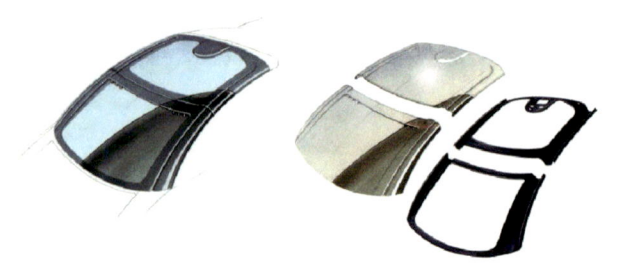

Figure 6-221 Concept of roof system with polymeric glass and frame (source: Freeglass)

6.4.2 Polycarbonate Automobile Glazing: Automotive Industry Requirements and Solutions

Erich Lehner, Gerald Aengenheyster

Automobile glazing made of polycarbonate has been used successfully for many years at DaimlerChrysler for fixed side windows and roof applications. The list of advantages (such as weight, styling, safety) that polycarbonate offers in comparison with mineral glass, however, also involves new technical challenges relating to both material and processing. We will provide a detailed treatment of automotive industry requirements, the latest developments by raw material and machine manufacturers, as well as current and future solutions from Freeglass.

6.4.2.1 Introduction

The first vehicle glazing at the beginning of the 20th century, when vehicles were open to the air, was used for protecting the driver against headwind. These windshields were initially made of flat glass, which could seriously injure the passengers if it were to break. Later on, laminated windows on the basis of nitrocellulose came into use. Today, laminated safety glass consists of sheets bonded together by an interposing layer of polyvinyl butyral film.

The single-pane toughened safety glass predominantly used for side and rear windows is based on a patent granted back in 1930 for pre-stressing glass to improve its resistance to flexural failure and to temperature changes and, above all, its fracture behavior. If this single-pane toughened safety glass breaks, it shatters into innocuous, tiny particles of glass without forming dangerous sharp edges. The first vehicles in Europe with plastic glazing were the Messerschmitt KR bubblecars during the 1950s, which had a laterally folding roof made of PMMA. However, its inadequate scratch-resistance and impact strength prevented this development from becoming more widespread.

The breakthrough for transparent plastic in the automobile came with the headlight lens made of polycarbonate (PC) with a scratchproof coating. As early as the 1980s, GE Bayer Silicones was marketing scratchproof paints specially designed for use with polycarbonates. The first applications in Europe were in 1993 with the lenses manufactured by Hella for the Opel Omega. There were further niche applications for polycarbonate in the field of special protective measures, in government agency vehicles (such as border guard vehicles) and, of course, in motor racing.

The first vehicle glazing made of PC to enter the market was the Smart City in 1998 (Figure 6-222 (left)). 2003 saw the launch of the Smart Roadster with a fixed quarter window (Figure 6-222, (right)). Even prior to this, the Mercedes-Benz C class sports coupe had been given a transparent styling panel at the

Figure 6-222 Smart Fortwo and Smart Roadster with polycarbonate quarter windows

rear [1] (Figure 6-223 (left)). All of these panels have a polysiloxane hard-coat and their shapes could not have been produced if glass had been used instead. The black frame accommodates mounting elements and gives a strong impression of quality in all three vehicles.

The Smart Forfour, introduced in 2004, not only has a fixed glazing pane of polycarbonate in the rear doors but also features a PC roof element (Figure 6-223 (right)). The fixed side window, which is encapsulated in TPE, exhibits a high level of functional integration feasible only in plastic. The polycarbonate roof element represents weight savings located high above the center of gravity of the vehicle.

It was also in 2004 that the A class with a transparent slatted roof entered the market (Figure 6-224). Here, too, the possibility of saving weight was a strong motivating force in the development.

Figure 6-223 Left: Mercedes-Benz C class sports with PC panel
Right: Smart Forfour with polycarbonate roof and quarter window

Figure 6-224 Mercedes-Benz A class with panorama slatted roof made of polycarbonate (source: Bayer)

Vehicles already on the market and those that will follow in the future exploit not only the material-specific but also the process-specific advantages of the plastics. Thanks to its density, which is 50% less than that of glass, and its very high impact strength, polycarbonate offers considerable weight-saving and safety potential. In addition, injection molding makes it possible to make parts that are geometrically very complex with regard to both esthetic design and functional integration [2]. In addition, polycarbonate is an interesting alternative for large-area roof glazing, considering the additional costs for laminated safety glass or toughened safety glass with protective film.

The prerequisite for all applications is, of course, a coating suitable for the component or vehicle in question. The comparatively sensitive surface of the polycarbonate and its considerably lower rigidity in comparison with glass are disadvantages which can, nevertheless, be overcome by the use of coatings and by constructive design measures.

Beyond this, there are still more challenges, some of which are solved or will be solved over the next few years. In all cases, every new development is measured against mineral glass. Not only light transmission and reflection optics, but also the integration of infra-red absorption/infra-red reflection or of a heating element should be mentioned in this regard. Even for a rear window with wiper system, very high quality coatings are required. Since these windows cover large areas and in some cases require large-area in-mold lamination, their process technology will be no trivial matter.

6.4.2.2 Process and Material Technology

Major players in plastic glazing are the companies Exatec [3] (a joint venture of Bayer and GE), Freeglass (a joint venture of Saint Gobain-Sekurit and Schefenacker) as well as Bayer and also Dynamit Nobel Plastics, who manufacture the transparent polycarbonate panel for the Mercedes-Benz sports coupe. All other polycarbonate windows mentioned are manufactured by Freeglass or Schefenacker in Schwaikheim.

Important developments in the raw materials sector include special polycarbonate types for automotive glazing (AG types from Bayer Material Science [4]), specially matched PC/PET types for large-area in-mold lamination of the transparent components, and new coating systems based on plasma CVD with considerably higher scratch and abrasion resistance (such as Exatec500 from Exatec). But even classic polysiloxane coatings with improved weather resistance are available, such as AS4700 from GE Bayer Silicones.

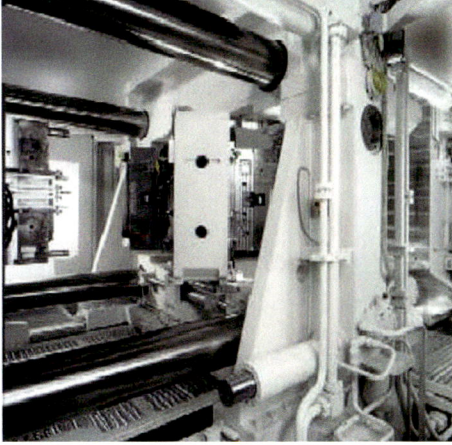

Figure 6-225 Two-component injection molding machine using the Krauss-Maffei swivel-platen technique (source: KM)

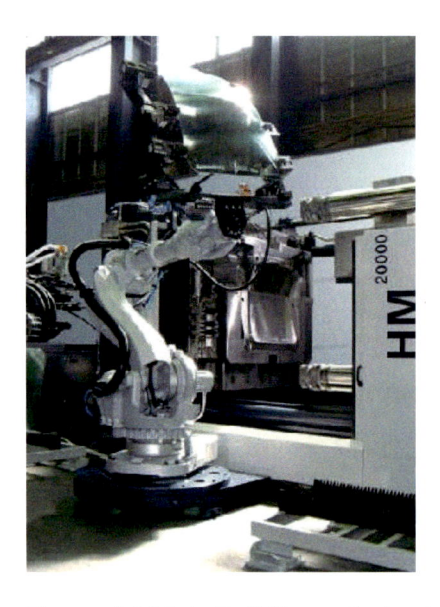

Figure 6-226 Left: IMPmore technology (source: Summerer) Right: GLAZEMELT technology (source: Engel)

Advances in process technology include swivel-platen machines (Figure 6-225) for manufacturing two-component parts and the adaptation or development of various compression molding methods, such as the clamping, breathing, expansion and wedge versions of compression molding for manufacturing extremely thin-walled and low-stressed parts [5]. The IMPmore compression molding process developed by Battenfeld and Summerer (Figure 6-226 (left)) applies a comparatively complex mold technology to produce a single-component polycarbonate window by wedge- or tilting-type compression molding [6]. Another technology developed especially for large-area car windows is Engel's GLAZEMELT (Figure 6-226 (right)). Particularly noteworthy here is its innovative compression and mold/gating technology [7].

The goal behind these developments is to achieve in-mold lamination with as high a degree of control as possible. This is of particular importance considering warpage, dimensional stability and surface quality and it allows the manufacture of parts with the lowest internal tension possible. This is necessary not only for a low deformation potential during the coating process – which takes place at temperatures close to the glass transition temperature – but also to prevent the risk of primers (e.g., AS4700) triggering stress-cracking.

6.4.2.3 Measurement Techniques and Delivery Specifications

The requirement for parts to exhibit good reflection properties and low internal stress calls for measurement techniques during both development and sampling that allow an objective, reproducible and at least semi-quantitative determination of these properties. At DaimlerChrysler, the Ondulo procedure by Techlab (F) is used together with comprehensive photoelastic analysis equipment by Strainoptics (US). The Ondulo method is a fringe projection method for determining the surface topology of components which correlates well with the impression formed by the human observer. Defects in the black component manifest themselves clearly in the reflective optics of the component surface (Figure 6-227).

The polarization optics of separate transparent components and of the finished two-component parts before and after coating supply very useful information about the quality of the injection molding process and additional information about the non-transparent component, since the latter forces a deformation or stress state on the transparent component. Figure 6-228 shows the polarization optics of the lower half of the part section shown in Figure 6-227.

6.4 Automotive Glazing

This and other analytical methods are currently beginning to appear in the delivery specifications for vehicle safety glazing. At DaimlerChrysler in the future, mineral glass glazing and polycarbonate glazing will be dealt with uniformly in a single delivery specifications sheet. The measurement technology required is currently being set up in an "optical measurements" laboratory at DaimlerChrysler.

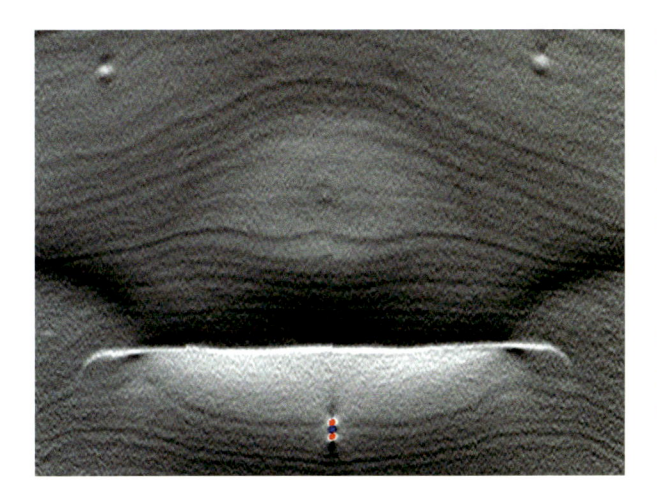

Figure 6-227 Ondulo measurement for a section of glazing

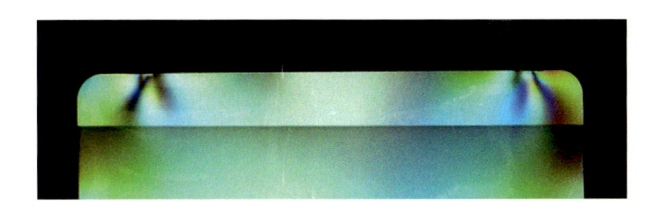

Figure 6-228 Polarization in a section of glazing

In addition, test programs are being developed at external institutes to determine the long-term behavior of different substrate coating combinations, including the influence of internal stress states in the specimens.

6.4.2.4 Process Chain for Manufacturing Plastic Glazing

In the manufacture of plastic glazing, every single process step has a decisive influence on the product characteristics and quality. Figure 6-229 shows the principle steps in the manufacturing process. Even during material storage or delivery to the production line, it is possible for the plastic granules to become contaminated, e.g., due to abrasion in the conveyor system, which can lead to a high proportion of rejects in the final product. A plasticizing unit suitable for processing polycarbonate must be used for the injection molding process.

At Freeglass, the great majority of plastic glazing panels are manufactured by the two-component method on swivel-platen machines. IN the following we will discuss the advantages of this method and the special features of the mold technology required.

If requirements for scratch resistance for vehicle windows and adequate ultraviolet protection are to be met, the polycarbonate panels always have to be coated on both sides. To do so, the product is first masked off according to its particular application, e.g., in areas that

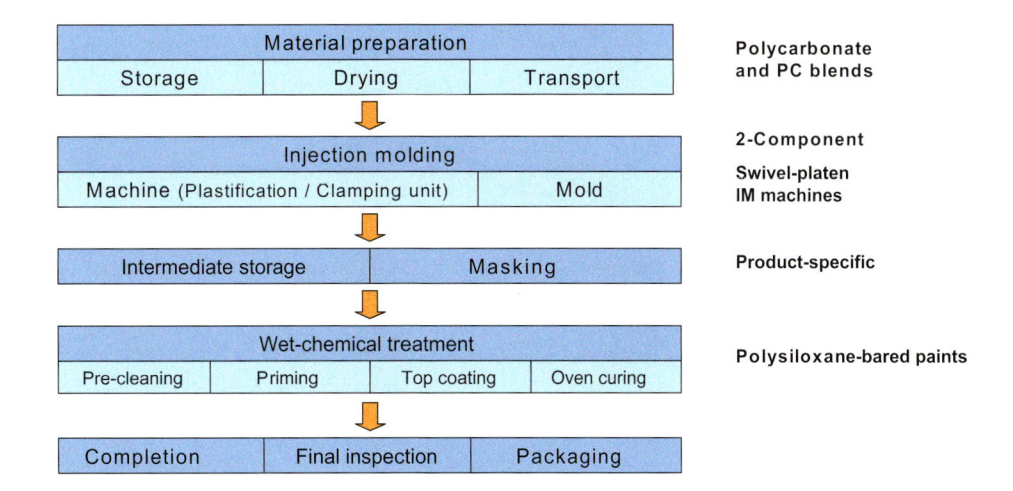

Figure 6-229 Process chain for manufacturing vehicle windows from polycarbonate

Material preparation			
Storage	Drying	Transport	Polycarbonate and PC blends

Injection molding		
Machine (Plastification / Clamping unit)	Mold	2-Component Swivel-platen IM machines

Intermediate storage	Masking	Product-specific

Wet-chemical treatment				
Pre-cleaning	Priming	Top coating	Oven curing	Polysiloxane-bared paints

Completion	Final inspection	Packaging

subsequently will be glued and therefore must remain free of coatings. State-of-the-art is a wet-chemical flow coating with polysiloxane paints. Systems with and without an intermediate primer layer are used. How the coating is applied is of decisive importance for the results (prevention of curtaining, homogeneous distribution of coating thickness, etc.) as well as for preventing dirt particles from being incorporated in the clear coating. This is ensured by pre-cleaning the polycarbonate panels and then painting them under clean room conditions.

6.4.2.5 Advantages of the Two-Component Method

When plastic panels are made by the two-component method, the printing common with vehicle windows is replaced by a (black) decorative frame.

First, the transparent (tinted) window is injection-molded (first component) and then the decorative frame is molded-on (second component) (Figure 6-230).

Here, the decorative frame not only serves as a cover for the connection to the body flange but it also has the following important advantages:

- Reinforcement of the plastic window module

- Integration of functional elements is possible (e.g., positioning aids, water drainage, mounting clips for the bodywork or for add-on parts, such as interior lighting (cf. Figure 6-230).

- Direct glazing by bonding onto the black component is possible with virtually any adhesive system normally used with glass

6.4.2.6 Machines and Mold Technology

At Freeglass, swivel-platen machines with locking forces of up to 4000 t are used for manufacturing plastic windows by two-component injection molding. The underlying principle of the machine configuration is shown in Figure 6-231. In these horizontal machines, actually two separate molds are used for the first and second plastic component. By rotating the central platen 180°, the transparent premolding can be in-mold laminated on the cycle 2 side.

This technique offers advantages in the comparatively small size of the platens necessary compared with the product size and in good clamping unit balance despite markedly different sizes of the projected areas of the first component (the panel) and of the second component (the frame). The swivel-platen technique is thus particularly suitable for two-component injection molding of large-area plastic windows.

Special requirements also apply to the injection molds used for plastic windows. In addition to the correct balancing of the two mold blocks used in the swivel-platen technology, the high-quality high-gloss polish of the large-area cavities is of decisive importance. Temperature control is also complex and must provide, among other things, a homogeneous temperature distribution as well as pre-defined local heat dissipation with high temperature control performance. At least for the manufacture of the transparent panel (the first component), the mold should also be suitable for an injection-compression process. For the frame component, which is frequently relatively thin-walled, the design of the hot runner is of decisive importance.

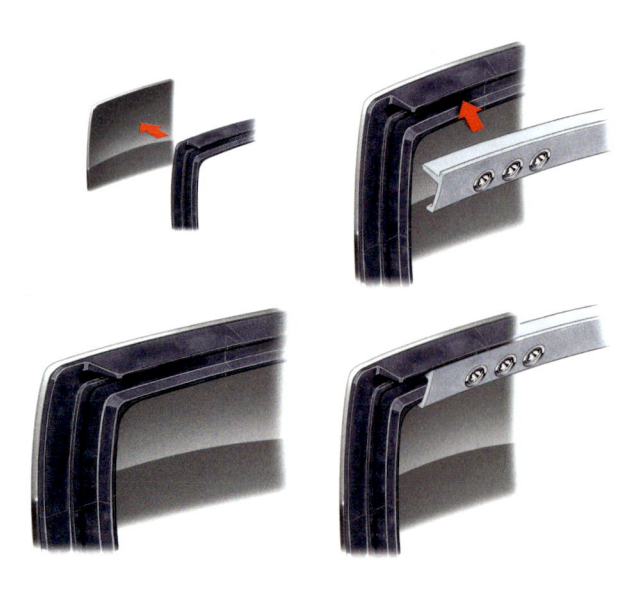

Figure 6-230 Two-component plastic window with clips for add-on parts

6.4 Automotive Glazing

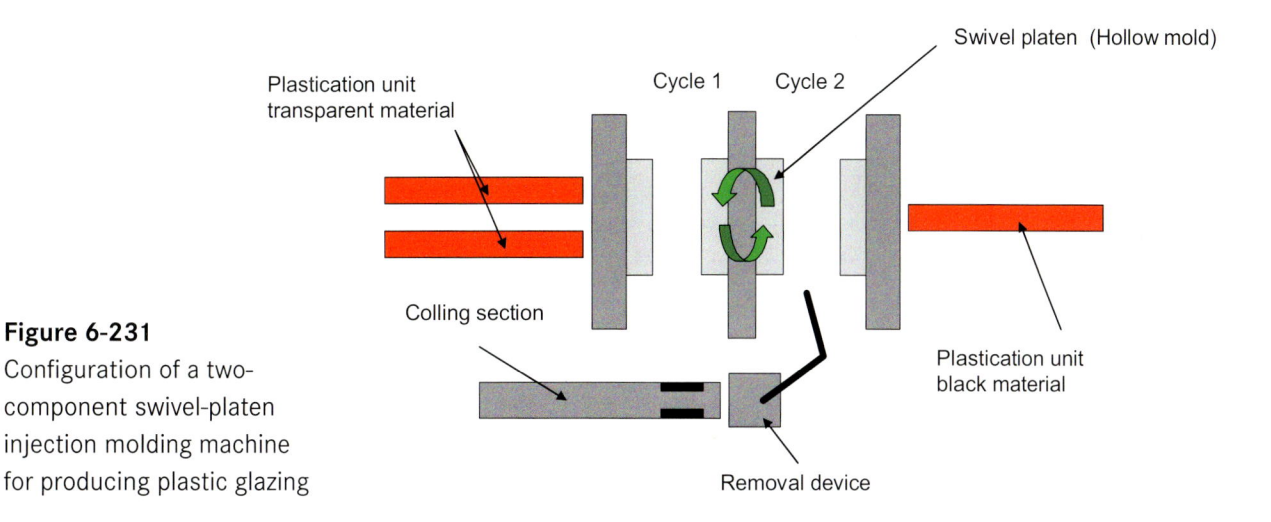

Figure 6-231
Configuration of a two-component swivel-platen injection molding machine for producing plastic glazing

6.4.2.7 Product Development and Process Optimization

The use of CAE systems is indispensable in product and process development. FE simulations and rheological calculations are performed for cavity filling of the first and second components in the injection molding or injection-compression method, as well as for the design of the hot runner.

By means of mechanical FE analyses, collision and wind-load simulations are performed and the joining plan (e.g., bonded joints) is laid out. In addition, simulation calculations provide important information on the thermal expansion of the polycarbonate window in its installed state, as well as on its dynamic properties (e.g., natural frequencies).

Informative results relating to the relationships between the product quality and the parameters of the production process can also be obtained with the aid of designed experiments (DOE).

Figure 6-232 shows the influences of mold and melt temperature of component 1 on warpage and on the quality of the reflective optics of a large-area window.

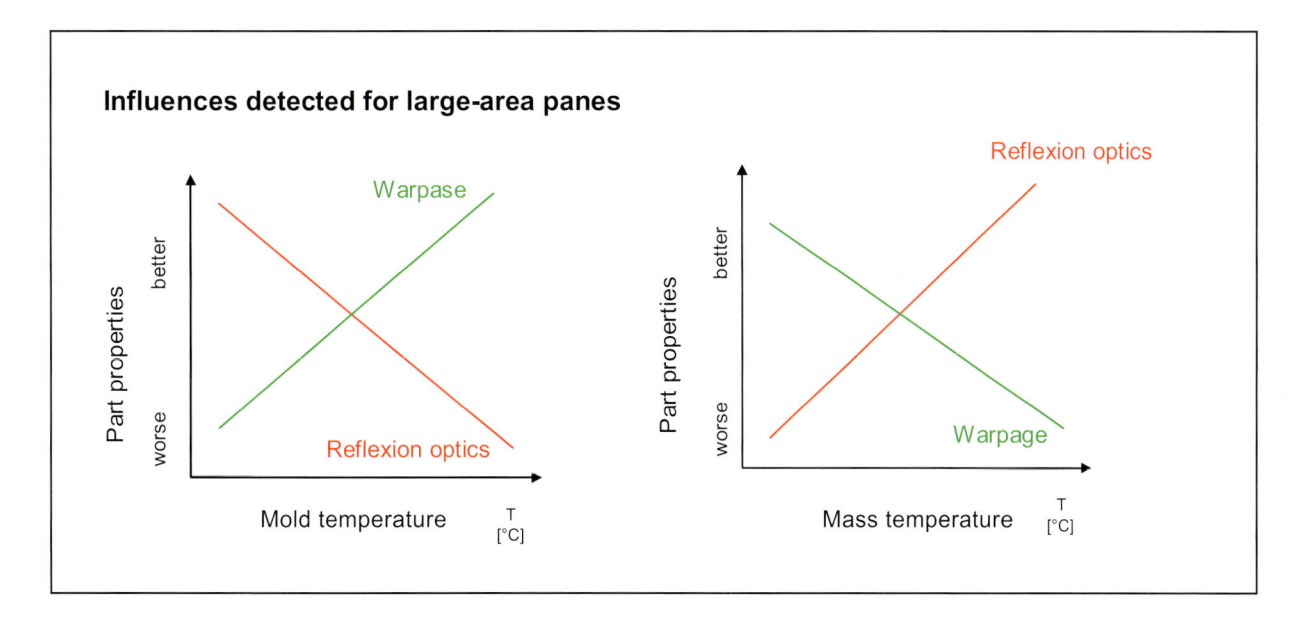

Figure 6-232 Results of design of experiments (DOE)

6.4.2.8 Additional Product Functions

The future range of applications for plastic windows does not depend solely on the optical quality and scratch resistance obtained, but also on the size of window that can actually be manufactured. Also of importance is the implementation of functions already available with other glazing solutions.

It is possible, for example, to heat plastic windows by means of heating wires embedded in the polycarbonate. This manufacturing process was developed by Freeglass and is characterized by the following steps:

- Hot embedding of heating wires in a transparent polycarbonate film
- Back-injection molding of the film with transparent (tinted) polycarbonate
- If necessary, back-injection molding of the window with a decorative frame by the two-component process
- Incorporation of electrical contacts for the window and additional final details depending on the application

Figure 6-233 shows thermal images of windows manufactured by this method after a heating-up period of 5 minutes. This meant that the specifications of the automobile manufacturer were met. Other advantages of this technology are that the heat conductors are virtually invisible and that a comparatively homogeneous temperature distribution is achieved over the surface of the window. The latter is particularly important due to the low thermal conductivity and higher thermal expansion of the plastic material.

Other requirements on vehicle windows apply for the field of thermal comfort – in other words, energy transmission which is particularly influenced by transmission values in the infrared range.

Table 6-15 shows the optics-related transmission data for different window materials. For clear polycarbonate or clear glass, energy transmission reaches approx. 82%. Due to the dark tinting of the plastic (residual light transmission approx. 21%) in roof windows, for example, the energy transmission can be reduced to approx. 46%. When special infrared absorbers (with sufficient UV stability) are added to the polycarbonate, the energy transmission can be further reduced to approx. 29%.

Table 6-15 Comparison of optics-related transmission data

	Clear glass	Green glass	SOLARIT Venus 10 grey	Glass IR-reflecting	PC dark pigmented	PC with IR-absorbers	PC clear
T_L	90%	71%	10%	77%	21%	26%	87%
T_{IR}	77%	22%	5%	15%	80%	29%	82%
T_{UV}	59%	18%	1%	1%	0%	0%	0%
T_E	83%	44%	8%	46%	46%	29%	82%
Dicke	4 mm	4 mm	4 mm	5 mm	5 mm	5 mm	5 mm

T_L: Light transmission; T_{IR}: Infrared transmission; T_{UV}: UV transmission; T_E: Energy transmission

Heating wires in series   and   in parallel

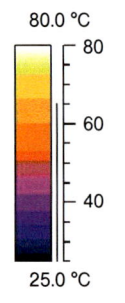

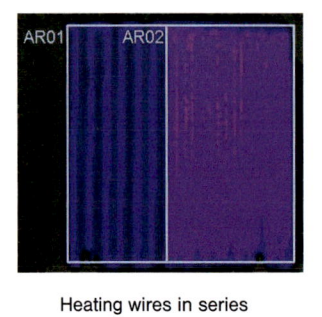

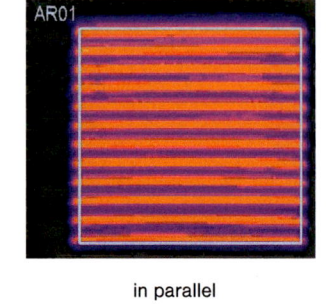

Figure 6-233 Thermal images of heated polycarbonate windows (heating time: 5 min)

6.4.2.9 Coating Methods

The wet-chemical coating method used in current volume production applications provides the following special characteristics:

- Coatings are transparent (dirt particles are not hidden)
- The coating thickness ranges from only approx. 4–9 μm
- Higher proportion of solvents in the coating
- Coatings are in some cases applied in two layers (a primer and a top coat)
- Coatings are applied to both the interior and the exterior of the products
- Coatings are thermally cured at approx. 130 °C (within the range of the heat resistance of polycarbonate)
- No subsequent coating is possible
- Products are sometimes of large dimensions (order of magnitude of 1 m^2)

As an alternative to this scratch-resistant coating, development work is in progress at Freeglass to vacuum-coat polycarbonate windows.

The underlying principle of this method is shown in Figure 6-234. Here, a precursor gas is introduced into a vacuum chamber with oxygen also being introduced from time to time. The precursor gas is excited to a plasma and a film is deposited on the polycarbonate surface, so that a graduated layer builds up, initially with an organic component but then increasingly of a purely inorganic, glass-like character.

Freeglass operates a pilot plant using this technique that can produce substrates measuring approx. 900 mm × 350 mm.

Figure 6-235 shows the homogeneity of the coating thicknesses achievable by this process. Similarly to polysiloxane coatings, the average coating thickness is 6 μm, with a thickness tolerance of only ±0.25 μm.

Figure 6-234 Principle underlying plasma CVD

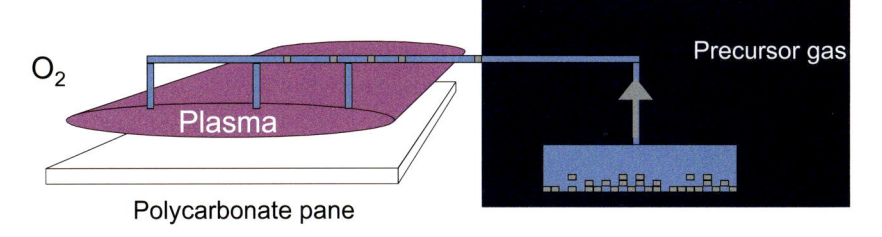

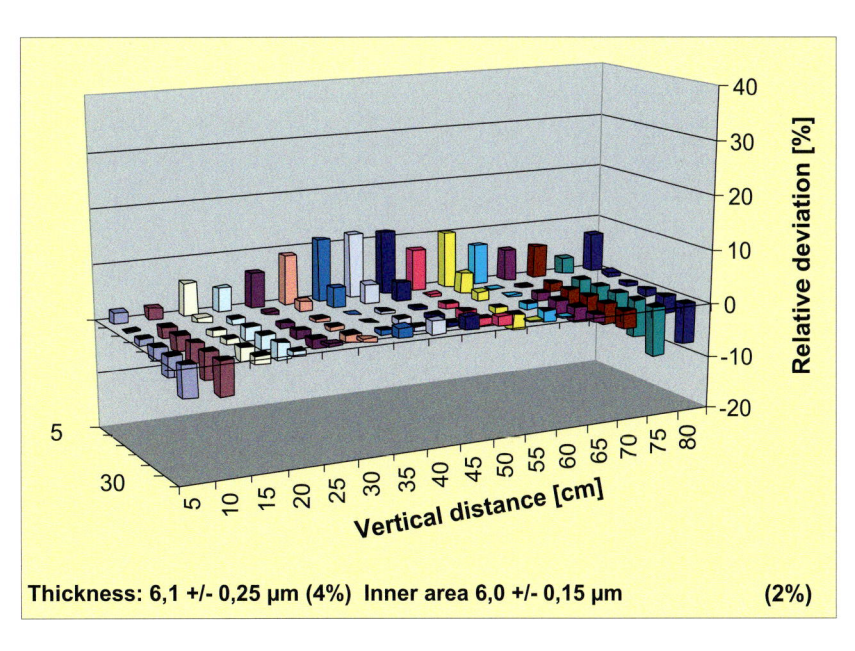

Figure 6-235 Coating thickness distribution in plasma CVD coating

Thickness: 6,1 +/- 0,25 μm (4%)   Inner area 6,0 +/- 0,15 μm   (2%)

Table 6-16 shows an overview of the properties of the plasma CVD layer. Particular emphasis should be given to the achievable scratch resistance: After 1000 rotations in the Taber abrasion test, an increase in opacity of less than 4% occured. The corresponding values for polysiloxane coatings range from 6–10% increase in opacity after 500 rotations in the Taber abrasion test.

Table 6-16 Properties achieved with plasma CVD coating

Test	Result
Thermal cycles (wet and dry)	passed
Cataplasma Test	14 days, passed
Storage at elevated temperatures (90 °C)	4 weeks, passed
Taber Test: increase in opacity after 1000 rotations	< 4%
Boiling water test	2 hours, passed
Chemical resistance	passed
WOM-Test (Florida)	1000 hours passed
Layer adhesion (grid-cut-test)	GT 0 according to all tests

Table 6-17 State of coating techniques

Application	Coating technology	Status
Fixed side windows	Wet chemical: coatings: PHC 587, AS4000	In mass production
Roof modules	Wet chemical: coatings: AS 4000, AS 4700	In mass production
Rear modules	Wet chemical	In mass production
	Plasma-CVD	Preliminary development (first series applications in approx. 3 years)
Windshields, moving side windows	Plasma-CVD?	Still under research

The plasma CVD method also offers additional advantages:

- It is possible to create multifunctional layers (IR-absorbent, water-repellent, etc.)

- Compared to painting lines, the installations are smaller in size while offering the same capacity

- No solvents are used.

The first mass production applications for the plasma CVD process are expected within the next three years (see Table 6-17). This method is, for example, suitable for rear window applications in hatchbacks that are subject to additional abrasive stress from the rear window washer.

6.4.2.10 Summary

Based on the already numerous applications in automotive volume production for polycarbonate windows, future focus for development of plastic glazing will be on:

- Further development of injection molding (optical quality, product size, etc.)

- Further development of coating technology (scratch resistance, ultraviolet protection, economic efficiency, etc.)

- The implementation of functions already available with other glazing solutions (heatability, infrared reflection, etc.)

- Implementation of new functionalities (such as integration of lighting)

This will open up an increasingly broad spectrum of applications for plastic windows in vehicles, which in the future may also result in a considerable contribution to the reduction of vehicle weight.

References

[1] Ullmann, F., Dannenberg, M., Grevener, C.: Mercedes-Benz C-Klasse Sportcoupé-SMC-Kennzeichenblende mit integrierter PC-Scheibe. VDI-Reihe Kunststofftechnik, 2001, Kunststoffe im Automobil, pp. 173-184

[2] Ullmann, F.: Kunststoffverscheibung im Automobil. VDI-Reihe Kunststofftechnik, 2004, Kunststoffe im Automobil, pp. 275-289

[3] Orth, P., Rappelt, T., Stein, F.: Für das Auto von morgen. Automobilverscheibung aus hartbeschichtetem Polycarbonat. *Kunststoffe* 90 (2000) 8, pp. 90-92

[4] Anon.: Makrolon AG2677. technical bulletin Bayer (2004), KU28062-412.de

[5] Bürkle, E., Klotz, B., Lichtinger, P.: Durchblick im Spritzguß. Das Herstellen hochtransparenter optischer Formteile – eine neue Herausforderung. *Kunststoffe* 91 (2001) 11, pp. 54-60

[6] Hof, K., Sauer, R.: Groß dimensionierte Fahrzeugscheiben aus Kunststoff. In-Mold-Pressing. *Kunststoffe* 93 (2003) 8, pp. 91-93

[7] Giessauf, J., Kralicek, M., Pitscheneder, W., Steinbichler, G.: Große Autoscheiben aus der Spritzgießmaschine. *Kunststoffe*, 94 (2004) 10, pp. 164-166, 169-170

[8] Götz, W.: Großer Markt für große Scheiben – Polycarbonat-Fahrzeugverscheibungen in der Serienfertigung. *Plastverarbeiter*, 55 (2004) No. 10, pp. 146-149

[9] Aengenheyster, G.: Fahrzeugscheiben aus Kunststoff – Fertigung im Spritzgieß- und Spritzprägeverfahren. 8. Darmstädter Kunststofftag, 18th June 2004

6.4.3 Plastic Automobile Windows

CHRISTIAN HOPMANN

6.4.3.1 Process Technology

Plastic is finding increasing use as a replacement for glass in automobile windows. An extremely dynamic market is developing, but one that places special requirements on materials, designs and production techniques. Success depends on an efficient logistical chain that addresses the product-specific requirements and the complex manufacturing processes.

With the introduction of the first Smart car in 1998, plastic debuted as a material for glazing in the mass market for automobiles. Since then, plastic has made significant inroads into the domain of glass, which previously had no competitors for automotive glazing. Today, three new product groups for plastics are taking shape in this field:

- stationary side windows, as was mentioned in the above example,
- transparent roof systems (Figure 6-236) and
- rear window modules (e.g., C-Class Sport coupé) and body attachments with the appearance of glass that impart an exclusive sense of depth (Figure 6-237).

Figure 6-236 In the panorama roof of the Smart Forfour glass and plastic elements complement one another ideally (photo: Bayer MaterialScience)

6.4.3.2 Reasons for the Use of Plastic in Automobile Windows

While the original motivation for the use of plastic in place of glass was to reduce vehicle weight in order to reduce fuel consumption (Figure 6-238), today numerous other factors have gained in importance and eclipsed this original idea. The optimization of vehicle wind resistance and improved efficiencies mean that weight-reducing measures no longer have as great an effect on the vehicle [1]. The added value of the weight loss of 30 to 40 kg per car resulting from the use of plastic is found rather in the lowering of the center of gravity of the entire vehicle and in an optimized load distribution on the front and rear axles. This provides new resources in the search for the best possible vehicle dynamics and makes a substantial contribution to improved safety as well as increased driving enjoyment.

Figure 6-237 The B-pillar of the Mercedes CLS-Class is constructed of two layers.
The combination of a black and a transparent PC creates a glass-like appearance with great depth (photo: Bayer MaterialScience).

6.4 Automotive Glazing

Figure 6-238 In the prototype of the 1-liter-car, VW has consistently pursued a lightweight design (photo: Volkswagen)

In addition, there is tremendous design freedom compared to glass, so that geometries are conceivable in plastic that are unimaginable in glass for the foreseeable future. Because of their lower stiffness, plastics are better suited for applications with complex geometries that provide a pronounced structural stiffness. With regard to safety, plastic is superior to glass in spite of the lower stiffness. In the event of a crash, the plastic structure remains intact due to its better impact strength, so that, in contrast to glass (which fails spontaneously), it contributes to maintenance of the survivability envelope.

The high degree of functional integration made possible by the use of plastic is also valued by the automotive sector, which usually calculates with a very sharp pencil. Here, through plastic-specific joining techniques such as multi-component injection molding and snap fits, assembly steps can be saved all along the production line to final assembly. In this respect, the use of plastic still offers considerable potential through integration of spoilers and lamps.

To fully exploit the potential of the material requires – as with all applications – optimal interplay between the areas of materials development, design and process technology. All participants have done a lot in this regard in past years, so that, in the mean time, plastic glazing is used in volume production that is ECE R43-certified und thus meets all the requirements that must be met by glass. On the material side, polycarbonate has proven itself, with special grades (e.g., [2]) being offered to the market for such applications that satisfy particular requirements such as stability when exposed to weather, for instance, exhibit compatibility with commonly encountered adhesive systems, are available with a high degree of purity and process easily. The drawbacks of the lower stiffness of plastic and the greater thermal expansion compared to the steel body are addressed through use of innovative designs.

6.4.3.3 Coordinated Mold and Machine Technology

Demanding requirements are faced with regard to both mold and machine technology when producing automotive glazing. To create an optically high-quality window surface requires that large segments of the mold surface be polished (Figure 6-239) and also kept free of deposits during the molding process.

The gate is often designed as a film gate and, for reasons of functionality, located on a long edge of the part. Ample dimensions assure a stable process and high part quality. In the mean time, there are also solutions in which several valve-gated pinpoint gates are actuated sequentially. The use of telescoping edges in injection compression molds assures that the molded part is always in contact with the mold, thus contributing to achievement of good surface quality

Figure 6-239 Polished and absolutely clean mold surfaces are essential for high-quality molded parts (photo: Engel)

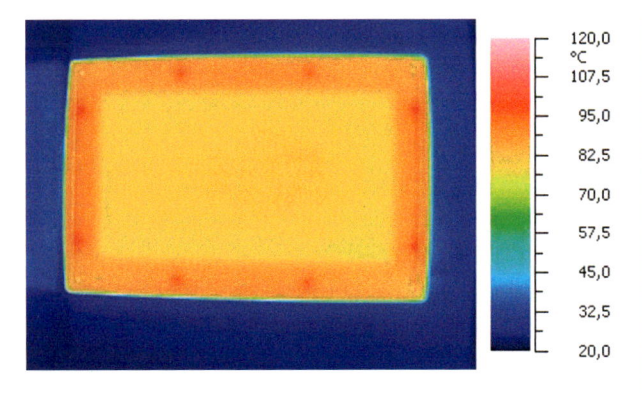

Figure 6-240 A uniform temperature distribution in the molded part shortly after part removal from the mold indicates a uniform mold temperature and is very important for molded part quality (photo: Engel)

on the molded part. Precise and very uniform mold temperature control also assures uniform cooling of the molded part (Figure 6-240), which is a necessary prerequisite for dimensionally accurate products with low molded-in stress.

Because of the size of the parts to be produced, it is necessary to consider the mold and machine in their entirety and match them to one another precisely from the processing standpoint. The degree of interdependence depends, of course, on the specific injection molding process employed. For the usual large-area parts with a corresponding large flow length/wall thickness ratio, various injection compression techniques are currently employed. On the one hand, this greatly reduces the pressure needed to fill the cavity; on the other, the shear stress on the melt is lowered, which results in less molecular orientation. In view of the fact that, because of its toughness, polycarbonate is currently favored for large windows, minimal orientation is especially important, since this material exhibits noticeable birefringence as the result of molecular orientation, which can give rise to undesirable optical effects.

Likewise, absence of molded-in stress is required in optical parts, since these reduce the adhesion of the scratch-resistant coating that is absolutely essential. Optical investigations show the positive effect of the injection compression molding process over conven-

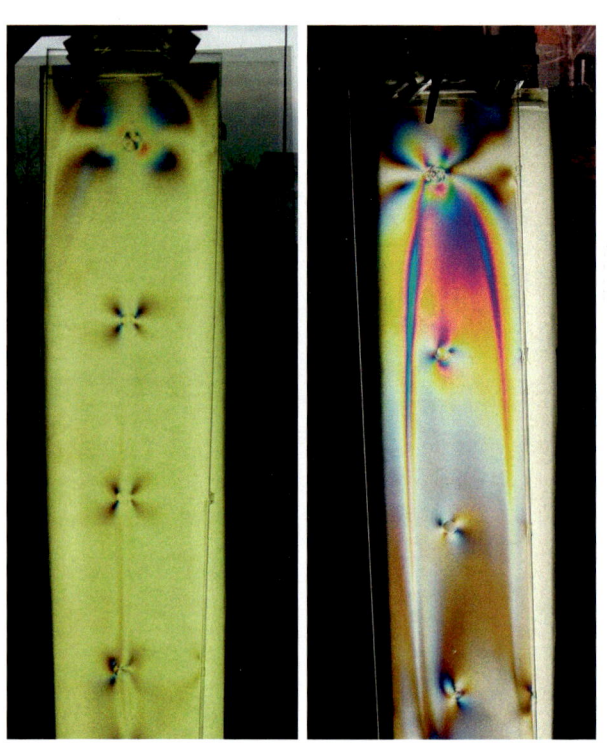

Figure 6-241 Internal stresses made visible: the injection-compression molded test specimen is almost stress-free (left) and thus demonstrates the superiority of this process over conventional injection molding (right) (photo: Engel)

tional injection molding (Figure 6-241). This explains why, at present, only injection compression molding processes requiring exactly parallel platens during the compression step with no pressure fluctuations are employed for large windows. Similarly, all machine manufacturers rely on the two-component rotary platen technique. This permits economical production of parts with a high degree of functional integration by means of multi-component injection molding [3]. At the same time, the associated molds, with weights of 15 to 20 t, are still relatively easy to handle.

6.4.3.4 Injection Compression Molding Techniques Set the Scene

Currently, there are basically three different injection compression molding techniques in use. Most closely related to the classical injection compression technique employed, for instance, for thin-walled articles

6.4 Automotive Glazing

Figure 6-242 With the Glazemelt technique, the two injection units are mounted horizontally opposite one another (photo: Engel)

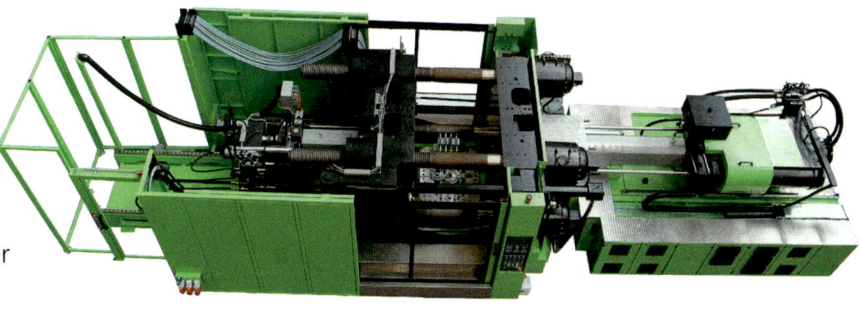

is the so-called long-stroke compression technique utilising the clamping unit. Engel Austria GmbH, Schwertberg/Austria, demonstrated this process [4] at the K 2004 in a two-component variation with two horizontal injection units and a stack mold with a rotary center section (Figure 6-242).

In the long-stroke compression technique, the approximately 4 mm thick and almost 1000 mm wide cavity of the injection compression mold is opened by an amount about equal to the part thickness; as a result, the pressure required during filling drops drastically. Subsequently, the clamping unit of the machine closes the enlarged cavity and the compression motion distributes the melt. Because of the large part dimensions and the edge gating required for windows, asymmetrical loads that must be counteracted are created in the mold during the filling and compression phases. With the technology described here, this is accomplished by means of platen parallelism control for the moving platen that regulates the pressure in the hydraulic cylinders on the four tie bars individually.

Another variation of injection compression molding that is employed to produce windows with low molded-in stress is the expansion compression technique offered by Krauss-Maffei Kunststofftechnik GmbH, Munich/Germany (Figure 6-243).

The mold is completely closed at the start of injection; it is also possible to close the cavity to below the final part thickness desired. The benefit of this measure: it addresses the jetting phenomenon that causes problems with edge gating. After the cavity is filled completely, the part thickness is increased by enlarging the cavity, under controlled pressure, by the amount of the com-

Figure 6-243 Krauss-Maffei is pursuing two-component (2K) technology with a rotary platen, but mounts the injection units in parallel. The expansion compression technique requires especially precise control of mold motions (photo: Krauss-Maffei).

pression stroke through the introduction of additional melt. Once the desired compression stroke is reached, the gate is closed and the compression force is applied and counteracts the shrinkage of the plastic like a holding pressure distributed over a large area. For precise control of part formation, the stroke of the clamping unit is monitored very carefully. Using this technique, Freeglass GmbH & Co. KG, Schwaikheim/Germany, is already producing stationary side windows and roof modules in series.

For long flow paths, Battenfeld GmbH, Meinerzhagen/Germany, in cooperation with Summerer Technologies, Rimsting/Germany, has developed the "IMPmore" process, which it demonstrated in public for the first time at the NPE 2003 (Figure 6-244).

Figure 6-244 With the IMPmore technology, Battenfeld has introduced a new injection compression molding principle with a wedge-shaped compression gap (photo: Battenfeld)

The basic idea here is likewise a phased enlarging of the flow cross-section that, in this case, is provided by a wedge-shaped opening of the cavity [5]. This is achieved by tilting the moving platen of the clamping unit with the cavity surface by a few degrees prior to closing of the mold. Here, too, closing of the machine platens is delayed until after injection has started. By narrowing the wedge-shaped compression gap, the cavity fills at very low pressure. As shown at the K 2004 with an approximately 1 m^2 demonstration window, this process permits injection molding of even large windows with little molded-in stress (Figure 6-245).

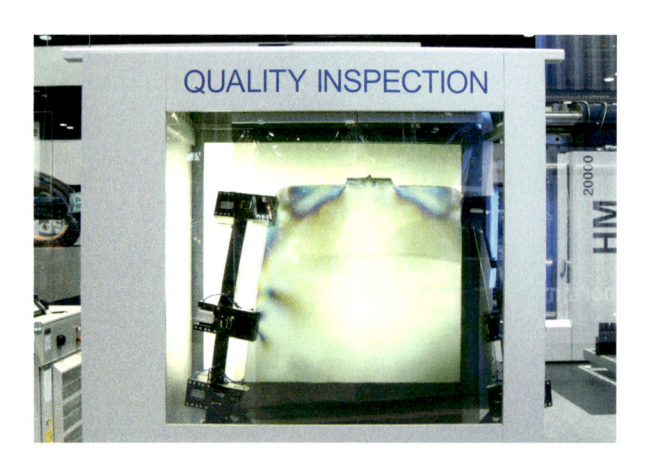

Figure 6-245 Windows with low molded-in stress and orientation result from an optimal combination of material, design and process know-how (photo: Battenfeld)

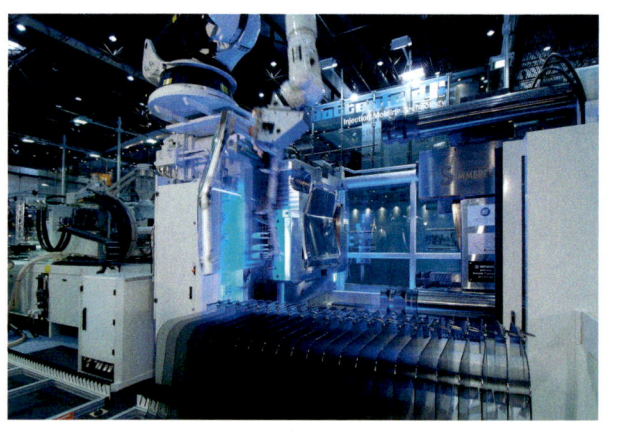

Figure 6-246 Retraction of the tiebars from the stationary platen increases access to the mold and creates space to permit entry of the part removal robot (photo: Battenfeld)

In conjunction with the edge gating, closing of the wedge-shaped compression gap also creates asymmetrical opening forces in the mold. These are counteracted with the aid of hydraulic cylinders incorporated into the mold, so that the machine is subjected to an apparently uniform opening force. Especially good access for robotic removal of the molded part is provided as the machine tie bars retract from the stationary platen during motion by the mold (Figure 6-246).

6.4.3.5 Melt Quality as a Factor for Success

Preparation of the melt is extremely important for achieving high-quality molded parts. Black specks, in particular, must be avoided in the melt. Selection of a suitable screw geometry and an appropriate non-return valve are especially important in this regard [6]. Because of the tendency of polycarbonate to adhere to the machine screw and subsequently damage it, special attention should be devoted to the screw material. Chrome and hardfaced screws as well as screws with surface coatings have been used successfully.

As a research project, Battenfeld GmbH, Meinerzhagen, Metaplas Ionon Oberflächenveredelungstechnik GmbH, Bergisch-Gladbach/Germany, Ticona GmbH, Frankfurt/Germany, and Wahl optoparts GmbH, Triptis/Germany, together with the Materials Science Department and the Institute for Plastics Processing (both at the RWTH University in Aachen/Germany), have been investigating various surface coatings for components of the plasticizing unit as to their suitability for injection molding of optical parts. The goal of this project is to identify the most suitable coating system for any given plastic resin and at the same time find a coating system with the best all-around properties. To this end, the behavior of various PVD coating systems is being characterized during processing of optical-grade resins. In particular, the tendency to adhere to the screw surface is being critically evaluated.

6.4.3.6 Types of Surface Treatment

The requirements with regard to stability when exposed to weather (UV stability) and chemicals (car wash detergents, salt, etc.) as well as with regard to abrasion resistance of plastic windows can, at present, only be met through application of an additional surface coating. To apply the coating, there are, in principle, three different methods, which can also be combined with one another. Protective coatings can be applied as a solution (liquid), be deposited from the gas phase or be applied as a film.

At present, application as a solution is the most commonly employed technique. These find use, above all, for mass production of headlamp lenses and can be applied to the substrate by the usual methods (e.g., flooding, immersion, spraying). Curing occurs either thermally or via exposure to UV radiation. For extremely stringent requirements regarding the optical quality of the coating, cleanroom conditions are necessary.

Primarily polysiloxane coatings are employed to provide scratch resistance. Application involves a precursor solution with a very high proportion of solvent that evaporates very quickly at room temperature. The coating is then subsequently cured to a glass-like film. Hydroxyl and alkoxy groups on the polysiloxane chains condense to form a three-dimensionally cross-linked methyl-functionalized SiO_2 film. With a typical thermally cured system, the coating is cross-linked at 120–130 °C for about 30 to 60 minutes. UV-curing systems are also available on the market, but their scratch resistance and stability when exposed to weather (weatherability) are inferior to those of thermally curing systems.

The typical coating starts with a thin film of primer that assures good adhesion of the brittle scratch-resistant film to the soft PC. The considerably thicker scratch-resistant coating is applied as a second layer. Both the primer layer and the scratch-resistant top coat contain UV absorbers and UV stabilizers that assure stability of the system upon exposure to weather. Film thicknesses lies in the range of 5–15 μm. To reduce process times, primerless scratch-resistant coatings have also been developed in recent years. Since one curing step is eliminated, considerable cost savings result.

In industrial practice, the coating systems PHC587, AS4000 and AS4700 from GE Bayer Silicones GmbH & Co. KG, Leverkusen/Germany, have found the greatest acceptance (e.g., for headlamp lenses, triangular side windows in the Smart car, rear window of the C-Class). These coatings meet the requirements of the automotive industry and have proven their long-term stability in actual use, even though the scratch resistance and UV stability are less than those of glass.

The scratch resistance of such coatings can be increased considerably through addition of nanoparticles (diameter < 40 nm) and then almost equals that of glass. The nanoparticles involved are metal oxides (e.g., SiO_2, Al_2O_3) that are dispersed in organic compounds. Since the particles are very small, they do not refract light; the coatings remain transparent. The greatest difficulty encountered to date has been achieving a stable dispersion with a narrow size distribution of the nanoparticles. Recently, it has become possible to accomplish this under economical conditions, so that, at present, development of scratch-resistant coatings containing nanoparticles is being pursued actively. For instance, Hanse Chemie AG, Geesthacht/Germany, mixes up to 50 wt.-% of nanoparticles in a UV-curing acrylic coating system. The coating can incorporate UV protection and be processed quite normally. In the near future, several interesting developments are to be expected.

Scratch-resistant films can be deposited not only from solution, but also from the gas phase (PVD: Physical vapor deposition; PECVD: Plasma-enhanced chemical vapor deposition). With these techniques, metal oxides, silicone or organosilicone compounds are vaporized in a low-pressure process and deposited from the gas phase as a scratch-resistant coating. In the case of PECVD, a chemical reaction takes place in the gas phase, creating the scratch-resistant coating. The PVD technique involves only a condensation process. The scratch resistance and other surface properties that can be achieved via these coating techniques are comparable to those of pure glass. However, these coatings are UV-transparent, making an additional UV protective coating necessary. To date, it has not been possible to create such a coating economically via the PVD or PECVD process.

Exatec GmbH & Co. KG, Bergisch-Gladbach/Germany, has solved this problem with its Exatec 900 glazing system through use of a UV protective coating that is applied to the PC in combination with a primer prior to the PECVD scratch-resistant coating. According to information from the manufacturer, this coating is stable for more than ten years even when exposed to weather. The equipment costs for this system are relatively high, since both coating and vacuum technology are needed. For this reason, this process can be operated economically only when other functional elements are integrated into the plastic window, thus reducing the system costs. To date, only a few prototype vehicles have been equipped with this system.

6.4.3.7 Functional Integration by Using Film

Application of a film to the PC window represents the final possibility for creating a scratch-resistant surface. Use of a PMMA film, for instance, increases the scratch resistance noticeably. If the film contains UV absorbers, it can also protect the underlying PC from UV radiation. However, the scratch resistance of PMMA is not yet adequate for such applications. For this reason, it is necessary to apply an additional scratch-resistant coating to the film.

The possibilities for functional integration were demonstrated by KRD Coatings GmbH, Geesthacht/Germany, in a prototype for an automobile rear window. An IR-reflecting film, a thin-wire heating system and a UV-absorbing film coated to be scratch-resistant were applied to a PC window. At the moment, economical production of this window is not possible, but it shows clearly where future development potential lies.

For plastic windows, which are still more expensive than glass windows in simple applications, to become economically viable, the greatest possible additional utility value must be integrated. It is not enough for a film to meet the minimum requirements such as scratch resistance, chemical resistance and UV stability; it must also provide additional functions such as IR reflection, window heating and EM shielding. The challenge, then, is to incorporate these numerous properties in the fewest possible steps on three-dimensional geometries.

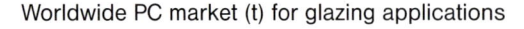

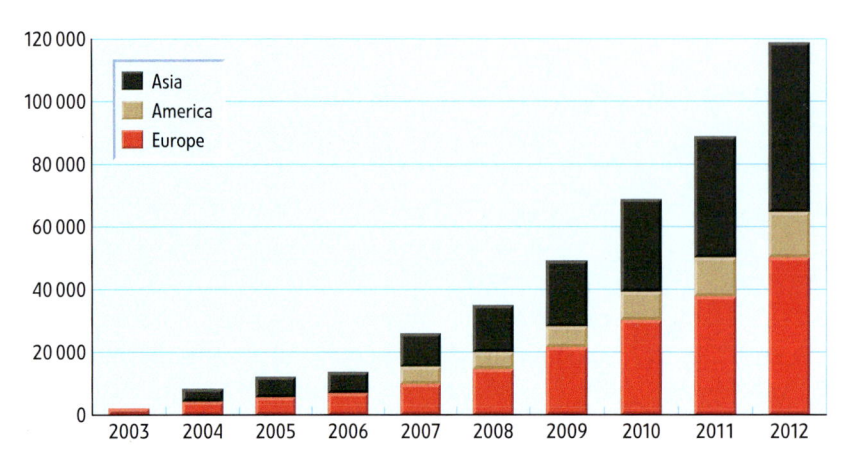

Figure 6-247 The prognosis shows a strong growth for car windows made from PC (figures in t)

6.4.3.8 Efficient Logistics Chain Decisive for Success

An efficient and well thought out logistics chain is of great importance for the economics of the entire, usually multi-step manufacturing process and the quality of the parts produced. This includes, in particular, product-specific factors such as the cleanliness of the entire manufacturing process, part removal from the mold and subsequent handling, for instance. According to an analysis by Krauss-Maffei, less than 5% of all defects when processing polycarbonate for optical applications result from the injection molding process. Considerably more important are factors such as drying and material conveying (33%) as well as material manufacturing and packaging (62%) [7]. Regardless of whether the concrete numerical values apply to the special application of automobile windows, there is clearly a multitude of defect sources that must be eliminated. Whoever is up to this challenge is in a position to exploit the enormous market potential with its relatively high growth rates (Figure 6-247, [8]).

References

[1] Kircher, W.: Stagnation statt Zuwachs – Trends von Kunststoffen im Automobil. *Kunststoffe* 92 (2002) 10, pp. 153–156

[2] N. N.: Makrolon AG 2677. Technical Information, Bayer Material Science, 2004

[3] N. N.: Großer Markt für große Scheiben. *Plastverarbeiter* 55 (2004) 10, pp. 146–149

[4] Gießauf, J., Kralicek, M., Pitscheneder, W., Steinbichler, G.: Large Car Windows from an Injection Moulding Machine. *Kunststoffe plast europe* 94 (2004) 10, pp. 164–170

[5] Hof, K., Sauer, R.: Large Automotive Windows Made from Plastic. *Kunststoffe plast europe* 93 (2003) 8, pp. 42–43

[6] Berthold, J.: IMPmore – Innovative Verfahrenstechnik mit Werkzeug und Maschine. In: Spritzgießen von Kunststoffkomponenten mit optischer Funktion, Seminare zur Kunststoffverarbeitung, IKV Aachen, 21./22.10.2003

[7] Bürkle, E., Klotz, B., Lichtinger, P.: The Production of Highly Transparent Optical Mouldings – A New Challenge. *Kunststoffe plast europe* 91 (2001) 11, pp. 17–21

[8] Bangert, H.: Bedeutung und Potenzial von Polymeren in Marktsegmenten für optische Anwendungen. In: Spritzgießen von Kunststoffkomponenten mit optischer Funktion, Seminare zur Kunststoffverarbeitung, IKV Aachen, 21./22.10.2003.

6.5 Acoustics and Aerodynamics

6.5.1 Specific Requirements on Aeroacoustic Development for Convertibles

PETER KALINKE

6.5.1.1 Introduction

Today, the buyers of convertibles demand a higher level of comfort. Excessive wind noise is no longer accepted in a high to mid-range convertible. This is in particular due to the fact that such vehicles frequently provide high engine performance, allowing for high driving speeds. A current J. D. Power study on customer satisfaction registers complaints by owners of 12 out of 14 premium sports cars regarding wind noise that is too high. For six of these, wind noise is the major cause for complaint [1].

In the case of high to mid-range sedans (between 100 and 130 km/h), wind noise dominates interior noise, thus masking noise coming from the running gear and the drive train. Wind noise primarily enters through the windows and seals due to their lower insertion loss as compared with steel [2]. In a soft-top convertible, even at low speeds, wind noise is the dominant noise source. There are several reasons for this:

- Due to its convertible top system, a convertible has more seals than a sedan (such as between the convertible top and the windshield frame, the top retaining clip and the lid of the convertible top stowage)

- The fabric of the convertible top provides an insertion insulation that is considerably lower than steel

- Parts of the convertible top (such as the rear window) can be excited to vibration and generate drumming noise

- In contrast to most sedans, convertibles have frameless door windows.

In Figure 6-248, the internal noise spectrum for a Mercedes CLK convertible and its corresponding coupe is compared at 160 km/h. In the range of measurement level between 200 and 1000 Hz, level differences up to 7 dB(A) can be noted. However, the range above 1 kHz, which is important for the subjective impression, exhibits a continuously decaying spectrum, as is the case with the coupe. No conspicuous leaks could be detected in this convertible (compare competitor vehicles in Figure 6-250). In the range below 100 Hz, the spectrum of the convertible is comparable with that of the coupe. Excessive levels in this frequency range, such as are found with most convertibles, would be perceived as booming or drumming noise. In summary, we may say that the driver of a Mercedes CLK Convertible traveling at 160 km/h would have a similar, if somewhat louder subjective impression of noise, than if he were driving the coupe.

Figures 6-249 and 6-250 present a comparison of the internal noise level, speech intelligibility and the internal noise spectra of a Mercedes CLK convertible and competitors at 160 km/h. In all three categories of interior noise, the Mercedes CLK, developed by Karmann according to Mercedes Benz specifications, emerges as the best in its class.

The Mercedes CLK convertible is a good example for possibilities for wind-noise optimization in convertibles. We will describe general measures for optimizing aeroacoustic noise in a convertible and indicate how this information was applied in the design and seals used for the convertible top in the Mercedes CLK convertible.

6.5 Acoustics and Aerodynamics

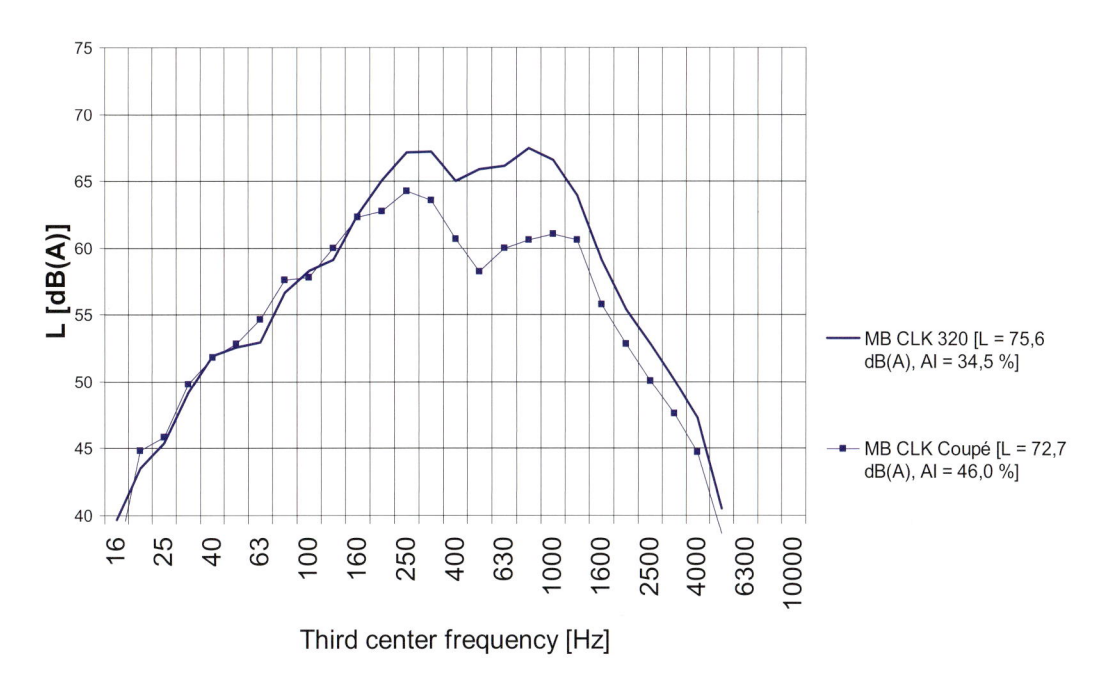

Figure 6-248 Comparison of wind noise at 160 km/h (measured on road) between a Mercedes CLK convertible and a coupe built on the same platform

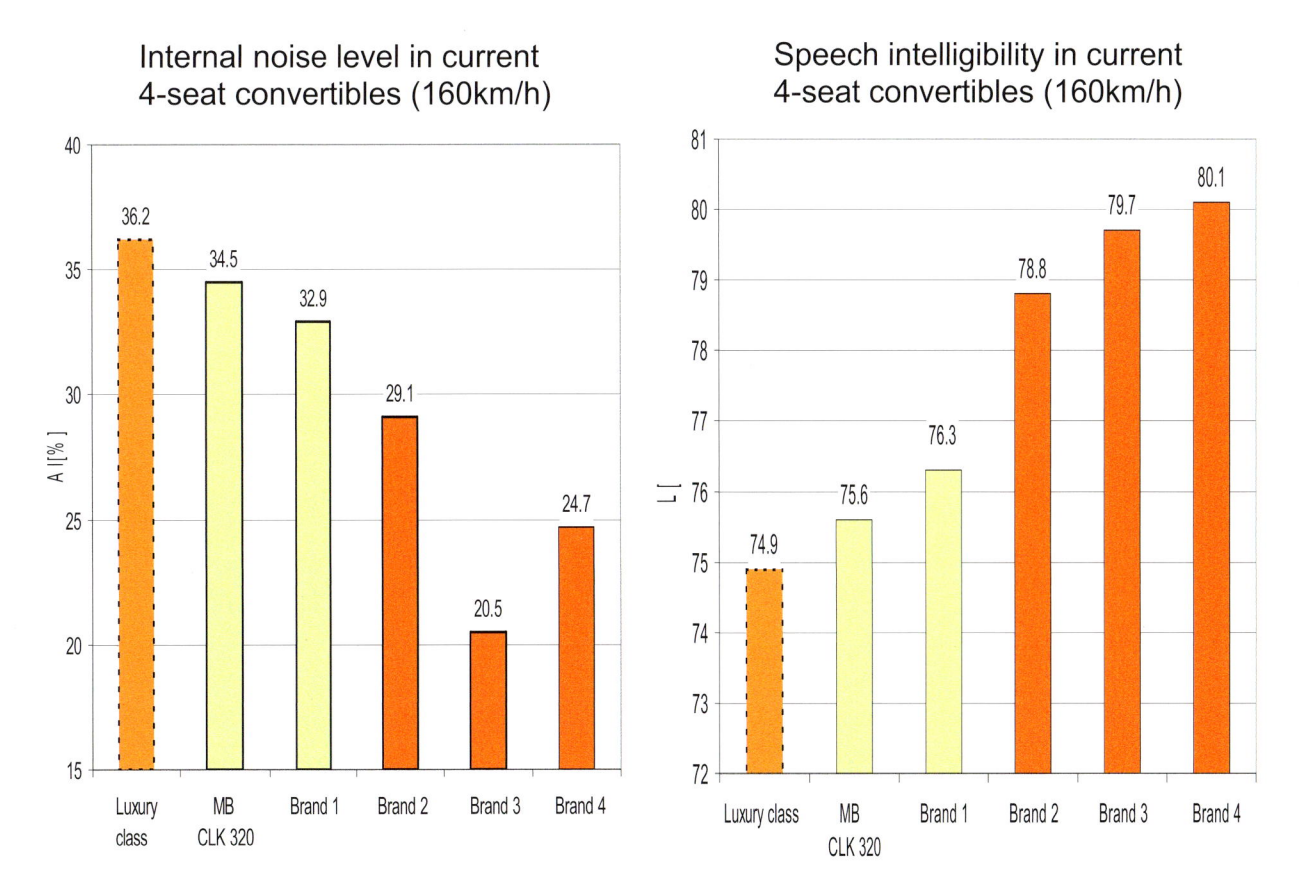

Figure 6-249 Comparison of internal noise level (left) and speech intelligibility (right) in the Mercedes CLK convertible and competing current four-seat convertibles at 160 km/h

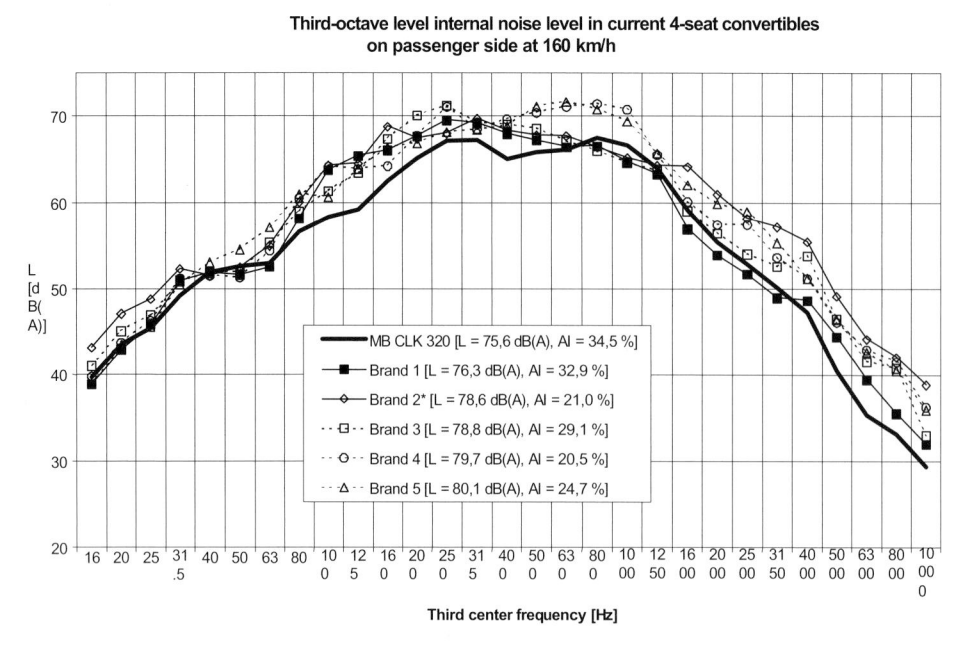

Figure 6-250 Third-octave level internal noise level in current 4-seat convertibles on passenger side at 160 km/h

6.5.1.2 Aeroacoustics

Mechanisms Causing Aeroacoustic Problems

Basically, three causes of aeroacoustic problems can be distinguished that subdivide interior noise into three different frequency ranges:

Aeroacoustic phenomenon	Typical frequency range
Airflow noise	200–1200 Hz
Leaks	> 800 Hz
Vibration excitation of components which in turn radiate noise into the vehicle interior	< 200 Hz

Airflow Noise

Airflow noise arises in particular from flow separation and vortex formation. The highest levels are to be found with three-dimensional vortex shedding [3]. The airflow noise is transmitted from outside into the vehicle. Here, the insertion loss of the outer vehicle skin is the variable that determines interior noise. In the case of a convertible, the material of the convertible top in particular determines the amount of insertion loss (cf. Section 6.5.1.2).

Leaks

Points of leakage on the one hand allow sound to penetrate directly from outside into the vehicle interior and also allow the air volume present in hollow spaces to be excited to resonant vibration, i.e., whistling noises.

Vibration Excitation of Component

Due to the formation of vortices, parts of the vehicle can be excited to vibration. In the case of convertibles, the area of the rear window and C pillar are particularly critical in this regard. Airflow around the vehicle can excite not only rigid-body vibration in the rear window in the cover material, but also higher vibration modes in the window itself. Such vibrations can excite unwanted drumming and booming in the vehicle interior.

Test Technology in Aeroacoustics

Test analyses of aeroacoustic problems are made in the aeroacoustic wind tunnel or on the test track with the aid of measurements using a dummy head. In order to optimize or reduce expensive testing time in the aeroacoustic wind tunnel, various states of a vehicle can be prepared beforehand on the testing grounds when the test track is dry and there is no wind. Measurements are then taken in the wind tunnel on the basis of product design specifications or measurements/tests are conducted that are possible only in the wind tunnel:

- Analysis of airflow with the smoke lance
- Measurement of vibrations in the convertible top fabric or rear window by laser vibrometer
- Intensity measurements in the passenger compartment in order to localize noise sources
- Measurement of the airflow noise, for example, using the concave-mirror method
- Analysis of influence from underbody airflow.

Aeroacoustic Simulation

Since the origin of aeroacoustic phenomena may be found in pressure oscillations, a transient simulation is an absolute necessity. Two possible approaches in flow simulation suggest themselves here:

- The implicit solution method which calculates the unsteady flow field on the basis of a steady-state solution.
- The explicit solution method which calculates a transient flow on the basis of a small time-step size.

Since aereoacoustic phenomena are caused by pressure oscillations, the logical approach is to calculate such oscillations directly. An absolute requirement for this is transient oscillation simulation of the type available in commercial software programs.

At Karmann, Powerflow software is employed: this software uses an explicit solution process which is intensive as regards computing time. The software delivers a good quality of results as regards aeroacoustics and even force coefficients (resistance and lift). The time-step size in a simulation, depending on the resolution, is in the range of $1-3 \times 10^{-5}$ s. This results in a theoretical confidence capability greater than 20 kHz. In practice, however, it has turned out that wind noise on the surface of the vehicle can be satisfactorily predicted to 1–2 kHz. For simulating the wind noise in the driver's ear, the first software tools already exist which use statistical energy analysis (SEA). However, such calculations are still the subject of research work and have not yet become standard in the aeroacoustic development process.

Factors Influencing Wind Noise

Shape of the Convertible Top

An aeroacoustically favorable convertible top shape enables air to flow around the vehicle with little turbulence formation. For this to succeed, transitions must be as smooth as possible from the convertible top to the body and windows. In [4] it is demonstrated that even small flow separations induce marked pressure fluctuations in the boundary layer; due to their "long life" downstream from the separation area, they subject large areas of the body or convertible top to high alternating pressures.

To avoid unsteady flows with intense formation of vortices in the region of the rear window, the angle of the rear window should be kept away from the vicinity of 30° [3]. Favorable from an aeroacoustic point of view is a steeper angle of the rear window with a defined cut-off line at the convertible top bow or, alternatively, a shallow angle with air flow contact across the window. From the point of view of the platform-derivative developer, however, it should be noted that shape of the convertible top is predetermined to a considerable extent by the styling department and that aeroacoustically significant add-on components, such as the door mirror or the A-pillar drip molding, are usually components that have been taken over from the platform vehicle.

Cover Fabric

To minimize the airflow noise for the passengers, the insertion loss of the outer vehicle skin must have as high a value as possible.

Of course, the insertion loss of the fabric roof of a convertible is less than that of a sheet-metal roof. In

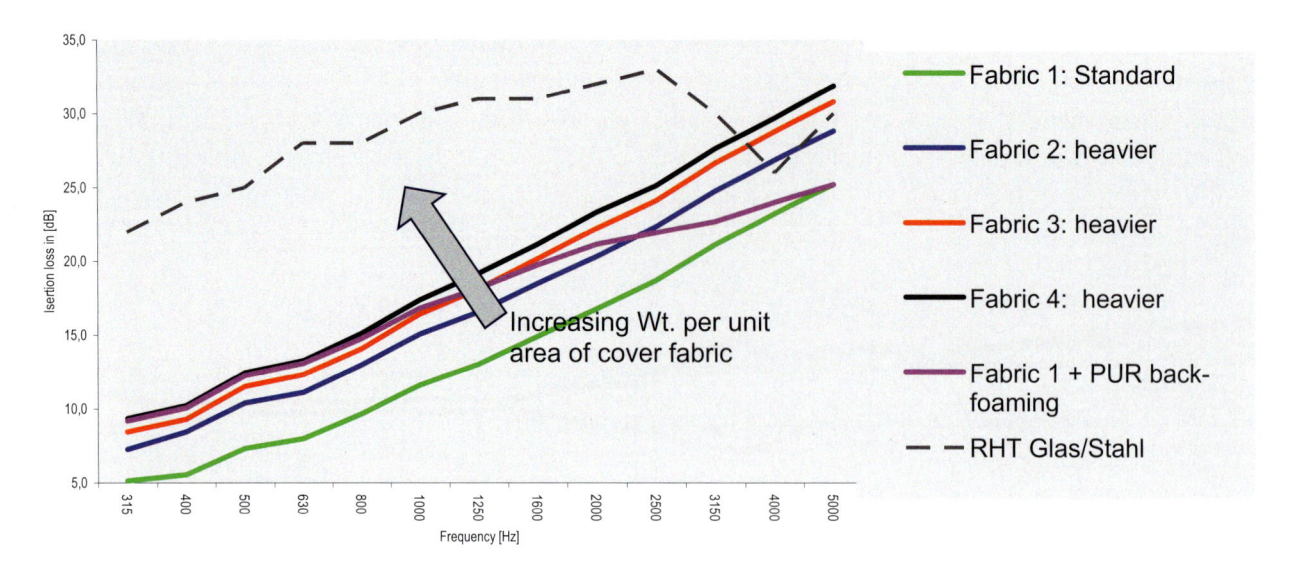

Figure 6-251 Insertion loss of different cover fabrics with differing weights per unit area in comparison with an RHT: steel and glass roof

Figure 6-251, data for the insertion loss of different fabrics are compared with those of a steel and glass roof. In the frequency range shown here, the insertion loss increases with frequency and is proportional to the weight per unit area of the fabric (this complies directly with the mass law). Damping from back-foamed fabric falls off at high frequencies due to its open porous structure. The dip in insertion loss at 4 kHz for the steel and glass roof is due to coincidence effect (flow velocity = flexural wave velocity of the RHT structure).

In the frequency range up to 4 kHz, the insertion loss of the steel and glass roof differs from that of fabric by up to 15 dB. Since at sufficient speeds, flow noise in the range between 200–1200 Hz is usually the one that determines the noise level, the low insertion loss of fabrics is the main reason why soft-top convertibles are subject to higher wind noise than sedans or hard-top convertibles.

Figure 6-253 shows that the effect of the weight per unit area of the fabric on the wind noise can be documented even when the vehicle is being driven. The diagram shows the interior noise of a convertible with two cover fabrics with different weights per unit area. Particularly at around 1 kHz, the level can be considerably reduced by using heavier cover fabric.

The increased insulation resulting from a heavy cover fabric is broadband in its effect.

Padding Mat

Despite its fabric roof, the convertible is expected to give the impression of a coupe. Especially unacceptable is the visible impression made by the convertible top bow in the top fabric. To prevent this happening, a padding mat is usually inserted between the crossmembers and the cover fabric. This padding mat can be given an acoustically absorbent design. Figure 6-252 shows the absorption coefficients measured for various padding materials in an alpha cabin. Material 5 in particular has been optimized for its absorption properties.

The effect of a padding mat during vehicle travel is shown in Figure 6-253. In the range above 500 Hz, the interior noise level is reduced by the presence of the mat.

Seals

The influence of seals on interior noise is shown in Figure 6-255. To determine this, interior noise was measured in a wind tunnel for the convertible in its initial state and also in the state with "all convertible-specific

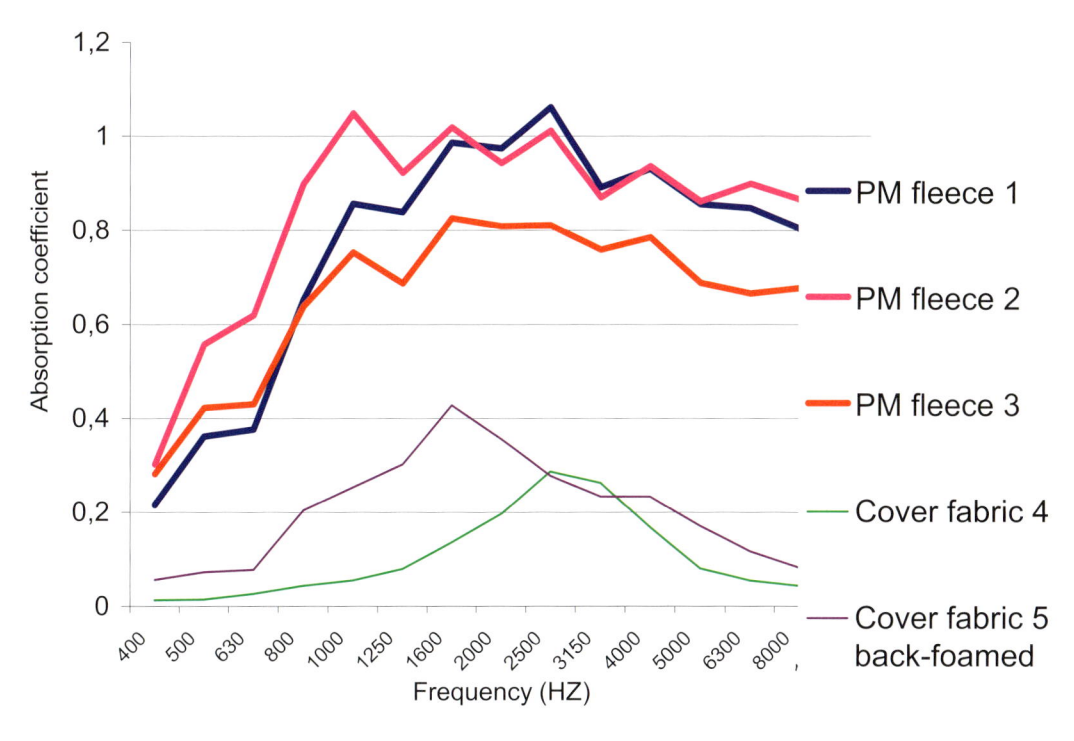

Figure 6-252 Absorption coefficient of different padding materials as measured in the alpha cabin

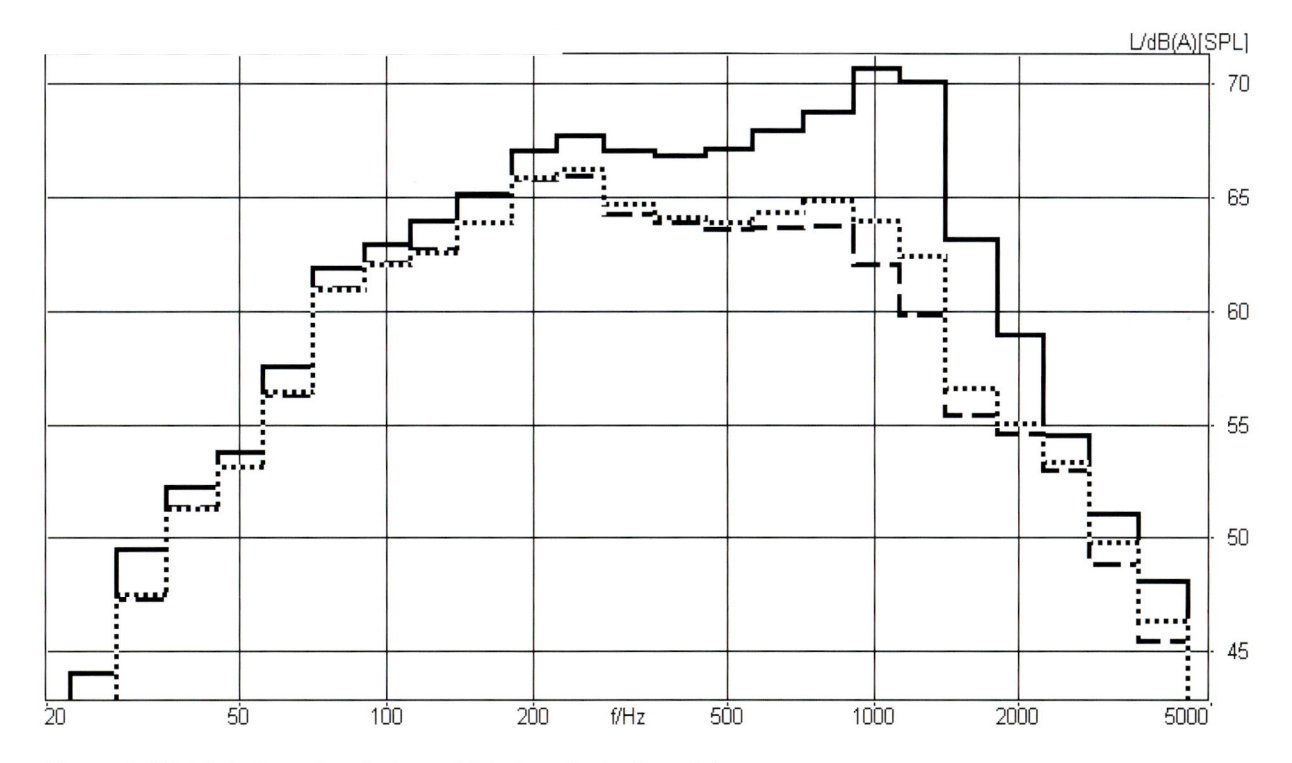

Figure 6-253 Interior noise during vehicle travel, starting state.
Solid line = cover fabric with heavier weight per unit area;
dashed line = cover fabric with heavier weight per unit area and padding mat.

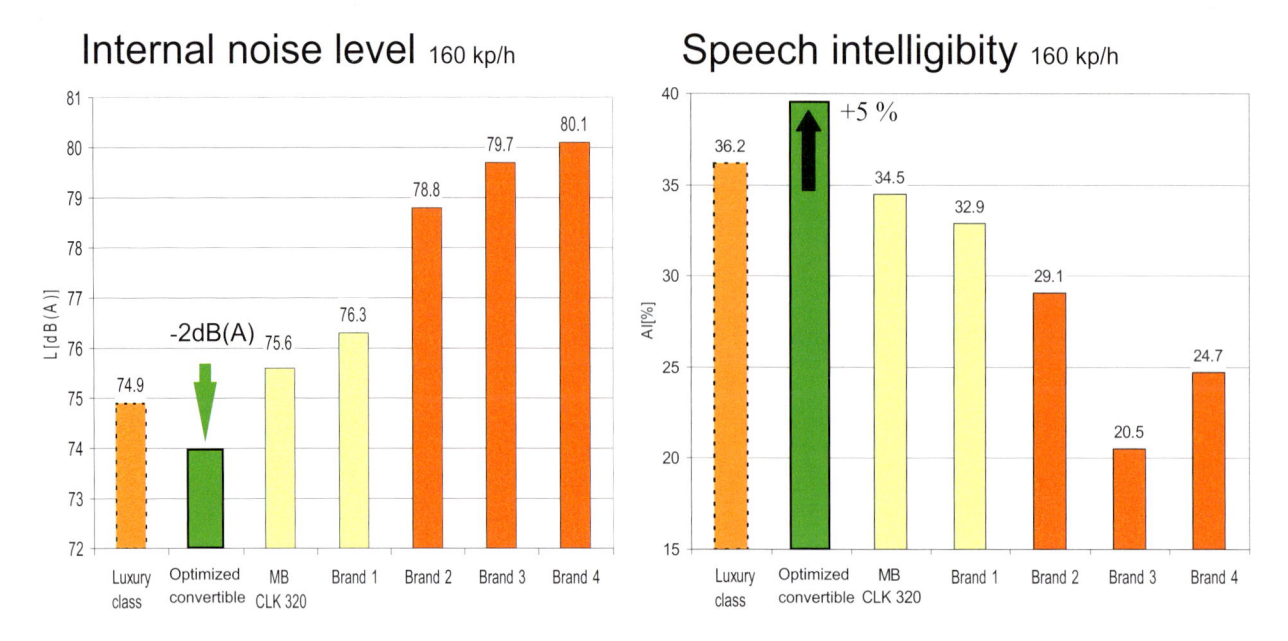

Figure 6-254 Comparison of an optimized mid-range convertible with its competition and a current high-end convertible for interior noise level (left) and intelligibility (right) at 160 km/h

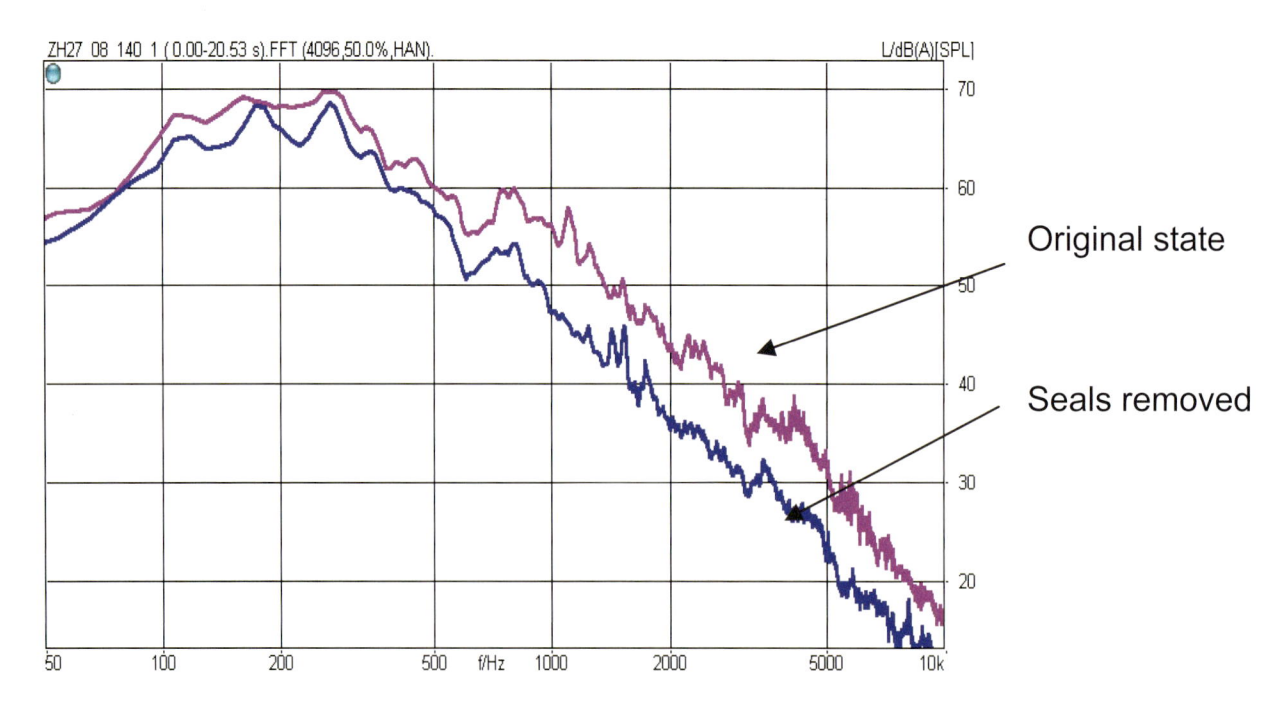

Figure 6-255 Influence of seals on interior noise level at 140 km/h

6.5 Acoustics and Aerodynamics

seals removed". Considerable level drops can be seen starting at approx. 800 Hz due to removal of the seals.

Rear Window and Cover Material Vibrations

The influence of air flow on rear window and cover material vibrations was analyzed as part of a wind tunnel test. At an airflow arrival speed of 140 km/h, the vibration velocities at the convertible top and windows were measured, and the acoustic pressure impacting the left ear of the driver as well.

The results are shown in Figure 6-256. Up to frequencies of around 200 Hz, marked vibration velocity peaks can be observed at the convertible top, at the B pillar (the results are based on a two-seater) and at the rear window; these usually lead to sound pressure peaks in interior noise. The vibrations in the windshield and side window are smaller by one order of magnitude and in the frequency range up to 500 Hz do not have any influence on interior noise.

Figure 6-257 shows the effect of anchoring the convertible's rear-window on the interior noise level during vehicle travel. Reductions in interior noise of up to 4 dB(A) are detected at approx. 100 Hz due to dampened rear window vibration – subjectively perceived as a reduction in drumming noises. Anchoring the rear window the roof frame can dampen additional natural vibrations in the rear window, whereby irritating booming below 50 Hz is eliminated.

In summary it may be stated that preventing the cover material and rear window from vibrating can considerably improve the interior noise in the frequency range up to 200 Hz (drumming and booming).

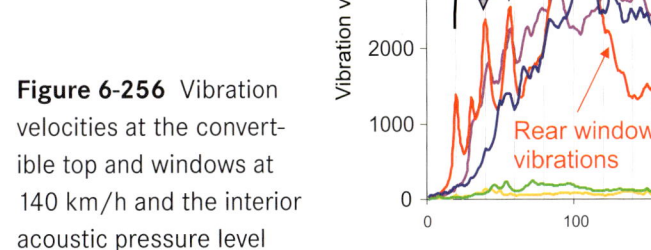

Figure 6-256 Vibration velocities at the convertible top and windows at 140 km/h and the interior acoustic pressure level

Figure 6-257 Interior noise at 180 km/h with glass rear window (solid line) and with a molded part that prevents the rear window from vibrating (dotted line)

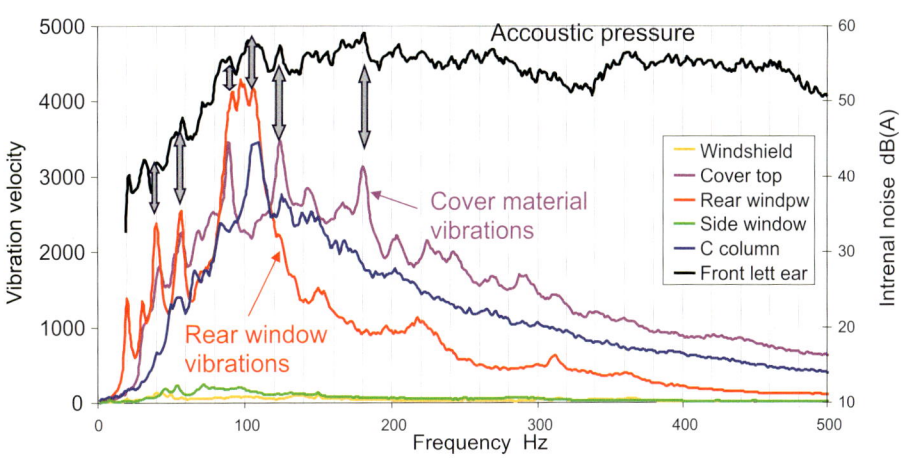

6.5.1.3 Convertible Top

Structure of the Convertible Top of the Mercedes CLK Convertible

The convertible top developed by Karmann for the Mercedes CLK convertible according to Mercedes Benz specifications is characterized by a particularly harmonious shape and a C pillar that is extraordinarily narrow for a fully-automatic convertible. The use of a padding mat means that the cross-members of the convertible top do not show through. This padding mat has been acoustically optimized as regards selection of materials, material thicknesses and material structure (see Figure 6-259). In contrast to most of the competition's vehicles, the padding mat is brought over the C pillar as far as the convertible top stowage lid and as far as the rear window. This solution on the one hand results in an especially large absorption area and on the other hand reduces through-transmission in the regions of the C pillar and rear window. This effect is backed up by the provision of an additional support in the region on the C pillar whereby the cover material is supported on a belt.

The rear window of the Mercedes CLK convertible is not supported in the cover material – as is the case with its competitors – but is held in a frame by the convertible top frame (cf. Figure 6-260). This prevents rigid-body movements of the rear window within the cover material. Due to the frame, even local vibration modes of the rear window are higher than in the case of "free – free bearing" in the cover material. These higher local rear-window vibration modes are not excited by the air flow around the vehicle. This means that the Mercedes CLK convertible behaves acousti-

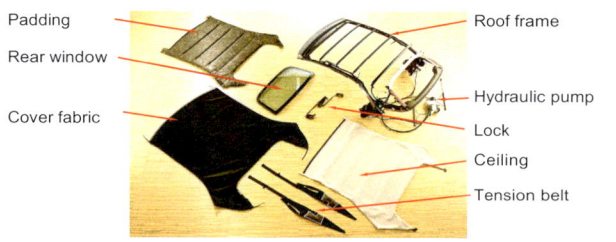

Figure 6-258 Side view of the Mercedes CLK convertible

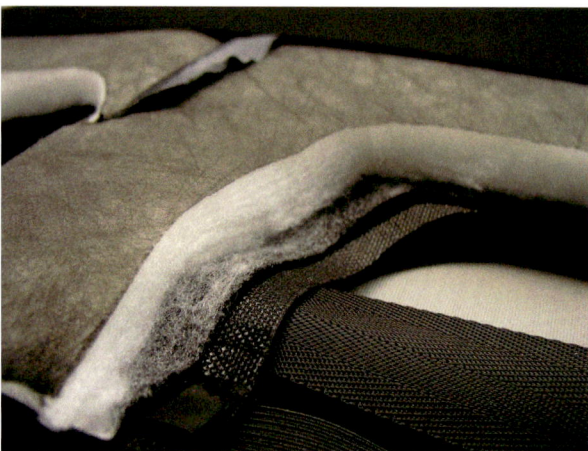

Figure 6-259 Shape of the padding mat in the region of the rear window; structure of the padding mat

6.5 Acoustics and Aerodynamics

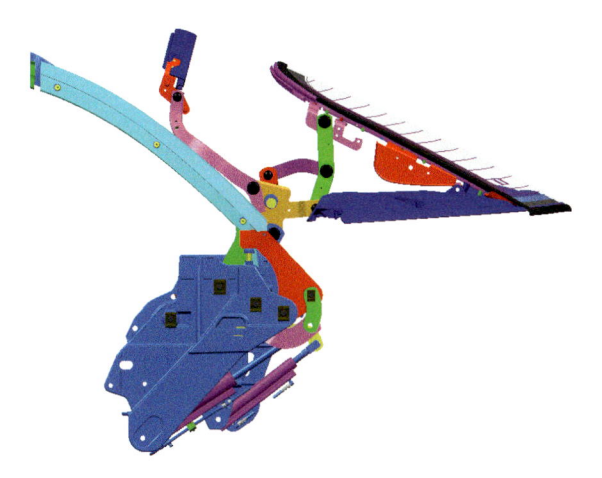

Figure 6-260 Side view of the connection between the rear window and the convertible top frame

Figure 6-261 Seals particularly critical as regards wind noise

cally up to around 150 Hz almost the same as a coupe and is thus free of booming or drumming noises (cf. Figure 6-248).

Sealing Concept of the Mercedes CLK Convertible

An independent sealing concept is a must for convertibles. The seals must ensure that the convertible can use a car wash and they also determine its aeroacoustics to a major extent. When the convertible top is open, a large number of seals can be seen, i.e., they also have to satisfy visual (esthetic) requirements.

Particularly critical, as far as the aeroacoustics is concerned, are sealing lines which abut onto each other, such as those marked in Figure 6-261:

Transition 1:
A pillar – mirror triangle – front side window

Transition 2:
A pillar – windshield – front of convertible top

Transition 3:
front roof frame – rear roof frame – B pillar

Transition 4:
B pillar – window-channel weatherseal of the front side window – window channel weatherseal of rear side window

Transition 5:
rear side window – convertible top at C column – window-channel weatherseal of rear side window.

By way of example, two convertible-specific seal cross-sections of the Mercedes CLK convertible are shown in Figures 6-262 and 6-263. The windshield frame seal is shown in Figure 6-262. To balance out tolerances between the convertible top and the body, large-volume soft seals are used. The aeroacoustic effect of this seal is optimized by a second sealing line which increases the insulation of the seal.

Figure 6-263 shows the seal at the front roof frame. The Mercedes CLK convertible has a short-travel lowering feature in the side windows which makes it possible to see the seal shown with its double sealing line at the side windows from the passenger compartment. In the same manner, the seal between the roof frame and the cover material has a double sealing line.

6.5.1.4 Summary

We have described how development aspects specific to the convertible can have a decisive influence on the aeroacoustics of the convertible. In the Mercedes CLK convertible many design measures have been implemented that have a positive influence on the aeroacoustics. The result is that even when the Mercedes CLK convertible is driving at high speeds, a

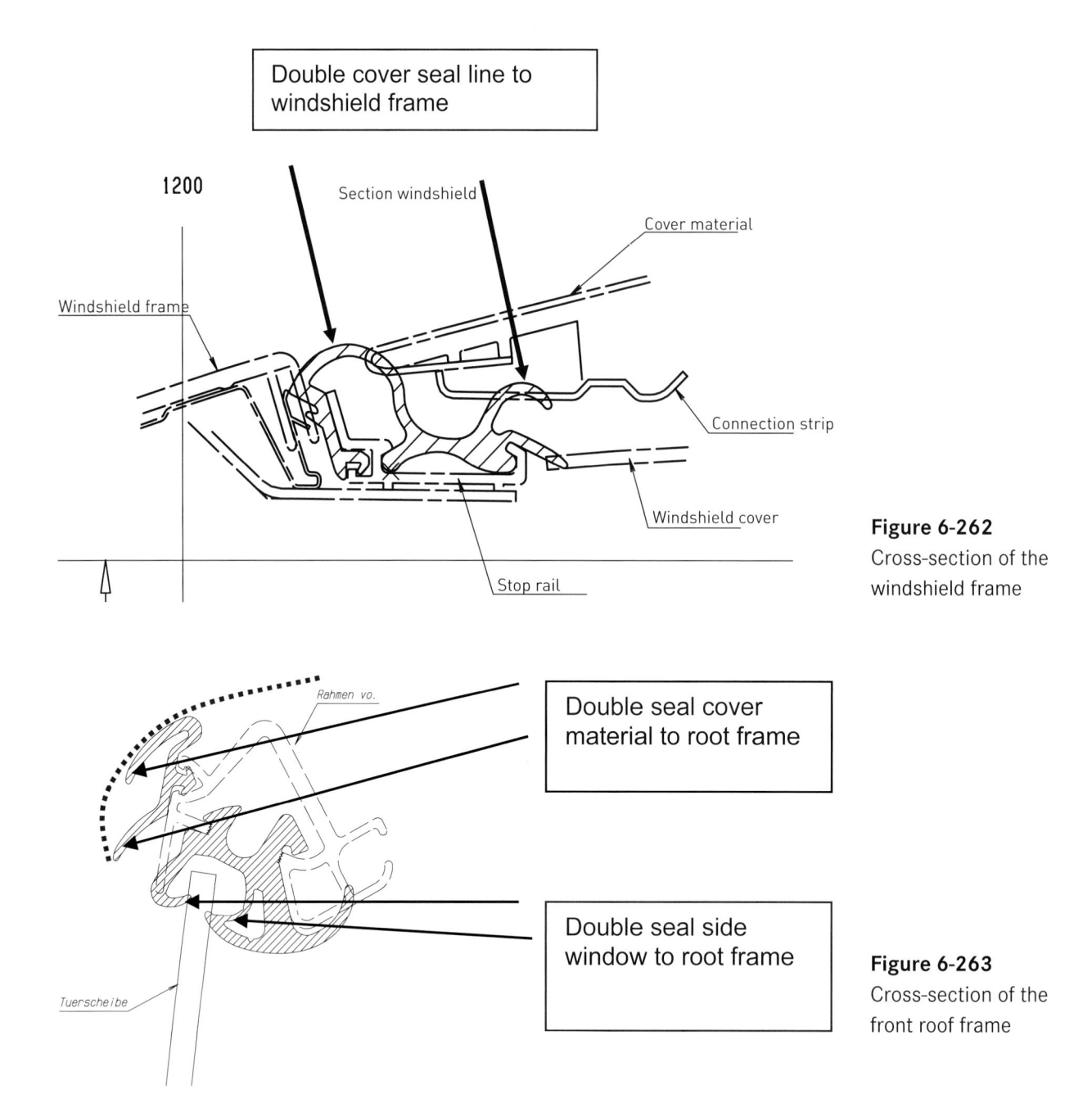

Figure 6-262 Cross-section of the windshield frame

Figure 6-263 Cross-section of the front roof frame

sound impression has been created that is similar to that of the coupe. If the cover fabric and padding mat are further optimized, a level of interior noise can be achieved in the upper mid-range that is even better than that of current high-end models.

References

[1] J. D. Power IQS-Study "Premium Sporty cars" April 2004
[2] George, A. R.: Automobile aerodynamic noise. SAE Paper 1990, pp. 1–24
[3] Hucho, W.-H.: Aerodynamik des Automobils. Düsseldorf: VDI Verlag 1981
[4] Dobrzynski, W.: Windgeräuschquellen am Kraftfahrzeug. Akustik und Aerodynamik des Kraftfahrzeugs. ed. S. Ahmed, Expert Verlag, Renningen 1995

6.5.2 Automotive – Compression Molding – LWRT Technique for Car Underbody Covers

EGON MOOS

6.5.2.1 Lightweight Design

The material concept of low weight reinforced thermoplastics (LWRT) can yield lower weight per unit area and increased flexural strength compared with glass-mat-reinforced thermoplastics (GMT) and direct long-fiber-reinforced thermoplastics (D-LFT). LWRT technology has been developed by Seeber AG to suit mass-production and will be used during the production of the new BMW 5 and 6 models.

The reduction of thickness has reached its limits for components made from glass-mat-reinforced thermoplastics (GMT) and direct long-fiber-reinforced thermoplastics (D-LFT). Further wall-thickness reductions would require an insupportable molding pressure, resulting in unstable components. Therefore, the ever more exacting requirements of OEMs can only be met by introducing a completely new technology.

Considering the homogeneous density of GMT structures, the significantly lighter sandwich design sets the target. However, this new method dispenses with material blends, the optimum objective being to use layers of the same pure material grade. The conflicting goals of pure material grades and layer structure have now been united within one process. Easy recycling and lightweight construction are characteristic features of the LWRT (Low Weight Reinforced Thermoplastics) material concept.

6.5.2.2 Ecological Balance Sheet as a Guiding Principle

Conventionally, car underbodies were protected by a liquid PVC underseal. But a growing market has begun to emerge for alternative concepts. So far, only GMT-based and D-LFT-based underbody covers have become established in the market. While they offer

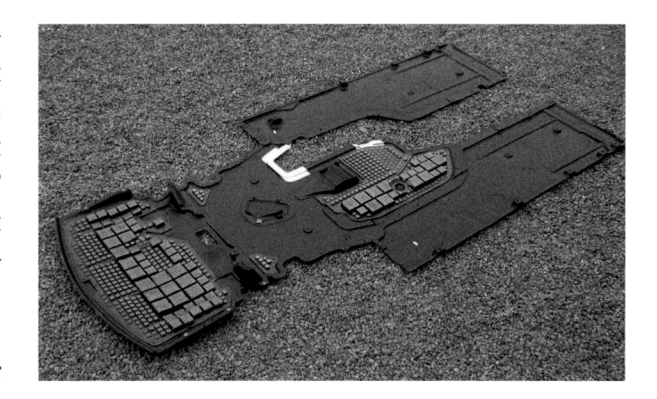

Figure 6-264 Only four component parts cover the main area of the underbody of the BMW 5

advantages over PVC coatings in terms of weight and recycling, they have already reached their limits as far as weight reduction is concerned. However, the new BMW 5 model manages the impossible: bar its equipment, the new model weighs 75 kg less than its predecessor and is already enjoying major success in the market. For car owners, lightweight design offers cost advantages across the entire vehicle lifecycle.

The LWRT underbody cover is an integral constituent of the new 5 model's intelligent lightweight design (Figure 6-264).

BMW drew up an "ecological component balance sheet" for the entire vehicle in general and its underbody in particular. This balance sheet accounts for aspects such as the car's "design for recycling", the reduction of driving noise and weight, as well as easy disassembly. These aspects are not only relevant for ELV's but also for routine inspections. Hence, the new model is equipped with special access points to facilitate the extraction of fuel and other service operations.

6.5.2.3 LWRT is Suitable for Mass Production

The LWRT production technique is very similar to GMT methods: a semi-finished product is heated and subsequently formed by applying pressure to make the product conform to the shape of a mold (Figures 6-265 and 6-266). The base material is produced along the principles of paper making and airlay processing.

Figure 6-265 The semi-finished product is heated and subsequently formed by the mold in a compression molding machine

The length, orientation and interaction of the glass fibers determines the structure of the mat when it is reheated. The development of a skin on the porous middle layer requires a multi-layer discharging method. This method is prerequisite for the required sealing properties of the finished product. The sealing film is an important constituent of the semi-finished product.

Another important aspect is the profile of the heating process, which shows the time progression and the relevant temperatures of the process. The swelling causes a defined porous texture. This so-called lofting is determined by the needle structure of the glass fibers.

Thermoforming is as complex as producing semi-finished components and heating up with subsequent lofting. Similar to the tailored blanks used in car body design, the final product exhibits varying wall-thickness depending on the local stress profile.

In a downstream water jet chamber, the thermoformed and cured sheets are cut precisely along their contour. The final product is devoid of any sharp edges. Hence, there is no risk of injury during assembly or disassembly. During the following processing stages, the product is equipped with heat protection and acoustic insulation elements.

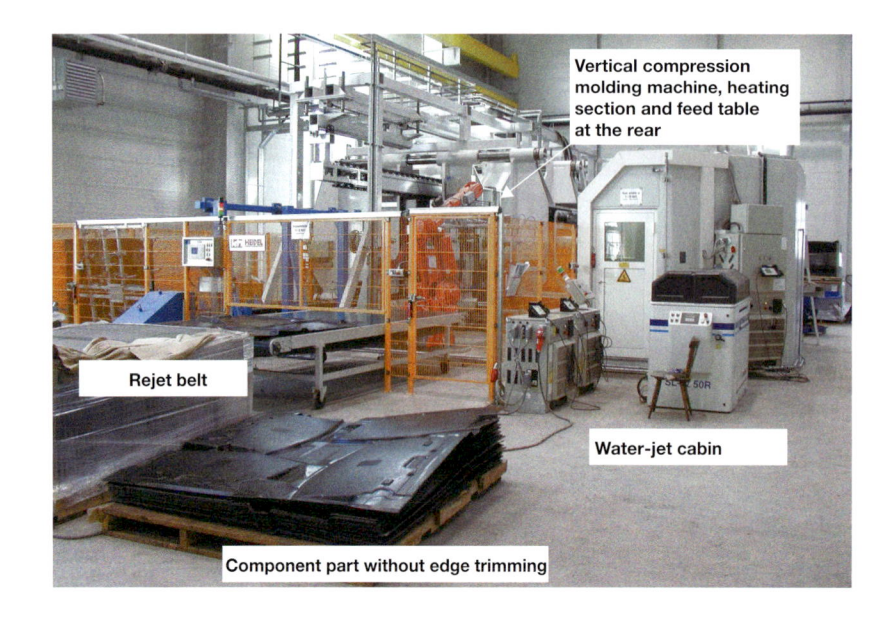

Figure 6-266 Production line for underbody covers made from LWRT

6.5.3.4 Component Properties and Requirements

While the density of GMT and D-LFT materials ranges between 1.04 g/cm^3 and 1.13 g/cm^3, the porous core of LWRT materials has a density of only 0.4 g/cm^3 (Figure 6-267).

The sealing film has a density of 0.9 g/cm^3 and the edges have a density of 1.0 g/cm^3. The total weight of the underbody cover amounts to only approx. 5 kg for a surface area of more than 3 m^2. Moreover, despite a 30% reduction of the weight per unit area, the product's flexural strength is approx. 130% higher than that of BMT or D-LFT products (Figure 6-268). The decisive design parameter is the wall-thickness of the finished component. It affects the flexural strength by the power of three. A slight increase in wall-thickness to 4 mm, i.e., a weight still below that of standard materials, is sufficient to deliver the required flexural strength.

Some applications may require a wider distance between two attachment points. Accordingly, the flexural strength can be increased to cover wider distances between two attachment points without a corresponding increase in the wall-thickness or weight. Lock-beading, a prerequisite for GMT designs, is widely dispensable for LWRT processing. The underbody of the new BMW 5 was smoothed across a large surface area. This is beneficial for both the car's aerodynamic and aeroacoustic properties.

The variation of the material density reduces the car's noise emission. The LWRT layer construction reduces the emission of air-borne sound while the porous core layer of the material has solid-borne sound dampening properties. The reduced wall-thickness on the edges and attachment points reduces the propagation of solid-borne sound even further. This dampens both interior noise at high speed and noise caused by flying stones or splashing water.

In addition to the flexural strength of the large surface area, the tear strength of the material near the attachment points requires special attention. In this area, it would not be sufficient to combine a porous middle layer with a thin sealing film. Therefore, more

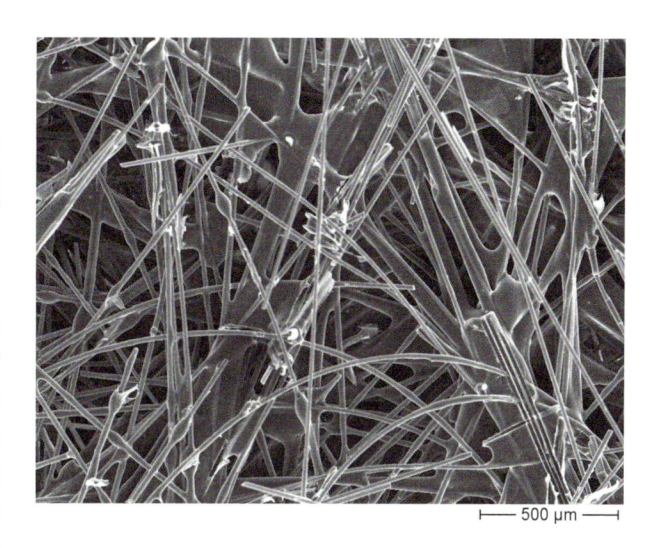

Figure 6-267 Scanning electron micrograph of a glass-fiber-PP structure (after removal of the sealing film)

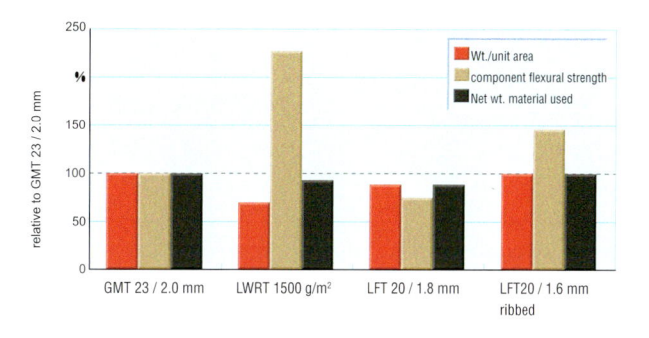

Figure 6-268 The flexural strength of LWRT products is more than double than that of GMT or D-LFT-components of largely similar material weight

intense compression is applied to increase the density of the area in question. The creation of a highly compressed, circumferential edge has the beneficial side-effect of creating a welded trimming effect along the entire component edge, preventing lateral moisture absorption.

The remaining surface area is protected by a thin film cover. It is important that this outer skin has sufficient impact strength. Adequate testing equipment was used to assess the strain caused by stones (Erichsen 508 based on DIN 55 996-1, under tougher conditions).

These tests showed the skin's sufficient impact resistance even at very low temperatures. The adhesion between the sealing film and the core layer is a crucial criterion for the assessment of the product's resistance against damage caused by loose stones.

However, the most trying situation is a critical angle of incline often approached in underground car parks. Snow drifts or deep tracks on country lanes pose a similar challenge. Due to the elasticity of the material and an adequate distance to the vehicle body, the underbody cover will show a certain deflection. In this situation, both the tear strength and the tear propagation strength of the sealing film and the entire layer construction are of crucial importance. The results of the simulation were verified by numerous practical tests. In addition to resistance to extreme stress discussed above, the product is also resistant to every-day abrasion and ageing.

6.5.2.5 Summary

"The development of a smooth underbody achieved an improvement of 40% over the original design." This is BMW's summary with respect to alternative underbody protection concepts [1]. This result affects a multitude of areas in automotive engineering: the attainable maximum speed and the longitudinal dynamics (acceleration in the area of between 100 and 200 km/h) are higher.

The underbody cover has a crucial effect on the front lift coefficient of 0.11 and the rear axle lift coefficient of 0.06. This represents a maximum stress removal from the rear axle of less than 5% at a speed of 250 km/h.

The air stream skips over the complex structure of the wheel suspension, carriers, steering equipment, bearings and pushrods, and the more homogeneous flow reinforces the efficiency of the diffuser at the rear of the vehicle. The car retains a smooth performance even in extreme situations and is less sensitive to crosswind. The low aerodynamic drag coefficient of the predecessor was maintained in spite of the extension of the projected surface area. This reduces fuel consumption particularly at high speed.

The one-piece underbody cover protects the car from contaminants, stones and corrosion. One of the four component parts encases the engine from the front axle onwards, which dampens noise from the outside and reduces the front axle lift coefficient. The cover also accelerates the heating-up of the engine during the cold start phase. These components and the controllable vent flaps located behind the BMW kidney grille, which are also supplied by Röchling Automotive, are all part of the car's active engine cooling system.

The second-generation of LWRT has been successfully launched into the market. Further weight reduction, improved mechanical and acoustical performance as well as cost are the key factors for the success of LWRT in underbody applications.

Reference

[1] Krist, S., Mayer, J., Neuendorfer R.: Aerodynamik und Wärmehaushalt der neuen BMW 5er, *ATZ* and *MTZ* Special edition, 8 (2003)

6.5.3 Noise-Reducing Coatings in Buses: the Mercedes-Benz Citaro

EDMUND SIENER

6.5.3.1 Introduction

There is a wide variety of potential uses for this two-component coating process in the vehicle interior or underfloor area, some of which are described as they are applied in Mercedes Benz city buses. Also described are the basic dependencies involved with measures for noise insulation and damping, respectively.

In the case of external applications, damping properties in the engine compartment and of the entire vehicle substructure are a focus of development. At the same time the new material can replace conventional underbody coating as corrosion and stone impact protection.

Inside the vehicle, functional damping can be supplemented by a two-component polyurethane decorative coating on floors and platform. This offers a higher degree of freedom for more sophisticated and functional geometries without having to cut complex shapes for the floor covering.

In the future, we can look forward to additional fields of application such as the prevention of damage from vandalism and simplification of repair measures.

Among environmental aspects, we will also describe a method and its process environment, which meets in particular the high load requirements on low-floored vehicles.

6.5.3.2 Sound Emissions in General

More and more people are annoyed and bothered by traffic noise, with road traffic noise being the major contributor. Aircraft noise stands in second place, followed by train noise. Noise from shipping is almost negligible as an irritant.

When sound is experienced as bothersome or a disturbance, we speak of noise rather than sound. In other words, noise is unwanted sound.

Noise can damage people's health and sense of well-being. The effects of noise depend on the sound level (especially in the case of loud sounds) and on its information content (disturbance and annoyance), which may result in:

- Hearing damage
- Sleep disturbance
- Impairment of mental activity due to distraction and interference with concentration
- Interference with speech intelligibility
- Impairment of leisure activities, including rest and relaxation.

With regard to the overall significant disturbance of the population by road traffic noise, reduction of sound emissions from motor vehicles must take on great importance. This must take into account not only the restriction of vehicle sound emissions due to design, but also the prevention of excessive design-related sound emissions over the course of the vehicle's service life.

Here, it should be noted that changes in the rating level of traffic noise in a range of 3 dB(A) are barely perceived by human hearing. The noise level has to be cut by 10 dB(A) before the sound level is perceived as "halved". This corresponds to a 90% reduction in traffic – in other words, from 20,000 to 2000 vehicles or from 100,000 to 10,000 vehicles ([1], [2]).

6.5.3.3 General Noise Reduction Measures

In general, sound can either be insulated or absorbed (damped). Since both of these approaches can be used not only on air-borne sound but also on solid-borne sound, this means that we have four ways of reducing noise [3]:

Solid-Borne Sound Absorption (Damping)

Solid-borne sound damping:
Decay of sound waves (oscillations) in the material

Characteristic:
Loss factor d, values between 0.0001 and 1.0

Solid-borne sound absorption is achieved by converting part of the acoustic energy into thermal energy as it penetrates homogeneous layers or coatings permanently connected or glued to the vehicle body. Solid-borne sound is thus absorbed before it can generate air-borne sound. Absorption of solid-borne sound will increase with a higher ratio between the modulus of elasticity of the coating material and that of the carrier material, multiplied by the ratio between the layer thickness of the coating material and that of the carrier material.

Solid-Borne Sound Insulation

Solid-borne sound insulation:
Preventing sound propagation in components

Characteristic:
Sound transmission loss R, values in dB

Solid-borne sound insulation is achieved by sound propagation being reflected by sound insulation (soundproofing) at a resilient intermediate layer. The softer and bulkier this layer is, the better the solid-borne sound insulation.

Air-Borne Sound Absorption (Damping)

Airborne sound absorption:
Friction at boundary surfaces or other surfaces

Characteristic:
Absorption coefficient α (alpha), values between 0.001 and 1
Frequency-dependent!

With air-borne sound absorption, part of the energy of the air-borne sound is converted into thermal energy as it penetrates into fibrous or foamed materials, thus absorbing the air-borne sound. The thicker the fibrous or foamed (open-pored) material layer, the better the air-borne sound absorption.

Air-Borne Sound Insulation

Air-borne sound insulation:
Preventing sound from passing through separating or partitioning components

Characteristic:
Sound transmission loss R, values in dB

With air-borne sound insulation, some of the acoustic energy is reflected from a wall. The rest of the acoustic energy is then emitted at the other side of the wall as air-borne sound. The heavier and more flexible the wall is, the better the air-borne sound insulation.

In practice these measures usually need to be combined to achieve the best results.

6.5.3.4 Statutory Requirements and Guidelines

In 1970, uniform legislation relating to restrictions on the noise produced by motor vehicles came into force throughout the European Community (EC). In subsequent years up until the present, noise limits have been lowered by 8 to 12 dB(A) for cars, buses and trucks [4].

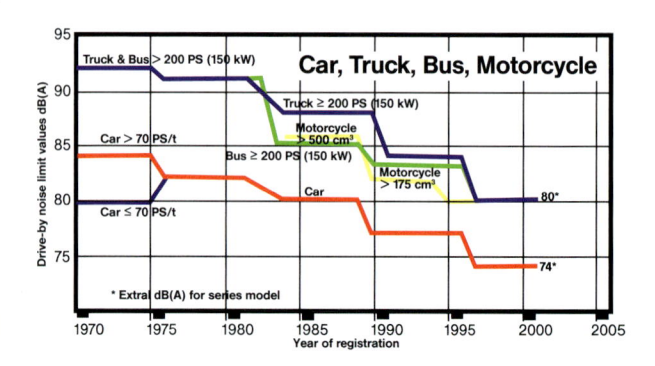

Figure 6-269 Legal limits of noise emission [4]

The legal limit values as defined in EU 70/157 for noise generation during drive-by under acceleration recorded at a distance of 7.5 m from center of the vehicle have recently been markedly lowered throughout the EC; the following limits currently apply to buses:

- Buses > 3.5 t and < 150 kW   78 db(A)
- Buses > 3.5 t and ≥ 150 kW   80 db(A)

Noise emission is taking on ever greater importance in environmental discussions. For this reason, in order to achieve a reduction in the sound emission burden on bus passengers and local residents that goes beyond legal requirements, sound levels should not exceed the following values:

Exterior Noise

Measurement method as defined in DIN ISO 362 and DIN ISO 5130

The sound levels specified are definitive limit values!

Driving noise
(accelerated drive-by at 7.5 m distance)   80 dB(A)

Requirements expected in the future
(EURO IV, for example)   78 dB(A)

Air compression noise   72 dB(A)

Exterior noise of supplementary
heating unit   65 dB(A)

Interior Noise

Measurement method as defined in DIN ISO 5128

Specified sound level at 50 km/h   ≤ 72 + 2 dB(A)

Measuring points in the vehicle rear section (center aisle) at 1.50 m height between drive axle and rear bench seat; at full acceleration on the flat up to 60 km/h (without kick-down); vehicle empty   ≤ 81 dB(A)

Stationary noise in the center aisle at 1.50 m height between the front and rear axles, with ventilation fan at lowest setting
(without air-conditioning)   ≤ 64 dB(A)

With fan at maximum setting and
air-conditioning switched on   ≤ 68 dB(A)

If so required, the manufacturer must supply documentary evidence of the values actually reached (exterior and/or interior noise).

Special demands apply to the new generation of city buses as regards thermal and acoustic insulation. In particular the input of heat into the vehicle interior from the drive unit must be kept low. In addition, both noise generation and engine vibrations must be dependably decoupled from the vehicle interior. Here, special attention must be given to the design of the maintenance access hatches in the floor and rear panel to ensure that they consistently meet the requirements mentioned.

The engine and gearbox must be provided with the corresponding engine compartment encapsulation in the underbody. It must be possible for a single person to remove or install the encapsulation (this also applies to inspection pits). The engine enclosure must not cause any thermal overload in the engine compartment [5].

6.5.3.5 Implementation of a New Noise Reduction Concept in the Citaro City Bus

Situation in 1995/1996

Corresponding to the state of the art at the time, the approach to this problem was characterized as follows:

- Heavy-spar mats were glued in with solvent-based adhesive, in some cases overhead; contact-gluing meant separate precoating of the mats, which were also difficult to cut

- Cutting flat material to shape manually involved high risk of injury

- Problems with work hygiene, work protection, environmental protection

- Large variety of parts with approx. 70 different part numbers; expense involved in stock control and storage of some 30 mats per vehicle with 3 different thicknesses

- Acoustic bridges caused by gaps between the several pieces in a layer

- Detachment due to thermal stress in engine compartment area

- Sub-surface migration of the glued-on mats by dampness, salt, dirt leading to corrosion in the engine compartment area

- Employment of 130 °C PVC underbody sealing – environmental relevance of corrosion damage due to PVC – embrittlement – loss of plasticizer

- High level of complaints due to damage to the floor surface, welded seams opening up, blisters form beneath the floor covering, shrinkage of floor coverings, delamination of the plywood sheets

- Example of expense: additional noise optimization for O 404 SHD coach

- Additional production time of 1400 minutes per vehicle

Trigger for the development of the complete two-component polyurethane system based on Desmodur/Desmophen (DD) were these additional requirements:

- Quality improvement – rationalization

- Optimization of production flow via cycle times rather than a location-based approach

- Innovation for the new generation of city buses

- Meeting requirements set by clean air guidelines and the VOC guideline

6.5.3.6 Specifications for Two-Component Polyurethane Functional Coatings

On the basis of the problems described and in connection with the development of the new Citaro city bus, a concept was drawn up in November 1995 for the development not only of a functional coating based on two-component polyurethane (solvent-free) to replace the PVC underbody sealing and heavy-spar mats, but also of a decorative coating – also solvent-free – based on two-component polyurethane for the vehicle interior.

In general, the effectiveness of noise-reduction coatings on components depends among other things on:

- The material thickness of the coating underlay (thin carriers: 0.88–1 mm thick) behave considerably better than thick underlays (1.5–2 mm thick)

- Conditions of fitting or clamping for the coating underlay

- Design of the underlay (single-shell, double-shell, sandwich construction, etc.)

- Temperature range to which the component is exposed (–40 °C – +80 °C, under certain circumstances more)

- Glass transition zone of the coating or insulating material (transition of polymer materials from the elastic-plastic state to the glass-like state and vice versa, with change in their physical properties)

- Glass-transition zone can be adjusted to suit the requirements applicable to the coating or insulating material (within certain limits)

- Density of the coating

- Coating thickness – compliance with physical and economic limits

- Single-sided or double-sided application to the substrate

- Double-sided application is in most cases better than one-sided

- Structure of the coating material (packed, closed, open-cell, porous)

- Flexural strength of the coating material (modulus of elasticity) depends on temperature

- Frequency of the noise source(s) – with superposed frequencies

- Points of leakage or openings providing sound with unimpeded access to components

Replacement of PVC – Underbody Sealing/Stone Impact Protection on Desmodur/Desmophen Basis

Research and development was successfully completed by the paint manufacturers Groß & Perthun of Mannheim [6, 7] in collaboration with Bayer AG of Leverkusen [8].

It had the following objectives:

A single material with at least two functions: Underbody sealing, stone impact protection and noise damping, applicable on all vehicle models

Heat resistance from $-40\ °C$ to $+140\ °C$

Low-temperature flexibility down to $-40\ °C$
(DIN 53 152 mandrel flex test)

Fire behavior rating of Class B or C
actual value: Class B (DIN 75 200)

Shore hardness A desired > 60
actual value: Shore A 63 (DIN 53 505)

Density approx. $1.70\ g/cm^3$
(similar to heavy-spar mats)

Stone impact test according to DBL OK
actual value: 82 min at 560 μm

- Spraying on of a two-component polyurethane insulating material by airless method onto electrophoretic-dip-primed substructure and rear areas, wheel housings and wooden floor, both inside and outside (from below)

- Can be sprayed on in coating thicknesses up to 4 mm on one side as a function coating, with continuous coating throughout

- Damping effect as with heavy-spar matting of comparable layer thickness (loss factor d)

- System is solvent-free (maximum solvent content of 3%), recyclable for possible return of end-of-life vehicles as per EU regulations

- System suitable for air-drying and oven-drying application

- Also suitable as repair material

Replacement of PVC Material on Rolls by a Sprayable Plastic Coating Based on Desmodur/Desmophen

Materials used must be dirt-repellent, easy to clean, water resistant, non-slip and operationally safe. In particular, the floor and the lower part of the side walls must be designed such that no water or cleaning agents can enter the vehicle substructure (a closed pan, for example) [9].

Use of solvent-free coating systems for floor coating

- Material is air- and oven-drying:
  60 minutes at 90 °C
  (6–8 h at room temperature)

- Fire behavior:
  Type B according to DIN 75 200
  (self-extinguishing)

- Surface slip properties according to BAM Berlin
  (slip resistance as per Road Licensing Regulations, Article 35 d [4])

- Decorative coating dry: markedly anti-slip
  Actual value: 68 cm

- Decorative coating wet: markedly anti-slip
  Actual value: 88 cm

- Surface slip properties as per DIN 18 032 Part 2
  Coefficient of sliding friction µm

- Dry min PVC series old (PVC 0.58) or better
  Actual value: 0.76 µm

- Wet min PVC series old (PVC 0.41) or better
  Actual value: 0.59 µm

- Wear resistance: 80 mg/1000 revolutions
  (Taber abrasion test)

- High-temperature light-fastness: Level 7
  (ultraviolet resistance DIN 75 202)

- Shore hardness A: > 60
  Actual value: Shore A 63

Taking these criteria into consideration, the effective loss factor actually detected can only be measured, evaluated and optimized under actual installation conditions. However, on the basis of the tests carried out, conclusions may be drawn regarding the comparative basic suitability of different materials for the end use specified.

Since it is a very difficult task involving a great deal of effort to register the frequencies of individual sources of acoustic disturbance in the bus, determined as they are by the superpositioning of the most varied frequencies when the vehicle is stationary or under way, and with further ambient influences also playing a part (engine cold or hot, road surface, and so on), the insulating or coating material selected should be effective over a wide frequency spectrum from about 200 Hz to 5000 Hz.

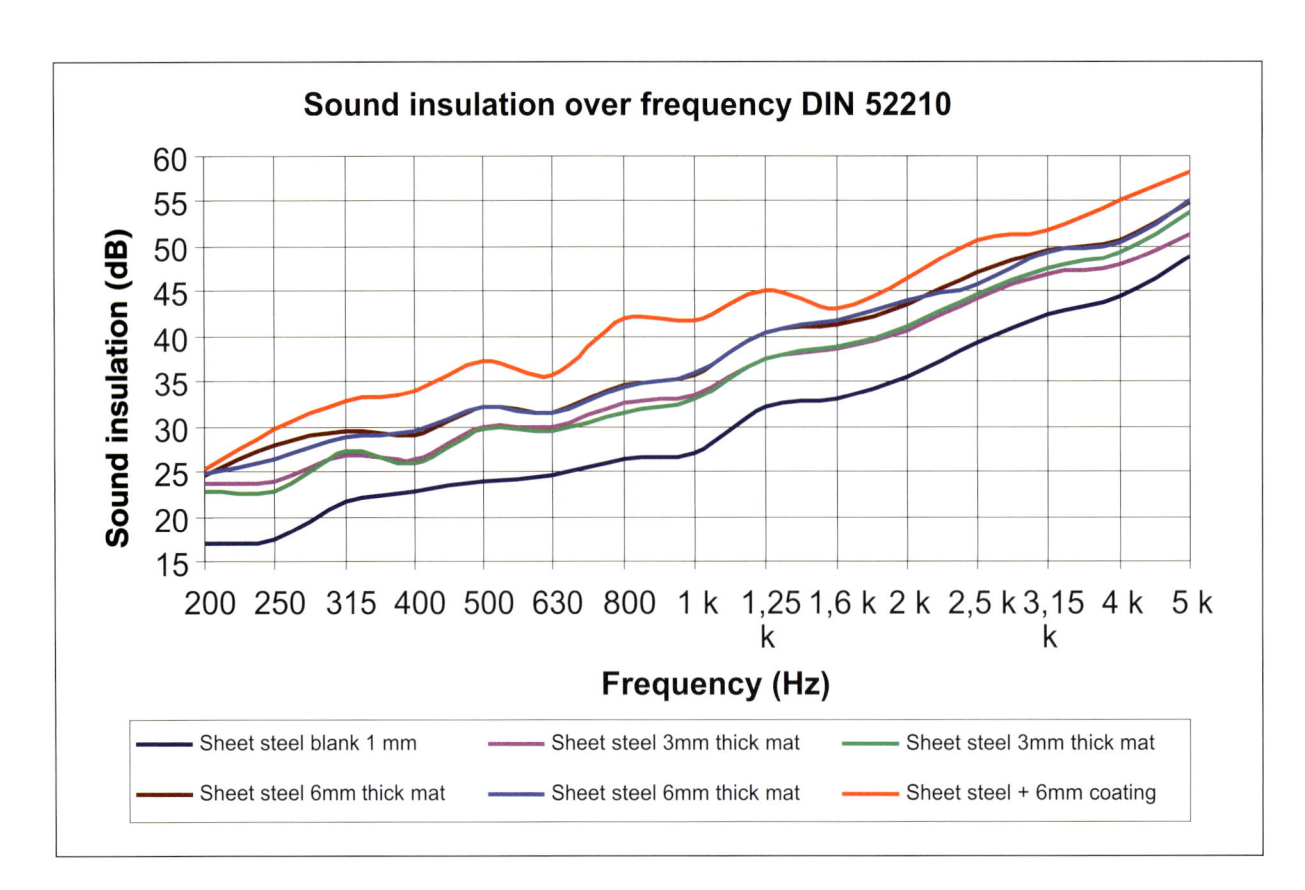

Figure 6-270 Sound insulation of various sound-absorbing materials/layers

6.5.3.7 Technical Information and Measured Gradients

In order to determine the general, and in particular the layer-thickness-dependent, effectiveness of coatings, different sheet metal thicknesses, temperature dependence and possible materials that could be used in addition to further reduce interior noise in the Citaro city bus, various tests were carried out, including determination of the loss factor d as per DIN 53 440.

Specimens were prepared for this purpose with a sheet thickness of 0.88 mm and 1 mm, each coated with different materials with different coating thicknesses. Measurements were carried out over a temperature range of 0 °C to +60 °C; measurement frequency: 200 Hz.

The acoustic values of this coating sprayed onto both sides shows the same damping values compared to a heavy-spar mat of the same thickness; in other words, 2 + 2 mm of a bilaterally sprayed-on coating corresponds in its damping effect to a 4 mm thick heavy-spar mat glued onto one side.

The measurement results show clearly that under the test conditions selected a thicker steel sheet does not behave as well as thinner steel sheet. In the case of all materials tested and the formulations based on chemical polymers suitable for car or truck use, all have the common disadvantage that their ability to reduce noise diminishes as the temperature of their surroundings or of the coated support material rises.

This effect is due to the chemical formulation and to the coating requirements profile of each case. Accordingly, the best damping properties can be achieved only over a limited temperature range.

However, if the coating is to be approximately equally effective over the specified temperature range – under certain circumstances up to more than +100 °C – it is recommended that it is protected against such heating or suitable steps taken (such as covering with heat-reflecting foils, e.g., aluminum) to allow it to satisfy both requirements, namely low-temperature flexible underfloor protection and temperature-dependent solid-borne sound absorption to approximately the same extent.

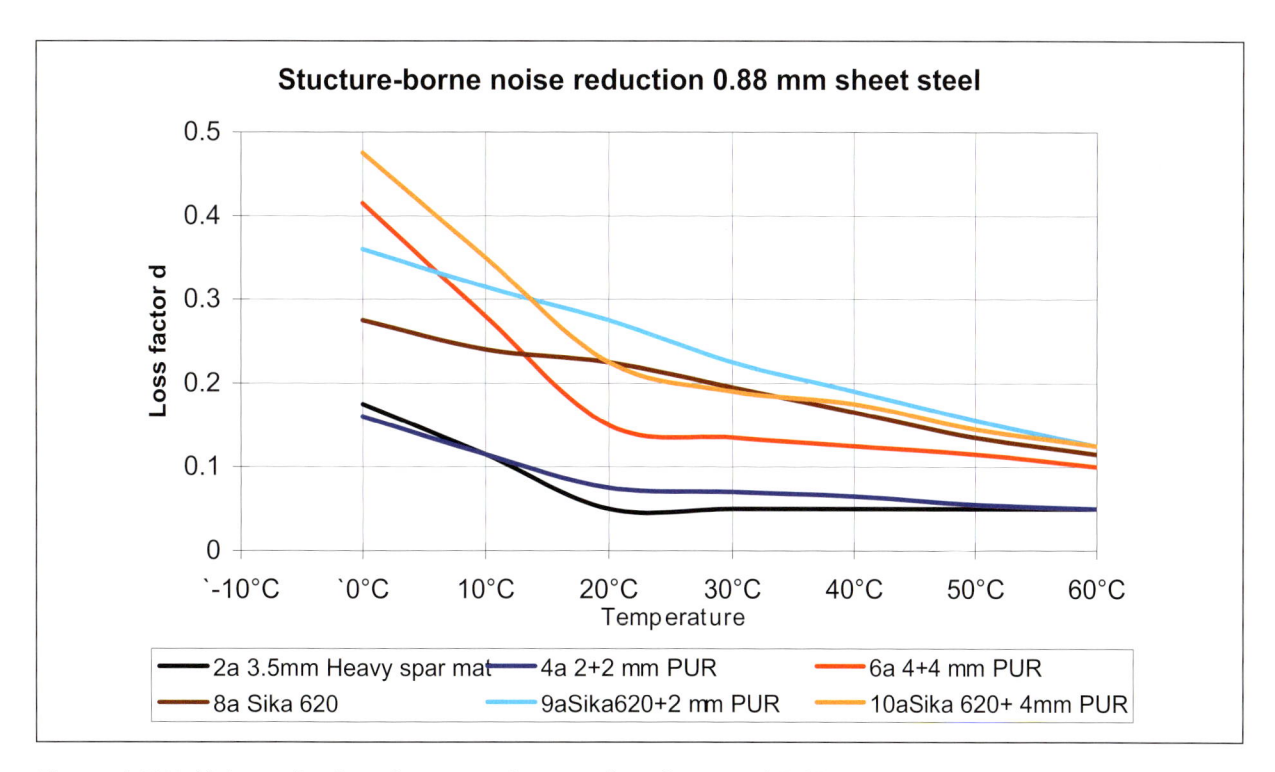

Figure 6-271 Noise reduction of structure-borne noise, sheet steel 0.88 mm

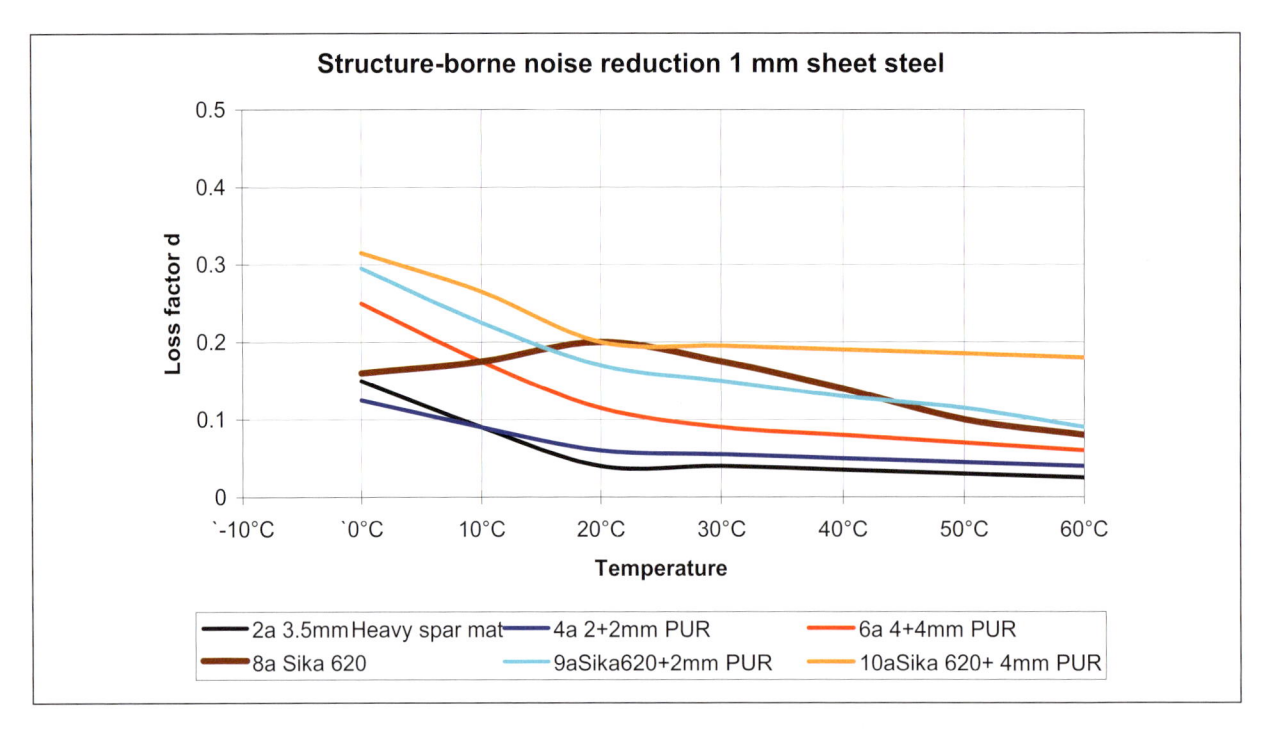

Figure 6-272 Noise reduction of structure-borne noise, sheet steel 1 mm

6.5.3.8 Implementation in Volume Production

Stone Impact Protection – Underbody Sealing – Solid-Borne Sound Absorption

The PVC underbody sealer coating used for years in our vehicles has been replaced by a completely newly developed two-component polyurethane functional coating. The term "functional coating" was selected in light of the fact that it was now possible to solve a number of different requirements in the vehicle with one coating material and in one operation. What should be emphasized here is the considerable improvement in production-related environmental protection. Moreover, introducing this system into volume production without any loss of quality meant that it was possible to dispense completely with coating the entire substructure with two-component epoxy high-build filler as was necessary after electrophoretic dip coating. This corresponds to about 10 kg of paint per vehicle of which solvent makes up approx. 35%.

Basically, the entire underbody area was given this functional coating – following partial caulking of seams – to a level corresponding to the requirement profiles

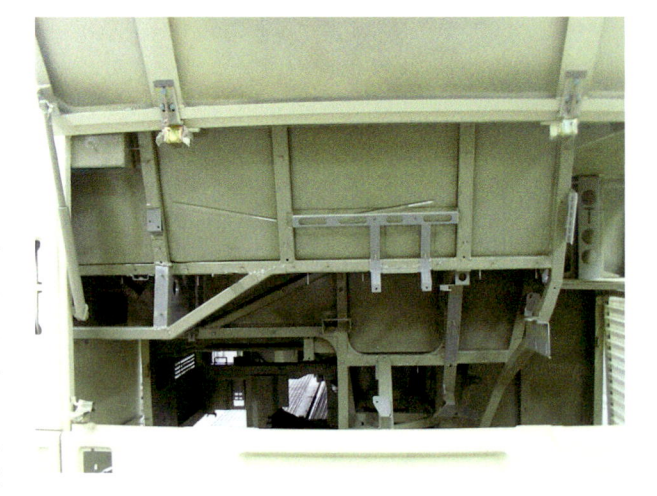

Figure 6-273 Sound-absorbent layer in underbody, rear engine compartment

for the particular loading case. In those areas requiring stone impact protection alone, coating thicknesses measured between 500–800 μm. In those parts of the vehicle where additional steps needed to be taken for solid-borne sound absorption (normally from the central entrance to the engine compartment bulkhead), coatings with thicknesses of up to 4 mm were applied, to one side of both horizontal and vertical faces, depending on design-related requirements.

The two-component material is applied manually by means of an airless spraying method not only to the underbody (primed by electrophoretic dip coating) but also to the wooden floor in a single, continuous coating at the thicknesses required. Since the film was continuous it prevented sound bridges, in contrast to when heavy-spar mats were used.

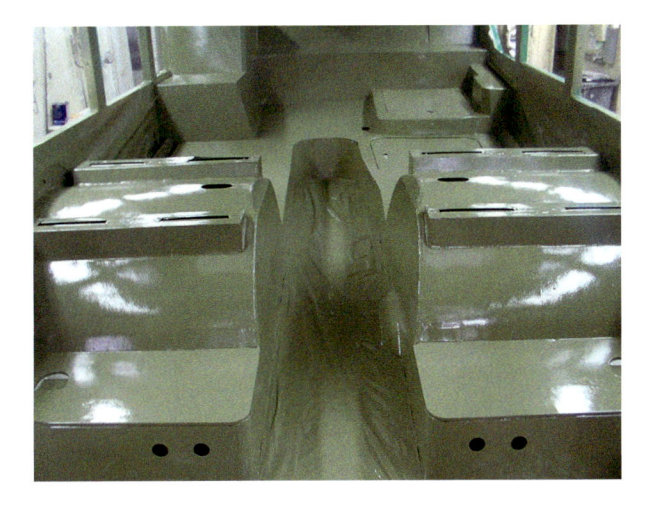

Figure 6-274 Sound-absorbent layer, rear interior of bus

Solid-Borne Sound Absorption in the Passenger Compartment

Particularly in the engine compartment, the coating thickness of up to 8 mm on one side of the metal paneling required by the vehicle design reaches the limits of the application technology: That would also excessively restrict the space left for subsequent installation work. In the light of these facts, it was decided that a 4 mm layer of the same two-component polyurethane material should be applied on each side of the metal panels, thus still resulting in an overall thickness of 8 mm.

This approach resulted in a sandwich structure (preferable from an acoustic point of view), which had the added advantage that the full-area coating of these parts created a pan structure which is sealed both from the underside and also from the inside of the vehicle.

This also meant that there was now no need to glue heavy-spar mats into the passenger compartment using adhesives with a high solvent content.

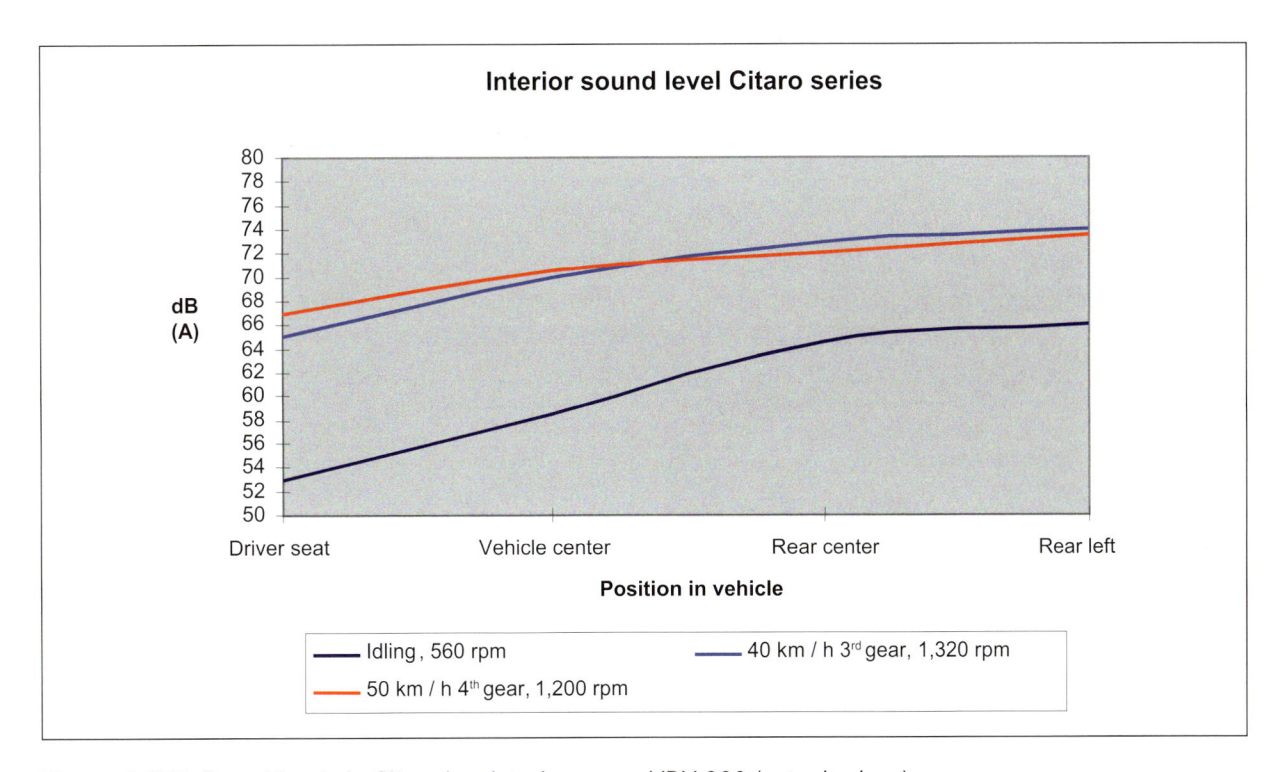

Figure 6-275 Sound levels in Citaro bus interior as per VDV 230 (actual values)

The success of this innovative approach can be seen from the sound curves recorded in a vehicle manufactured on the new production line (measuring points as per VDV 230).

6.5.3.9 "Sealed Pan" Concept – Passenger Compartment Coating

Decorative Coating System in the Passenger Compartment – Floor Functionality

Once the solid-borne sound absorption had been applied to the vehicle interior and was cured, the substrate was prepared. Unevenness and transitions were filled with a special two-component polyurethane and smoothed down and, in the case of the complete coating of the floor, the transitions from floor to raised areas, wheel arches and so on were chamfered. This ensured that a jointless and seamless transition from the horizontal to the vertical coated areas.

With this procedure the goal of a sealed pan in the passenger area of buses was finally achieved.

Spray application also used the airless method, with a coating thickness of 2.5 to 3 mm, depending on the useful layer thickness of the rolled PVC. This coating is characterized by providing a seamless, jointless, sealed, wear-resistant and resilient surface and satisfying the relevant requirements made by mass transit companies. Since 1996, approx. 2500 buses have been built to this standard of quality.

Since 1998, the standard production state has been the so-called mixed design, whereby the complex shapes in the rear component and elsewhere are sprayed with two-component polyurethane material produced by the paint manufacturers Groß & Perthun of Mannheim. The result is a basic color and design virtually identical to PVC rolled material with an overall smooth surface area of the vehicle previously achieved using glued down sections of rolled PVC.

The decorative floors in the case of both the mixed design and the complete coating can have virtually any color – depending on the customer requirements – combined with a colored flecking strewn over it when the film is still wet. The size of the flakes can be selected from between 2–3 mm and 3–4 mm as can the percentage of different colored flakes in the mixture.

To satisfy one customer's requirements for even greater slip-proofing for a ski bus, various dulling additives were tested, such as fine corundum, plastic fibers, spherical pellets, and others. The final decision was to use fine-grained plastic pellets which can be applied at the same time as the flakes. These pellets did not impair the coloring of the floor design either, nor did they increase the risk of grazes or abrasions in summer operation as would have been the case with corundum, for example.

Protection against Vandalism – Anti-Graffiti Properties

The word "graffiti" is Italian and means "a drawing or writing scratched on a wall". It is now applied to anything undesirable applied by whatever method to walls and other surfaces in public areas. Usually, graffiti employ very bright colors and frequently express political, racist, or sex-related sentiments and millions are spent each year in an effort to remove them. Even though the so-called anti-graffiti systems are not yet able to prevent graffiti being applied in the first place, it should nevertheless be possible to make their removal considerably easier by prophylactic measures.

Figure 6-276 Production state – 2-component PUR paint + floor covering in rear compartment

In Berlin, the current fad is to spray graffiti on the floor. In this case, the standard two-component polyurethane direct coating has the advantage over roll PVC in that, once the area in question has been partially ground, it is a relatively easy matter to re-apply the original color with a roller (including flakes) – and overnight the graffiti are gone.

In another application, the two-component polyurethane decorative coating in the region of the rear bench seat and the radiator tower is protected against graffiti by an additional, two-component polyurethane clear lacquer film applied wet-on-wet.

For one major customer in Germany and in Austria, and also for various customers in France, an additional low-solvent system was developed on a two-component polyurethane basis to protect side wall panels by coating them. Instead of so-called print laminates, anti-scratching films, or the familiar clear lacquer systems, a colored two-component polyurethane primer paint was applied wet-on-wet, followed by the droplet-shaped application of a different colored two-component polyurethane paint, resulting in an appealing dotted effect on the ABS side panels.

This dotted effect painting – also available in a single-color fine structure without the so-called dotted effect – can also be easily changed to suit the color scheme desired by the customer for the passenger compartment.

In trials and during practical operation with the customers, a drop in the incidence of graffiti was recorded, probably due to the structured, tough, resilient and hard surface. Cleaning has also been easier, as is any recoating necessary following contamination of the surface.

6.5.3.10 Application Techniques – Application Equipment

The complete application method (airless hot spraying process) was designed and developed since 1996 by EVOBUS Mannheim in collaboration with Wagner-OTEG of Grünstadt and material manufacturers, Groß & Perthun of Mannheim.

Over the years, experience in volume production has been fed back into further development of the application system.

The complete system, that is, two-component polyurethane underbody sealing/absorption coating and decorative coating on a DD basis in the passenger compartment, is standard for all Citaro vehicles, with the optimized spraying facilities of Wagner-OTEG also in service at the sister plants at Ligny and Samano.

In the Hosdere/Davutpasa plant in Istanbul, a similar type of facility is also in use. However, only the two-component polyurethane underbody sealing (a version of the two-component polyurethane functional coating) is currently being used there.

Spraying parameters:

Material application by the airless method	150–180 bar
Material temperature at nozzle outlet	40 °C, nozzle 525
Material output	approx. 1800–2500 ml/min

Figure 6-277 Two-component mixing and proportioning unit for applying decorative coating

6.5.3.11 Ecological Aspects and Environmental Protection

The two systems used at EVOBUS Mannheim – the functional coating and the direct floor coating on the basis of DD – were dubbed the currently "best technique available" for this sector of industry in 2002.

This award was presented as part of the research project "Best techniques available in the field of paint and adhesive processing in Germany" by the DFIU in Karlsruhe (French-German Institute for Environment Research) as commissioned by the German Federal Bureau of the Environment [1]).

6.5.3.12 VOC Guideline

The aim of the VOC guideline is to prevent or reduce the effects of emissions of volatile organic compounds into the environment, especially into the air, and to prevent or reduce possible risks to human health.

Here, the European directive 1999/13/EG of the Council dated 11 March 1999 applies; this concerns the restriction of emissions of volatile organic compounds which occur with certain activities and in certain installations when organic solvents are used.

The VOC limit of 150 g/m^2 – in other words, solvent usage for painting buses (in force as of 31 October 2007) – is already met by Evobus Mannheim in the case of single-color vehicles thanks to the use of the above-mentioned materials.

Advantages and relevance to the environment:

- Both systems are standard in production with the Citaro at the Mannheim plant

- Both systems are solvent-free; solvents for cleaning or rinsing the application equipment pass via a solvents management system to a service provider for recycling

- Recycling solvents are set to values of less than 5% xylene

- Both systems have been pigmented without using hazardous substances
  (no lead, chromium or cadmium)

- Both systems replace PVC materials
  (PVC underbody sealing, PVC rolled material)

- Replacement of PVC seam-caulking material by single-component polyurethane material
  (solvent-free)

- Both systems are fully recyclable
  (possibly with return of end-of-life vehicles according to EU law)

- Neither system generates any paint or dye sludges or hazardous waste requiring special monitoring; cured material can be disposed of as domestic waste

- Dry deposition in the spraying booth meets the requirement of the Clean Air Guidelines with values of less than 3 mg particles/m^3 waste air

- No emissions from a drier; two-pack polyurethane functional coating is not annealed, meaning no decomposition products
  (polyurethane addition reaction)

- Both systems replace the heavy-spar mats previously used

- Reduction in solvent-based adhesives from approx. 15 kg to approx. 2 kg per vehicle.

6.5.3.13 "Blue Angel" Environmental Label

Vehicles painted as described with the new technology can receive the "Blue Angel" environmental label (for example, Gasbus Mühlhausen, Zwickau) because:

All products, insofar as they meet the prerequisites for awarding the label, can apply to RAL for permission to use the environmental emblem on the basis of a label use agreement [11].

6.5 Acoustics and Aerodynamics

Figure 6-278
German "Blue Angel" environmental acceptability certificate

This environmental label is intended to help reduce the considerable burden of hazardous substances and noise caused by commercial vehicles, distribution vehicles, communal vehicles and buses particularly in city centers, conurbations and areas requiring protection. It takes into consideration the requirements and recommendations regarding emissions of hazardous substances and noise, regarding painting of vehicles, as well as insulation and refrigerants in refrigerated vehicles.

6.5.3.14 Economical Aspects

Statements about the economic efficiency of components and/or processes in the field of commercial vehicles – and buses in particular – are always characterized by two criteria:

- A wide variety of models must be offered in order to meet customer demands; which means that

- Many versions will have only very small production numbers, which in turn means high charges for each part or for each vehicle.

A cost comparison between the predecessor model, the O 405 N2 (with its riveted floor, primered substructure, PVC seam sealing and PVC underbody sealing, insulation by means of heavy-spar mats, full lining with PVC off-roll material in the passenger compartment, structural paintwork in the front end area and U spaces) and the systems first used in the Citaro with mixed design and sprayed two-component polyurethane solid-borne sound insulation and a sprayed floor on DD basis is difficult, because this would involve comparing quantities that resist comparison technically, functionally and qualitatively. Nonetheless, a higher level of economic efficiency derives from:

- A marked reduction in the time required for painting and gluing

- Inter-departmental optimization of process sequences between floor assembly and painting

- Cost reductions in energy and disposal

- No expensive investments to meet environmental protection requirements.

References

[1] Lärmschutz im Verkehr, Technische und rechtliche Grundlagen, Lärmschutzverordnung, Bundesministerium für Verkehr, 2nd edition, January 1998
[2] K. Egenschwiler, EMPA Düsseldorf, Grundlagen der Akustik und Lärmbekämpfung, ERFA Seminar 25th February 2002
[3] Bayer AG Leverkusen, Dr. Sägmühl, Schmick, TE/PAT Schall- und Schwingungstechnik, March 98
[4] Dr. Reiner Stenschke, Umweltbundesamt Berlin, LFUG workshop 'Road traffic noise', May 2004
[5] VDV publications September 2002, VDV Publication 230 Rahmenempfehlungen für Stadt-Niederflur-Linienbusse (SL III), Section 3.2.4. Wärme- und Geräuschisolation
[6] Lackfabrik Groß & Perthun, Mannheim, Bälz, Lütz, Frey, Preissig, material data sheets and technical information; development meetings
[7] Lackfabrik Groß & Perthun, material datasheets and technical information
[8] Bayer AG Leverkusen, Dr. Bock, Dr. Casselmann, Dr. Petzold, Dr. Fleck, Müller, Kobelka, Schmidt, various test reports 1995–2003
[9] VDV publications as at September 2002, VDV Publication 230, Rahmenempfehlungen für Stadt-Niederflur-Linienbusse (SL III), Section 6.5. Fussboden
[10] University of Karlsruhe, Institute DFIU/IFARE, Peters, Nunge, Geldermann, Rentz; Report on best available techniques in the field of paint and adhesive processing in Germany; Vol. I: Paint processing, Vol. II: Adhesive processing, August 2002
[11] Jürgen Trittin, Federal Minister for the Environment, Protection of Nature and Reactor Safety at www.BlauerEngel.de, December 2004

Appendix

A World Wide Web References Related to Plastic Part Design[1]

Note: ▶ symbol represents MUST SEE World Wide Web sites.

Associations/Organizations	
ABIQUIM – Associação Brasileira da Indústria Química	http://www.plastivida.org.br/
American Chemical Society	http://www.acs.org
American Plastics Council	http://www.americanplasticscouncil.org/
ASM International	http://www.asm-intl.org
Asociación Nacional de la Industria Química, Mexico	http://www.aniq.org.mx
Associação Brasileira de Polímeros	http://www.abpol.com.br
Association of Plastics Manufacturers, Germany	http://www.vke.de/de/index.php
Association of Plastics Manufacturers Europe	http://www.apme.org/
British Plastics Federation	http://www.bpf.co.uk/
Canadian Plastics Industry Association	http://www.plastics.ca
Food Packaging Institute	http://www.fpi.org
Instituto Technologico del Plastico	http://aimplas.es
Plastic Industry in South America	http://www.cosmos.com.mx/pla.htm
Plasticx Universe	http://www.plasticx.com
Polystyrene Packaging Council	http://www.polystyrene.org
Product Design & Development	http://www.pd3.org
Plastics And Chemicals Industries Association, Australia	http://www.pacia.org.au/
Plastics Institute of New Zealand	http://www.plastics.org.nz/
Rubber and Plastics Machinery Sector Association, Germany	http://www.vdma.org/english/guk/index.htm
▶ SAE (Society of Automotive Engineers)	http://www.sae.org
SAMPE	http://www.sampe.org
Syndicat des Producteurs de Matières Plastiques (SPMP)	http://spmp.sgbd.com/
▶ SPE (Society of Plastics Engineers)	http://www.4spe.org
SPI (Society of Plastics Industry)	http://www.plasticsindustry.org/
Swiss Plastics Association	http://www.kvs.ch/
The All India Plastics Manufacturers' Association (AIPMA)	http://www.aipma.org/

---

[1] Tres, P. A., Designing Plastic Parts for Assembly, 6th ed. (2006), Hanser Publishers, Munich

Books/Magazines	
Carl Hanser Verlag	http://www.hanser.de
▶ Chemical Week	http://www.chemweek.com
▶ Design News	http://www.manufacturing.net/magazine/dn
▶ Injection Molding Magazine	http://www.immnet.com
Hanser Gardner Publications	http://www.hansergardner.com
Japan Chemical Week	http://www.chem-edata.com/index_e.html
▶ KunststoffWeb	http://www.kunststoffweb.de/
▶ Machine Design	http://www.machinedesign.com
Materie Plastiche ed Elastomeri	http://www.ovest.it/mpe/
Plastics News	http://www.plasticsnews.com
▶ Plastics Technology	http://www.plasticstechnology.com

Forums	
Composites	http://www.advmat.com
Forum PET	http://www.forum-pet.de
▶ Plastics Engineering Resources	http://www.plastics.com/forums.php
▶ Plastics Network	http://www.plasticsnet.com
▶ PolySort News & Information	http://www.polysort.com

Material Databases	
▶ CAMPUS	http://www.campusplastics.com
▶ IDES	http://www.idesinc.com
▶ Moldflow	http://www.moldflow.com
Material Property Data	http://www.matweb.com
▶ Plaspec Network	http://www.plaspec.com
▶ Rapra Technology	http://www.rapra.net/

Material Suppliers	
Asahi Kasei Corporation	http://www.asahi-kasei.co.jp/asahi/
Asahi Glass Company	http://www.agc.co.jp/english/index.htm
Ashland Chemical	http://www.ashchem.com
Atofina	http://www.atofina.com/groupe/gb/f_elf_2.cfm
Ausimont	http://www.ausiusa.com
Akzo Nobel	http://www.akzonobel.com
Basell Polyolefins	http://www.basell.com
BASF	http://www.basf.de
Bayer	http://www.bayer.de
Borealis	http://www.borealis.com
British Petroleum	http://www.amoco.com or http://www.bp.com
Celanese	http://www.celanese.de
Chevron Phillips Chemical Company LP	http://www.cpchem.com/index.asp
Degussa	http://www.degussa.de
DuPont	http://www.dupont.com
▶ Eastman	http://www.eastman.com
ExxonMobil	http://www2.exxonmobil.com/corporate/
Ferro	http://www.ferro.com
▶ GE Plastics	http://www.ge.com/plastics
Honeywell Plastics	http://www.alliedsignalplastics.com
ICI	http://www.ici.com
Kyocera	http://www.kyocera.co.jp
LNP Engineering Plastics	http://www.lnp.com
Owens Corning	http://www.owenscorning.com
PolyOne	http://www.polyone.com
▶ PPG Industries	http://www.ppg.com
Radici Plastics	http://www.radiciplastics.com
Rhodia	http://www.us.rhodia.com/
Rohm and Haas	http://www.rohmhaas.com
RTP Company	http://www.rtpcompany.com
Sumitomo Chemical	http://www.sumitomo-chem.co.jp

Museums

- Plastics Museum, Leominster, MA — http://npcm.plastics.com
- Sandretto Plastics Museum Italy — http://www.sandretto.it/museonew/default.htm

Rapid Prototyping

- Worldwide Rapid Prototyping Site — http://www.biba.uni-bremen.de/groups/rp/index.html
- Rapid Prototyping Journal — http://www.emeraldinsight.com/rpj.htm

Testing/Research

- Akron Rubber Development Laboratory — http://www.ardl.com/
- Plastics Technology Laboratories — http://www.ptli.com

Tips on Design and Processing

- Bad Human Factors Designs — http://www.baddesigns.com
- Guide to Thermoplastics — http://www.endura.com
- Software for Snap-Fits, Press-Fits, etc. — http://ets-corp.com/tools/development/

Tooling

- D-M-E Company — http://www.dmeco.com
- Foboha — http://www.foboha.com
- Harbec — http://www.harbec.com
- Mold-Masters — http://www.moldmasters.com/

Universities

- Brown University — http://www.chem.brown.edu
- Cornell University Materials by Design — http://www.mse.cornell.edu
- Delft University — http://www.io.tudelft.nl
- University of Hannover — http://www.imc.uni-hannover.de
- University of Leeds Polymer Science — http://www.materials.leeds.ac.uk
- University of Massachusetts – Lowell — http://www.uml.edu/
- University of Wisconsin – Milwaukee — http://www.uwm.edu/UniversityOutreach/catalog/ENG/index.shtml

B Plastic Materials for Automotive Applications

JULIUS VOGEL

Application	Processing	Material
1. Headlamp front lens	Injection molding	Polycarbonate (PC)
2. Headlamp casing	Injection molding	Acryl butadien styrene (ABS)
3. Air ducts	Blow molding, injection molding, combined injection – blow molding	Polyamide 6 (PA6)
4. Instrumental pnels	Injektion molding	Acryl butadien styrene (ABS), Polypropylene (PP)
5. Engine Covers	Injektion molding	Polyamide 6 (PA6)
6. Seat backs and valances	Injection Molding, Blow Molding, Painting	Urethane foams (PUR)
7. Steering column cover	Injection molding, 2-K-Technology	Polyamide 6
8. Pillars trim	Injection molding, 2-K-Technology, Ultrasonic welding, Lamination, Overmolding	Polyamide 6
9. Door trim	Injection molding	Polypropylene (PP) – reinforced with glass fibers
10. Cockpit modules hard panels	Injection molding, painted, soft painted Injection molding, soft painted	Polypropylene (PP) Acryl butadien styrene (ABS)
11. Cockpit modules soft panels	Foam in place, positive/negative forming, powder slush molding, leather wrap Thermoforming, Powder slush molding	Styrene maleic anhydride (SMA) Polyvinyl chloride (PVC)
12. Rear light lens	Injection molding, multi color injection molding	Polymethylmethacrylate (PMMA)
13. Water tank	Blow molding, welded together from injected halve shells	Polypropylene (PP)
14. Fuel tank	Blow molding	High Density Polyethylene (HDPE)
15. Bumper	Injection molding	Polypropylene (PP), ethylene propylene rubbers (EPDM)
16. Tubes	Extrusion through tubing die	Polyamide-TPE (thermoplastic elastomers)
17. Fender	Injection molding	Polycarbonate-blends (PC), Polyamide-blends (PA)
18. Pedal	Injektion molded	Polyamide (PA)
19. Air manifolds	Injektion molded, welded	Polyamide 6, 66 (PA6, PA66)
20. Airbag container	Injektion molding	Polyamide 6 (PA6)
21. Door handles, fuel caps	Injektion molding	Polyamide 6 (PA6)
22. Exterior mirror	Injektion molding	Polybutylenterphtalate (PBT), Polyethylenterphtalate (PET)
23. Hydraulic hoses	Extrusion	Copolyester (TPC)
24. Seals	Extrusion	Polyamide-TPE (thermoplastic elastomers)
25. Steering wheel	Injection molding	Polyurethane (PUR), vinyl resins

C   Plastics Acronyms[1]

In the plastics industry it is common to define a polymer by the chemical family it belongs to, and assign an abbreviation based on the chemistry. However, many times instead of using the standardized descriptive symbol, often engineers use the tradename given by the resin supplier.

This book uses the standardized notation presented in Table C-1. The symbols which have been marked with an asterisk (*) have been designated by the ISO standards, in conjunction with the material data bank CAMPUS.

The plastics presented in the table are presented in detail in Chapter 6 of this handbook. Furthermore, the acronyms presented in Table C-1 may have additional symbols separated with a hyphen, such PE-LD for low density polyethylene, or PVC-P for plasticized PVC. The symbols for the most common characteristics are presented in Table C-2.

Table C-3 presents the most commonly used plasticizers and the symbols used to describe them. Plasticizers are also covered in detail in Chapter 6 of this handbook.

Table C-1 Alphabetical overview of commonly used acronyms for plastics

Acronym	Chemical notation	Acronym	Chemical notation
ABS*	Acrylonitrile-butadiene-styrene	CO	Epichlorhydrine rubber
ACM	Acrylate rubber, (AEM, ANM)	COC*	Cyclopolyolefine-Copolymers
ACS	Acrylonitrile-chlorinated polyethylene-styrene	COP	COC-Copolymer
AECM	Acrylic ester-ethylene rubber	CP	Cellulose propionate
AEM	Acrylate ethylene polymethylene rubber	CR	Chloroprene rubber
AES	Acrylonitrile ethylene propylene diene styrene	CSF	Casein formaldehyde, artificial horn
AFMU	Nitroso rubber	CSM	Chlorosulfonated polyethylene rubber
AMMA	Acrylonitrile methylmethacrylate	CTA	Cellulose triacetate
ANBA	Acrylonitrile butadiene acrylate	DPC	Diphenylene polycarbonate
ANMA	Acrylonitrile methacrylate	E/P*	Ethylene-propylene
APE-CS	see ACS	EAM	Ethylene vinylacetate rubber
ASA*	Acrylonitile styrene acrylic ester	EAMA	Ethylene acrylic acid ester-maleic acid anhydride-copolymer
AU	Polyesterurethane rubber		
BIIR	Bromobutyl rubber	EB	Ethylene butane
BR	Butadiene rubber	EBA	Ethylene butylacrylate
CA	Cellulose acetate	EC	Ethylcellulose
CAB	Cellulose acetobutyrate	ECB	Ethylene copolymer bitumen-blend
CAP	Cellulose acetopropionate	ECO	Epichlorohydrine rubber
CF	Cresol formaldehyde	ECTFE	Ethylene chlorotrifluoroethylene
CH	Hydratisierte cellulose, Zellglas	EEAK	Ethylene ethylacrylate copolymer
CIIR	Chloro butyl rubber	EIM	Ionomer Copolymer
CM	Chlorinated polyethylene rubber	EMA	Ethylene methacrylic acid ester copolymer
CMC	Carboxymethylcellulose	EP*	Epoxy Resin
CN	Cellulose nitrate, Celluloid	EP(D)M	see EPDM

---

[1] Osswald, T. A. et al., International Plastics Handbook (2006), Hanser Publishers, Munich

Acronym	Chemical notation
EPDM	Ethylene propylene diene rubber
EPM	Ethylene propylene rubber
ET	Polyethylene oxide tetrasulfide rubber
ETER	Epichlorohydrin ethylene oxid rubber (terpolymer)
ETFE	Ethylene tetrafluoroethylene copolymer
EU	Polyetherurethane rubber
EVAC*	Ethylene vinylacetate
EVAL	Ethylene vinylalcohol, old acronym EVOH
FA	Furfurylalcohol resin
FEP	Polyfluoroethylene propylene
FF	Furan formaldehyde
FFKM	Perfluoro rubber
FKM	Fluoro rubber
FPM	Propylene tetrafluoroethylene rubber
FZ	Phosphazene rubber with fluoroalkyl- or fluoroxyalkyl groups
HIIR	Halogenated butyl rubber
HNBR	Hydrated NBR rubber
ICP	Intrinsically conductive polymers
IIR	Butyl rubber (CIIR, BIIR)
IR	Isoprene rubber
IRS	Styrene isoprene rubber
LCP*	Liquid crystal polymer
LSR	Liquid silicone rubber
MABS*	Methylmethacrylate acrylonitrile butadiene styrene
MBS*	Methacrylate butadiene styrene
MC	Methylcellulose (cellulose derivate)
MF*	Melamine formaldehyde
MFA	Tetrafluoroethylene perfluoromethyl vinyl ether copolymer
MFQ	Methylfluoro silicone rubber
MMAEML	Methylmethacrylate-exo-methylene lactone
MPF*	Melamine phenolic formaldehyde
MPQ	Methylphenylene silicone rubber
MQ	Polydimethylsilicone rubber
MS	see PMS
MUF	Melamine urea formaldehyde
MUPF	Melamine urea phenolic formaldehyde
MVFQ	Fluoro silicone rubber

Acronym	Chemical notation
NBR	Acrylonitrile butadiene rubber
NCR	Acrylonitrile chloroprene rubber
NR	Natural rubber
PA	Polyamide
PA11*	Polyamide from aminoundecanoic acid
PA12*	Polyamide from dodecanoic acid
PA46*	Polyamide from polytetramethylene adipic acid
PA6*	Polyamide from ε-caprolactam
PA610*	Polyamide from hexamethylene diamine sebatic acid
PA612*	Polyamide from hexamethylene diamine dodecanoic acid
PA66*	Polyamide from Hexamethylene diamine adipic acid
PA69*	Polyamide from hexamethylene diamine acelaic acid
PAA	Polyacrylic acid ester
PAC	Polyacetylene
PAE	Polyarylether
PAEK*	Polyarylether ketone
PAI	Polyamidimide
PAMI	Polyaminobismaleinimide
PAN*	Polyacrylonitrile
PANI	Polyaniline, polyphenylene amine
PAR	Polyarylate
PARA	Polyarylamide
PARI	Polyarylimide
PB	Polybutene
PBA	Polybutylacrylate
PBI	Polybenzimidazole
PBMI	Polybismaleinimide
PBN	Polybutylene naphthalate
PBO	Polyoxadiabenzimidazole
PBT*	Polybutylene terephthalate
PC*	Polycarbonate (from bisphenol-A)
PCPO	Poly-3,3-*bis*-chloromethylpropylene oxide
PCTFE	Polychlorotrifluoro ethylene
PDAP	Polydiallylphthalate resin
PDCPD	Polydicyclopentadiene
PE*	Polyethylene

Acronym	Chemical notation
PE-HD	Polyethylene-high density
PE-HMW	Polyethylene-high molecular weight
PE-LD	Polyethylene-low density
PE-LLD	Polyethylene-linear low density
PE-MD	Polyethylene medium density
PE-UHMW	Polyethylene-ultra high molecular weight
PE-ULD	Polyethylene-ultra low density
PE-VLD	Polyethylene-very low density
PE-X	Polyethylene, crosslinked
PEA	Polyesteramide
PEDT	Polyethylenedioxythiophene
PEEEK	Polyetheretheretherketone
PEEK	Polyetheretherketone
PEEKEK	Polyetheretherketoneetherketone
PEEKK	Polyetheretherketoneketone
PEI*	Polyetherimide
PEK	Polyetherketone
PEKEEK	Polyetherketoneetheretherketone
PEKK	Polyetherketoneketone
PEN*	Polyethylenenaphthalate
PEOX	Polyethylene oxide
PESI	Polyesterimide
PES*	Polyethersulfone
PET*	Polyethylene terephthalate
PET-G*	Polyethylene terephthalate, glycol modified
PF*	Phenolic formaldehyde resin
PFMT	Polyperfluorotrimethyltriazine rubber
PFU	Polyfuran
PHA	Polyhydroxyalkanoate
PHB	Polyhydroxybutyrate
PHFP	Polyhexafluoropropylene
PI*	Polyimide
PIB	Polyisobutylene
PISO	Polyimidsulfone
PK*	Polyketone
PLA	Polylactide
PMA	Polymethylacrylate
PMI	Polymethacrylimide
PMMA*	Polymethylmethacrylate
PMMI	Polymethacrylmethylimide

Acronym	Chemical notation
PMP	Poly-4-methylpentene-1
PMPI	Poly-m-phenylene-isophthalamide
PMS	Poly-α-methylstyrene
PNF	Fluoro-phosphazene rubber
PNR	Polynorbornene rubber
PO	Polypropylene oxide rubber
PO	General notation for polyolefins, polyolefin-derivates
POM*	Polyoxymethylene (polyacetal resin, polyformaldehyde)
PP*	Polypropylene
PPA	Polyphthalamide
PPB	Polyphenylenebutadiene
PPC	Polyphthalate carbonate
PPE*	Polyphenylene ether, old notation PPO
PPI	Polydiphenyloxide pyromellitimide
PPMS	Poly-$para$-methylstyrene
PPOX	Polypropylene oxide
PPP	Poly-$para$-phenylene
PPQ	Polyphenylchinoxaline
PPS*	Polyphenylene sulfide
PPSU*	Polyphenylene sulfone
PPTA	Poly-p-phenyleneterephthalamide
PPV	Polyphenylene vinylene
PPY	Polypyrrol
PPYR	Polyparapyridine
PPYV	Polyparapyridine vinylene
PS*	Polystyrene
PSAC	Polysaccharide, starch
PSIOA	Polysilicooxoaluminate
PSS	Polystyrenesulfonate
PSU*	Polysulfone
PT	Polythiophene
PTFE*	Polytetrafluoroethylene
PTHF	Polytetrahydrofuran
PTT	Polytrimethyleneterephthalate
PUR*	Polyurethane
PVAC	Polyvinylacetate
PVAL	Polyvinylalcohol
PVB	Polyvinyl butyral
PVBE	Polyvinyl isobutylether

Acronym	Chemical notation
PVC*	Polyvinyl chloride
PVC/EVA	Polyvinyl chloride-ethylene vinylacetate
PVDC*	Polyvinylidene chloride
PVDF	Polyvinylidene fluoride
PVF	Polyvinyl fluoride
PVFM	Polyvinyl formal
PVK	Polyvinyl carbazole
PVME	Polyvinyl methylether
PVMQ	Polymethylsiloxane phenyl vinyl rubber
PVP	Polyvinyl pyrrolidone
PVZH	Polyvinyl cyclohexane
PZ	Phosphazene rubber with phenoxy groups
RF	Resorcin formaldehyde resin
SAN*	Styrene acrylonitrile
SB*	Styrene butadiene
SBMMA	Styrene butadiene methylmethacrylate
SBR	Styrene butadiene rubber
SBS	Styrene butadiene styrene
SCR	Styrene chloroprene rubber
SEBS	Styrene ethene butene styrene
SEPS	Styrene ethene propene styrene
SEPDM	Styrene ethylene propylene diene rubber
SI	Silicone, Silicone resin
SIMA	Styrene isoprene maleic acid anhydride
SIR	Styrene isoprene rubber
SIS	Styrene isoprene styrene block copolymer
SMAB	Styrene maleic acid anhydride butadiene
SMAH*	Styrene maleic acid anhydride
SP	Aromatic (saturated) polyester
TCF	Thiocarbonyldifluoride copolymer rubber
TFEHFPVDF	Tetrafluoroethylene hexafluoropropylene vinylidene fluor
TFEP	Tetrafluoroethylene hexafluoropropylene
TM	Thioplastics
TOR	Polyoctenamer
TPA*	Thermoplastic elastomers based on polyamide
TPC*	Thermoplastic elastomers based on copolyester
TPE	Thermoplastic elastomers
TPE-A	see TPA

Acronym	Chemical notation
TPE-C	see TPC
TPE-O	see TPO
TPE-S	see TPS
TPE-U	see TPU
TPE-V	see TPV
TPO*	Thermoplastic elastomers based on olefins
TPS*	Thermoplastic elastomers based on styrene
TPU*	Thermoplastic elastomers based on polyurethane
TPV*	Thermoplastic elastomers based on crosslinked rubber
TPZ*	Other thermoplastic elastomers
UF	Urea formaldehyde resin
UP*	Unsaturated polyester resin
VCE	Vinylchloride ethylene
VCEMAK	Vinylchloride ethylene ethylmethacrylate
VCEVAC	Vinylchloride ethylene vinylacetate
VCMAAN	Vinylchloride maleic acid anhydride acrylonitrile
VCMAH	Vinylchloride maleic acid anhydride
VCMAI	Vinylchloride maleinimide
VCMAK	Vinylchloride methacrylate
VCMMA	Vinylchloride methylmethacrylate
VCOAK	Vinylchloride octylacrylate
VCPAEAN	Vinylchloride acrylate rubber acrylonitrile
VCPE-C	Vinylchloride-chlorinated ethylene
VCVAC	Vinylchloride vinylacetate
VCVDC	Vinylchloride vinylidenechloride
VCVDCAN	Vinylchloride vinylidenechloride acrylonitrile
VDFHFP	Vinylidenechloride hexafluoropropylene
VF	Vulcanized fiber
VMQ	Polymethylsiloxane vinyl rubber
VU	Vinylesterurethane
XBR	Butadiene rubber, containing carboxylic groups
XCR	Chloroprene rubber, containing carboxylic groups
XF	Xylenol formaldehyde resin
XNBR	Acrylonitrile butadiene rubber, containing carboxylic groups
XSBR	Styrene butadiene rubber, containing carboxylic groups

Table C-2 Commonly used symbols describing polymer characteristics

Symbol	Material characteristic
A	Amorphous
B	Block-copolymer
BO	Biaxially oriented
C	Chlorinated
CO	Copolymer
E	Expanded (foamed)
G	Grafted
H	Homopolymer
HC	Highly crystalline
HD	High density
HI	High impact
HMW	High molecular weight
I	Impact
LD	Low density
LLD	Linear low density
(M)	Metallocene catalyzed
MD	Medium density
O	Oriented
P	Plasticized
R	Randomly polymerized
U	Unplasticized
UHMW	Ultra high molecular weight
ULD	Ultra low density
VLD	Very low density
X	Cross-linked
XA	Peroxide cross-linked
XC	Electrically cross-linked

Table C-3 commonly used plasticizers and their acronyms

Acronym	Chemical notation
DODP	Dioctyldecylphthalate
ASE	Alkylsulfone acid ester
BBP	Benzylbutylphthalate
DBA	Dibutyladipate
DBP	Dibutylphthalate
DBS	Dibutylsebacate
DCHP	Dicyclohexylphthalate
DEP	Diethylphthalate
DHXP	Dihexylphthalate
DIBP	Diisobutylphthalate
DIDP	Diisodecylphthalate
DINA	Diisononyladipate
DMP	Dimethylphthalate
DMS	Dimethylsebazate
DNA	Dinonyladipate
DNODP	Di-n-octyl-n-decylphthalate
DNOP	Di-n-octylphthalate
DNP	Dinonylphthalate
DOA (DEHA)	Dioctyladipate, also diethylhexyladipate, DEHA no longer used
DOP, DEHP, DODP	Dioctylphthalate, dioctyldecylphthalate
DOS	Dioctylsebacate
DOZ	Dioctylazelate
DPCF	Diphenylkresylphosphate
DPOF	Diphenyloctylphosphate
DPP	Dipropylphthalate
ELO	Epoxidized linseed oil
ESO	Epoxidized soy bean oil
ODA	Octyldecyladipate
ODP	Octyldecylphthalate
PO	Paraffin oil
TBP	Tributylphosphate
TCEF	Trichlorethylphosphate
TCF	Trikresylphosphate
TIOTM	Triisooctyltrimellitate
TOF	Trioctylphosphate
TPP	Triphenylphosphate

Subject Index

A

abrasion resistance   351
ABS/ASA blends   207
ABS/PC blend   55
absorber tube   286
absorption coefficient   359
accelerators   84
acoustics   16, 306, 354
– bridges   372
– glazing   329
– mapping   224
– test   223
adapter frame   291
adhesive   187, 198, 203, 303
– applications   173, 174
– bonding   22, 175, 178, 181, 192, 195
– bonds   83
   – off-line   188
   – on-line   188
– chemical thixotropic   197
– cold-curing   183
– collision-optimized   176
– humidity-reactive fusion   198
– one-pack   183
– shrinkage   176
– single-component   198
– structural   176
– systems
   – non-reactive   198
   – reactive   198
– tapes   173
   – two-component   198
– usage   22
– warm-hardening   201
adult head impactor   11
advanced SMC   90, 94
aeroacoustic   356
– aerodynamics   354
– improvements   245
– simulation   357
aftermarket versions   107
aftershrinkage   231
ageing   154, 156, 180

air-borne sound   307, 367, 369
– absorption   370
– insulation   370
air compression noise   371
airflow   357
– noise   356, 357
Allianz Crash Test (AZT)   49
aluminum   20, 21, 59, 68, 182
– /CRP combinations   189
– space-frame   316
– wrought alloys   12
angle of inclination   159
anisotropic   146, 155, 156, 257
– failure   257
– failure model   258
anisotropy   154, 171
– due to glass fibers   257
antenna   107
– covers   43
– integration   107, 211
anti-graffiti-properties   378
antireflectiveness   329
anti-slip   374
A pillar   235, 363
A-pillar deflector   249
appearance   246
– harmony targets   26
aramid fiber laminate   189
armored cars   328
assembly time   103
atmospheric pressure plasma   200
attachment concept   35
autoclave technology   91
automobile design   19
automotive glazing   291, 347
AZT test   284

B

back-foaming process   297
back-injection   99
– molding   342
back-molding   44, 104

barrier layer   186, 300
basecoat   45, 212
batch consistency   84
behavior
– linear-elastic   169
– plastic   169
– temperature-dependent   170
– yield-rate-dependent   169
belt pulley   163
– thermoplastic   164
– thermoset   164
benchmarking   251
biaxial tension   268
birefringence   348
blackout printing   46
black specks   350
blind rivet
– nuts   185
– rust-resistant steel   187
Blue Angel   380
BMC   248
body   176, 189, 192
– spot-weld-glued   176
– spot-welded   176
body components
– films used   206
– horizontal   207
– vertical   207
body construction   19
body design   182
Bodyflex   105
body-in-white (BIW)   62, 65, 89
body panels   205, 217
body rigidity   178
body shell   13, 15, 234, 236
– module   240
body side molding   206, 207
body structure   176
– dynamics   292
body tub assembly   21
bodywork
– plastics   234
bolts   185
bonding   173, 194
– joint profile   222
– of SMC
– primerless   195
– process   222

– technology   193
bonds
– backing   175, 176
– folded   175, 176
– in the structure   175
bottom load system   290
B pillar   346, 363
braiding machine   275
braiding ring   275
brake system   244
brittle fracture   267
buckling resistance   58
bulkhead   189
bumper   207, 209, 227, 228
– fascia   130, 256, 261, 280
bus   234
– body shell   235

C

cab   315
CAD data, 3D   137
CAD synchronization points   265
CAE exchange points   265
CAE systems   341
Californian Air Resources Board *see* CARB
CAMPUS   168, 171
CARB   86
car body
– panels   208
– temperature distribution   88
carbon-fiber-reinforced plastic (CRP)   59, 272
carbon fibers   71, 89, 90
CARB test   75
cargo roofs   109
carrier films   212
car underbodies   365
cascade-controlled injection molding   43
catalysts   84
central locking   106
CFD simulation   316
CFRP   59, 61, 192, 193
– applications   59
– body applications   62
– cost   63
– cost/performance trade-off   64
– structural parts   62
chemical foam   119

Subject Index

chemical resistance   216, 296
child head impactor   11
chrome-plate   302
circulation method   331
class A   210
- body parts   206
- component   302
- finish   43, 44, 48, 109, 219
- paintability   75
- requirements   130
- surface   178, 246
clean room   99, 297, 351
- conditions   45
clearcoat   45, 207, 212
clinching   183
C-Mold   146, 148
coating   337, 375
- noise-reducing   369
- systems   330, 351
- techniques   344
coefficient of linear thermal expansion   219
coefficient of sliding friction   374
coefficient of thermal expansion   221, 227, 232
coextruded films   208
collision
- behavior   276
- characteristics   175, 292
- high-speed   283
- low-velocity   283
- safety   30
color-matching   26, 215, 230, 246
comfort   53
component
- design   34, 273
- failure   156
- performance   270
- properties   127
- rigidity   236
- tests   270
composite   5
- body sides   22
- fiber boxes   280
- materials   20, 116
- strength   287
compression   157
- gap
- wedge-shaped   350
computer-aided engineering (CAE)   49

computer tomography   278
concealed joints   173
concept
- holistic   1
- jointing   2
conditioning   37
conductive components   229
conductive primer   229
conductivity   40
continuous filaments   117
continuous operating temperature   75
convertible top   306, 309, 354, 362
- frame   362
- shape   357
coolers   273
cooling time   118
corona   246
corrosion   180, 185
- bimetallic   186
- electrochemical   186
- proof   173
- protection   12, 273, 278
cost   245
- benefit analysis   128
- comparison   317
- cutting   102
- effectiveness   53
- free take back   27
- reduction   245, 248
- savings   224
cover shield   198
cowl grille   226
crack   212
- propagation curve   163
crash
- analysis   172, 257
- box   280, 284
- efficiency   283
- element   253, 271
- management   280
- simulation   57, 167, 264, 267, 268
- simulation models   264
crazing   212
creep behavior   154
creep tendency   86
crevice corrosion   178
critical section plane   159
CRP component   183, 186

cure-shrinkage   195
curing   83, 308
cutting element
– property profile   288
cutting plate   282
– encapsulated   286
cutting speed   283
cutting unit   302
cyclical stress   145, 156
cyclic crack propagation   162
cyclic loading   162
cyclic tensile/compressive stress   150

D

damage accumulation   145
damping properties   369, 375
data mining   70
deflashing   302
deformation behavior   268
delamination   156, 189, 276
demolding   302
design
– development   315
– for recycling   365
– freedom   347
– of experiments (DoE)   69
development
– costs   264
– phase   274
– process   135
– vehicles (DVs)   124
dilatant polymer dispersions   54
dimensional stability   75, 338
dimensional tolerances   193
dimensioning   220
direct floor coating   380
direct glazing   173, 340
direct long-fiber-reinforced thermoplastics (D-LFT)   365
dirt repellence   246
dismountability   222
disposability   19
disposal   249
dissipation factor   16
door   43, 226
– locks   103
– modules   103

– panels   207
– systems   103
double-shell design   194
driving noise   371
drumming noise   361
dry paint film   207
dual barrel cartridges   197
ductile fracture behavior   79
dynamic collision computation   273
dynamic flexural rigidity   175
dynamic impacts   253
dynamic stress   146
dynamic test   282

E

ECE R21   167
elastomer   268
electrical conductivity   278
electrochromatic   107
electroluminescence   110
electrolytic solution   278
electron beam   137
electrophoretic coating dryer   228
electrophoretic dip coating   87, 183, 188, 231
– bath   88
elongation at break   171
emission   75, 300
– limits   302
EM shielding   352
encapsulation   105
End of Life Vehicle Directive   26
energy
– absorber   285
– absorbing components   272
– absorption principles   281
– management system   284
– transmission   342
entropy elastic deformation   267
environmental resistance   246
epoxy   193
– one-component systems   84
– resin   82, 140
– two-component systems   84
EPP   92
– content   76
esthetic properties   246
EuroNCAP   253, 254, 264, 280

Subject Index

European End of the Life Vehicles Directive (ELVD) 19
expanded films 104
expanding polypropylene *see* EPP
expansion 83
– compression technique 349
exterior
– body
– body applications 42
– body panels 192
– noise 16
– painted 13
external body panels 21

F

fabrication process
– influences 159
facelift 219
failure
– analysis 172
– behavior 258
– criteria 156, 157, 158
– mechanisms 156, 157
– potentials 157
falling dart impact
– test 222
fast closing 302
fastening 185
– system 35
fatigue 156
– behavior 161
– crack growth 162
– endurance 163
– life 13, 14, 143, 146, 148
– life calculation 147, 151
– life estimation 145
– life prediction 143
– strength 159, 193, 273
– tensile stress 150
FE analysis 155, 260
FEM 35
– analysis 169
– calculations 176
– programs 167
FEMFAT 158
FE model 260
fender 13, 68, 207, 232

fender module 225
FE simulations 341
fiber
– detachment 156
– distribution 145, 146
– failure 156
– -composite plastic 240
– -matrix adhesion 162
– -matrix system 154
– -reinforced composite 189, 234
– -reinforced composite plastics 182
– -reinforced plastic components 189
– -reinforced plastics (FRP) 153, 154, 156, 161, 171, 181, 185, 192
– -reinforced PU 48
– inorganic 154
– local 154
– natural 154
– organic 154
– orientation 145, 146, 147, 148, 154, 155, 160, 162, 171, 260, 275
– synthetic 154
– tear 195, 196
– tearing 156
fill time 260
film 205, 206
– gate 347
– scratch-resistant 352
– technology 25
fine-celled foam 84
finite element method 146
finite element network 146, 149
fixed side window 109, 332, 336
flame treatment 105
flaming 246
flange plate 280
flax fibers 99
fleet business 312
flexibly bonded 235
flexion 219
floor functionality 378
flowability 72, 75
flow-coating 331
flow paths 309
flow properties 54
fluid injection 117
– technique 104
fluorination 246

FMVSS 301   10
foam   280
– cores   322
– curing   87
– -backing   207
– -laminated fabric   324
– -molding   309
– incursions   309
foaming   118
– mold   308, 322
foam structure   83
foil technology   25
folding boxes   280
folding top   306
folding without fold bed   176
force-deformation characteristic   168
force-to-area ratio   283
fracture-mechanical   162
framework construction   130
freedom of design   227, 328
frequency-dependence   154
frontal collision   272, 273, 292
front bumper   273
front-end   21, 253, 272
– components   239
– module   68, 104, 105, 225, 240, 253
  – inside   241
– structure   49
front floor   189
front mudguards   192
front side window   363
FRP   193
– components   190
fuel
– cell   19, 56
– consumption   244
– door   229, 230, 231
– economy   19
full-scale production   128
functional coating   372, 376, 380
functional integration   330, 332, 347, 352
functional properties   246
function integration   218, 219

G

galvanic corrosion   278
gas counterpressure method   121
gas internal pressure technique   117
gas strut   220
gearshift linkage   147, 148
geometric stability   218
glass-fiber-PP structure   367
glass-mat-filled composites   115
glass-mat-reinforced thermoplastics (GMT)   192, 365
glass microspheres   93
glazing   108, 235, 329, 335
– automotive   327, 337, 347
– mineral glass   339
– organic   327
– plastic   337, 339, 346
– polarization   339
– polycarbonate   339
– safety   339
glazing components
– injection-compression molding   108
gloss measurement   295
glue selection   222
gluing   173, 182, 187
– surfaces   303
– unit   303
gravimetric feeding   197
grille opening panel   24
grinding   195

H

Haigh-Goodman diagram   144, 147
handling strength   201, 202
hardtop   289, 299, 301, 303, 304
HC emissions   9
head impactor   253
headlamp lens   42
headlight   111, 242, 273
– bonding   198
– lenses   328
heating element   337
heating technologies   222
heat resistance   75, 194, 227, 296
higate hybrid
– assessment   224
highroof   319
high-speed collision   284
high-speed frontal impact   253
high-temperature light-fastness   374
holistic analysis   69, 70

hollow section   82
homogeneity   75
homogenization   258
honeycomb structures   280
hood   210
– hinge   273
– systems   226
horizontal components   210
horizontal panels   211
hot embedding   342
hot melts (RHM)   198
– reactive   199
hot-runner nozzles   103
housing   198
hybrid
– carrier   105
– design   316
– part   22
– tailgate   219
    – comparison   224
– technology   2
hybrid aluminum-FRP designs   182, 187
hydroforming   182
hydrogen storage   56

I

IHS test   284
impact simulation   219
impact strength   232
in-line painting   229, 246, 247, 251
in-mold film lamination   110, 217
in-mold laminating   103, 331, 333, 337, 338
in-mold pressing   331
indicator   35
induction heating   176
infra-red
– absorbers   342
– absorption   337
– reflection   337
injection-compression molding   103, 112, 113, 115, 348, 349
injection-compression molds   347
injection-compression process   107, 221
injection molding   4, 36
injection-molding simulation   258
inner panel
– molded in color   222

insertion loss   358
insertion pins   185
instant-fix RWM   201
instrument panel   238
insurance classification   253
– low speed   257
– tests   262, 264
integral
– components   2
– foam structure   103
– front-end   234, 235, 238
– nose   236, 239, 240
– rear-end   236, 237
integrated functions   248
integrating hybrid   2
integration   340
– of simulation   264, 266
– potential   227
integrative crash simulation   261
integrative simulation   57, 258, 259, 260
intelligent damping device   55
inter-fiber failure (IFF)   156
interior noise   354, 358, 359, 361, 371
interior trim
– glued-in   324
interlayer resin technology   25
internal
– noise   355
– stresses   348
– tension   338
IR reflection   46, 352
isotropic   155

J

joining   173, 182
– methods   22
– mold   303
– techniques   347
– technology   181
joint
– blind rivet   187
– blind-riveted   184
– blind-riveted metal/CRP   184
– clinching   184
– location   184
– punch-riveted   184
– strengths   196

just-in-sequence delivery   325

K
keyless entry   106
keyless go   106
knee-bending angle   49, 280
knee shearing displacement   280

L
laminated film   45
laminate structure   274
landfill directive   27
large-area components   331
laser
– beam   136
– light   137
– powder sintering   140
– soldering   178
– technology   99
– welding   178
layer   72
LD-SMC microstructure   93
leak   356
leakage test   198
LED light sources   111
legform impactor   255
legform impact test   255, 262
leg impact kinematics   256
leg impactor   11, 280
life-cycle   249
– properties   246
life estimation   159
light   111
– transmission   337
light-emitting diodes (LEDs)   111
lightweight bodywork   82
lightweight design   1, 30, 72, 102, 173, 181, 190, 205, 293, 347
lightweight door designs   104
lightweight vehicle design   153
linear thermal expansion coefficient   75, 77
liquid-crystal films   109
load capacity   155, 157
load cycles   157
loading state   2
load transmission   196

long-fiber
– composite   99
– reinforced thermoplastics (LFT)   115
long-haul operations   313
long-haul segment   312
long-stroke compression technique   349
loudspeakers   103
low-density (LD) SMC   70
low density sheet molding compound (LD-SMC)   90, 93
lower bumper stiffener   253, 254, 255, 256, 257, 258, 261
– design   262
lower impactor   280
lower leg crash calculations   261
lower legform   253, 255
low impact
– resistance   221
low-stressed   338
low-temperature viscosity   75
low weight reinforced thermoplastics (LWRT)   365

M
macrocracks   156
magnesium   59
– alloys   12
– die castings   12
MAH (maleic anhydrid)   99
manual lamination processes   91
material
– behavior   145, 258
  – under cyclic loading   143
– data   172
  – acquisition   259
– fatigue   156
– model   172, 268
– modeling   259
– properties   13, 155, 259, 267, 268, 295
– restrictions   27
– structural ranking   252
maximum acceleration   280
maximum strain   157
maximum stress   157
mean stress   144, 154
mechanical behavior   155
mechanical fastening   22
mechanical joining   182, 183

mechanical properties   75, 170, 196, 222
– testing   302
melting stability   75
melt viscosity   118
Mercedes-Benz Development System (MDS)   124
mesh size   168
metal-organic framework (MOF)   56
meter-mix-machines   197
methacrylate-base formulations   193
microcracking   156
micro-mechanical model   258
microscratches   296
microstructure foaming   106
milling adapter   303
Miner calculation   147
Miner rule   145
mirror galvanometer   136
mirror triangle   363
mixed materials   22
modular concept   240
module   22, 23
– carriers   103
modulus of elasticity   155
moisture absorption   37, 39, 75
mold design   308
molded-in stress   348, 350
molded-in color   219
mold film technology   206
mold flow   146, 151
molding principle   261, 262
mold temperature   300, 302
molecular orientation   348
molecular structures   53
morphology   160
moveable cores   103
MuCell®   120, 121
– technique   119
multiaxial loading   154, 156
multiaxial stress   157
– states   156
multi-component   116
– injection molding   347
multifunctionality   173
multipurpose specimens   161

N
nano particles   330, 351

nano technology   25, 53, 54
necking   167, 268
– zone   168
nickel shell   320, 321
noise   369
– damping   373
– emission
  – buses   371
– optimization   306
– spectrum   354
nominal strain at break   167
non-fuel emissions   79
non-linearity in stress-strain   257
non-woven   275
normal stress   147
nose
– cross-member   235
– module   238
– panel   241

O
oblique-angle impact   273
occupant safety   253
ODB crash   284
OEMs   19, 61, 63, 65
off-line coating   45
off-line painting   205, 246
on-line paintability   39, 227
– thermoplastics
  – material properties   40
on-line painting   205, 229, 231, 246, 247
optical components   111
organic coatings   186
orientation distribution   258
Original Equipment Manufacturers (OEM's)   19, 61, 63, 65
orthotropic   155, 156
outer body panels   68, 234
outer skin   192, 194, 294
outer skin bonding   179
overlap   273

P
package situation   314
packaging   308

paint
- polysiloxane-based   331
painted
- in-line   32, 228, 230
- off-line   228
- on-line   32
painted film   210
paint film   212, 214
painting   304, 324
- preparations   303
- process   248
paint systems   330
panorama roof   46, 107, 108, 109, 346
part cost reduction   219
partial reinforcements   117
parting line   221
part marking   27
PA/SAN blend   55
passenger protection   167
payload   244
PC copolymer   207
pedestrian
- crash model   261
- impact   49
- lower leg protection   253
- protection   10, 105, 130, 253, 254, 256, 257, 264, 280, 281
- safety   253
peeling stress   187
pendulum impact   284
perceived quality   246
performance properties   246
physical foam   119
physical vapor deposition (PVD)   352
plasma   246
- CVD   337, 343
    - coating   344
    - layer   344
    - processes   330
- generator   200
- -enhanced chemical vapor deposition (PECVD)   352
- pretreatment   199
- technology   109
- treatment   105
plaster printer   141
plastic
- body applications   7

- utilization   20
plastic-based inserts   117
plastic-metal hybrid   104
Pliogrip 7770   188, 189
PMMA   207
Poisson's ratio   155, 260
polarization optics   338
pole test   284
polycarbonate   108, 109, 332, 339, 347, 348, 350
- automobile glazing   335
- glazing   45, 114
- trunk-lid panel   333
polymer-blend technology   100
polyolefinic EEA   207
polypropylene   207
- long-fiber reinforced   103
polypropylene-based films   207
polysiloxane   109, 330
- coatings   337, 351
- paints   340
polyurethane   46, 47, 193
- adhesive   291, 303
- foam system   307
- insulation   306
- part
    - process chain   141
- primer   379
post-treatment   229
powder coating   232
power windows   103
PP catalysts
- metallocen-based   81
PP continuous-strand   103
PPE+PA
- heat resistance   33
- mechanical characteristics   34
PP/EPDM compounds   75
PP/EPR   75
precision backstroke technique   120
preform   72, 73
- production   274
- technologies   91
pre-gelling adhesives   87
prepreg/autoclave technology   60, 90
pre-treatment   27
pressure loss   260
pressure oscillations   357
primer   351

- application 246
- coat 105
primerless SMC adhesives 196
priming 36, 195
process ability 75
process control 278
processing window 232
process simulation 257
product brief 129
product design 248
production technology
- evaluation 317
product validation 222
prototype
- assembly 140
- components 123, 125, 126, 130, 132
- costs 125
- molds 128, 131, 132
- production 123, 274
- requirements 125
- suppliers 129
- tools 127, 131
PU LFI structure
- material properties 293
pumping equipment 203
punch riveting 178
PUR-LFI 210
PUR-SRIM 210
PU sandwich technology 47

Q

quality 102
- assurance 278
- gates 124
quick-fix 199
- RWM 201

R

radiator grilles 228
rain flow
- counting procedure 146
- matrix 146
rapid manufacturing (RM) 138
rapid prototyping (RP) 135, 136, 138
- methods 137
rapid tooling (RT) 135, 138

reaction injection molding *see* RIM
reactive blending 100
reactive warm melt (RWM) 200, 201
rear
- cover 192
- door 192
- impact at 80 km/h 10
- paneling 236
- window 357, 361
  - modules 346
recovery 249
recycling 10, 27, 104, 249
- material 11
- status 28
- thermal 11
- targets 27
re-engineering 245
reflection optics 337
reinforced reaction injection molding *see* RRIM
reinforcement 340
- continuous strands 104
- fabric 104
relative movements
- thermally induced 176
release agents 84
reparability 219
required pressure 75
requirement profiles 3
resin 84
- film infusion (RFI) 90
- infusion (RI) 90
- injection processes 92
- transfer molding (RTM) 46, 47, 276
re-use 249
reverse engineering 168
reversible deformations 268
rigid components 12
rigidity 12, 218, 235
rigid pole collision 292
RIM 46, 47, 276
rivet 192
- coated carbon-steel 187
riveting 173
- self-pierce 184
rocker panels 206, 207, 228
roll bar 235
roof 226
- antennae 292

- box
  - external 289
- fabric 358
- frame
  - front 364
- module 107, 289, 290, 291, 292, 294, 312
- panel 210
- rack 290
- spoiler 192
- strip 290

rotary platen 349
RRIM 47
RTM 5, 90, 316
- mold 92
- setup 277
running gear 147

S

safety 53, 251
- concepts 253
- glazing 328, 339

sag resistant 197
samples
- close-to-production 171

sandblasting 195
sanding 246
sandwich
- construction 103
- design 235
- injection molding 43, 44, 116
- structure 377

scale factor 258
scratchproof 332
- coating 335
- paints 330

scratch resistance 75, 216, 246, 296, 330, 343, 351, 352
screw 192
- connection 278

sealed joints 193
sealed pan concept 378
sealing film 367
seals 358, 360, 363
seamless surfaces 75
seat pans 57
secondary finishing 302
selective laser sintering (SLS) 136, 137

self-cleaning 329
- surfaces 54

self-extinguishing 373
semi-automated sequential preforming 91
semiconductor light sources 111
sequence effects 154
sequential preform process 92
sequential valve gate injection 221
service life 13, 14, 153, 180, 242
- calculation 158

service temperatures 196
shear
- boxes 280
- load 287
- modulus 155
- stress 348

sheet metal 316
sheet molding compound see SMC
shore hardness 374
short-term dynamic testing 86
shrinkage 75
side deflectors 245
side panel 31, 43
- attachment 38

silicate coatings 25
sill panel 13
simulation 232, 253
- non-linear behavior 169
- parameters 172
- techniques 53

single-step process 116
sink holes 176
sink marks 118
sinusoidal loading 161
skin 295
sliding sunroof 108
slip resistance 373
smart-surface technology 110
smart windows 329
SMC 5, 24, 25, 68, 93, 95, 192, 218, 248
- blank 302
- class A 299, 300
- components 302
- emission-optimized 301
- emissions 299
- fiber failure 195
- formulations 299
- headlight housing 194

- material properties 32
- production 301
- rear license-plate panel 194
- technology 68
  - lightweight potential 70
- visible components 299

SMI technology 316
snap fits 192, 347
solid-borne sound 307, 369
- dampening 367
- damping 370
- insulation 370
- sound absorption 376, 378

solvent-free coating 373
sound
- airborne 16
- structure-borne 16

sound insulation 103
space-frame concept 182
split control 302
split markings 232
spoilers 43
spot-weld bonding
- collision strength 177
- comparison 177

spot welding 178
- collision strength 177
- comparison 177

stable dispersion 351
standard mold unit 131
standard rivet nuts 185
static strength 156
static stress analysis 156
static tension 144
statistical orientation distribution function 259
steel 15, 20, 224
- hot-formed 15
- low-alloy deep-drawing 15
- rust-resistant 186
- stainless 15

steel/aluminum plastic 316
steel-plastic composites 16
steel shell design 182
steering column 235
stereolithography unit 140
stiffness 75, 222, 250
stone impact protection 373, 376
stone impact resistance 246

stowage boxes 292, 298
stowage facilities 289
strain-rate dependence 257
strength 250
stress
- amplitude 147
- capacity 155
- concentrations 196
- distribution 146, 193
- ratio 144, 147

stress-strain
- characteristics 267, 269
- curves 168
- hystereses 156
- relation 155

structural
- behavior 219
- body panels 68
- bonding 176, 193
- components 13
- foam 82, 83, 87, 88
  - mechanical characteristics 85
  - components
  - fabrication 83
- modules 95
- plastics 21
- rigidity 20, 287, 292
- thermoplastics 250
- validation vehicles (SVVs) 124

structure-borne noise
- noise reduction 375, 376

styling 220, 248
- freedom 218, 220

styrene copolymers 207
sun blinds 291
sun test 222, 223
support frame 293, 294
surface 75
- coating 351
- decoration 104
- defects 213, 214
- finishes 24
- preparation 195, 222
- pretreatment 199
- quality 232, 338
- topology 231

swivel-platen 330, 331, 338, 340, 341
- system 116

synchronization process 266
syntactic core 60
system suppliers 129

T

tailgate 68, 218
– hybrid 218
– panel 109, 207, 210
tapering tubes 280
telescopic tubes 280
telescoping edges 347
temperature 156, 283
– distribution 348
– range 170
– stability 296
tempering 37
– step 229
tensile
– modulus 167
– strength 144
– test
  – simulation 169
  – curve 172
tension 157
– -compression asymmetry 257
test 80
– compression 160
– cyclic 161
– monotonic tension 160
testing 124, 178, 287
– noise 356
testing machine
– servo hydraulic 160
testing of 'Class-A' surfaces 129
test methods 40
test setup 282
test specimen 77, 160
testworthiness 125
textile preform technologies, 3D 90
thermal
– behavior 75
– comfort 342
– conductivity 342
– expansion 75, 182, 183, 193, 218, 294, 342
– insulation 306, 329
– loads 251
thermoforming 251

thermography 278
thermojet 3D printer 141
thermoplastic 249
– chopped-strand-reinforced 143
– material properties 32
thermoplastic absorber 286
thermoplastic composites
– recycling 250
thermoset 224, 249
thermoset composites
– recycling 250
thermoset materials 162, 220
– short-glass-fiber-reinforced 163
thermoset technology 218
thermosetting materials 245, 248
thigh impactor 11
thin film composites 44
thin-walled 338
tiger skin pattern 77
time-critical 131, 132
time-to-market 264
titanium alloy 186
top-load system 107, 290
torsion 219, 221, 222
torsional rigidity 175
total lifecycle 19
toughness 75
towing loads 273
TPO 79, 81
– key properties 78
track control arm 149
transparent roof systems 346
transverse stiffness 221
trimming 276
trim panels 207
trim parts 209
Triple C 69
truck cab 243, 244, 245, 246
– sun-visors 249
trunk lid 210, 218
tufting 276
tufting head 276
tufting threads 276
two-component
– injection molding 82, 330
– method 340
– parts 338
– polyurethane 372

- technology 349
two-pack 173
type approval 27

U
ultrasound 278
ultraviolet
- light 137
- resistance 75
underbody
- airflow 357
- covers 366
- sealing 373, 376
underseal
- liquid PVC 365
- panels 68
uniaxial loading 268
unidirectional reinforced material 155
upper legform 253
utilization scenarios 126
UV 246
- absorption 46
- protection 351
- stability 342

V
vacuum-coat 343
vacuum deposition methods 330
validation 270
valve-gated pinpoint gates 347
vapor deposition 25
vehicle development 270
- process 265
vehicle life-cycle
- CFRP 64
venting 309
vertex load cycle 159
vertical
- flash face clearance 302
- panels 209, 210
- process chains 65
vibration 357
- excitation 356
virtual development 264
virtual tools 136
virtual vehicle assessment gates 264

viscoelasticity 267
viscoplasticity 267
viscosity 55
visual properties 246

W
wall impact 284
warm melts (RWM) 198
- reactive 199
warm reactive bonding 198
warpage 75, 105, 118, 221, 231, 260, 338
wash resistance 246
water absorption 232
- /ambient medium 154
water-based paints 229
water content 267
water injection molding 117, 118
water tightness 222
wear resistance 374
weather resistance 330
weight 244
- reduction 7, 9, 19, 175, 219, 244, 248, 365
- saving 103, 218, 224
wet-chemical coating 340, 343
wettability 199
wind noise 354, 355, 357
window heating 352
windshield 329, 363
- frame 364
- wiper 235
  - panel 241
Wöhler
- curves 158, 159, 161, 162
- master curve 144, 145, 147
- tests 144, 161
wrought steel alloys 15

X
XRE technology 95, 96
XRI technology 95, 96

Y
yield rate 169
yield strain 167
yield stress 167

One-stop Solutions
for your Lightweight Structures

The **Institut für Leichtbau und Kunststofftechnik** (Institute of Lightweight Structures and Polymer Technology, ILK) of the Technische Universität Dresden with its **Lightweight Structures Innovation Center** and the **Leichtbau-Zentrum Sachsen GmbH** stand for first-class research and development in the field of load-adapted lightweight structures.

With a staff of more than 150 engineers, a comprehensive approach is pursued with respect to both materials and products throughout the total engineering chain – material, design, simulation, manufacturing, component, quality assurance.

A special focus is set on multi-material design with plastics, composites, hybrids, metals and ceramics.

Selected research topics:
- Lightweight components for high-performance purposes
- Ultra-lightweight structures and adaptive lightweight structures
- Design rules and optimisation criteria
- Innovative material concepts with high lightweight potential
- Analytical and numerical (FE) modelling, simulation and optimisation of compounds and composites

Our services for you:
- Material characterisation and failure analysis
- Design and engineering
- Modelling and simulation
- Prototype manufacturing
- Component testing, quality assurance

You are interested? Contact us!

Prof. Dr.-Ing. habil. Werner Hufenbach
Technische Universität Dresden
Institut für Leichtbau und Kunststofftechnik
01062 Dresden
Germany
Tel.: +49 351 463-38142
Fax: +49 351 463-38143
eMail: ilk@ilk.mw.tu-dresden.de
www.tu-dresden.de/mw/ilk/

Dr.-Ing. Dipl.-Math. Martin Lepper
Leichtbau-Zentrum Sachsen GmbH
Marschnerstraße 39
01307 Dresden
Germany
Tel.: +49 351 463-39477
Fax: +49 351 463-39476
eMail: info@lzs-dd.de
www.lzs-dd.de